Risk and uncertainty in dam safety

CEA Technologies Dam Safety Interest Group

Desmond N. D. Hartford
Gregory B. Baecher

Thomas Telford

CEA Technologies

Published by Thomas Telford Publishing, Thomas Telford Ltd,
1 Heron Quay, London E14 4JD.
URL: http://www.thomastelford.com

Distributors for Thomas Telford books are
USA: ASCE Press, 1801 Alexander Bell Drive, Reston, VA 20191-4400, USA
Japan: Maruzen Co. Ltd, Book Department, 3–10 Nihonbashi 2-chome, Chuo-ku, Tokyo 103
Australia: DA Books and Journals, 648 Whitehorse Road, Mitcham 3132, Victoria

First published 2004

Reprinted 2007

Also available from Thomas Telford Books
Interim guide to quantitative risk assessment for reservoirs. A. J. Brown and J. D. Gosden. ISBN 07277 3267 6
Long term benefits and performance of dams. The British Dam Society. ISBN 07277 3268 4
Engineering judgement and risk. J. Parkin. ISBN 07277 2873 3
The philosophy of risk. J. Chicken and T. Posner. ISBN 07277 2666 8

A catalogue record for this book is available from the British Library

ISBN 10: 0 7277 3639 6
ISBN 13: 978 0 7277 3639 0

Typeset by Apek Digital Imaging Limited, Bristol
Printed and bound in Great Britain by
CPI Antony Rowe, Chippenham and Eastbourne

Contents

Dedication

This book is dedicated to all those who lost their lives or suffered harm or loss from dam failures in the hope that it will lead to safer dams.

Sketch by W. Nicholson of the failure of the Dale Dyke Dam, Sheffield, 1864. Sheffield Libraries, Archives and Information retain rights to the use of this image. The image was provided specifically for use in the book Risk and Uncertainty in Dam Safety, *with Thomas Telford named as the publisher.*

CEA Technologies Dam Safety Interest Group

The preparation of this book, *Risk and Uncertainty in Dam Safety*, has been sponsored by the Dam Safety Interest Group (DSIG) under the auspices of CEA Technologies Incorporated of Montreal, Canada (http://www.ceatech.ca). The DSIG is a group of power utility owners interested in collaboration for the sharing of technical information strategic to their business, and for the combining of resources (both staff and fiscal) to applied research topics of mutual interest. Risk is one such topic.

At the present time there are thirteen DSIG members in North America, four in Europe and one in Australia. Not all members fiscally support each initiative that is proposed, but those members and their representatives that supported this book on risk are shown here in Table 1.

Table 1. Sponsors and their representatives

Sponsor	Representative
BC Hydro & Power Authority	Ray Stewart
Elforsk	Urban Norstedt/Anders Isander
Great Lakes Power	Harvey Walsh
Hydro Quebec	Tai Mai Phat
Manitoba Hydro	Gordon Schellenberg
Ontario Power Generation	Tony Bennett
Scottish and Southern Energy	Neil Sandilands
US Bureau of Reclamation	Bruce Muller
US Army Corps of Engineers	Charles Pearre
Gary Salmon, Technology Coordinator, CEA Technologies	

The support came in the form of monetary contributions made on an annual basis by the participating sponsors during the period 1997–2003, and from technical input and discussion during the Spring and Fall Business Meetings held each year. In addition BC Hydro and the US Bureau of Reclamation contributed considerable in-kind funding (i.e. staff time not charged during this same six-year period) and, in doing so, helped spearhead the initiative and kept it moving to a successful conclusion.

Risk and Uncertainty in Dam Safety is aimed at providing a contemporary description of evolving techniques for risk-based dam safety management. The principal focus is risk analysis, with the material presented representing the collective views of the contributors, reviewers and sponsors. The book is intended to provide the sponsors and the dam safety community with a basis for establishing their own Standards of Practice for risk-based management of the safety of their dams, and concurrently provide a focus for ongoing research and development into new and improved risk management practices.

Acknowledgements

BC Hydro served as Contractor for preparation of the book. Technical acknowledgements (and there are a considerable number) are included in the Preface to the volume as written by Desmond Hartford and Gregory Baecher, primary contributors to the volume. Alan Imrie has been Project Manager for the period 1999 to date.

The Project Manager would like to thank Josef Lampa and Tonia Jurbin for their careful editing of earlier drafts of the book and Donna Reedy for her patient and competent approach to text processing and formatting. Final technical editing was carried out by the sponsors' Working Group (see Preface). The Project Manager would also like to thank Gary Salmon who took on the challenging job of DSIG Technology Coordinator in 1998 and used his persuasive powers to pave the way for the sponsors and Contractor to move ahead in concert on the business side of the initiative. Finally, it was a pleasure to work with Kaare Høeg who used his vision to write a thoughtful and meaningful Foreword to the book, describing its usefulness to the dam safety community. His articulate critique of an earlier draft helped move the Project Team to the final product.

The sponsors' Working Group

Desmond N. D. Hartford, BA, BAI, MA, PhD (Dublin), CEng FICE, CEng FIEI, EurIng, PEng is Specialist Engineer – Dam Safety Risk Assessment at British Columbia Hydropower Authority. Specialising in dam safety and catastrophic loss risk management of large dams, hydroelectric facilities and water resource infrastructure, he is primarily responsible for the development and implementation of risk management solutions for BC Hydro's portfolio of 43 dams. Dr Hartford also advises dam and water resource agencies across Canada and internationally on dam safety and risk management. Dr Hartford is one of five principal authors of the forthcoming International Commission on Large Dams (ICOLD) bulletin on Risk Assessment for Dams. He is active in the training and professional development of engineers in risk management of dams and has presented workshops and seminars on dam risk management to government agencies and professional dam safety organisations around the world. Prior to joining BC Hydro, Dr Hartford held senior positions on international infrastructure projects. He was Research Fellow in Civil Engineering at Trinity College (1984–87) and subsequently Research Associate (1987–present).

Gregory B. Baecher, BSCE (Berkeley), MSc, PhD (MIT) is Professor of Civil and Environmental Engineering at the University of Maryland at College Park (UMCP). Dr Baecher is a consultant to government and industry on risk and reliability of civil infrastructure, especially in water resources development, and has served on a wide range of government and related committees. A member of the National Research Council's Water Science and Technology Board since 1998, Dr Baecher also served on the NRC's Board on Infrastructure and the Constructed Environment (1999–2002) and the Consultative Council of the National Institute of Building Sciences (1998–2002). Prior to joining UMCP he was President and co-founder of the information technology company ConSolve Inc., providing software solutions for environmental engineering. Earlier, he was Professor of Civil Engineering at MIT. He is recipient of the Thomas A. Middlebrooks and State-of-the-Art Awards of the American Society of Civil Engineers, and is Lilly Teaching Fellow and Freeman Foundation Fellow at the University of Maryland. He is co-author, with John T. Christian, of *Reliability and Statistics in Geotechnical Engineering* (Wiley, 2003).

Karl M. Dise, BSc (Cornell), MSc (Colorado), PE is a Civil Engineer and Technical Specialist with the US Bureau of Reclamation in Denver, Colorado. He is a principal investigator, consultant, advisor, and reviewer for extensive or complex geotechnical engineering matters encountered in current reclamation work. Mr Dise is a member of Reclamation's Risk Analysis Cadre, responsible for writing reclamation guidelines on risk assessment methodology and for facilitating risk analyses. Mr Dise has served as Principal Designer for dam safety deficiency modifications, has designed and supervised many field exploration programmes and has written numerous modification decision analyses, performance parameter reviews, and comprehensive facility reviews. He is a member of the American Society of Civil Engineers and the Society for Risk Analysis.

Marc Smith, BSc (Montreal), MSc (Laval), Ing is a geotechnical engineer with 18 years experience in the Dams and Civil Works division of Hydro Quebec. His practical experience in dam safety covers the design and construction of many large embankment dams as well as the surveillance and monitoring of dams during impoundment and during operation. He has performed a variety of safety assessments and risk analyses for dams in the province of Quebec, Latin America, Africa and South-East Asia. He is currently conducting embankment dam safety analyses in Hydro Quebec and carrying out research at the Center for Risk and Performance at École Polytechnique de Montréal aiming for the resolution of practical problems posed by dam safety risk evaluation and minimisation.

Przemyslaw A. Zielinski, MSc, M. M (Warsaw Technical University), PhD (University of Warsaw), with specialist expertise in probability, statistics, stochastic hydrology and water management, is Senior Engineer, Hydrotechnical Studies and Risk Assessment, Ontario Power Generation. Since the late 80s Dr Zielinski has been involved in over 30 dam safety assessment studies of OPG river systems, including all aspects concerning structure classification, watershed hydrology and hydraulics, dam breach and inundation modelling. He is actively involved in the development and adaptation of risk assessment techniques for dam safety purposes. Dr Zielinski is presently a Chairman of the ICOLD Committee on Dam Safety.

Foreword

Kaare Høeg

"Risk" is a combined measure of the probability and severity of an adverse event, and is often estimated by the product of the probability of the event occurring and the expected consequences. The public (our society) now demands, more than ever before, transparency in decision-making regarding safety issues and risk levels associated with dams and other forms of infrastructure. The dam owner/operator is required to document that the dam is safe or to improve its safety, as specified by regulations or dictated by societal standards or expectations. The courts in many countries are leaning towards strict liability. If the dam fails, for any reason other than acts of war or terrorism, the owner is liable for the consequences. Owners may even be found criminally liable if they cannot demonstrate due diligence in keeping their dams safe.

The safety evaluation of aging dams is an essential but complex undertaking, especially when the original foundation investigation, dam design and construction details are not known or are associated with significant uncertainty. The cornerstones of good dam safety practices are surveillance, periodic dam safety reviews, tested operation procedures, regular maintenance and emergency preparedness. Risk-based dam safety management represents an enhancement of the traditional dam safety practices through an integrated process that affords explicit recognition and analysis of risk and formal treatment of the uncertainties.

The book consists of three integrated parts, each covering basically the same topics with an increasing degree of detail. Part I "Guiding principles" refers to Part II for guidance and illustrative examples which, in turn, refers to Part III for theoretical background.

Risk analyses can be qualitative or quantitative and with varying levels of detail as required. A qualitative analysis can be extremely valuable although it stops short of quantitative risk estimation. Quantitative analyses have up until now (2003) not been extensively used in dam safety practice mainly because the profession has not known how to get meaningful and reliable estimates of the probability of an event occurring (e.g. dam failure). The book brings the reader up-to-date on the state-of-the-art and philosophy regarding the assignment of probability values to be used in risk analysis procedures.

It is widely recognised that dams seldom fail owing to a simple flaw in design or construction. A failure is generally a result of a complex linking of circumstances, conditions, flaws or errors that, when combined, result in failure (a failure mode). Failure mode identification is an essential step in the risk estimation process as it lays the foundation on which the remainder of the study is built. The identification is best accomplished through a team of experienced people who cover all aspects of the dam and its appurtenant structures. Knowledge, perspectives and opinions are synergistically combined in a focused team-work exercise to identify the possible failure modes.

Part II of the book presents the procedure for the failure modes and effects analysis (FMEA) and influence diagrams which are used to show the interrelationships among initiating events, states of nature, and conditions of the system. FMEA may be extended by a "criticality" analysis (FMECA) to provide a means of ranking the failure modes in terms of an index of risk that incorporates representations of probability and consequences.

Event tree analyses (ETAs) represent the logic of how an initiating event, like flood, earthquake or internal erosion, can lead to various types of damage and failure. Fault tree analyses (FTAs) are sometimes used within an event tree to assist in assigning probability values to different branches in the event tree. FTA is common in industrial applications to mechanical and electrical systems. For dams, FTA is therefore often associated with tree branches having to do with spillway gates and turbines. The sections treating the construction of event trees in Part II, supported by the more theoretical development in Part III, represent very valuable contributions to the future correct use of this powerful tool in dam safety practice. The text is supplemented by several illustrative examples to assist the reader.

The authors present new approaches and comprehensive sections on consequence analysis and assessment which are necessary for the estimation of risk and for the planning of emergency preparedness. The modelling of dam breaching and downstream inundation presently has many inherent uncertainties and requires further research. The warning time provided for the people downstream is essential for the outcome of an evacuation.

The challenging task of assigning probability values, for instance for use in an event tree analysis, is devoted a lot of attention and discussion. The book advocates that branch probabilities be estimated in one of four ways or in a combination of the following:

1. statistical estimates based on empirical data;
2. engineering models based on physical processes;
3. fault tree analyses based on logical constructions;
4. judgement by experts.

The second and third of these methods may be grouped together under the name, engineering reliability models.

Statistical estimates are characterisations or summaries of past observations. *Engineering models* are constructs based on reasoning from first principles of physics. Uncertainties in the model parameter values and in the model itself are propagated through the calculations to establish probabilities that the dam can resist the required loading conditions. *Fault trees* differ from engineering models in that they model the logic of a system rather than the physics of the system. *Judgement* is based on experts' intuition and reasoning that reflect a base of knowledge and evaluated experience. Collective judgement of experts, structured within a process of debate, may yield as good an assessment of probabilities as can be obtained by mathematical analyses. Some would claim a better assessment (e.g. Vick, 2002).

The authors provide insight as to the relative merits among the use of judgement (degrees of belief) and statistical or reliability analyses. They conclude that the approaches provide equally important information, but usually different information, which can be applied to different parts of a risk analysis.

The main purpose of a risk analysis is to provide support for decision-making. The integrated dam risk management process requires risk analysis, risk evaluation, risk communication, risk assessment and risk control. Part I of the book puts the process in perspective and clearly points out the value of risk information in decision-making.

The International Commission on Large Dams advocates the use of risk assessment in dam safety management (ICOLD, 2003). The book in front of you, *Risk and Uncertainty in Dam Safety*, is an authoritative, comprehensive, valuable and welcome contribution to dam safety practices. Through the presentation of a systematic and integrated process, it assists the dam owner in evaluating the needs for dam safety improvement, selecting and prioritising remedial and corrective actions, and improving the operation, maintenance and surveillance procedures. The book is a result of the unique cooperation among experienced and knowledgeable dam owners, dam safety managers and engineers, and experts in the theoretical basis for risk assessment. It contains a thorough review of the state-of-the-art, provides lessons from first-hand practical experience, and gives significant new contributions that will enhance understanding of the risk assessment and management process and how to apply it effectively.

I hope the book will be widely distributed and read. It will increase awareness and reduce complacency regarding dam safety issues. It should also appeal to readers outside the dam engineering community due to its general and excellent treatment of the various topics in the integrated process of risk assessment.

Kaare Høeg
Past President of ICOLD
June 2003

Reference

International Commission on Large Dams (ICOLD) (2003) *Bulletin on Risk Assessment in Dam Safety Management* (ICOLD).

Preface

This book is designed to provide all those interested in the safety of dams with knowledge and insight into the nature of risk and uncertainty as they arise in dam safety decision-making. It also provides details of the analysis methods available to characterise and quantify these risks and uncertainties. While the book is primarily intended to be a source of knowledge for dam owners, their engineering managers and engineering analysts, and dam safety regulators, it is structured and written in a way that is also relevant to other branches of civil engineering.

The underlying philosophy is to present a wide-ranging view of risk and uncertainty, in a practical way, as they arise in dam safety. The approach combines the philosophical underpinnings of risk and uncertainty with the techniques available to characterise their engineering and scientific dimensions in the physical performance of dams operating in the natural environment. The notion that $Risk = Probability \times Consequences$ is constant for all views. The adherence to mathematical and logical principles throughout the book is intended to ensure that the representations of risk and uncertainty are consistent across the full range of views from that of the detailed analysts to the ultimate decision-makers who may not be engineers and may well be public representatives.

At a fundamental level, the book holds as a premise that dams, like other civil structures, operate under the influence of human and natural forces in accordance with the laws of physics, the language of which is mathematics. The book attempts to strike a balance between the mathematical, scientific and engineering analysis elements of the characterisation of risk and uncertainty, as they apply to engineered structures in the natural environment. This approach does not diminish the importance of qualitative reasoning and judgement in characterising risk and uncertainty, rather it is intended to provide structure and context to their utilisation. Throughout, the use of the term science pertains to its meaning in the sense of systematically acquired knowledge.

The approach recognises that dams are not fully engineered structures in the sense that buildings or even nuclear power plants are engineered and, as such, the techniques available to characterise risk and uncertainty in fully engineered systems are not sufficient or adequately structured for direct application to dams. At the outset it was recognised that writing the book would not simply be a matter of adapting existing techniques for characterising risk and uncertainty in engineered systems, and that it would be necessary to go beyond conventional engineering analysis.

To achieve this wide-ranging perspective, the book is divided into three parts, with the bulk of the mathematics in Part III. The first part is intended for dam owners and dam safety policy makers; the first and second parts for engineering managers charged with ensuring that the necessary information about risk and uncertainty are available to inform decision processes; all three parts are relevant to dam risk analysts. This design is intended to ensure that all those involved have a common understanding of what the decision-makers require, why risk analysis projects are structured the way they are, while properly accommodating the needs of the analysts. This said, there is no reason why any reader might not want to be familiar with all aspects of the characterisation of risk and uncertainty in dam safety, and the writing style is intended to foster such permeation of knowledge. The objective is to ensure that those responsible for risk analysis projects structure them to permit the analysts to reveal the nature and extent of risk and uncertainty and to appropriately inform the decision process.

This book is the end result of an initiative that arose out of an industry desire for information on risk analysis for dam safety as identified in a number of surveys in Canada and the United States in 1996. In early 1997, the Canadian Electricity Association's (later CEA Technologies Inc.) fledgling Dam Safety Interest Group (DSIG) sought a proposal from BC Hydro to develop guidance on risk analysis for dam safety. The original proposal envisaged a three-phase multiyear project that would first develop a dam safety specific version of the Canadian Standards Association's (1991) *Risk Analysis Requirements and Guidelines*. The second phase, initiated in 1998, was to involve preparation of a *Commentary on the Guidelines* based on contemporary practices. This was to be followed by a third phase that would involve the development of two examples specially designed to illustrate the application of the principles described in the *Guidelines* and procedures described in the *Commentary*. The result was intended to be a three-part book on risk analysis in dam safety. A fourth phase to maintain and enhance the products of the first three phases completed the project. Considering the evolutionary state of risk analysis for dam safety, the BC Hydro proposal was designed to permit changes of scope and objectives in response to changing needs of the industry, and advances in the theory and practice of risk analysis.

The proposal envisaged that the *Guidelines* would be prepared by the project leader (Dr. D. Hartford), as guided by the sponsors and reviewers, and that they would reflect contemporary understanding of the endeavour of risk analysis in dam safety practice. The *Commentary* was to comprise submissions by dam safety practitioners working in the field of risk analysis for dam safety, edited by D. Hartford to ensure consistency. The reviewers for Phases I and II were; Professor C. A. Cornell, Professor R. Fell, Dr. H. Kreuzer, Professor N. C. Lind, Mr J. L. Von Thun and Professor R. V. Whitman.

Given the level of interest in quantitative risk analysis for dams and a growth in the practice of risk analysis across the industry at that time, the objectives appeared modest, especially considering the nature of the first phase of the proposal. During the first phase a need to adjust the objectives to re-focus the Phase I effort on development of Guiding Principles for Risk Assessment in Dam Safety was identified. This re-focusing was due, in part, to genuine differences in opinion among practitioners and between the project sponsors concerning the confidence that dam safety decision-makers could have in the results of contemporary (late 1990) practices. It was also, in part, in response to the advice of the project reviewers, and advances in knowledge and understanding of the characterisation of risk and uncertainty among dam safety engineers. The imbalance between the states of knowledge across the various analysis methodologies, as revealed by the submissions for the *Commentary*, was the third factor that influenced the decision. The re-focusing of the effort overcame these difficulties and Part I of this book (Chapters 1–3) is the result of this change in focus and subsequent work.

As Phase II was restricted to analysis methods, the preparation of these contributions continued during the re-focusing of Phase I of the project. The aspects of practice that had been selected and the invited contributors were: failure modes and effects analysis – Mr J. W. Findlay; event tree analysis – Dr. D. S. Bowles; fault tree analysis – Dr. M. W. McCann Jr; probability assignment – Mr L. A. McDonald; and dam failure consequence analysis – Mr W. J. Graham. These submissions were completed and reviewed in early 1999.

Critical review of the results of Phases I and II of the project by the project team and the sponsors, that incorporated the advice of the reviewers, concluded that despite meeting the objectives of the original scope, these Phases of the project could not be brought to a satisfactory conclusion. On reflection, "the bits did not fit together very well" and despite representing the state of the art, "there were holes and inconsistencies throughout". The overall conclusion was that the project could not be brought to a satisfactory conclusion without first preparing a suitable treatment of the philosophical underpinnings of the analysis of risk and uncertainty, and then revising the *Commentary* on analysis methods in the light of the theory.

This meant that the Phase III focus shifted from examples to theory with a Phase IV introduced to revise the *Commentary* and integrate the entire effort. The bulk of the effort between 1999 and 2002 was in writing Part III, Theory, with the major revisions to Part II following completion of the various elements of the theory. This very significant increase in effort and direction required a complete re-thinking of how the project would be staffed to write the theory and complete Phase IV.

The strategy to complete Phases III and IV was to establish a sponsors' Working Group to prepare the necessary material. The sponsors' Working Group initially comprised the authors supported by Dr. D. S. Bowles (Utah State University), Mr K. Dise (US Bureau of Reclamation) and Dr. P. A. Zielinski (Ontario Power Generation). Subsequently, Mr M. Smith (Hydro Quebec) replaced Dr. Bowles in the sponsors' Working Group upon completion of the chapter "Essential elements of probability theory". We split the writing of Part III (Baecher) and the re-writing of Part II (Hartford) between us with the other members of the sponsors' Working Group providing review, critique and guidance as to technical content.

The sponsors' Working Group met for intensive working sessions four times a year and twice a year with the sponsors themselves, during the period June 1999 through to July 2003, with the writing and review carried out between meetings as the material evolved. Parts II and III are the result of this four-year effort.

Part III, Theory, was really a matter of starting from scratch, with meetings focused on content and direction whereas Part II, Commentary, required generalisation and supplementation of the previous submissions based on the material in Part III.

There was no simple formula for the process of generalising and supplementing the original submissions on analysis methods. The process ranged from re-organisation of the contribution at one end of the spectrum (fault tree analysis), to creation of a completely new treatment using some of the ideas presented in the original contributions to varying degrees (event tree analysis and probability assignment). Intermediate between these extremes the contributions were subsumed to varying degrees into a more general discussion (FMEA and consequence analysis). The need to variously revise and re-write these contributions does not detract in any way from the usefulness of these contributions. Rather it was a necessary result of the re-focusing of the project as originally envisaged and in response to advances in knowledge and understanding.

The broad spectrum of interests that the book seeks to address necessarily means that there is some limited repetition across the parts. This limited repetition is intended to provide sufficient overlap between the three parts to ensure continuity of understanding across the audience spectrum.

This book had its origins in a desire for knowledge by the founding sponsors: BC Hydro, Hydro Quebec, Manitoba Hydro, Ontario Power Generation, Scottish and Southern Energy (formerly Scottish Hydro Authority), US Bureau of Reclamation and Elforsk. We thank the sponsors for their support and we greatly appreciate their commitment to this long-term project. Section 3.3 reflects the views of Bruce Muller of USBR, and Ray Stewart of BC Hydro as Part III approached completion and we would like in particular to thank them for bringing a unified sponsors' perspective to the book.

This book evolved through various phases, the course of which was strongly influenced by the views of the reviewers and contributors. We greatly appreciate the advice and guidance provided by Professor C. A. Cornell, Professor R. Fell, Dr. H. Kreuzer, Professor N. C. Lind, Mr J. L. Von Thun and Professor R. V. Whitman. The contributions to the original Part II concept by Dr. D. S. Bowles, Mr J. Findlay, Mr W. J. Graham, Dr. M. W. McCann Jr and Mr L. A. McDonald provided the platform for debate and subsequent advancement of the knowledge contained in Part II.

The formation of the sponsor's Working Group and the active participation of our colleagues David Bowles, Karl Dise, Marc Smith and Andy Zielinski, was pivotal to the completion of the book. The sponsor's Working Group was, in our view, one of the great successes of this project as it provided a successful model for collaborative research in dam safety. In addition to facilitating the completion of the book, our collective knowledge and understanding of the nature and characterisation of risk and uncertainty continued to develop throughout the life of the project. Strong friendships and deep respect for each other evolved to the extent that what were initially quite diverse perspectives changed to a degree and evolved to a broad and deep appreciation of each other's perspective.

This is the first edition of *Risk and Uncertainty in Dam Safety.* Whether or not we are fortunate enough to prepare the second edition, we are convinced that risk and uncertainty will always be a part of dam safety, be it during normal operation, or as a result of extreme natural events or human causes. We are privileged to be part of this effort and hope that it will be a useful contribution to dam safety.

D. N. D. Hartford
G. B. Baecher

List of figures

Chapter 1

Chapter 2

Chapter 3

Chapter 4

Chapter 5

Chapter 6

Chapter 7

Chapter 8

Chapter 9

Chapter 10

Chapter 11

List of tables

Chapter 9

Chapter 10

Chapter 11

Part I
Guiding principles

1 Dam risk management

This volume describes the fundamental aspects and benefits of risk analysis, risk evaluation, risk assessment and risk management approaches to dam safety. The guiding principles of Part I of this volume provide a framework for risk-based dam safety management which is founded on tested dam safety practices, the cornerstones of which are surveillance, periodic dam safety reviews, emergency preparedness (planning and testing), and operation and maintenance procedures. They also illustrate the broad applications of risk analysis in dam safety practice. While the focus of this volume is the safety of existing dams, the concepts are equally applicable to the design of new dams.

In managing dam safety, it is necessary to strike a balance between the need to protect against the risks posed by dams and the consequent benefits of those same dams. The dam risk management process is aimed at supporting the preparation of safety reports which demonstrate:

- How risks are identified and assessed.
- What has been done to reduce risks to appropriate levels.
- What systems are in place to ensure that risks are properly controlled.

The guiding principles of Part I are intended to provide owners, who may or may not have dam safety engineering expertise, with information as to the form of a risk-based dam safety management framework to assist them in discharging their responsibilities of managing the risks posed by their dams, and engineers, responsible for assessing the safety of dams, with information on the implementation of a risk-based dam safety management process.

The guiding principles are not intended to be a standard of practice. The onus is on owners and their engineers to:

- demonstrate that the guiding principles as outlined are appropriate for application to their situation;
- obtain regulatory approval on the application of these guiding principles; and
- ensure that in utilising these guiding principles all legal, professional and regulatory requirements are met.

In addition to the normal professional responsibilities of engineers to act at all times in accordance with Codes of Ethics and Codes of Professional Practice, some professional licensing bodies recognise that engineers have a peculiar responsibility when dealing with risk issues, especially when forming a judgement about the tolerability of societal risk.

Although few professional licensing agencies (in North America) have established formal practice for engineers regarding risk issues, risk is becoming recognised as a matter worthy of its own professional practice policy. Therefore, in addition to awareness of compliance with the normal Codes of Ethics and Professional Practise of the engineering profession, users of this volume should be aware of any policies specific to risk issues adopted by their licensing body or association, plus any regulatory body which may be involved.

1.1 Dam risk management process

The risk-based dam safety management process is an enhancement of the traditional dam safety management process in that the three fundamental components – surveillance, periodic dam safety reviews and operation and maintenance procedures – are central to the procedure. The enhancements may be achieved through an integrated process that affords explicit recognition and analysis of risk in the system, formal treatment of the uncertainties that are ever present in dam safety practice, and a defined risk evaluation process leading to the risk assessment processes. Together they permit informed decision-making concerning the manner in which the risks are being controlled. The purpose of this section is to provide an overview of the various aspects of dam risk management. A comprehensive dam risk management process is illustrated in Figure 1-1.

Consistent with traditional dam safety decision-making, the dam risk management process requires five supporting processes:

1. The generation and analysis of information about individual dam systems.
2. The establishment of criteria with which the information on the individual dam systems can be assessed.
3. A decision-making process which leads to the most appropriate course of action.
4. A control process to ensure adequate control of the risks.
5. A periodic audit to continually monitor the scope and suitability of the risk controls.

This requires in addition that; (a) all activities are carried out within a framework that addresses the safety management needs of the dam system as a whole as well as the individual components, and (b) all information, either gathered or generated through some form of analysis at any step in the process, is formatted in a manner that permits it to be used in any subsequent steps.

This process is based on a top-down approach whereby the dam system is systematically and successively decomposed into elemental components to determine the fundamental causes of risk in the system.

Risk assessment is central to the dam risk management process, the essential features of which are illustrated in Figure 1-2. In risk assessment, the results of the risk analysis and risk evaluation processes are integrated and recommendations are made with respect to the risk control process.

A schematic diagram is often developed as a tool for comprehending the system under consideration, and its boundaries (Figure 1-3). System component diagrams (Figure 1-4) form the basis of risk analyses processes and have application at all levels of a risk analysis from scoping, through the various levels of iteration, to the final decision recommendation.

Performance-based surveillance processes which incorporate risk-based methods, including failure modes and effects analysis, are gradually emerging and have two principal objectives: to identify the significant failure modes that are unique to the structure, and to ensure that the surveillance pertains to all potential failure modes so that advance warning of deteriorating performance can be obtained and intervention actions initiated.

System component diagrams provide a particularly useful means for communication, ensuring transparency, and testing the logic of the analysis process. They assist in maintaining focus during a risk analysis, and provide a useful check that an appropriate amount of effort is afforded to each part of the analysis.

Risk analysis leads to an understanding of the sources and magnitude of risk. Risk analysis for dam safety is a structured process aimed at identifying both the likelihood of failure of the dam or dam components, and the extent of the consequences of failure. Risk analysis processes reveal the fundamental performance characteristics of the dam and its components. It generates information about the risk in the system and the contributors to that risk.

Risk analysis processes are finding increasing applications in the management of the safety of dams including, but not restricted to, determining the consequence (hazard in traditional terminology) category of a dam, assisting in developing more effective

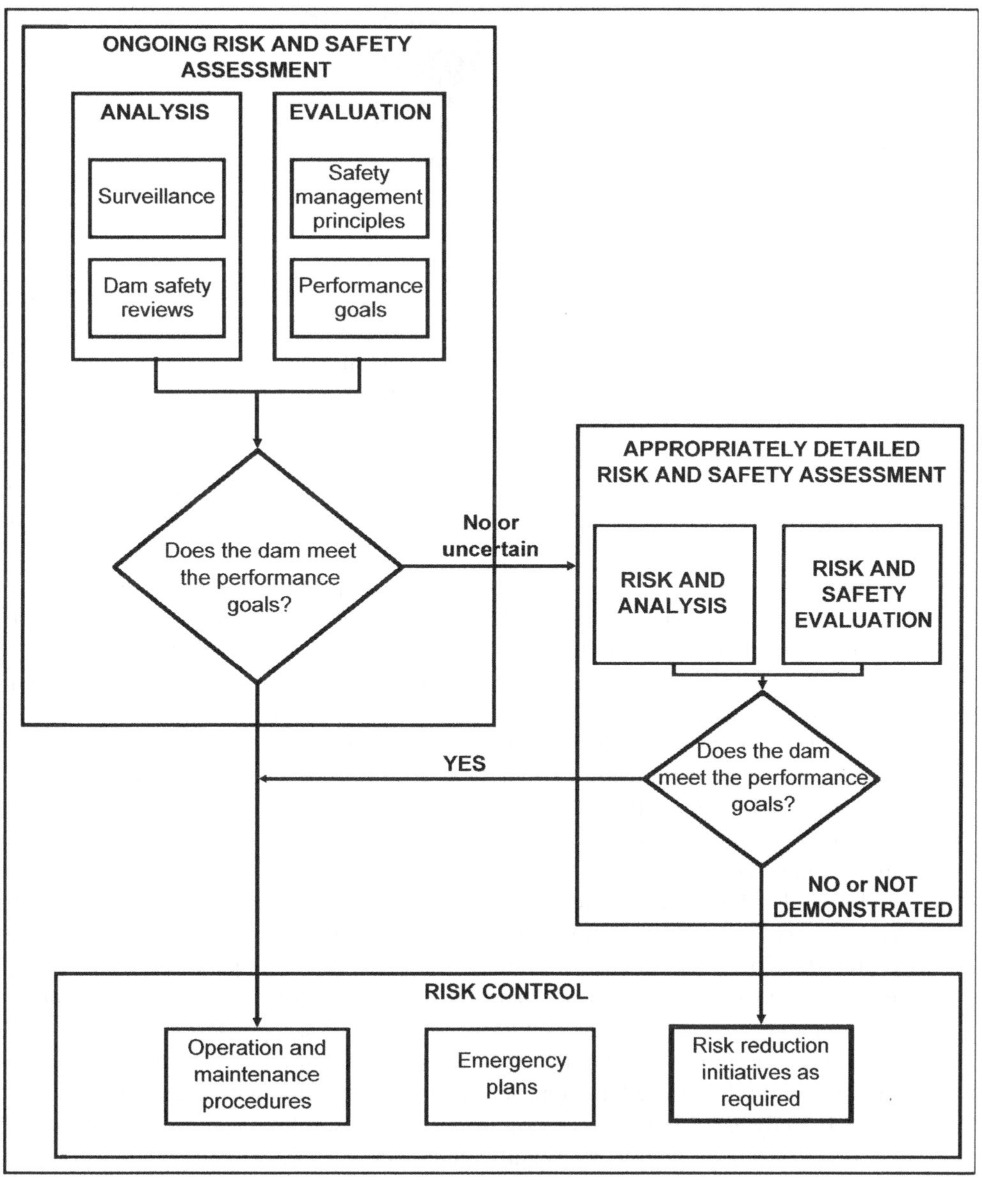

Figure 1-1. Dam risk management process

Figure 1-2. Risk assessment

surveillance processes, improving conventional dam safety assessment processes, demonstrating a sound understanding of the sources of risk and their relative contributions, determining resource requirements for investigation, analysis and surveillance and how they should be allocated to a particular dam or across a portfolio of dams, identifying modification alternatives that most specifically address the major risk contributors for a particular dam, prioritising dam safety modifications and improvements across a portfolio of dams, and communicating dam safety recommendations and/or decisions to financial planners, senior management, regulatory bodies and the public.

Risk analysis techniques enhance traditional processes by providing, a comprehensive integrated analysis of all of the components of the dam system and how they interact, a clear understanding of how the dam might fail (i.e. how the adverse consequences might materialise), an estimate of the probability of failure (if required), the associated consequences of the failure, and an estimate of magnitude of the risk (if required). Risk analysis processes can assist in all aspects of dam safety management which lead to the conclusion that, given that certain performance requirements are met, the dam can be operated safely.

To ensure that surveillance and periodic dam safety reviews fully support the dam risk management process, risk analysis methods may form the basis of surveillance activities and periodic dam safety reviews. Risk analysis can be either qualitative or quantitative and to varying levels of detail (i.e. decomposed or de-aggregated to the extent that is required). Thus, risk analyses can take several forms depending on the scope and intended end use. One form of analysis is not necessarily superior to another, if the requirements of the scope are met and the method is appropriate for the problem in hand. It is recommended that the end use be defined at the outset of the risk analysis so that the methods selected are appropriate and compatible with the intended application.

Risk analysis of dam safety requires a multidisciplinary approach as it may cover areas of science, engineering and social science. Expertise ranging from hazard analysis, to dam response analysis, to consequence analysis involving economics, sociology, and psychology may be required.

Risk evaluation, the process of understanding and judging the significance of risk, is fundamental to risk assessment and risk-based decision-making. The principal role of risk evaluation in risk assessment is the generation of decision guidance against which the results of a risk analysis can be assessed. The process of generating risk-based decision guidance requires a statement of the owners' safety management principles and of the values and preferences of the public, which include consideration of prevailing financial, legal and regulatory conditions.

Definitive statements concerning the significance of a particular risk require definition of the background against which the significance of the risk is being described. There must be a general understanding of all risks within the system under consideration. In some cases the system will be limited to the dam itself, in others the system could include local downstream, regional, national and even international elements.

In principle, data and information generated during the surveillance activities should be formatted in a manner that permits this knowledge to be incorporated directly in more detailed risk analyses. This form of surveillance is simply an enhanced version of conventional check list type surveillance activities, in that the generic check lists are

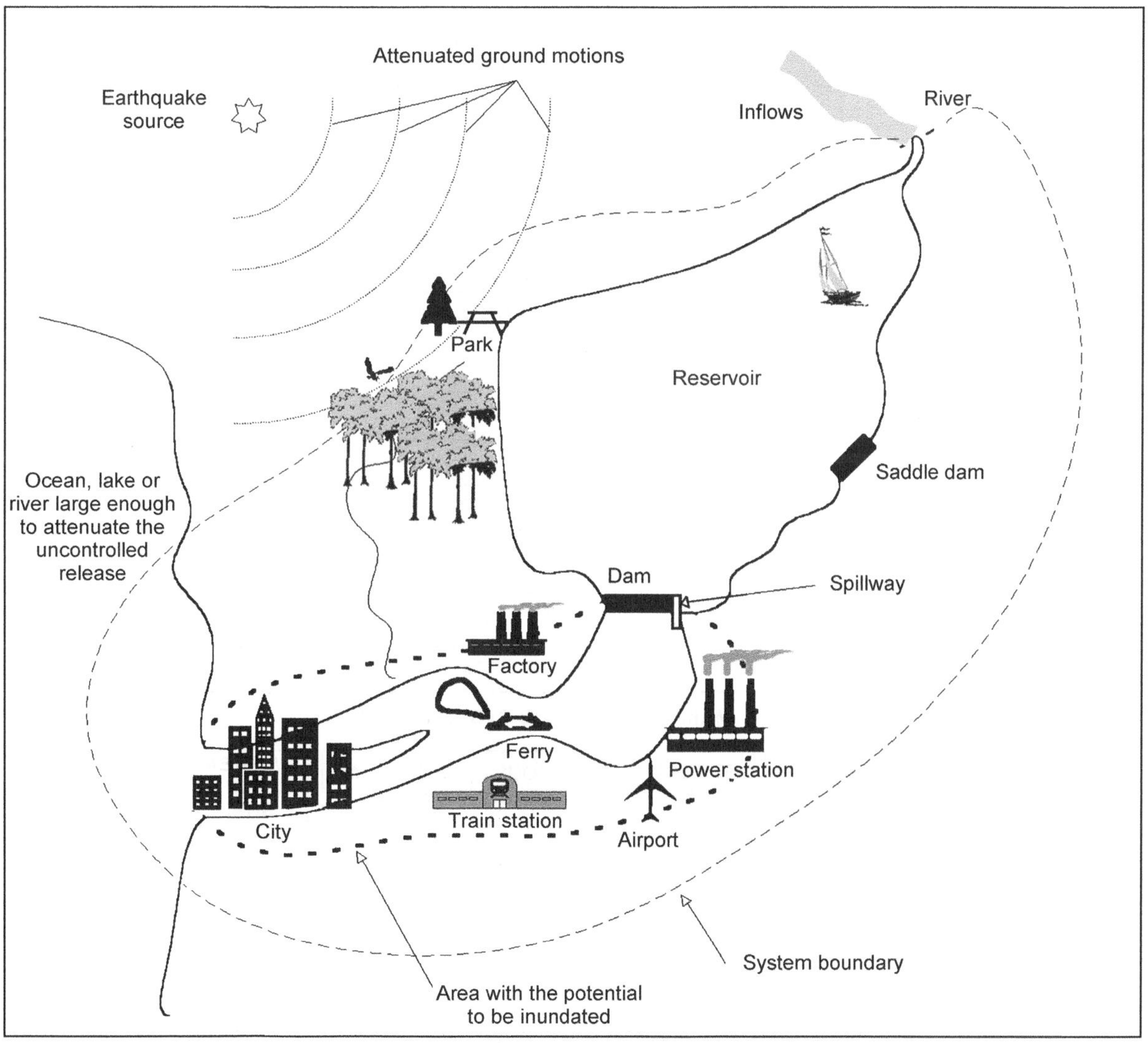

Figure 1-3. System schematic diagram

retained, but the primary focus is on the performance of the dam and the detection of signs of adverse performance of the individual components of what is a unique structure.

Performance-based surveillance processes have the advantage of transforming surveillance to a proactive and failure mode/failure consequence focused activity. This type of surveillance can be considered as an ongoing risk assessment process. The adoption of risk-based surveillance processes is compatible with and supports the guiding principle described in Section 1.2 on an ongoing basis.

The introduction of risk-based operation and maintenance procedures appears to be a logical next step in the development of a comprehensive risk management process for dam safety. While risk-based operation and maintenance processes are still at a conceptual stage, there does not appear to be any reason why these activities should not be based on sound risk management principles. It is envisaged that, in addition to

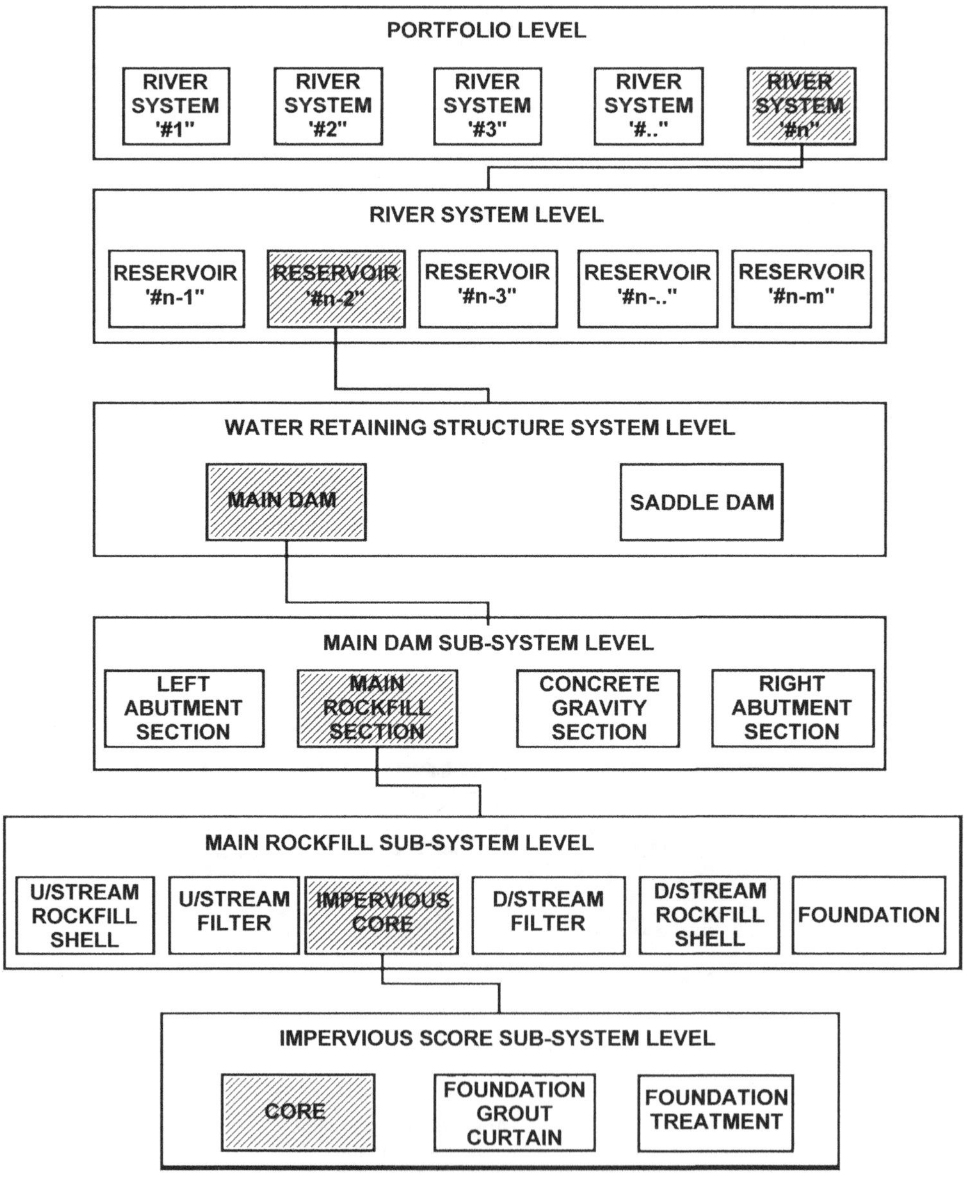

Figure 1-4. System component diagram

Table 1-1. Potential benefits of risk management of decision-makers

Aspect of risk management	Potential benefits to decision-makers
Risk analysis	• Identification of potential hazards or failure-causing conditions • Identification of potential failure modes • More focused gathering and analysis of data • Improved understanding of the dam and its performance • Qualitative and quantitative statements of risk in the system • Identification of the important contributors to risk and the vulnerabilities of the system • Identification of uncertainty in the analysis and quantification of the degree of confidence in the result
Risk evaluation	• Clearer description of the nature of risk in the system, and the magnitude of the consequences should they materialise, enabling the decision-maker to establish the extent to which he/she is averse to the risk • Clear statements of the values and philosophies of the owner and regulatory authorities
Risk communication	• Consistent framework for communication between owners, designers, surveillance engineers, dam safety review engineers, operations and maintenance staff, emergency planners, decision-makers and regulators • Comparison of risks posed by one dam with similar dams, and other facilities • Communication of uncertainties and engineering/science-based policy considerations that are generally implicit in dam safety decision-making
Risk assessment	• Statement that the dam can or cannot be operated safely and under what conditions • Assessment of options and alternatives to reduce the risk • Level or risk during any remediation programme • Definition of the post remediation residual risk and identification of residual risk control strategies • Answer to the question "Is the dam safe enough?" • Demonstration of due diligence
Risk control	• Aid in establishing priorities and consistency for expenditures on safety improvements for individual dams and between dam systems • Plans for surveillance procedures • Improvement of emergency planning

statements of the conventional requirements that are to be met, the risks involved in operating and maintaining the dam in accordance with these requirements would be described, as would be the increased risk associated with not meeting the requirements.

1.2 Potential benefits of integrated risk management

The results of dam safety risk analyses can be assessed and used in a risk management program by a decision-maker to help judge the tolerability or acceptability of risks and to aid in choosing between potential risk-reduction measures. From a decision-maker's perspective, the potential benefits of an integrated risk management approach are summarised in Table 1-1.

All of the above are important in the effective risk management of dams, whether the objective is improved public safety, prevention of economic loss, or compliance with government regulations.

2 Guiding principles for risk analysis

Risk analysis for dam safety seeks to address several fundamental issues including: what can happen to cause the dam to fail or be damaged? How probable is it? What are the various consequences and their associated probabilities? What are the probability weighted consequences, or risks?

The risk analysis process generally involves the following activities (Figure 2-1):

- Scope definition.
- Hazard and load identification and definition.
- Probabilistic analysis of hazards and loads.
- Failure mode identification.
- Dam response and failure probability analysis.
- Estimation of consequences corresponding to each failure event.
- Risk estimation.
- Uncertainty and sensitivity analysis.
- Documentation.
- Expert review and/or verification (if possible).
- Analysis update (if required).

Risk analysis incorporates failure probability and failure extent along with consequence magnitude and associated probability. Expert or peer review verification and analysis update provide quality control and a process to permit the analysis to be used in on-going risk management.

2.1 Risk analysis methods

The principal methods available for conducting a risk analysis for dams are introduced here and explained in more detail in Part II. The following has been adapted from the Canadian Standards Association's (1991) *Risk Analysis Requirements and Guidelines*. The principal methods are: (1) failure modes and effects analysis (FMEA); and associated methods; (2) event tree analysis (ETA); and (3) fault tree analysis (FTA).

Failure modes and effects analysis (FMEA)

FMEA (including such variants as failure modes, effects and criticality analysis (FMECA) and hazard and operability (HAZOP) studies, is a method of analysis whereby the effects or consequences of individual component failure modes are systematically identified and analysed.

While the actual analysis is inductive (i.e. is based on the question "What happens if a component or element fails?"), it is first necessary to break the dam system down into

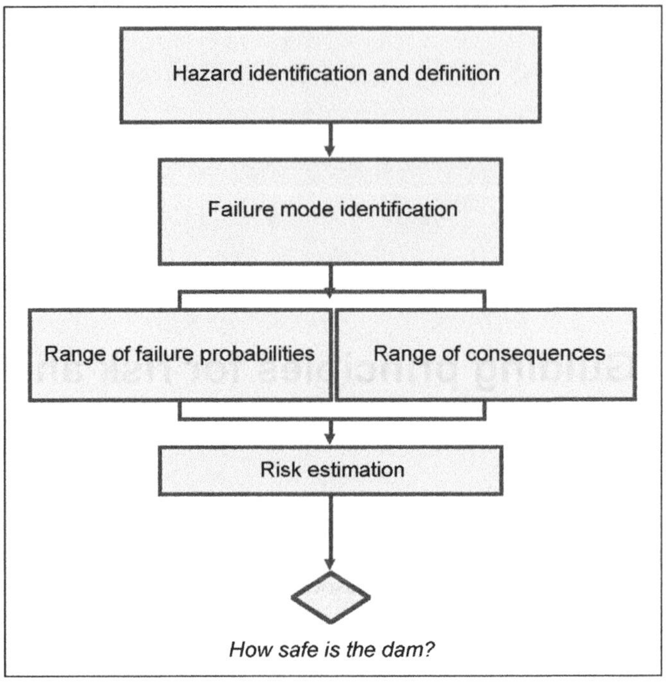

Figure 2-1. Risk analysis process

its individual components or elements. A logic diagram of the type commonly used in reliability centred maintenance can often be of assistance in carrying out this task. Once the system has been de-aggregated the failure modes of each of the fundamental elements can be identified.

Some aspects of failure mode identification that are well established in the mechanical and electrical engineering disciplines, and in the aerospace, petrochemical and nuclear industries, are directly applicable to some aspects of dam safety risk analysis. However, these techniques are less well suited to certain aspects of the analysis of "continuous sub-systems" such as dams and it may well be necessary to modify the process and perform a root cause analysis.

Once the failure modes and/or root causes have been identified, the effects of the failure mode on other components of the sub-system and on the system as a whole are systematically identified. The analysis is usually descriptive and information is normally presented in tabular and/or spreadsheet form. FMEA clearly relates component failure modes and their causative factors to the effects on the system and presents them in an easily readable format. The hierarchical nature of FMEA and the relationships between failure mode, failure cause and failure effect is illustrated in Figure 2-2.

A FMEA can be extended to include criticality considerations (FMECA). In a FMECA each failure mode identified is ranked according to the combined influence of severity and its probability of occurrence.

Event tree analysis (ETA)

Event tree analysis (ETA) is a technique, either qualitative or quantitative, that is used to identify the possible outcomes and, if required, their probabilities given the occurrence of an initiating event. ETA is an inductive type of analysis where the basic question that is addressed is "What happens if . . .", e.g. what happens if there is an earthquake of magnitude M 6.5 and peak ground acceleration of 0.27 g? An example of an event tree for one failure mode for a dam subjected to an earthquake is illustrated in Figure 2-3.

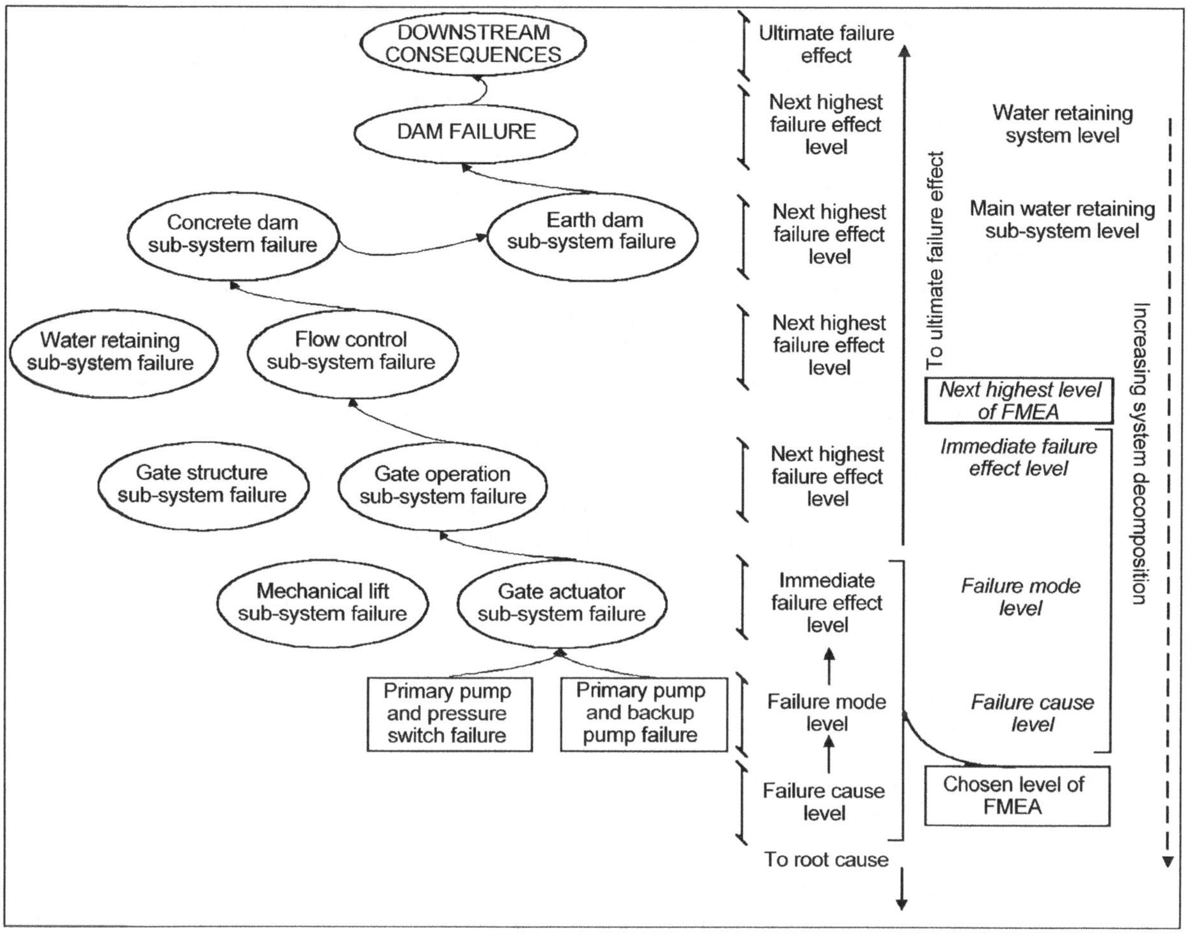

Figure 2-2. Hierarchical nature of FMEA

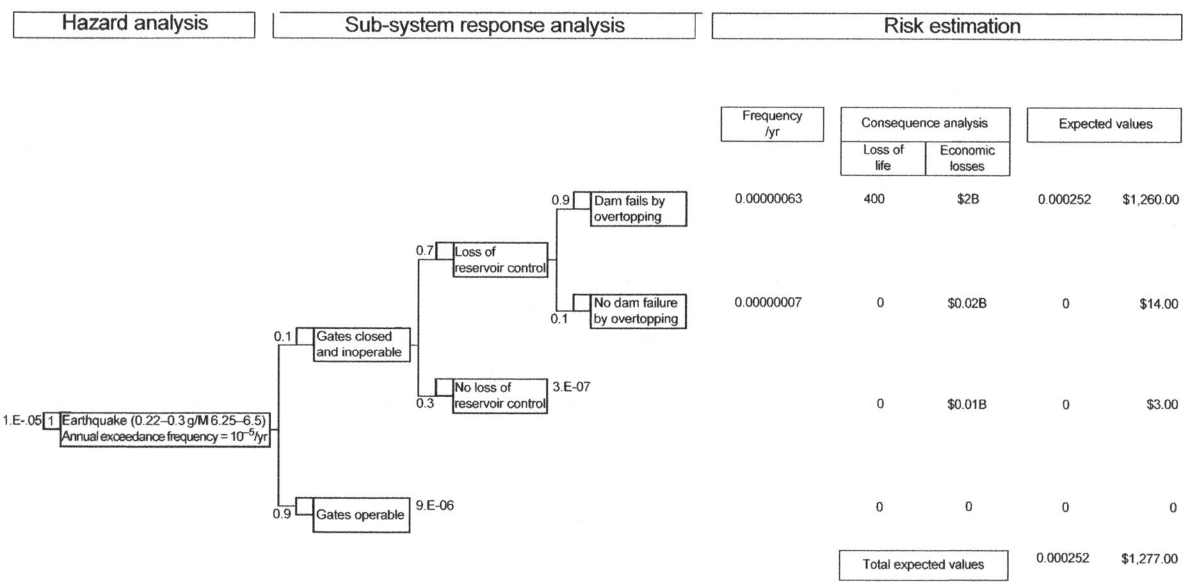

Figure 2-3. Example of event tree analysis

ETA is widely used for facilities provided with engineered accident-mitigating features to identify the sequence of events that follow the initiating event and produce specified consequences. In dam safety applications, ETA reveals the relationship between the functioning or failure of various mitigating systems and it is useful for identifying events that require further analysis using fault tree techniques (i.e. individual event tree branches become the top events of the fault trees).

Fault tree analysis (FTA)

Fault tree analysis (FTA) is a technique, either qualitative or quantitative, by which conditions and factors that can contribute to a specified undesired event (called the top event) are deductively identified, organised in a logical manner and represented pictorially. The faults identified in the tree can be events that are associated with component hardware failures, human error or any other pertinent event that leads to the undesired outcome (e.g. dam overtopped).

Starting with the top event, the possible causes or failure modes on the next lower functional system level are identified. Following the step-by-step identification of undesirable system operation to successively lower system levels will lead to the desired system level which is usually the component or element failure mode.

An example of a fault tree for a failure of an emergency spillway generator to start is illustrated in Figure 2-4.

Consequence analysis

There are essentially two aspects to failure consequence analysis:

1. Dam breach flood definition.
2. Flood impacts analysis.

Computer programs for dam breach flood definition, developed for traditional dam safety applications, are commercially available. These programs provide estimates of flood arrival times and average flood depth and velocity with time at defined cross sections in the inundated area.

Flood impact analysis requires definition of the characteristics of the inundated area including details of the population at risk, property and environmental impacts, as well as the responses to the inundation conditions.

Assigning probabilities

Risk analysis for dam safety is fundamentally a characterisation of the uncertainties in the performance capability of dams under the loading conditions of interest. Risk analysis is useful because it provides a systematic structuring of uncertainty, and this structuring allows us to better understand how uncertainty arises and how information may lessen it.

The most commonly used measure of uncertainty in dam safety studies is *probability*. Probability is a mathematical construct used to express degrees of uncertainty about the occurrence of events, state of the world and truth of propositions. As discussed in Chapter 7, two principal interpretations of probability are common: probability as frequency, and probability as degree of belief. Because there is more than one interpretation of the meaning of probability, there is also more than one way to assign probabilities.

A review of the contemporary literature creates the impression that there is no unique way to assign probabilities in dam safety risk analysis, and such an impression is correct. That two distinctly different interpretations of probability exist make it necessary for the analyst to differentiate which is used for specific probabilities in the risk analysis. As a general rule, probabilities describing rates of occurrence are interpreted as frequencies, while probabilities describing states of nature (e.g. parameter

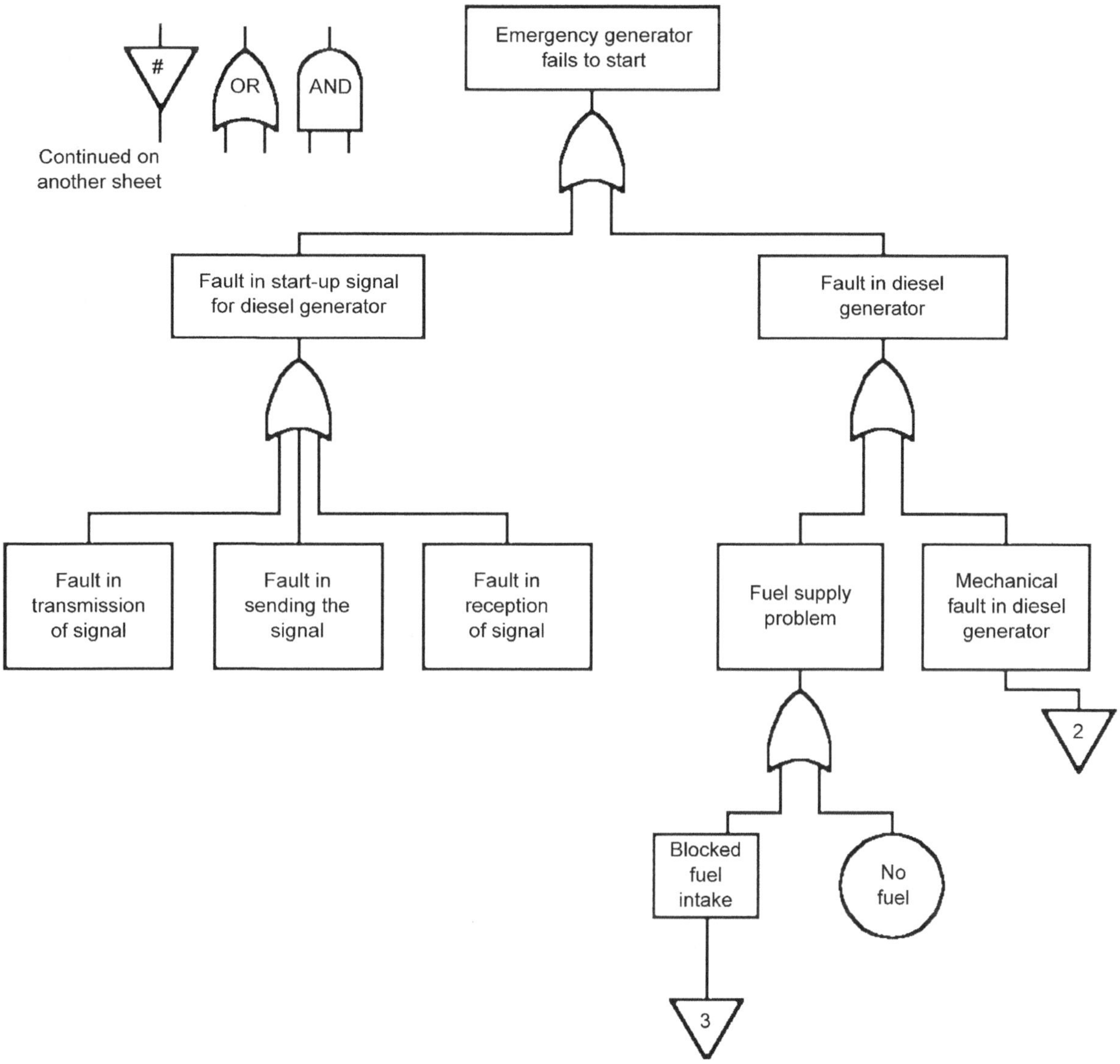

Figure 2-4. Fault tree for failure of an emergency spillway generator to start

values) or the truth of hypotheses are interpreted as degrees of belief. Philosophical issues concerning probability are discussed in detail in Chapter 7, while practical methods of assigning probabilities are outlined in Chapter 6 and discussed in detail in Chapter 11.

The most important components of uncertainty in risk analyses for dam safety include at least the following:

Natural variability associated with the "inherent" randomness of natural processes, manifesting as variability over time for phenomena that take place at a single location (temporal variability), or as variability over space for phenomena that take place at different locations but at a single time (spatial variability), or as variability over both time and space. Such natural variability is approximated using mathematical simplifications, or models. These models may or may not provide a good fit to natural phenomena. In the best of cases, they are close but only approximate fits.

Model uncertainty, reflects the inability of a model or design technique to precisely represent a system's true physical behaviour, or our inability to identify the best model, or a model that may be changing in time in poorly known ways (e.g. flood-frequency curve changing because of changing watershed). The models we fit to naturally varying phenomena need to be fit to natural processes by observing how those processes work, by measuring important features, and by statistically estimating parameters of the models. Model uncertainty is sometimes treated as uncertainty without assigning probabilities to alternatives.

Parameter uncertainties, result from our inability to accurately assess parameter values from test or calibration data, due to limited numbers of observations, and the statistical imprecision attendant thereto.

Data uncertainties, including: (i) measurement errors, (ii) inconsistency and inhomogeneity of data, (iii) data handling and transcription errors, and (iv) inadequate representativeness of data sample due to time and space limitations all of which reflect upon the information or knowledge available to describe historical system behaviour.

2.2 Risk analysis process

The risk analysis process, illustrated in Figure 2-1, generally involves the following activities:

1. Scope definition and selection of analysis methods.
2. Hazard identification and definition.
3. Estimation of probability of dam failure.
4. Estimation of consequences of dam failure.
5. Risk estimation.
6. Documentation.
7. Verification.
8. Analysis update.

These activities can be carried out using an iterative approach with increasingly detailed levels of analysis. At each level, a risk analysis should be documented and verified to an appropriate level of detail to facilitate subsequent review and revisions. Depending on the intended application, only certain elements of the process outlined below may need to be considered. However, the omission of these elements may have to be justified.

Scope definition

The scope of the risk analysis should be defined and documented in the risk analysis plan at the start of the project. This will include a thorough familiarisation with the dam and the various sub-systems that make up the dam system under consideration. Definition of the scope of a dam safety risk analysis involves the following steps.

1. Describe the reasons for carrying out a risk analysis, and formulate the objective of the risk analysis based on the main concerns identified.
2. Define the dam system being analysed, including:
   - general description of the dam system including the hazards being analysed.
   - physical and functional boundaries;
   - environment (e.g., physical environment);
   - all flows (e.g., reservoir inflows, outflows, debris) and influences (e.g., seismic shaking) that cross the boundaries of the dam system;
   - operating conditions covered by the risk analysis, and any relevant limitations;
   - identification of all the technical, environmental, organisational, and human circumstances that are relevant to the analysis.
3. State data availability assumptions and constraints governing the analysis.
4. Identify the decisions that have to be made, the decision processes to be used, the criteria and the decision-makers.

5. Identify suitable methods for analysing the risks.
6. Document the risk analysis plan, including all of the above.

Selection of methods

Once the objectives and scope of the risk analysis have been defined, the initial level of effort should be selected based on such factors as the following:

1. Role of the analysis in the decision-making process.
2. Objectives of the study.
3. Type of dam system and hazards being analysed.
4. Level of risk involved.
5. Personnel, level of expertise and resource availability.
6. Information and data requirements/availability.
7. Modification/updating of the analysis.

Hazard and load identification and definition

The hazards and loading conditions which generate risk in the dam system require identification. Risk analyses for dams can be categorised by the nature of the hazard leading to the consequences of concern.

The common usage of "load" relates to the forces on a dam, foundation and appurtenant facilities (e.g. spillway, outlet works) resulting from external loads: reservoir under normal conditions, reservoir under flood conditions, earthquake, and other natural hazards or special conditions (ice, debris, landslides, avalanche). In the broader sense, loading conditions should include any causative condition which acts on the dam and could, depending on the response of the dam and the action taken, result in dam failure. These broader "causative conditions" fall into three categories: (1) operational process (equipment response, human error, etc.); (2) internal process (alkali–aggregate reaction, corrosion, metal fatigue, seepage and internal erosion, etc.); and societal actions (war, sabotage, etc.).

Known hazards, such as floods and earthquakes, should be identified. To identify unknown hazards, formal methods can provide a structure for asking questions requiring foresight so that a system is fully analysed.

Probabilistic hazard analysis

For quantitative risk analyses, probabilities of hazards will be required. While detailed guidance on the estimation of probabilities of hazards is beyond the scope of Part I, guidance on process is included in later chapters.

Failure mode identification

Failure mode identification is an essential step in the risk estimation process as it lays the foundation on which the remainder of the study is built. The extent to which failure modes are defined may depend on the level of the analysis. Failure mode identification can be performed as part of, or as a precursor to, a failure modes and effects analysis.

Failure mode identification requires that the dam system be systematically reviewed to identify the manner in which the dam, foundation or appurtenant structures may fail under the imposed loading or causative conditions. This systematic review might include: dam safety reviews which provide basic dam-specific input to the process; consideration of appropriate case histories of dam failures and historical records of dam incidents, experience from previous risk analyses provides useful input to this process, or checklists of causative conditions and failure modes to assist in identifying potential failure modes for the dam under review; a formal process that provides a structure for raising issues and posing questions to a group of people familiar with all aspects of the project, so that the system is exhaustively analysed; event tree logic diagrams; or fault tree logic diagrams.

Dam response and failure probability analysis

Depending on the scope, the dam response analysis can take various forms including qualitative failure modes and effects analysis, various levels of event tree and/or fault tree analyses, and/or detailed quantitative analyses with formal treatment of uncertainty.

Dam response analysis involves modelling the response of the dam to the full ranges of loads due to hazards and/or operating conditions. The first stage of this process involves selection or development of a suitable model and identification of data requirements.

The second stage involves providing input data and running the model for the various conditions under consideration. A simplified example of a dam response model for one failure mode was illustrated in Figure 2-3.

Disaggregation or decomposition of the failure mechanism into its constituent parts is a key element of the analysis process. The extent to which this disaggregation or decomposition is required will depend on the complexity of the failure mechanism and the level of the analysis. Again, an iterative approach to disaggregation/decomposition is appropriate and in some cases it may be necessary to reduce the failure process to its fundamental component parts.

Consequence analysis

Consequence analysis involves estimating the direct and indirect impacts of the failure or incident. The consequence analysis should provide a clear picture of what emergency response personnel would be faced with should the failure occur as well as a picture of the long-term effects of the failure. Consequence analysis consists of identification of potential losses and loss magnitude estimation. In some cases it may be necessary to describe the estimate of consequences probabilistically and account for temporal variation in the characteristics of the inundated area. The essential components of the consequence analysis process involve considerations such as those in Table 2-1.

Risk estimation

Depending on the scope of the analysis, the output from a risk analysis for a dam will provide descriptions of the risk in one, some or all of the following categories.

- Individual risk to the public.
- Societal risk to the public.
- Occupational risk to workers.
- Environmental risk.
- Commercial risk to the owner.
- Social and economic risk.

The extent to which these categories are considered depends on the nature and level of effort of the risk analysis. The outputs of the risk analysis should be structured so as to be useful inputs to the risk control process. Graphical representations of frequencies and consequences are also useful.

In quantitative analyses, risk and how it accumulates can be represented in a summarised event tree based on condensed versions of the dam failure and failure consequences event trees. Risk is estimated by combining the probabilities of failure-initiating events obtained from the hazard analyses with the probabilities of dam failure obtained from the dam response analyses, and the magnitudes of consequences and their associated probability distributions from the consequence analysis phase. This can be achieved by multiplying the end-branch probability of each failure scenario by the associated consequences and, if the expected value is required, by summing the results. Summed over all hazards and failure initiating events, the risk in the system is:

$$Risk = \Sigma\{P(Events) \times P(Failures \mid Events) \times P(Consequences)\} \qquad (2\text{-}1)$$

Table 2-1. Essential components of consequence analysis

Component	Activity
Identification	• The extent of inundation associated with each failure mode • Estimating warning time and effectiveness of emergency plans including consideration of the time for failure detection, decision to warn, issuance of the warning and evacuation of the population • The forcefulness of the flood water and its destructive power • The population at risk • The inventory of properties that will be inundated • The potential for environmental damage • The socio-economic implications of dam failure including public outrage
Analysis	• Determining the temporal variation in the number of people located in the inundated area and estimating the number that might die or be injured as a result of the release of water • Estimating environmental damages • Estimating the total economic impact of the failure in terms of damage costs including compensation for loss of life, costs of environmental rehabilitation, fines, and economic losses • Estimating the incremental consequences of failure may be required if the context of the purpose of the risk analysis is the estimation of "incremental risk"

where the symbol "|" represents conditionality (i.e. failure conditional on the event). The estimate of risk can be presented in product form (expected value) or in one of several non-product forms. It may be appropriate to present the results of the analysis in several formats.

2.3 Iteration

An iterative process in a risk analysis with increasing levels of effort in subsequent iterations is expected and is generally appropriate. Such an iterative approach to risk analysis permits an initial and relatively inexpensive screening technique, such as checking conformance with established performance requirements. Since the risk is unknown at this stage, the analysis should guard against the possibility of understating the risk by screening out contributors to risk whose significance is unappreciated. The results of a screening analysis should be used to set priorities for conducting more detailed studies. The approach to these more detailed studies should also be iterative. This will permit sound decisions to be made for the next course of action.

Levels of analysis

Risk analyses can be carried out at different levels of detail and rigour. The risk analysis project plan should recognise the following issues: hierarchical structures for risk analyses; continuum of processes to ensure continuity in analyses from one iteration to the next; that uncertainty may be reduced and knowledge of uncertainty may be increased with improved data collection and analysis techniques; and identifying areas where lack of information and/or analysis techniques, and imbalances between them, may create difficulties or inconsistencies in conducting some forms of risk analyses.

A high level of effort is needed if a high degree of defensibility is required. However, the quality of the study and the ability to use the study to assist in dam safety decision-making does not depend directly on level of effort as defined below in the hierarchy of levels. Rather it depends on the complexity of the problem, the availability of data and models and the skills of the analyst(s).

The most important issue is to ensure that the method used is appropriate for the intended end use of the analysis. At each level of risk analysis, there needs to be an appropriate amount of supporting evidence and engineering input.

All levels of this iterative process are forms of risk analysis. At any level of the iterative process some form of decision is needed. Therefore, if the output of any stage of this iterative process can be used as input to a risk assessment (compatible with the end use) then there may be no need to continue with further iterations.

2.4 Uncertainty and sensitivity analysis

Ultimately, an estimate of risk should be accompanied by a statement of the degree of confidence in the estimate. The statement of the degree of confidence describes the extent to which the result of the analysis can be relied upon in the decision-making process.

For the purpose of this volume, a distinction is made between sensitivity studies and uncertainty analysis. Sensitivity studies involve determining the change in response of a model to changes in individual model parameter distributions or changes in the model. They provide an indication of how wide the spread of results can be if individual parameters are varied within realistic bounds.

Probability distributions of parameters to which the analysis is sensitive should be identified and their validity and accuracy verified to the extent that is realistic. Changes should be made to these probability distributions where justified.

Sensitivity studies are a powerful way of determining if some uncertainties significantly influence the estimate of risk.

In this volume, uncertainty analysis is an extension of sensitivity analysis where probability distributions are associated with the various parameters or models being varied. Thus, the output is in the form of a probability distribution which specifies the likelihood of each possible result across the full range of possible results.

The analysis of uncertainties associated with data, methods, and models used to estimate the risks posed by dams is important. Uncertainty analysis involves determining the variation or imprecision in the model conclusions resulting from the collective variation in the parameters used to define the model. Estimating uncertainty is done by translating the uncertainty in the analysis models and in the crucial model parameters into uncertainty in the outputs of the risk model.

2.5 Issues of good practice

A number of issues of good practice recommend themselves in dam safety risk analysis studies, having to do with documentation, review, and personnel.

Documentation

The extent of the report will depend on the objectives and scope of the analysis. The documentation should, however, contain the following sections:

- Title or cover page.
- Summary.
- Major findings and understandings achieved.
- Conclusions and recommendations.
- Table of contents.
- Objectives and scope.

- Limitations and assumptions.
- Dam system description.
- Sections covering all aspects of the study.
- Discussion of results, including a discussion of analysis challenges and degree of confidence in the results.
- References, tables, figures and appendices.

Review and verification

A formal review process provides a basis for confirming the integrity and correctness of the analysis process through internal or external review to an extent that is reasonable, at appropriate times, by someone other than the analyst(s). Review involves the following steps:

1. Check that the scope is appropriate for the stated objectives.
2. Review all critical assumptions and ensure that they are appropriate in the light of available information.
3. Ensure that the analyst(s) used appropriate methods, models and data.
4. Check that the analysis is comprehensive and the results are reasonable.
5. Check that the analysis is insensitive to the way the data and results are formatted.

For the best results, the review should be formally documented and all comments should be evaluated and incorporated into the risk estimation process if appropriate. Where adequate field experience is available, verification may be accomplished by comparing the results of the analysis with direct observations.

Analysis update

If the risk analysis is to form part of an on-going risk management process, the analysis should be performed and documented in a manner that permits it to be maintained throughout the life of the dam. The analysis should be updated as new information becomes available and in accordance with the needs of the management process.

Personnel

The working group carrying out a risk analysis should have a thorough understanding of the dam and the relevant failure modes as well as knowledge of the methods of risk analysis. Those responsible for the surveillance, monitoring and operation of the dam should be included in the working group. Risk analysis for dams may involve the analysis of several complex and interacting sub-systems and the necessary specialised knowledge of the behaviour of the sub-systems should be provided as required.

3 Risk in decision-making

This final chapter of Part I attempts to pull together the considerations that management must take into account in using the results of a risk analysis study of dam safety.

3.1 Significance of risk

While there are no firmly established principles for determining the significance of risk in any endeavour, general principles are evolving in different areas of society. Since dams generally impose risks on third parties and the environment, it is appropriate for risk evaluation for dam safety to be consistent with approaches to risk evaluation that are evolving for other societal activities. The basic principles of risk evaluation that appear to have become established are that risk evaluation processes be:

- Comprehensive.
- Fair and equitable.
- Transparent.
- Consultative.
- Defensible.

The risk evaluation process is illustrated in Figure 3-1.

The extent to which each of these basic principles applies depends on the nature of the risk and the objective of the risk assessment. Dam risk management that involves risk to third parties should incorporate all of these principles. On the other hand, an

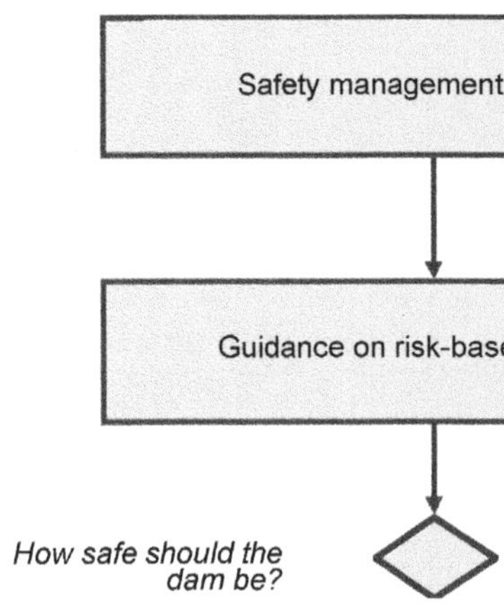

Figure 3-1. Risk evaluation process

individual owner could choose to manage commercial risk on the basis of instinct and experience. The former type of risk evaluation process should be as objective as possible, even though there will always be subjectivity present, and the latter will be entirely subjective. In general, those responsible for making decisions concerning risk should identify the extent to which the above principles apply in the risk evaluation process, as it will vary from owner to owner and within an owner's portfolio of risks to be managed. Depending on the risks, an owner could require all five principles considered for some risks and none for others. In other instances, regulatory and societal expectations might require the owner to ensure that all five principles apply. The statement of the extent to which any of the above principles apply is, in a way, a statement of some of the values and preferences of those responsible for making decisions concerning risk.

Categories of risk

Risks posed by dams can be categorised in the same way as risks in any other business activity. Although the following categories are not universally accepted, they are presented to promote consistent use of terminology in dam risk management, thereby avoiding the confusion and misunderstanding that often results from inconsistent use of terms.

These risks can be subdivided into two categories: commercial risk and societal or third party risk. Losses in either of these sub-categories constitute a loss to the business.

Commercial risk refers to those risks where the consequences of incurring the loss only appear on the company balance sheet. For dam owners, commercial risk may involve many of the diverse aspects of the process of storing and using water and ranges from reservoir inflow risks to market risks. Although diverse, this category of risk may only constitute a small component of the total risk associated with the design, construction, operation and decommissioning of dams.

Societal risk refers to all situations where the adverse consequences relate to life safety considerations and other impacts that will be borne to varying degrees by society. This is a broad category of risk and it comprises all risks where the consequences of the risk materialising result in some impact to people, property and the environment. It is possible that what appears to be a commercial risk, such as generator reliability, could in fact be a societal risk if, say, loss of the generation capability were to result in loss of control of the reservoir and dam failure. Operation and maintenance of dams is another function where what is perceived as a commercial matter may actually involve a societal risk. Life safety risk to individuals or "individual risk" is a subset of societal risk which should be considered.

Societal vs commercial risk

Management of societal risk and commercial risk may not be synchronised with each other, and risk evaluation for dam safety involves the often difficult task of rationalising these competing interests. Commercial risk involves risk-taking for benefit where the beneficiaries are the owner and shareholders and even the rate payers. Societal risk usually involves risk-taking for the benefit of society in general. However, there may be situations where the segment of society on which the risk is imposed does not gain any benefit from the taking of the risk. Such situations, which frequently arise in dam risk management, usually require careful consideration in the risk evaluation process.

Values, preferences and choices

In respect of risk evaluation, values are principles, standards or qualities considered worthwhile or desirable. While it is not the role of this volume to define what constitute "values" for dam safety, definition of owners' values in the case of commercial risk and those of society as represented by elected or appointed representatives in the case of societal risk are essential parts of the risk evaluation process. Statements of values are matters of policy for owners, regulators, and society in general. Similar remarks apply to preferences and choices.

One can reasonably expect values, preferences and choices to be different for each stakeholder group and even for similar stakeholder groups in different jurisdictions.

Risk aversion

Risk aversion describes the willingness or unwillingness of owners and stakeholders to take risks. It is a highly subjective concept where there is a trade off between maximisation of benefit and avoidance of loss. It could involve investment in dam safety at the expense of returns to the shareholders.

Methods have been developed to ascertain a decision-maker's aversion to commercial risks that are routinely taken in business and are within the experience of the decision-maker. Risk aversion can be determined from considerations as to how past losses have been financed and the extent to which steps were taken to reduce the possibility of that risk re-materialising.

Societal risk aversion is a difficult issue as different people within the societal group will have different values that result in different choices. In general terms, it appears that society is more averse to accidents that result in large losses per accident than it is to larger numbers of individual accidents that result in small losses per accident. This appears to be true even if the total loss from the serious accidents is less than the sum of the small losses from each of the large number of individual accidents; an example being a comparison between airline safety and motor vehicle safety. There are more deaths in road accidents per passenger-kilometre than there are deaths in aircraft accidents per passenger-kilometre; however, society is more averse to aircraft accidents.

Representations of risk

As described in the definitions, risk can be represented in either expected value (sum of the product of probabilities and consequences) or non-product (probability–consequence pairs). Two-dimensional graphical representation of expected value is generally not possible whereas probability–consequence pairs (frequency–number of deaths, etc., $ losses) can be plotted graphically in a Cartesian coordinate system, known as "f"-N or "F"-N plots.

Guidance to risk-based decisions

Risk-based decision guidance is an essential component of risk assessment. Several forms exist, including but not restricted to:

1. Benefit-cost type.
2. The expected value from where an upper limit to the expected value of the loss, as expressed in terms of lost "units of measure" per year, is set.
3. Graphical criterion where a boundary is established between two areas on an f-N plot, one side of the boundary constituting an undesirable level of risk. The cumulative form F-N may be an appropriate form in certain cases.
4. The decision guidance may require consideration of what constitutes appropriate levels of uncertainty.

ALARP principle

Regulations concerning societal risk may require that owners demonstrate that the risks posed by their facilities have been reduced to "As low as reasonably practicable" (ALARP), or the "As low as reasonably achievable" (ALARA) principle.

Multiple criteria
It may be necessary to establish several criteria, one for each of the principal considerations in the risk evaluation process. For instance, it may be necessary to have criteria for life safety, third-party socio-economic damage, environmental damage and corporate financial security. A hierarchy for the criteria and guidance as to which criterion governs in the event of differing outcomes for the various considerations in the analysis, should also be defined in the risk evaluation process.

Responsibilities

The responsibility to establish appropriate policies rests with industry, the creator of the risk and the regulatory authorities in consultation with the public where appropriate. It is essential for owners, to be pro-active in establishing policies for losses that are borne by others.

3.2 Risk assessment

Risk assessment is the process of deciding if the risk is sufficiently significant to require additional risk control measures. Risk assessment involves making statements of the form: "The risk is assessed as undesirable (or even unacceptable)." It brings together the measure of risk as determined in the analysis phase and the benchmark or yardstick from the evaluation phase against which the risk can be compared, to permit a decision to be made as to whether or not some risk management action is required.

Decision process

Risk assessment facilitates good dam safety decisions by achieving a better understanding of the dam and associated risk factors.

In those cases where benefit – cost, graphical or expected value criteria are used, the risk assessment process may be as simple as comparing the result of the analysis with the decision criteria. In general, however, one can expect the process to be rather more complex.

Generally, and unless otherwise specified in the risk evaluation process, all criteria should be met for the dam to be deemed safe enough. In some cases this might include conformance with the ALARP principle.

It is important to consider the sum of all risks and not just the risk from each hazard as the total risk must be less than the allowable risk.

Compatibility with deterministic approaches

Presently, risk assessment for decision-making is usually carried out in parallel with traditional assessment techniques. This is in part due to the fact that risk analysis processes are not yet established dam safety practices, and in part due to the fact that there are currently no accepted risk evaluation criteria for dam safety.

Inputs to the risk management process

The outputs from a risk assessment should be compatible with the risk management process. In the case of portfolio approaches, the risk assessment outputs could either lead to further detailed analyses or the implementation of risk reduction measures. However, do nothing can sometimes be a valid approach. For risk assessments used for decision-making, it may be necessary to format the outputs in a number of ways to be compatible with one or more of the evaluation criteria.

If the risk evaluation process provides a range of management options that meet the safety management principles for decision-making, the outputs of a comprehensive risk assessment may be used in deciding on the merits of the different options, and in choosing between those options.

3.3 The value of risk information in decision-making

Past experiences demonstrate that there is no simple recipe for evaluating and managing risks (Klinke and Renn, 2002). This is particularly true of dam safety where the use of risk assessment is in its infancy.

Dam safety decision-making generally falls into one or more of the following categories:

- Standards based decision-making.
- Decisions supported by risk analysis.
- Decisions based on risk assessments.
- Decisions dominated by company or societal values.

Decisions based on these processes tend to require an ascending order of sophistication from standards to values dominated decisions in terms of information, analyses, corporate level of decision and regulatory environment. If the decisions at lower levels are questioned by stakeholders, higher levels of sophistication may be required to address stakeholder concerns.

In reality, many complex and important dam safety decisions will involve portions of all four processes at different stages. A decision-making framework which explicitly incorporates and elaborates on these processes allows engineers, owners, regulators and stakeholders to determine the appropriate processes and tools to implement and sustain dam safety decisions given a wide variety of situations.

One example of such a framework was developed by the United Kingdom Offshore Operators' Association (UKOOA) (Brinded, 2000), and is shown in Figure 3-2.

Like other major decisions involving uncertainty that have the propensity to affect society, decisions regarding managing and controlling large dam risks should involve the risk bearers, other stakeholders and shareholders to ensure that the decision is widely

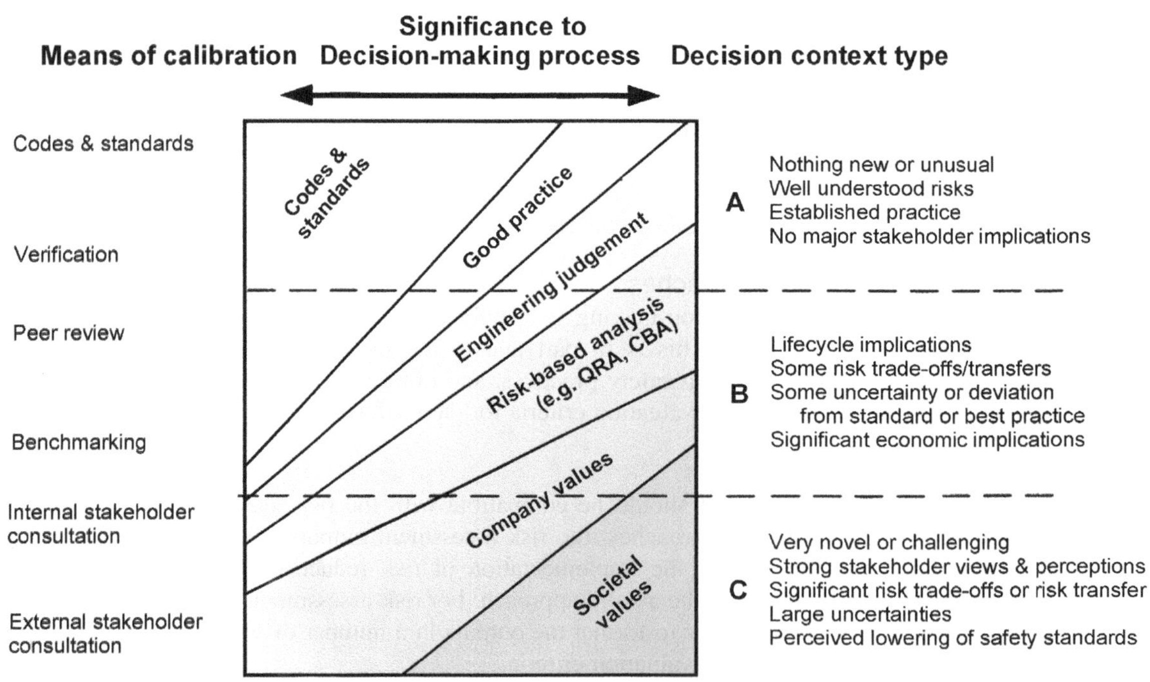

Figure 3-2. Risk decision framework (UKOOA), (Brinded, 2000)

Note: QRA – quantitative risk analysis
CBA – cost–benefit analysis

accepted, sustainable and implemented without protest. Risk information supporting the desirability for improvements and for comparing alternative concepts for reducing risk (increasing safety) may be a very valuable aid in communicating with all interested parties. It is common that the various parties have differing goals and priorities, and the ultimate decision may well involve some consensus regarding explicit risk trade-offs.

Standards based decision-making

Standards based decision-making is the simplest and most commonly applied approach to dam safety decisions. The term "standards" is used in the broader context of generally accepted "good practice", rather than the narrow definition of statutory codes of practice, because there are few of the latter in the dam safety discipline.

If a particular issue under consideration is clearly covered by a "standard", then the decision is simple, either the condition of the dam meets the standard or it does not. If it does not meet the standard, modifications to meet the standard would be warranted. Examples would include:

- The ability of dam discharge facilities to safely pass the inflow design flood "standard".
- The calculated factor of safety "standard" for a dam slope.

However, the decision is rarely that simple and in many instances judgement, including the judgements embodied in the engineering advice, is required. In effect, this complication is often a reflection of the interaction of various "standards" in a dam safety problem, and the ever-present uncertainties associated with unique structures such as dams that are built on unique foundations using materials that are unique to the site.

The advantage of being able to demonstrate compliance with a "standard" or authoritative good practice is that it is generally accepted as a threshold where the tolerable risk is "as low as reasonably practicable" and further consideration of risk is generally not required. This will almost always satisfy the expectations of regulators, the legal system, politicians and the public. One exception, however, might be where consequences of failure are very large; in which case simply meeting a "standard" may not be enough.

Two other potential shortfalls of a standards-based decision process are: (1) lack of adequate information for prioritising the multi-dam owner's resources for making dam safety modifications; and (2) the potential for increasing risk to the public at lesser events in order to provide protection for the dam at more significant events. For example, with flood-induced failure modes it is not enough to meet the standard (maximum design flood) without considering the reliability of the gates, and the power supplies.

Decisions supported by risk analyses

In the absence of a risk evaluation framework, a dam owner might consider using risk analyses to assist with decision-making. The presumption here is that there is no formal risk evaluation framework which would provide guidance as to how much risk might be considered to be tolerable. At the present time, this is the case for most jurisdictions. Risk analysis supported decisions could be considered under the following circumstances (McQuaid, cited in Stewart, 2000).

- Recognised good practice for a particular safety hazard does not exist.
- Past practice exists but may not have kept pace with technology or may be of doubtful quality in light of new ways of performing technical analyses.
- Good practice exists but is believed to be unbalanced, with too much attention in some areas and neglect in others.

- Good practice exists but the risk standards implied are believed to be too lax in the light of prevailing societal expectations.

Under these circumstances risk analysis may be valuable whether it is qualitative, semi-quantitative, or quantitative.

Risk analysis, at any level from qualitative to quantitative, is useful in the review of dam and reservoir operation, dam maintenance and the dam surveillance programme. A failure modes analysis will identify the modes of failure that generate risk in the dam system. Operation can be carried out in such a way that avoids identified failure modes and surveillance can be focused on detecting conditions that would lead to a failure mode such that it might be precluded.

The nature of the dam safety problem, the intended application, the form of the decision, the scientific or engineering knowledge of the mechanisms or processes, the availability of accepted models, and a reliable means of assigning probabilities will all contribute to the scope and degree of analysis to be used. Results from various levels of analysis will be sufficient under different circumstances. The important factor is for the owner (decision-maker) to make a determination of the required accuracy of risk estimates prior to embarking on the analysis. The scope of the analysis should be revisited during the process as information becomes available.

The value added to decisions supported by risk analysis can include:

- A structured framework to examine, display and debate all risks, their relative importance, and options for risk control.
- An open and transparent decision process for debate with regulators, stakeholders and the public on the nature of the risks.

In addition, risk analyses have been used in the dam industry to prioritise dam risk mitigation measures among a number of dams. This is akin to an indexing approach to aid decision-making.

Decisions based on risk assessments

Water resources' decision-making is recognised as a discipline that exists within a multi-objective environment. While there are no dam owners currently making decisions solely on the basis of risk, there are dam safety decisions being made by some organisations that are heavily influenced by the owners' evaluations of risk. Examples may be found among some owners in Australia, and within the USBR. Decisions based largely on risk assessment must be supported by a risk evaluation process and risk criteria/guidelines compatible with the risk analysis that has been performed. The evaluation process and criteria/guidelines provide two key elements needed for a dam owner to use estimates of risk in their decision-making processes. These elements provide the basis for the dam owner's decisions regarding appropriate actions necessary to address dam safety issues. Decisions may also require that a suitable regulatory framework exists to support such decisions.

While a decision at this level can consume significant resources, both in terms of the time required for the analysis and in terms of the development of processes and criteria/guidelines, the benefit comes in the form of risk comparisons to assist an owner in prioritising dam safety needs across an inventory of dams and selecting the preferred alternatives. Most applications of this type of decision-making make use of the computed risk results for prioritisation, while focusing on the specific conditions that cause the risks to be high when making a case for dam modifications to improve safety.

Value-driven decisions

Complex dam safety decisions which involve:

- strong stakeholder views or perceptions;
- significant risk trade offs or risk transfer;
- large uncertainties;
- perceived lowering of safety standards.

may require explicit consideration of company and/or societal values (see Brinded, 2000).

The experience of Shell UK with regard to the Brent Spar Offshore Oil platform disposal in the North Sea is a clear demonstration of the importance of "values" in driving a good and broadly acceptable decision (Brinded, 2000).

> [. . .] we were made painfully aware that our independently-verified technical assessment had produced a solution that was all too clearly not supported by everyone. We began to realize that some decisions need to be driven by values and not directed solely on a basis of sound science. No amount of additional analysis could rescue the original "Spar disposal strategy".

A clear, explicit decision-making framework which includes standards, risk analysis, risk assessment and company and stakeholder values is of great value to a dam owner because it provides a consistent approach to be used on multiple occasions. Such consistency provides the engineers, the public, the regulators and the risk bearers a degree of comfort and familiarity with how safety decisions are weighed and ultimately made.

Decision-making and communications

During the risk assessment and decision-making process, it is desirable and beneficial to implement a process for periodic communications to share information with stakeholders. This may include information notices, dialogue with community representatives, open houses, public meetings and focus groups. A participatory process allows development of a common understanding of relevant issues, perceptions, concerns and values which may influence the dam safety decisions, and which may change as more information becomes available during the assessment.

Once the assessment is complete, and the preferred risk management or control options identified, it is desirable to record this in a concise, non-technical document. The Safety Case concept as developed in the United Kingdom provides one vehicle for demonstrating the basis for the decision. The Safety Case must demonstrate that:

1. Either, engineering, operational, and managerial good practices are met and the facility is robustly fault tolerant; and risk estimates are below the basic safety objectives (BSOs)

or,

2. The risks are below the basic safety limit (BSL) and a sustainable argument can be made that it would be grossly disproportionate to take further measures.

The document which could be called the "dam safety case" (Hartford and Stewart, 2002) could be expected to include:

- dam safety issues addressed;
- nature of the risks and relevant uncertainties;
- comparison with well-established practices;
- nature and full scope of consequences;
- options to manage or control the risks;
- preferred course of action;
- description of how risks are to be mitigated;

- identification of any risk trading involved;
- summary of public consultation, and how issues have been addressed.

Such a safety case can be valuable to support funding requests for implementation for Boards of Directors, and funding agencies. Dam safety regulators and interested stakeholders would also derive value from such a document.

Thus there is a strong argument for extending the use of safety case regimes to other areas of government regulation where risks affect the public or give rise to public concerns, with full exposure of the principles of assessment and public availability of the safety case information. The ordered recording of the basis for decisions would do much to engender public confidence (McQuaid, 2002).

Part II
Commentary on risk analysis

4 Reliability analysis

This chapter provides an overview of the application of the three principal approaches to reliability analysis for dam safety studies: failures modes and effects analysis, event trees, and fault trees. Corresponding discussions at a more theoretical level are provided in Chapter 8 for event trees, and Chapter 9 for fault trees.

4.1 Failure modes, effects and criticality analysis

The basic features of the generally accepted form of failure modes and effects analysis (FMEA) are presented in this section. The technique, which is widely used in the discipline of system reliability, is encoded by national and international standards (British Standards Institution, 1991; Canadian Standards Association, 1993; International Electrotechnical Commission, 1985; US Department of Defense, 1980). The following material is based on these standards and while the basic technique is valid for dams, certain modifications may be required to apply the technique to dam safety practice. While a stand-alone method in its own right, FMEA can also be considered to be a precursor to more detailed event tree and fault tree analyses.

Overview

Failure modes and effects analysis (FMEA) is a form of reliability analysis that is used to map out the consequences of specific events that can occur during the operation of an engineered system, and to use this information to identify and prioritise necessary actions.

FMEA can be applied in several forms and for a number of purposes. In its simplest application it can be a free-standing technique to give a structured understanding of the failure modes applicable to the components of an engineered system, or it can be an integral part of a more comprehensive probabilistic analysis of the risks associated with multiple integrated systems.

Although there are many ways in which the basic technique can be applied to engineered systems and in which the output can be incorporated within wider risk and reliability studies, there are two clearly defined techniques in use. They are failure mode and effects analysis (FMEA) and failure mode, effects and criticality analysis (FMECA).

The essential difference between these two applications is that the first deals only with an understanding of the range and influence of a system's failure modes, while the second enumerates a representation of "frequency" of occurrence and consequence for each failure mode. This permits the ranking of the various events in order of the effect that they will have on the operation of the overall system. However, it should be noted that a FMEA is an essential first part of a criticality analysis.

FMEA is an inductive method of analysis where particular faults or initiating conditions are postulated and the analysis reveals the full range of effects of the fault or the initiating condition on the system.

In describing the two failure mode analysis techniques (FMEA and FMECA) reference has been made to the methodology set out in the above-mentioned standards. The terminology may differ slightly between various standards, however the basic principles are common to all of these standards.

Origins

FMEA is a technique developed originally for design purposes, which now finds application in the analysis of potential for failure of existing systems. It also finds wider application as part of a wider asset management process that deals with the ongoing satisfactory output of the system under consideration. The use of FMEA is no longer restricted to engineered systems and is now used in a diverse range of societal activities, healthcare management being an example.

FMEA is suited to the investigation of materials and equipment failures, and can be applied to categories of systems based on different technologies or combinations of technologies.

FMEA has grown from the needs of other branches of engineering, such as nuclear, mechanical, electrical and the space industry, primarily to help understand complex systems but specifically as one of a number of techniques used to assess system reliability. As such, its use in civil engineering applications has been limited in the past, but this is changing.

Fundamental concepts

A failure mode describes how component failures must occur to cause loss of the sub-system or system function.

Consider the functional failure "loss of pressure" in a hydraulic actuator of a spillway gate lifting mechanism with inherent sub-system redundancy (Figure 4-1).

One failure mode could be "loss of primary pump *and* pressure switch failure". Together these two component failures would cause the functional failure. A second failure mode could be "loss of primary pump *and* loss of backup pump".

Failure modes may refer to the root cause, depending on the level of analysis, but usually refer to the observed failure effect such as a seal failure or a motor failure coupled with switch failure, for example. Root cause implies that a systematic forensic analysis can determine the fundamental beginning of the failure. For example, a pump failure cause may be observed as a cracked impeller. The root cause is whatever caused the impeller to crack, e.g. metal fatigue.

The analysis process is of a hierarchical nature, with the important feature that all failure modes within a system are not uniquely defined. Rather, failure modes are defined in terms of the level in the system hierarchy at which the analysis is being carried out.

Basics

FMEA techniques are structured, logical frameworks that allow informed operatives and specialists to use available knowledge and information in a systematic way to lead to an understanding of the sources of risk in the system. The key benefits of the technique are transparency and amenability to audit. These can be more fully expressed as the ability to follow the process through and to understand the effects of varying the inputs without a mathematical or statistical background. It also provides the facility to allow third parties (perhaps without specialist engineering backgrounds) to review the process and, if necessary, to question the inputs and outputs.

Use of the technique follows a basic structure under the following headings:

1. Define the system.
2. Separate the system into its functional sub-systems.
3. Systematically separate the sub-systems into key elements and elemental functions.
4. Analyse element failure modes.
5. Assess direct failure effects and operational consequences.
6. Assimilate findings.
7. Respond.

The primary skill required is to understand the functional nature of the system being analysed. This requires knowledge of how the system was designed and built, how system function is achieved, and how it has been and is being operated. This gives a basic appreciation of the potential weaknesses within the system and forms the basis for carrying out a failure mode-type analysis.

Advantages and disadvantages

FMEA is a versatile design-based tool with significant scope for application in dam risk management. The process is aimed at systematically developing a picture of the dam system, its components and their interactions, and presenting details of how component failure could lead to system failure, the magnitudes of the failure effects and the criticality of the various components in preventing the failures from materialising.

FMEA provides a systematic procedure for analysing failure modes of dam systems by isolating each component and describing the effects of individual component failure modes on the system as a whole.

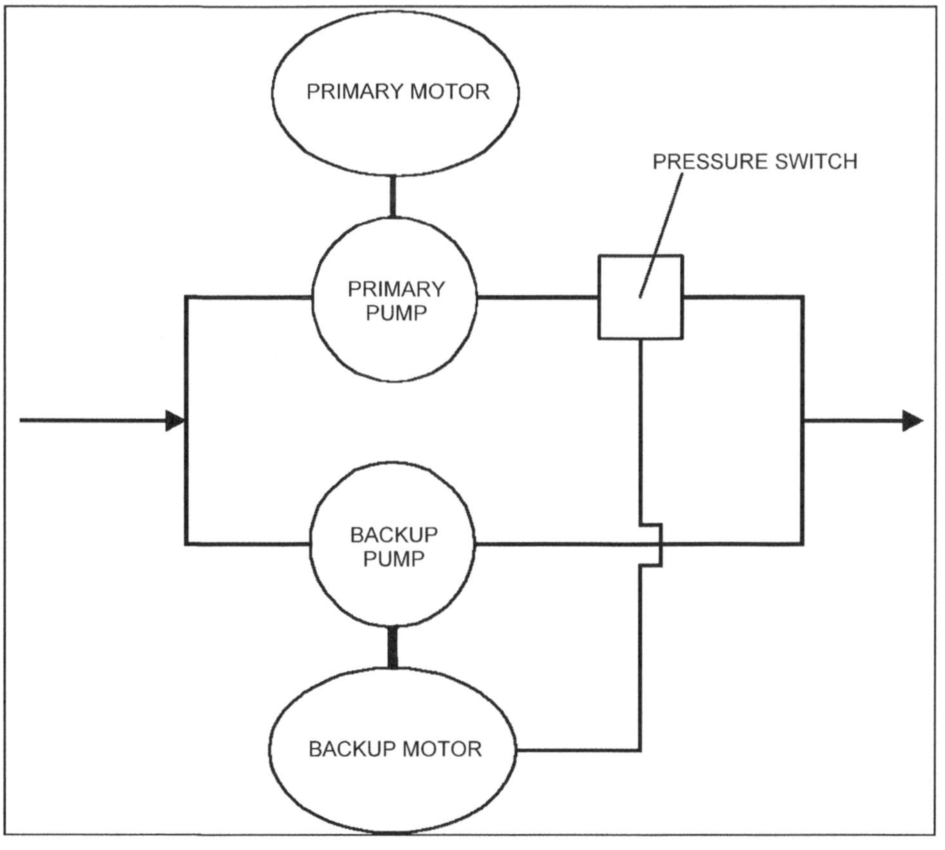

Figure 4-1. Hydraulic pump system with inherent system redundancy (after Jones, 1995)

FMEA is conceptually straightforward and readily applied to "discrete component" systems with little or no component redundancy. It clearly relates component failure modes to their effects on the performance of the system and presents them in an easily readable form.

FMECA extends FMEA to provide a means of ranking the failure modes in terms of an index of risk that incorporates representations of probability and consequences. This provides a sound basis for prioritising corrective or remedial actions.

A major disadvantage of the technique is the difficulty of modelling and analysing redundancy, as the individual component failure modes do not always result in observable system effects. The disadvantage of using the FMEA analysis tool becomes apparent when the technique is applied in dam safety applications where one component might compensate for the functional failure of another.

A second significant disadvantage of FMEA is the difficulty of applying the method to systems where the components can exhibit varying degrees of failure, i.e. the components do not necessarily go from a "functional" state to a "non-functional" state instantaneously.

While the general nature of the FMEA worksheet can be established and made transportable from one situation to the next, it may be necessary to tailor the generic worksheet for individual situations to incorporate the necessary detailed information.

A comprehensive FMEA/FMECA can be expected to generate a large number of potential failure modes. In a well-designed and maintained system, the analysis can be expected to demonstrate that a potential failure mode has been "designed out" of the system (inherently safe system) or controlled in some other manner.

From a technical perspective, the analysis can become extremely complex if the effects of multiple failures are taken into account. As a result, much time and effort might be expended on the analysis of failure modes that have a negligible effect on the performance of the system.

British Standard BS 5760: Part 5 (British Standards Institution, 1991) urges caution in the appropriate use of the technique and gives guidance on the circumstances under which it performs best and also those which other techniques such as FTA might provide better results. FTA is noted as being complementary to failure modes analysis in some circumstances, but not as well suited to the analysis of series failure logic. The converse is also true, that FMEA can provide insight that the FTA cannot. Experience has shown that FMEA can be used to efficiently analyse components, sub-systems, etc. that can cause global system failure, and it is most appropriate for use where detailed knowledge of the failure characteristics is required.

Since the analysis is required to first identify the significant potential failure modes and then identify compensating provisions, FMEA/FMECA may require a great deal of time and a very significant resource commitment. The more complex the system and the larger the number of components and component functions, the larger the FMEA and number of associated worksheets. The result is that application of the FMEA process can be highly resource intensive, the documentation voluminous and the review and interpretation of the worksheets tedious.

Failure modes and effects analysis

Although existing standards deal principally with design of new electrical or mechanical systems, the focus of this volume is the assessment of existing dam systems in service. Currently, the volume does not consider the use of failure mode analysis techniques as a design tool for new dams.

System structure and element definition

FMEA can be carried out to varying depths within a system, making it essential to define the "level" in the system for which the analysis will be carried out. Systems can be broken down into "sub-systems", "sub-sub-systems", etc., and then to an elemental (component) level. Complex systems can be broken down in a hierarchical manner whereby the failure effects at a lower level in the system become the failure modes at the next highest level in the system. In general, the system is broken down to a level where there is adequate understanding of the failure modes of the elements and the system.

At the outset, there should be a clear understanding of the hierarchical nature of the system and its elements, making it necessary for the analyst to transform the information about the system into some form of functional block diagram that also illustrates the system hierarchy (Figure 4-2). These diagrammatic representations of the system, sub-systems, components and elements should identify the functional structure of the various parts, the boundaries around and within the system and the influences crossing each boundary.

If the analysis is to include consideration of "criticality", the system must be broken down to a level where it is possible to obtain adequate and reliable failure-rate data for each failure mode, or to make reasonable assumptions of such failure rates.

Definition of the component level of the system, interaction between components and the relationship between the component level and higher levels is a critical step in the analysis process. Failure modes analysis is designed to be performed on a sequential level-by-level basis working from the lowest level of interest in the system upwards to the highest.

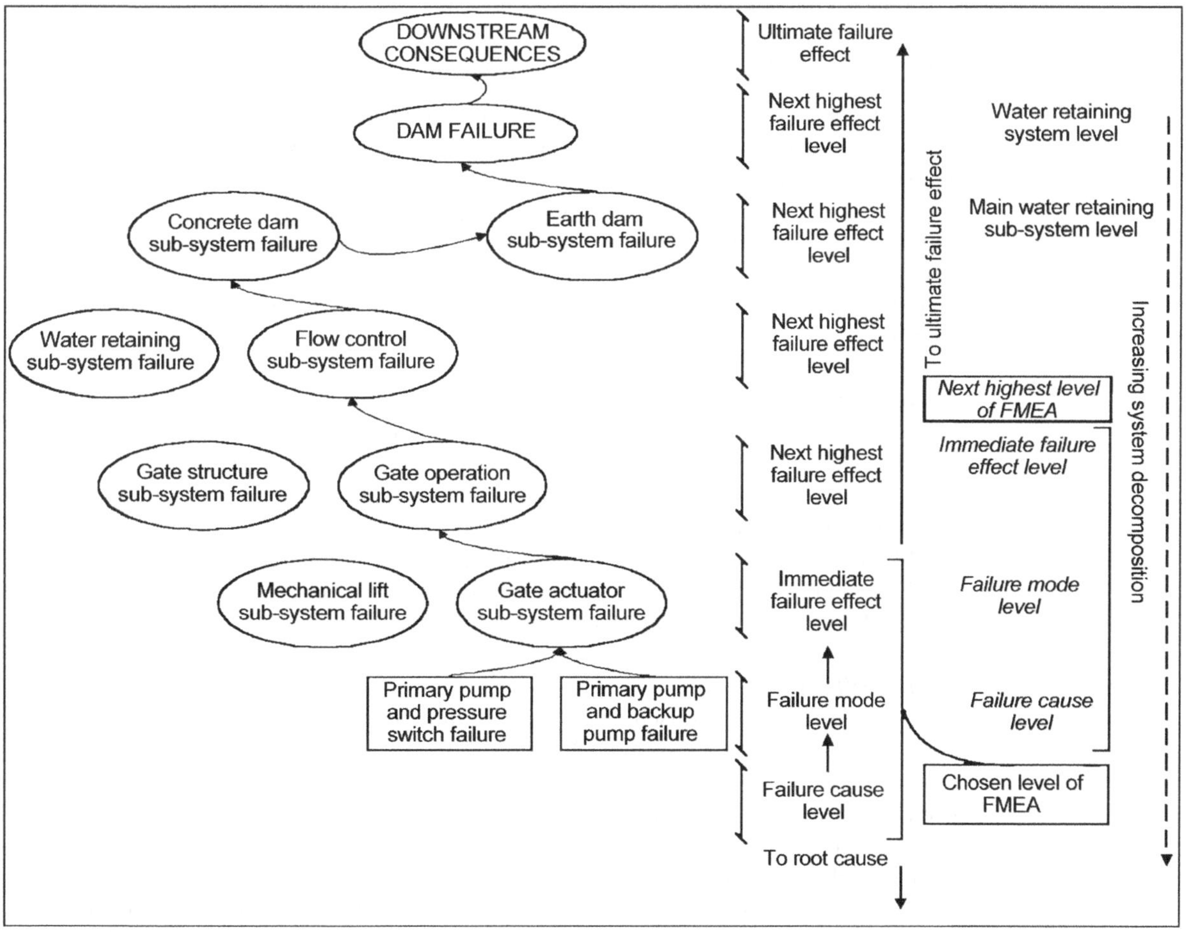

Figure 4-2. Relationship between failure modes and failure effects in a system hierarchy

Failure modes, causes and effects

As introduced in Figure 2-2 above, the definition of failure modes, failure causes and failure effects depends on the level within the system at which the analysis is carried out. Therefore, as the analysis progresses upward through the system, the failure effects identified at a lower level become the failure modes at the next highest level and so on. Furthermore, a failure of analysis mode at a lower level might be a failure cause at the next highest level of analysis, i.e. the effect of a failure mode at one level causes the sub-system at the next highest level to fail.

A failure mode is a descriptor of how component failures must occur to cause loss of the sub-system or system function. In this regard, failure modes are not unique features of the system but artefacts of how the system is modelled. Similarly, the term "component" is not unique, rather it is an artefact of the system model. For example, the entire spillway gate control sub-system would be considered to be a component if the FMEA was carried out at that level in the system (Figure 4-2).

Identification of "function critical" components is central to the analysis of system performance, and at the component level, it may be appropriate to define the failure mode of the component in terms of functional failure.

Many failure modes of systems and elements are often known in advance and are presented in tabular form as part of the preparation phase of the FMEA. General failure mode categories can be prepared for example for earth dams in terms of:

- overtopping;
- internal erosion/piping;
- mass movement;
- slope instability.

However, these general categories of failure modes are usually too broad for definitive analysis and should be expanded by going deeper into the system and carrying out the analysis at a more fundamental level. Again referring to Figure 4-2, these failure modes pertain to a failure mode and effects analysis carried out for the earth fill dam component of the sub-system for the "water retaining structures".

In principle, the analysis should be carried out for all items within the system boundaries at the lowest practicable level, so that all potential failure modes can be identified.

Failure causes

While, strictly speaking, it is not necessary for FMEA, it is often useful to identify and describe the cause(s) of each failure mode. However, if the analysis is to be extended to include consideration of criticality, identification and description of failure causes is necessary to estimate the "frequency of failure mode initiation".

FMEA can be used to qualitatively analyse common mode or common cause failures (CCFs), and the analysis should be structured to account for any common cause failures that might occur (see failure causes below).

Failure effects

A failure effect is a consequence of a failure mode in terms of the operation or performance of the dam system. A failure effect can result from one or more failure modes of one or more components or sub-systems. The effects of each failure mode need to be identified and recorded. Since the system is analysed in an hierarchical manner, provision should be made to be able to track failure effects at consecutively higher levels in the system.

Of all the "effects" of component failure, two categories of "effects" are of particular value: the "immediate" or "local" effects of component failure, and "global", "end" or "ultimate" effects of component failure. The "local" effects refer to the effects of the failure mode on the next highest level under consideration, whereas the "global" effects refer to the effects of component failure on the system as a whole should the failure sequence progress through the system. The "global" effects are defined by analysing and determining the significance of the failure effects at all intermediate levels in the system.

Description of how the effects of element failure at one level in the system propagate through intermediate levels can be of particular importance if detection and intervention considerations are a feature of the FMEA.

The worksheet

The worksheet (or spreadsheet) is the principal feature of a FMEA. The development of the worksheet requires a detailed understanding of the design function of the system and how the individual components contribute to system function.

As the FMEA is carried out at a component level in the system, the worksheet is a tabular representation of models of component functions and the effects of component function failure.

Considerable attention should be given to the setting up of the worksheets for the analysis and of the essential information that needs to be recorded. Worksheets tend to be highly project specific and their format generally depends on the objectives of the analysis and the available information. There are no hard and fast rules for setting up the worksheet.

The British and Canadian Standards identify ten pieces of key information (with slight differences), to be included in the worksheet.

1. Identification number of the element, which usually requires establishing a suitable referencing, scheme.
2. Name of the element: a concise statement of the function performed by the element.
3. A concise statement of the function performed by the element.
4. Failure modes of the element.
5. Failure causes and operating conditions under which failure can occur.
6. Effects of the failure on the system. It may be appropriate to consider local and global effects.
7. Failure detection methods.
8. Compensating and mitigating provisions, such as redundancy, alarms, warning systems, etc.
9. Severity classification which will require selection of a method to classify the various severity levels from minor to catastrophic.
10. Comments and remarks.

The essential elements of the FMEA process are illustrated in Table 4-1. When consideration of criticality is required, columns can be added to accommodate the criticality data.

Criticality analysis

Although FMEA is essentially a qualitative method of system analysis, interpreting the results of the analysis may require some representation that describes severity, importance, criticality, potential to occur, etc. Expressing the combination of frequency and severity as a "criticality" is one way to provide a metric.

Criticality analysis can be considered to be a second phase or extension of the FMEA process. It involves the ranking of each potential failure mode according to the combined influence of severity and probability of occurrence or failure rate.

This provides a management tool that allows the relative significance of particular events (occurrence of failure modes) to be assessed in terms of the performance

Table 4-1. Sample FMEA worksheet (from CSA, 1993)

System ____________________ Date ____________________
Indenture level ____________________ Sheet __________ of __________
Reference drawing ____________________ Compiled by ____________________
Mission ____________________ Approved by ____________________

Component identification number	Item/functional identification (nomenclature)	Function	Failure modes and causes	Failure effects			Failure detection method	Compensating provisions	Severity class	Remarks
				Local effects	Next higher level	End effects				

objectives of the system. As such it allows informed decision-making on priorities for action in terms of criticality, which is a representation of risk.

Severity

Extending the failure modes and effects analysis to consider criticality requires the introduction of probability/frequency concepts.

Criticality analysis can be either qualitative or quantitative. The qualitative approach is appropriate when detailed failure rate data are not available. The quantitative approach may be appropriate when detailed failure data are available.

Prior to carrying out a criticality analysis, it is useful to establish a set of "severity classes" ranging from catastrophic to inconsequential, with particular reference to the range of possible damage and loss. The FMEA provides a basis for allocating each failure effect to a severity class.

The severity classes should reflect the range of outcomes, from disastrous to insignificant, that might be due to the failure of a component. These classes will also reflect several different areas where the severity might vary with the attributes applied, such as death, injury, damage or loss of property, economic loss, etc.

The selection of severity class in FMECA may be more important than in FMEA and therefore should be chosen carefully to reflect those circumstances best ranking the combination of failure modes and effects. An example of how severity classes might be defined is presented in Table 4-2. These severity classes provide a basis for completing the "severity class" column (second from right) in Table 4-1.

Failure mode frequency and probability

In reliability engineering, probabilities of occurrences of failure modes are derived from existing data or synthesised using analytical techniques. In this regard, the frequency interpretation of probability provides the basis for assigning criticality values. In the context of the frequency interpretation of probability, estimates of probability of a particular failure mode in a particular operating environment require a statistically significant reliability database. In the context of dams, it may be necessary to derive the probability distributions through the use of alternative methods including the elicitation of expert opinions of frequency characteristics.

If event/failure frequency is available for mechanical and electrical equipment under the operational and environmental conditions that are envisaged, these data can be used directly. However, in reality, failure rates are usually available for system elements

Table 4-2. Sample severity classes

Severity class	Consequence			
	People	**Environment**	**Financial/damage**	**Political regulatory reputation**
A	Negligible	No impact	Low impact (€,$10K)	None
B	Minor first aid	Release to on-site environment	Medium impact (€,$10K–$100K)	Low
C	Minor injuries	Legal or permit violation	Med-high impact (€,$100K–$1MM)	Medium-low
D	Temporary disability	Detrimental impact on-site	High impact (€,$1 MM–€,$10 MM)	Medium
E	Permanent disability/fatality	Detrimental impact off-site	Major impact (€,$10 MM–€,$100 MM)	Medium-high
F	Multiple fatalities	Catastrophic off-site release	Massive impact (>€,$100 MM)	High

rather than for failure modes and the operating and environmental conditions under consideration.

A series of factors are used in reliability analysis to modify base-rate failure data to account for the differences between the situation under analysis and the database. BS 5760 recommends the following approach:

The modified frequency "λ" is given by:

$$\lambda = \lambda_b \times m \times a \times b \times r \qquad (4\text{-}1)$$

where: λ_b is the base failure rate of the element, m is a modifying factor for environmental loading and maintenance considerations, a is the failure mode factor, b is the conditional probability of system failure, and r is the time at risk proportion.

Guidance on the selection of the modifying factor m can be obtained from the specialist literature on reliability analysis. The value of m could be less than 1 or greater than 1 depending on if the operating conditions and operating history differ from that of the failure rate data source. The failure mode factor, a, accounts for the proportion of element failures attributable to the failure mode in question. If all failure mode factors are listed, they sum to 1 (100%).

The b factor, which represents the conditional probability of failure given the occurrence of the failure mode, may be available from case histories, alternatively a process of deductive or inductive reasoning may be applied to predict the outcome. The values recommended in the British and Canadian Standards are as follows:

Actual loss: $b = 1.0$.
Probable loss: $0.1 < b < 1.0$.
Possible loss: $0 < b < 0.1$.
No effect: $b = 0$.

The factor r is the proportion of the required operating time that the element under consideration is at risk.

The above material applies to FMECA as applied in the mechanical and electrical equipment analysis and is generally not directly applicable to dam structures, as the necessary failure rate data are usually not available. This is an area for future research and development in the field of FMECA for dams.

The idea of applying an equation such as (4-1) is not new for dam safety risk analysis as the "base rate" adjustment approaches to estimating failure probabilities for dams are essentially of this form. However, equations of this nature are not generally valid as they constitute *ad hoc* scoring rules which can lead to erroneous estimates of probability. The error is usually not noticeable for small failure rates ($\lambda_b \ll 0.01$), but can lead to absurd results for large failure rates. The matter of *ad hoc* scoring rules is discussed in some detail in Chapter 11. MIL-STD-1629A avoids this problem and any associated confusion about probability by computing a criticality number, C_m, instead of a modified frequency.

FMEA/FMECA process

The end result that is being sought from FMEA is a structured understanding of the effects on the dam system resulting from any failure of any element of the system for the pre-determined level of detail considered appropriate to the particular study. The FMEA/FMECA process is a structured systematic means to that end.

FMEA/FMECA is carried out in an iterative manner in a limited way during the early planning and scoping stages of the analysis. More detailed analysis is done at a later stage if necessary.

There are two primary approaches to conducting an FMEA/FMECA, the "component" or "hardware" approach and the "functional" approach. The "component" approach involves the listing of individual components and the analysis of their possible failure modes to identify the effects on the system. The "functional" approach is based on the premise that every element of the system is designed to perform one or more functions that can be considered as outputs. The functional FMEA/FMECA is performed by listing and analysing these outputs to determine their effects on the system.

Procedure

As described previously, a thorough understanding of the system under analysis is essential prior to undertaking an FMEA/FMECA. A flowchart of the process is presented in Figure 4-3. The essential building blocks of FMEA are:

- System elements.
- Functional diagrams.
- Failure mode concepts.
- Identification of additional features.

The Canadian Standard Q636-M93 (Canadian Standards Association, 1993) provides the following stepwise guidance for conducting an analysis:

1. Define the system and its performance requirements.
2. Define the assumptions and ground rules to be used in the analysis including the level of the analysis.
3. List all individual components and/or the various functions at the required level of the analysis.
4. Develop a simple block diagram or other simple model of the system.
5. Devise the analysis worksheet to provide the failure modes and effects information for each component, together with other relevant information to suit the specific objectives of the analysis.

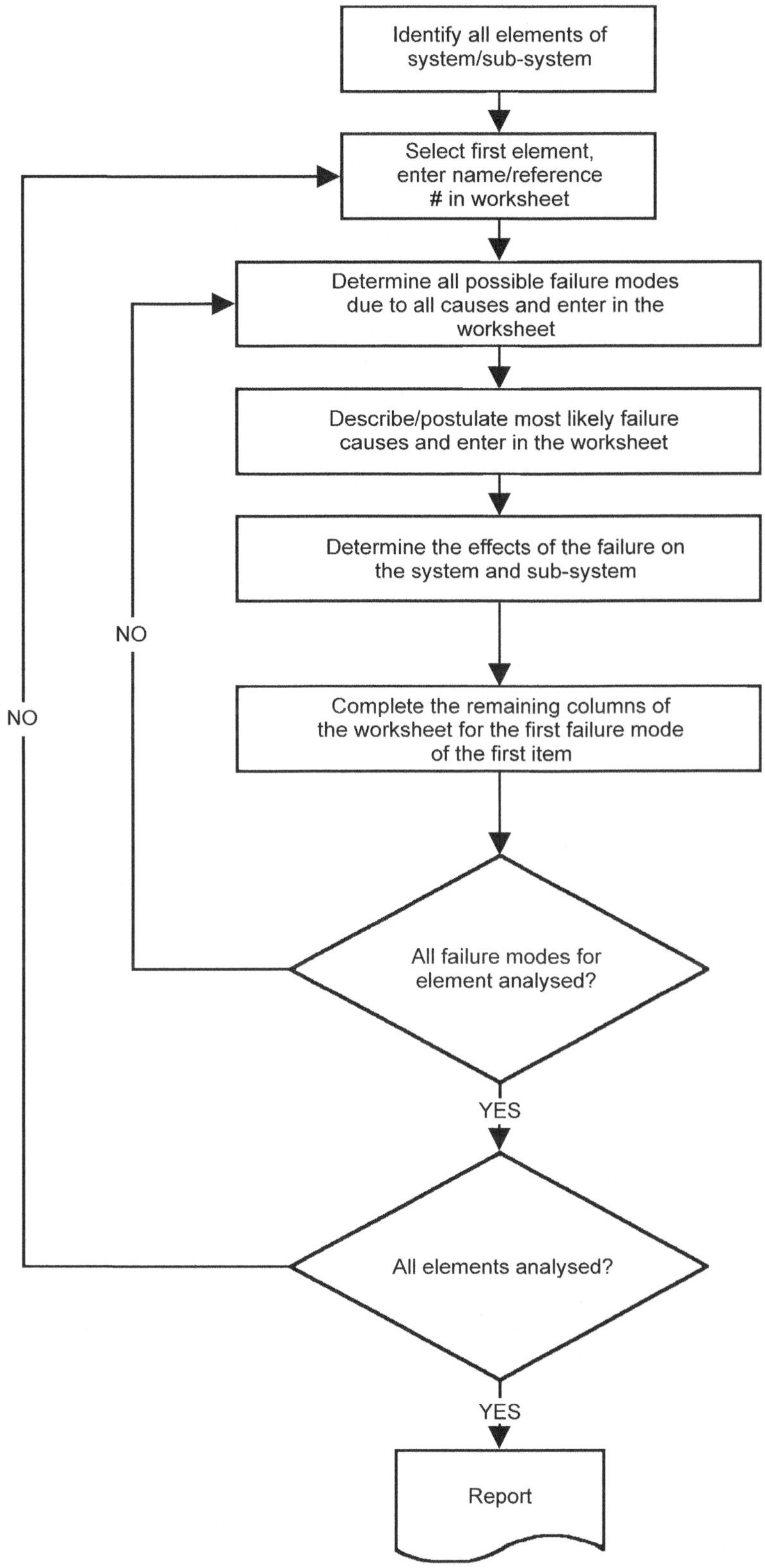

Figure 4-3. "Component" FMEA process

FMEA is not closely defined and regimented but depends on the ability and the experience of the analyst to deal intuitively with some of the concepts.

System definition and performance goals

As discussed previously, since systems can be broken down into sub-systems, and an analysis carried out at the sub-system level, clear definition of the system, sub-systems and elements or components is required. Performance requirements for the system as a whole and of all of the sub-systems and elements are also required.

In addition to design features, system definition should include functional, operational, maintenance and environmental requirements. For instance, a very high-consequence dam system would normally be required to safely pass a very large flood (in some jurisdictions it would be the probable maximum flood, in others it might be the flood with an annual exceedance frequency of 10^{-4}). The spillway sub-system may be required to have such capacity or it may be that a lesser capacity spillway together with operational restrictions (reservoir/operation sub-system), should large floods be seasonal, would meet the system requirement.

Assumptions, ground rules, and levels of analysis

The purpose and use of the FMEA should be clearly stated. One of the key considerations in the process is the decision on what is the appropriate level of detail at which to analyse the system. This can range from global analysis at the system level down to a very detailed analysis at the component or elemental level. The analysis is usually carried out at one particular level in a multi-level system. However, the process works best when working from one level up to the next. Failure effects from the lowest level of the system hierarchy form the failure modes for the next level up and so on progressively upwards through the system. Multi-level analysis may be possible but it is more complex than analysis performed one level at a time. BS 5760: Part 5 recommends that FMEA be limited to two levels in the system hierarchy.

System and sub-system boundaries should be clearly defined. The availability of information to define functions determines the lowest level within the system at which the analysis can be carried out. Assumptions must be clearly documented. Provision should be made to check the sensitivity of the analysis to the assumptions. An initial list of general hazards (or failure mode initiators) such as flood, earthquake, static, and operational, should first be developed. This list, which provides an initial focus, will in general be too broad for definitive analysis and a more detailed list should be prepared for each of the general failure modes.

It is important to ensure that component failure modes are not overlooked or omitted. Since failure mode data for dams and civil structures can not be obtained in the same way as failure data for mechanical or electrical equipment which can be obtained from suppliers, a thorough knowledge of the functional characteristics of all components and sub-systems of the dam system, as well as the causes of past dam failures and incidents are essential prerequisite for the analysis team. The implication here being that the quality of the analysis is directly related to the knowledge and experience of the analysis team in the analysis of failures and incidents.

The possible causes of each failure mode for each element should be identified and described. Since a failure mode may have more than one cause, all potential independent causes for each failure mode should be identified and described.

In addition to independent causes, the important, and sometimes dominant, condition of "common mode failure" (CMF) or "common cause failure" (CCF) can occur. Technically, a CCF is the result of a condition or event that, because of logical dependencies, causes coincidences of failure states in two or more components, excluding secondary failures caused by the effects of a primary failure.

Human error, such as spillway gate mis-operation or failure to respond to alarms, may be important and should be included in the analysis where relevant. If the safety of the dam depends on computers, the potential for software errors should be included in the analysis. Extreme natural events, such as unprecedented (for the location) earthquakes and floods, can pose particular difficulties for the human reliability elements of the analysis as the operating conditions and the response conditions are usually beyond experience.

Methods of detecting failure modes should be described while failure modes, other than the ones being considered, which manifest themselves in the same way should be considered separately. If methods of detecting failure modes are not provided or are considered to be less than adequate, FMEA can be used to identify what type of surveillance would be required to provide adequate indicators of failure mode development or occurrence. If there are redundant elements, separate methods of detecting failure may be required.

Failure effects refers to the consequence of the occurrence of the failure mode. The effects can be local or global, where the global effect is sometimes called total or end effect. Local effects refer to the effects on the sub-system at the particular level of analysis. Global effects refer to the impact at the highest system level. Global effects are determined through upward systematic analysis of all intermediate sub-systems. Local effects can be particularly important when FMEA is used to target surveillance needs, identify detection methods or to examine mitigation measures.

Good documentation is an essential component of FMEA. It should contain, as a minimum:

- summary;
- detailed record of the analysis;
- block and functional diagrams;
- references to drawings and data including the issue status on which the analysis is based.

The summary should include the principal aspects of the report, significant results, conclusions and recommendations.

Criticality analysis

The effect of a failure event, allocated to a class of severity, in the FMEA is combined with an assessment of frequency of occurrence to give the criticality of the effect.

$$Criticality = severity \times modified\ frequency \tag{4-2}$$

Table 4-3 illustrates a representative table of headings.

The steps in qualitative criticality analysis are as follows:

1. Extend the FMEA worksheet to include a qualitative description of probability, such as frequent, reasonably probable, remote, extremely unlikely, etc.
2. Combine the severity classes and the qualitative probability descriptions, possibly in matrix form.
3. Develop a method of ranking the severity–probability pairs.

The semi-quantitative alternative is to rank severity and probability descriptors on numerical scales of, say, 1 to 5 and combine them by multiplication. It is recognised that this approach may be overly simplistic and overlook many complex issues of system behaviour and the combinatorial mathematics of the associated physical behaviour. In some instances, it may be appropriate to include a rating for compensation or mitigation. The issue of incorporating human intervention in this way is discussed in Chapter 5.

Quantitative analysis requires additional treatment of the probability component and, depending on the nature of the problem, may also require consideration of exposure

Table 4-3. Sample criticality analysis worksheet

System ______________________________ Date ______________________________
Indenture level ______________________________ Sheet ______________ of ______________
Reference drawing ______________________________ Compiled by ______________________________
Mission ______________________________ Approved by ______________________________

Failure probability											
Identification number	Item/ functional identification	Function	Failure modes and causes	Mission phase/ operational mode	Severity class	Failure rate data source	Next higher level	End effects	Failure detection method	Compensating provisions	Remarks

time. The Canadian Standard Q636-93 describes an eight-step process. The British Standard describes a similar process. The appropriateness of these approaches for use in dam systems should be assessed on a case-by-case basis. They may find application in spillway gate reliability analysis but, for the most part, adequate failure-rate data are not available for dam safety applications.

FMEA/FMECA evaluation

Evaluation of the FMEA/FMECA is the determination of the significance of the knowledge gained through the analysis process and should not be confused with the risk evaluation phase of the risk assessment. In principle, FMEA/FMECA evaluation involves interpreting the analysis and expressing the results in terms of some metric. It does not involve determining the significance of the "metric". Determination of what constitutes an appropriate "risk metric" is carried out in the risk evaluation phase of the assessment.

An FMEA is essentially a self-explanatory statement of failure modes and effects. The significance of failure modes is determined initially by the severity class and subsequently by the measures that exist to prevent the initiation and/or progression of the various failure modes. The analysis is then transferred to an assessment phase where suitably qualified personnel assess the effectiveness of the risk control measures.

Criticality information, which can be considered to be a "risk metric" can be presented in several ways, including ordered lists, scales of significance or matrix representations. The ordered criticality list is simply a listing of failure modes from highest criticality to lowest. Once ordered, it is generally necessary to transfer the list to the assessment phase where the significance of the criticality index is assessed. The significance of these criticality numbers is determined in the risk evaluation phase of the risk assessment. For instance, simply ordering the criticality numbers may not be sufficient as it may be necessary to distinguish between a low frequency – high severity failure mode which generates the same criticality as a high frequency – low severity failure mode. Similarly, different criticality levels will, in general, lead to different courses of action and different rates of implementation of these actions. Again these considerations are dealt with in the risk evaluation phase. The matrix representation (Figure 4-4) is particularly useful as it provides a graphical representation of criticality upon which criticality bands developed in the risk evaluation process can give guidance on appropriate action can be superimposed. These critically bands can be superimposed on the critically matrix.

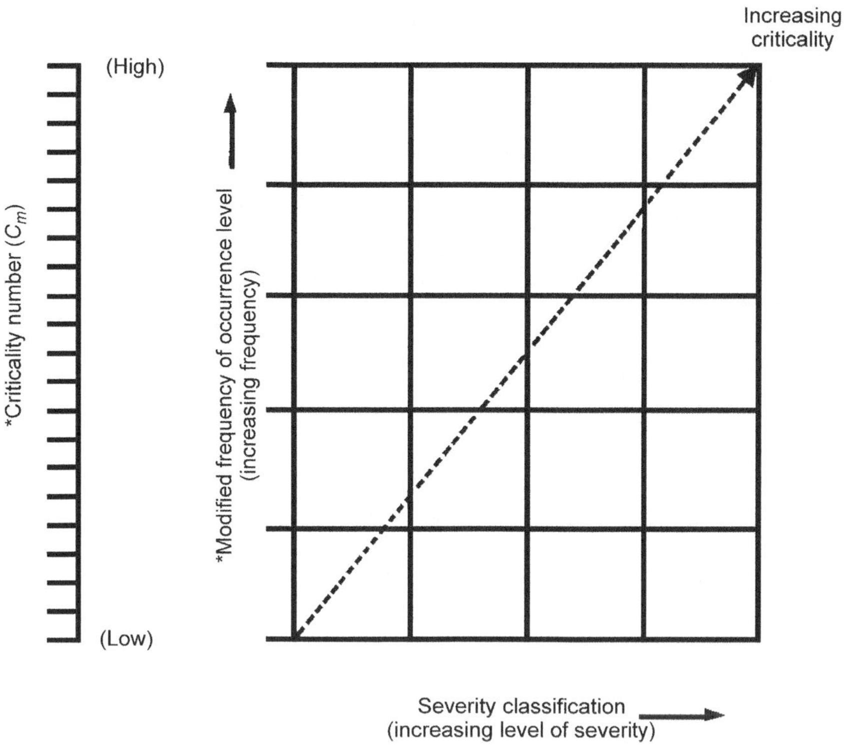

Figure 4-4. Criticality matrix (from CSA, 1993)

Note: * Both criticality number (C_m) and modified frequency of occurrence level are shown for convenience

4.2 Event tree analysis (ETA)

This section presents some of the basic features of the event tree analysis (ETA) and its application. Unlike FMEA and FTA the literature on ETA is sparse and this section should be read in conjunction with its companion sections in Chapter 8.

The description of ETA presented in this section is restricted to application of the technique to fully engineered systems where the sequence of events is known. When viewed in this way, the event tree can be considered as a deterministic model of the binary functional states of the system where probabilities are assigned in a conventional (statistical frequency) way. This discussion is directly applicable to those aspects of dam safety risk analysis that are amenable to analytical treatment in this way (e.g. spillway gate failure). It is also applicable to the analysis of dam safety. However, the apparently straightforward nature of ETA belies much more complex and subtle modelling concepts which may need to be engaged in a risk analysis of the safety of a dam. These complexities and subtleties, which include the inextricable links between the logic structure of the tree and the assigned probabilities, are considered in more detail in Part III.

The material presented in this section reflects standard practice in the nuclear and chemical process industries. Historically, this form of analysis was termed “accident initiation and progression analysis” (Jennings, 1974). The essential features of an event tree are illustrated in Figure 4-5 and discussed in more detail below.

Overview

Event tree analysis (ETA) is one of the techniques available to the engineer conducting a reliability or safety analysis for a dam. It is an apparently straightforward endeavour

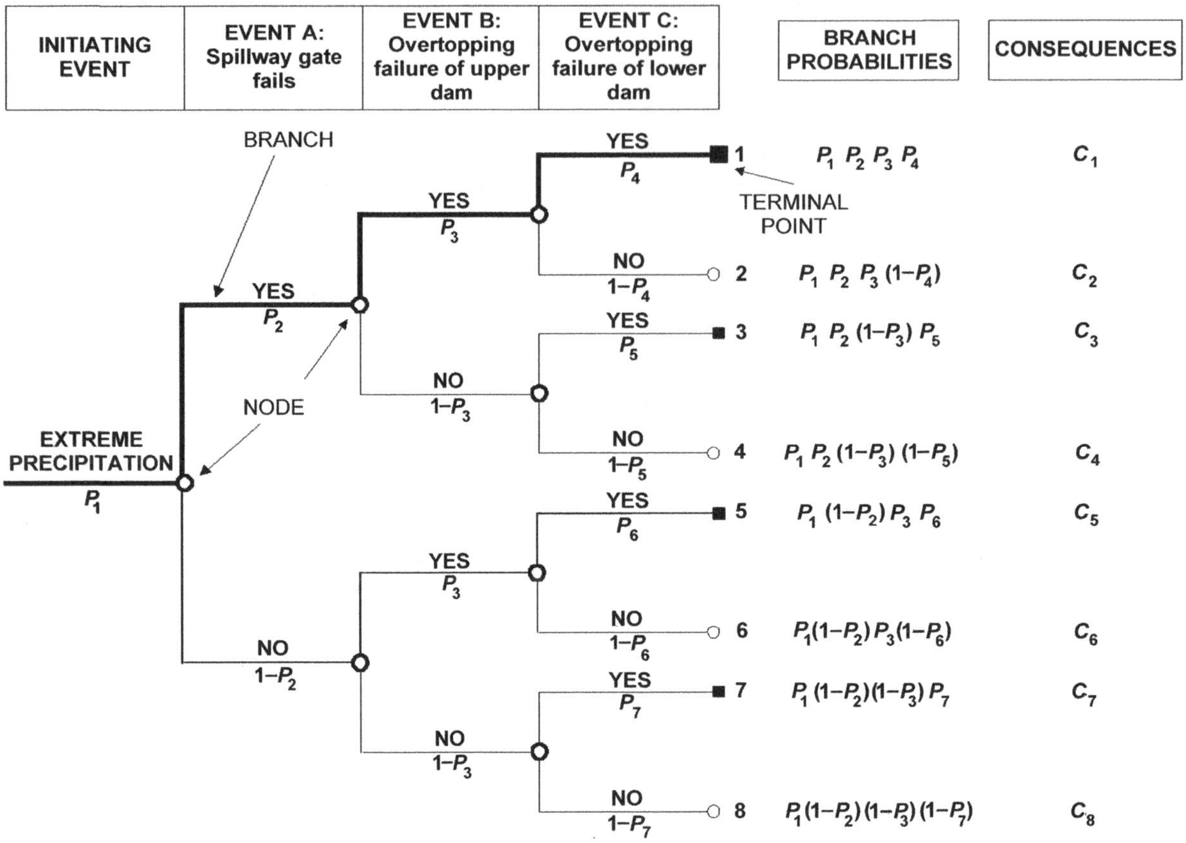

Figure 4-5. Essential features of an event tree as used in dam safety risk analysis

that finds widespread application in many industries and businesses. It is an inductive type of analysis that, unlike fault tree analysis, is not supported by an extensive theoretical basis. ETA is the most widely used form of analysis in risk analysis for dam safety, although the lack of theoretical basis means that the correctness of these constructs is difficult, if not impossible, to determine at this time. While this does not imply that ETA is not useful in safety assessment of dams, it does limit the extent to which it can be relied on in an absolute sense in decision-making.

An ETA is an analysis process of which the essential component is the event tree. The event tree is a graphical construct that shows the logical sequence of the occurrence of events or states of a system. Event trees offer the analyst the capability to construct a logic model of a system that is visual and, therefore, is easy to view and read, and that provides a qualitative and quantitative insight to the system's operations and reliability.

Although this section reflects the conventional view of ETA, recent developments in other areas, such as Bayesian networks and artificial intelligence, suggest that there may be a relationship between ETA for dams and causal analysis. More importantly, since ETA pertains to "cause and effect", it may be that the emerging science of "causality" with numerous analytical benefits that it brings might find direct application in ETA for dam safety.

Origins

The origins of ETA are not clear and, unlike FTA, there is no definitive time when the technique was first developed. Event trees are similar in form to tree diagram representations of decision problems, which predate 1970 (e.g. Raiffa, 1968).

Application of ETA to technological systems became more common in the 1970s (e.g. Rasmussen, 1975) and have continued to increase since then, particularly in the chemical, nuclear, and oil and gas industries. In this regard, ETA, as an analysis technique in its own right, appears to have evolved as an adjunct to decision analysis and system reliability analysis, particularly FTA and reliability block diagrams.

The differences in approach to ETA between the onshore and offshore oil and gas operation indicate the varying levels of detail that ETA can be carried out at. The onshore industry frequently employs relatively standardised event trees with little analysis effort required, whereas in offshore the analyses are platform specific and the ETA typically becomes the largest single risk analysis task (Pitbaldo and Turney, 1996). The first published application of ETA to dams for the purpose of estimating the risk was by Whitman (1984). Since then, ETA has emerged as the most common approach to risk analysis for dams.

Fundamental concepts

Event tree analysis (ETA) is a logic method, either qualitative or quantitative, that is used to identify the possible outcomes and, if required, their probabilities given the occurrence of an initiating event. ETA is an inductive type of analysis where the basic question that is addressed is "what happens if ...", e.g. "What happens if there is an earthquake of magnitude M6.5 and peak ground acceleration (PGA) of 0.27 g?"

ETA is widely used for facilities provided with engineered accident-mitigating features to identify the sequence of events that follow the initiating event and produce specified consequences. One use is as a "pre-accident" analysis technique that examines the systems that are in place that would prevent accident precursors from developing into accidents.

Another use is as a "post-accident" analysis technique that identifies consequences of an accident sequence. This is the most common application of ETA for dam safety where accident initiation is postulated and the event tree is used to illustrate how the various subsequent events and scenarios evolve. There are three attributes that make ETA a valuable engineering and safety analysis tool:

1. It is graphic.
2. It provides qualitative insight to a system.
3. It can be used to quantitatively assess a system's reliability.

Graphic tool. One of the attractive features of an ETA is the fact that an event tree is a graphic display of the sequence of system states (events) that lead to the occurrence of a specified undesired scenario; a failure consequence. Typically for engineered systems, event trees illustrate known "states" that the system can reside in given certain precursor conditions. When used in an inductive way, the event tree can illustrate the line of reasoning employed by the analyst in postulating a given outcome.

Qualitative analysis. When used in a qualitative way, an event tree provides a logical way of describing system performance in terms of a "what happens if . . .?" type of analysis. Typically, the event tree will be in binary form, where the various "events" of "system states" either occur or do not occur, with the outcome of each event sequence described at the end.

Quantitative assessment. The step from qualitative to quantitative ETA is straightforward in situations where the tree is precisely defined and probability of event occurrences are available. Typically, the probability of occurrence of an "event" is conditional only on the occurrence of the precursor event. The process of enumerating the event tree is simply a matter of multiplying the probabilities along individual branches. The result is a set of frequency–consequence pairs that are the fundamental components of a quantitative risk analysis.

Basics

As part of an ETA, a model or event tree is constructed that is designed to provide specific qualitative and quantitative information about a system, particularly its vulnerabilities and reliability. The model is the analytical means to obtain the information that is required in the risk assessment. Consistent with the Guiding Principles of Part I, before a model is constructed, that is before the ETA begins, the analyst will have addressed certain questions:

- What is the system (what is it comprised of, what are its bounds)?
- What information do we want to obtain?
- What is the environment the system will operate in or be subjected to?

These questions are vital to determining what will be analysed, the level of detail of the analysis, the modes of failure to be considered and the common cause effects that may exist. The analyst must also establish how a particular event tree fits into the scope of a risk analysis for a dam.

Precisely what the event tree represents must be determined at the outset, as must the nature of the probabilities to be assigned at the nodes. The meaning of the term "event" should be clearly defined and understood as should the description of system states. It will often be necessary to make a clear distinction between the state of the system and the state of the operating environment of the system as there are often important interactions between the two which can be come "mixed" in the event tree. For example, published event trees for dam internal erosion show nodes relating to whether or not internal erosion is detected, whether or not intervention is successful, etc. Clearly, the detection of internal erosion is not a functional state of the dam, and the probability of detection is not conditional only on the precursor "event" (e.g. "excessive seepage"). Similarly, the success of intervention is not solely conditional on detection.

Advantages and disadvantages

Like any analysis technique, ETA has its advantages and disadvantages. In many respects, the process is a craft that depends on the knowledge and depth of experience of the analyst as much as it requires sound engineering and scientific analysis techniques.

Building models that are either too detailed or too coarse for a particular application are obvious downsides to any analysis technique. As a result a premium is placed on experience, particularly when the systems to be analysed are large and complex.

Some of the recognised advantages of ETA include:

- It provides a logical and graphical means to illustrate the sequence of events from an initiating event to the complete set of possible outcomes, even for large systems. From the perspective of logic, the event tree illustrates the line of reasoning from a premise to the complete set of possible outcomes.
- In cases where the sequence of events is known, the process of event tree construction is simply a matter of graphically illustrating what is known about the possible ways the system can perform given the operating conditions and initiating event.
- It is oriented to illustrating necessary system conditions or states that have a bearing on the undesired consequence (e.g. system failure).
- Being an inductive modelling process, ETA provides the basis for the analysis of situations where the sequences of events are not known with certainty but which may be revealed by probing questions of knowledgeable experts. In such cases, the

event tree might be considered to be a knowledge structure or a belief structure (see Chapter 8).

- Once the tree is defined and the probabilities assigned, the enumeration of the tree is a straightforward matter (multiplication along each branch) as there are no logical gates (AND or OR etc.).

General principles

Event tree analysis is analogous to Decision Tree Analysis (DTA) (e.g. McNamee and Celona, 1991) and the same basic logical and mathematical principles apply to both endeavours. It is worth noting that the discipline of decision tree analysis is rather more extensive than that of event tree analysis for technological systems and the reader might wish to consider consulting the decision analysis literature.

As mentioned previously, the process of event tree analysis is straightforward at a conceptual level with few process considerations. In the process industries, the outcomes of an event tree analysis are governed by the laws of science (McCormick, 1981), unlike in DTA where human decisions influence the outcome. The focus here is on the former, with consideration of event trees as knowledge structures or belief structures discussed in Chapter 8. This is considered appropriate as, fundamentally, the performance of dams is governed by natural physical laws and if how these physical laws act were known precisely, as is the case for "engineered" systems, then principles described here could be expected to apply.

Event trees are usually applied to binary state systems but can also be used for systems that reside in multi-outcome states.

The event tree

The construction of an event tree is sequential with the usual convention being from left to right (a fault tree is constructed sequentially from top to bottom). The construction process begins with an initiating event, for example a large flood, and the sequences of occurrences of all necessary subsequent events are entered in the order that they occur.

Elements of event trees

While the basic structure of the event tree is simple, the actual form of the elements of the tree depends on what the tree is to represent as outlined below. Broadly speaking, the elements of an event tree can be considered to be the components of the lines of reasoning from the initiating event to the various outcomes.

Initiating event. An event tree begins with a defined initiating event. The initiating event could be a natural hazard, such as a large flood or a strong earthquake, or a failure internal to the system, such as spillway gate failure or the failure of some other internal function.

Branches. In their simplest form, the branches of an event tree are simply graphic links through the sequences of system states to the terminal point. In more advanced applications, the branches can be used to represent the processes whereby the system transitions from one state to another. Since event trees are cause–effect diagrams, it may be appropriate to consider the branches of event trees to be analogous to the directed arcs of Bayesian network analysis (Pearl, 2001). While each branch is unique, all branches have at least one common element: the initiating event.

Nodes. The nodes of an event tree represent transitions from one system state to one or more new states. Typically, for event tree models of functional states, outcomes of precursor events will be binary, Functional or Not Functional (Success/Failure, Yes/No). Nodes can also represent multiple outcomes such as 100%, 80%, 20%, 0% of functional capability. In addition, nodes can also represent continuous variables, which for computational and illustrative convenience are discretised into appropriately sized proportions of the continuum.

Terminal points (leaves). The terminal point is simply the end of a branch which defines a unique end-state, conditional only on all the preceding events (the intersection of these events). The terminal points of the "system" event tree may be interfaces with the analytical procedures that generate the consequences associated with the final system state.

System states

Once the initiating event is defined, the response of the system must be determined. For example, in the case of an electrical system, such as a spillway gate power system, where continued functional capability is maintained through the use of backup systems, (for example a stand-by generator), the analysis of system states will involve determining the response of the primary systems and backup systems to the influence of the initiating event.

Once the component and sub-system responses that are called into operation for a given initiating event are identified, the possible states of the system must then be defined. Success and failure states must be carefully defined to ensure that potential failure states are not included in the definitions of success states. Typically, the event tree will be in binary form, where the various "events" or "system states" either occur or do not occur, with the outcome of each event sequence described at the end.

Typically, the mechanical and electrical systems will be considered sequentially as "called into service" where the response of the primary system would be defined followed by the response of the secondary system. This is usually illustrated in the form of a block diagram as illustrated in Figure 4-6.

The ordering in Figure 4-6 does not necessarily mean that the primary system actually transitions in time to a failed state before the backup system, as both could be rendered inoperable at the same time by the same earthquake. However, they can be logically ordered in this way as they are called into service in this order. Figure 4-6 illustrates the basic first step in an ETA in that they provide the headings for the event tree.

Each branch of the tree yields a unique outcome sequence and the combination of events or states are normally identified.

Partial functional states

While ETA of mechanical and electrical systems of dams can be carried out by direct application of the general principles outlined above, the performance of water-retaining structures is not as amenable to representation in event trees. While there is no agreed upon way of dealing with this problem, the following suggestion is presented for discussion. One approach to ETA of a dam is to model the dam as a system of functional

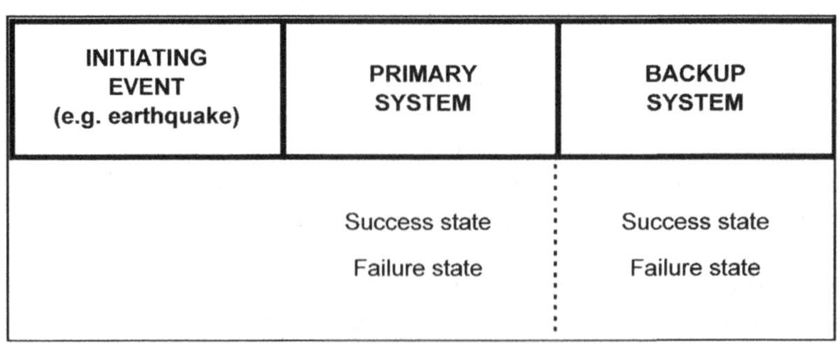

Figure 4-6. System state definition (based on Rasmussen, 1975)

sub-systems where system function is water retention, with a predictable design seepage rate. The various components of the dam – core, filters, drains, shells, etc. – each have defined individual functions that collectively provide system functionality.

The model of the functionality of the dam can be represented in event tree form where, instead of the event tree branches defining successful function or failed function, additional branches can be added to characterise the degree of functionality. Thus, instead of binary branching, there might be several branches for each functional sub-system denoting a different percentage of functionality (see Figure 4-7).

Alternatively, functionality might be represented by a continuous distribution over the possible range of functional states. Figure 4-8 illustrates the manner in which models of continuous variables or process are represented in event trees.

If the functional capability is measurable, and the event tree is being used to analyse some continuous process, then the event tree can be updated periodically as the deterioration continues.

Typically, one or more failure mechanisms (or failure processes) can be determined from the physics of the response of the dam. These failure mechanisms will normally be depicted as sequences or branches in the event tree. In developing the event tree all of the outcomes, successes and failures are depicted, although non-failure outcomes may later be pruned from the tree. The system states or events, once defined, are combined using decision-tree branching logic to obtain the various outcome sequences as illustrated in Figure 4-9.

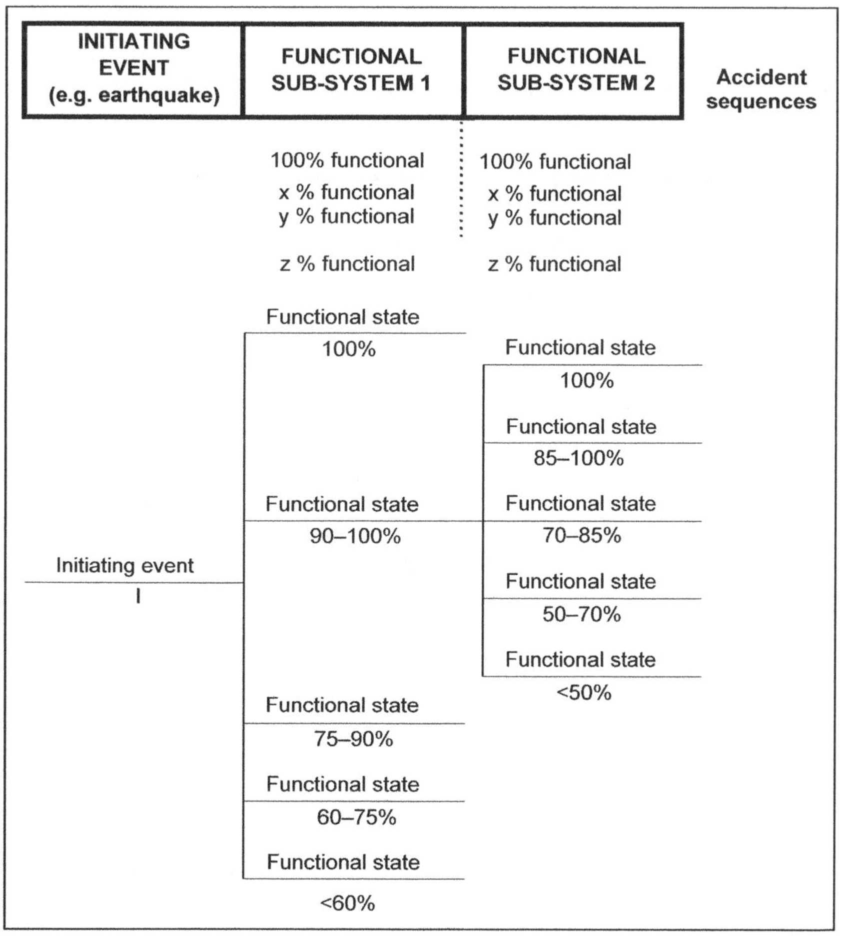

Figure 4-7. Discretised representation of functional states

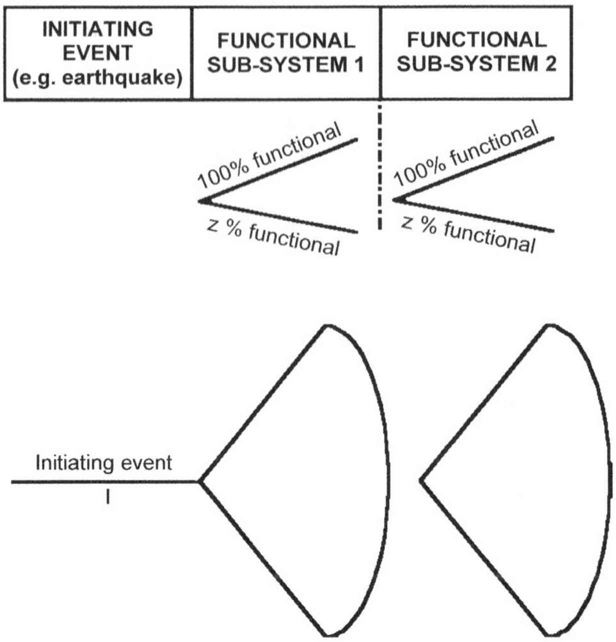

Figure 4-8. Continuous representation of functional states

The accident sequences or more generally the failure mechanisms that result from the tree structure represent the final state of the dam conditional on the initiating event. Each branch of the tree yields a unique mechanism or accident sequence. Typically, dam

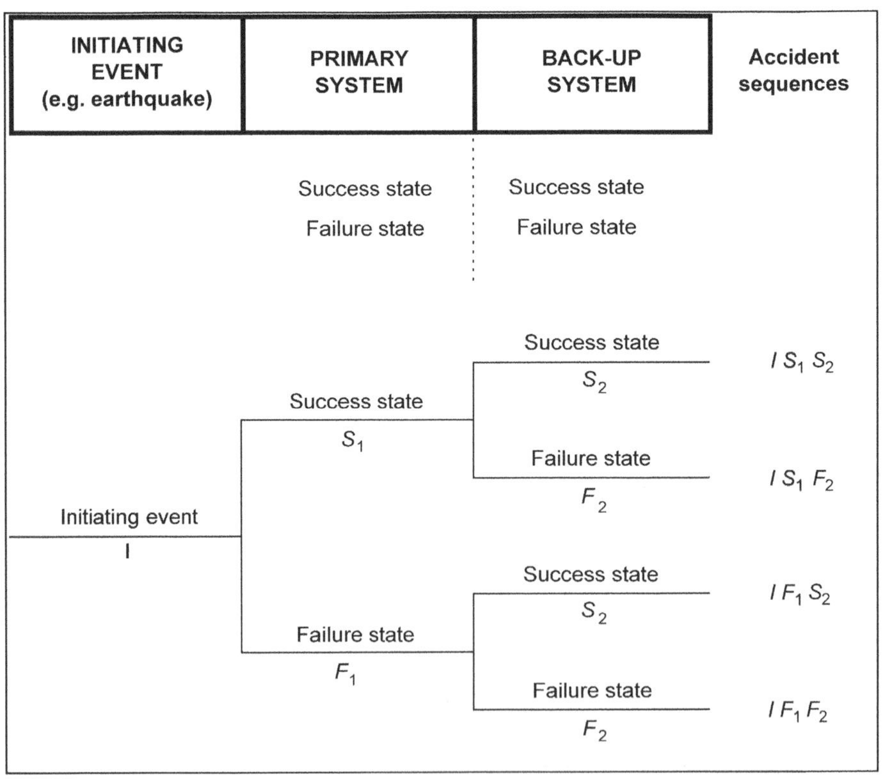

Figure 4-9. Illustration of event tree branching (based on Rasmussen, 1975)

failures can be broadly grouped into three principal failure modes (see Chapter 8, Event tree analysis):

- hydraulic failures due to abnormally high pool;
- mass movement due to extreme loads;
- deterioration and internal erosion.

A failure mode initiator (hazard) is associated with each of these modes. For example, an earthquake may cause dam foundation liquefaction leading to slumping of the dam or mass movement. These three failure modes can, in general, be initiated by more than one cause and, typically, for a dam an event tree will be constructed for every combination of failure mode and failure mode initiator

Failures: mechanisms, modes and effects

The consequences associated with each of the sequences are usually determined separately. Typically, for a dam, the final event in the event tree sequence would be the release of the reservoir. Consequences are then estimated from an analysis of the effects of the dam breach flood (see Chapter 10). It may be that the failure consequences are also estimated using an event tree.

Pre-existing system states

This volume takes the view that the state of the system should be fully defined by a logic tree prior to the initiation of the failure mode in order to ensure that the event tree is restricted to being a representation of the system response. This approach is somewhat different to the ETA practices for dams that emerged in the 1990s where pre-existing system states were often represented at nodes within the event tree.

At the outset, the state of the system prior to the occurrence of the initiating event must be defined. This means that the complete set of states that the system normally operates in, for example reservoir elevation, operating temperature, etc. are defined prior to constructing the event tree. It is also important that the knowledge (epistemic) uncertainties concerning the pre-existing state be represented if they influence the analysis of the system response. Logic trees provide a useful way of representing system states where each end branch of the logic tree describes a pre-existing system state (Figure 4-10).

When it is necessary to consider more than one precursor system state and/or uncertainty in the precursor state, it is necessary to condition each event tree for each failure mode by all possible system states.

The possible states α β, χ and the various state possibilities and/or state uncertainties ($i=1$ through N) are represented in the logic tree with an event tree for each failure mode being constructed at the end of each terminal point or leaf of the logic tree. Figure 4-10 also illustrates how an ETA can rapidly become very large and cumbersome.

Dynamic aspects of event tree analysis

Correctly accounting for the timing of the events is a significant issue in ETA, which is more amenable to representation of instantaneous state transitions rather than time-dependent changes in functional state. The analyst should be aware that, as is the case in some nuclear power applications, the failure logic sequence might change depending on the rate at which interdependent changes in states take place.

Event tree construction

The process of constructing an event tree depends on what the tree is supposed to represent and the nature and functional characteristics of the system under examination. As is explained in Chapter 8, event trees can be graphical representations of: models of

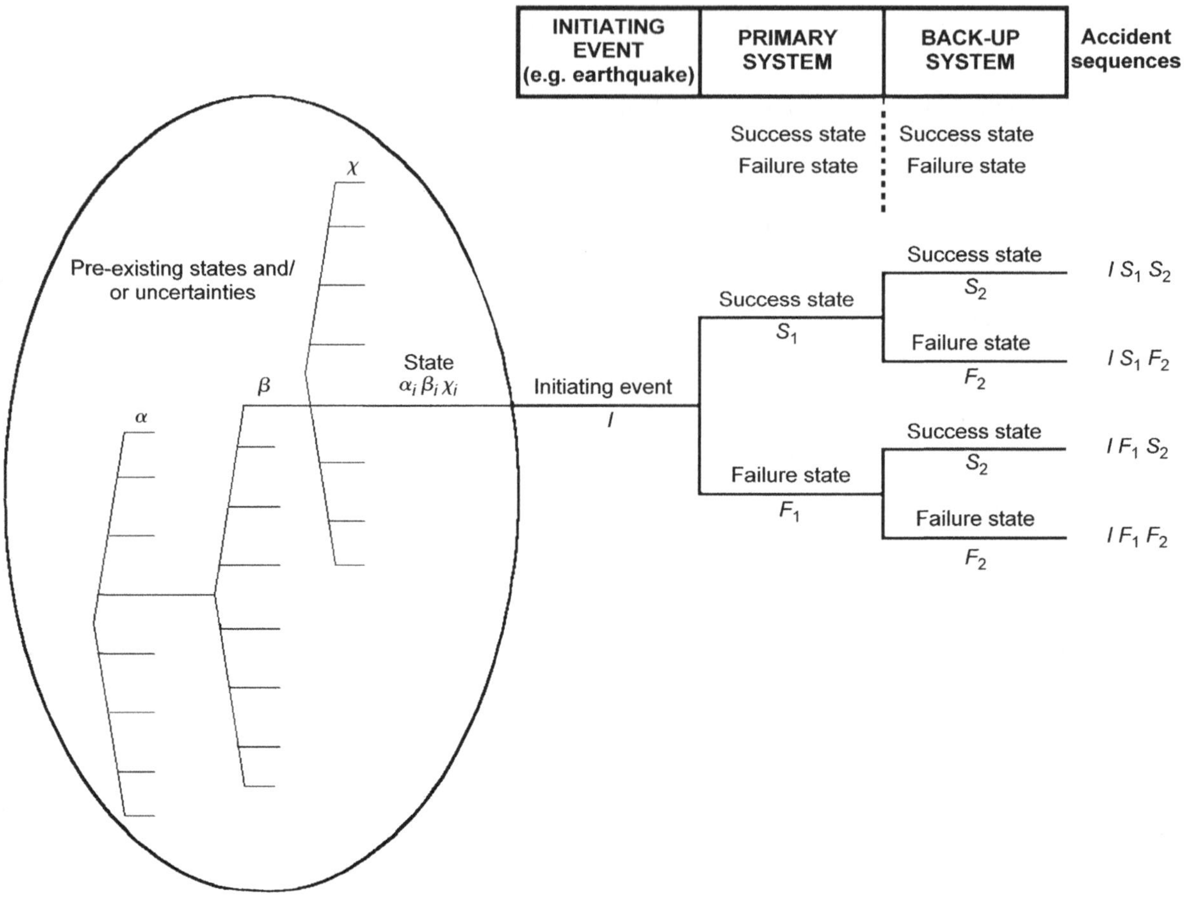

Figure 4-10. Logic tree of states that condition the event tree

physical systems; statements about the joint probabilities of random variables; or accounting schemes for information, knowledge, and beliefs.

If the event tree is a graphical representation of a model of a physical system, then it can be drawn directly from its influence diagram by ordering the nodes according to the way the directed paths of the influence diagram connect the events. The event tree is then generated with the event nodes as vertices. The only difference between the event tree and the influence diagram will be that in the event tree there will be no decision nodes.

If the event tree is a graphical representation of the statements about the joint probabilities of random variables then, from the total probability theorem, the order of the events is not a consideration.

If the event tree is a belief structure, then the event tree represents the lines of reasoning from the initiating event to all of the possible outcomes. It sets out a framework for characterising knowledge and uncertainties, and for analysing the implications of those uncertainties.

The above considerations are important because of the nature of the variables represented in the different types of tree. As explained in Chapter 8, what the event tree represents determines the types of variables in the tree. It also determines the considerations in the assignment of probabilities in the tree. The remainder of this discussion on ETA focuses on event trees as a physical model of the system response. The manner in which an event tree is constructed will depend on what it is intended to represent.

Influence diagrams

Influence diagrams (Figure 4-11) provide a useful way of characterising the risk analysis problem under consideration (Howard and Matheson, 1981) as an aid to construction of event trees.

When used in risk analysis, an influence diagram illustrates relationships between initiating events, states of nature, conditions of the system and consequences. The process of creating an influence diagram provides the fundamental model of the risk analysis problem from which event trees can be constructed and other reliability calculations developed. Design of influence diagrams is a craft as there is no rigid procedure other than that imposed by logic in the form of a generally structured procedure. Detailed treatment of influence line diagrams and their use in ETA are presented in Chapter 8.

Influence diagrams provide a powerful way of overcoming the problem of an event tree rapidly becoming an unmanageable "bushy mess" (Raiffa, 1968), because they are a means of representing the same analysis problem in a compact form. Every node in the event tree can be represented by a single node in the graph, and relationships between the nodes are represented by directed paths, sometimes called influence lines or directed acyclic arcs.

Influence diagrams are a particularly useful way of framing the risk analysis problem, structuring the analysis process, organising analysis tasks, managing the iterations of the analysis and presenting a summary overview of the problem. They provide a straightforward way of illustrating dependencies between nodes, be they deterministic or probabilistic.

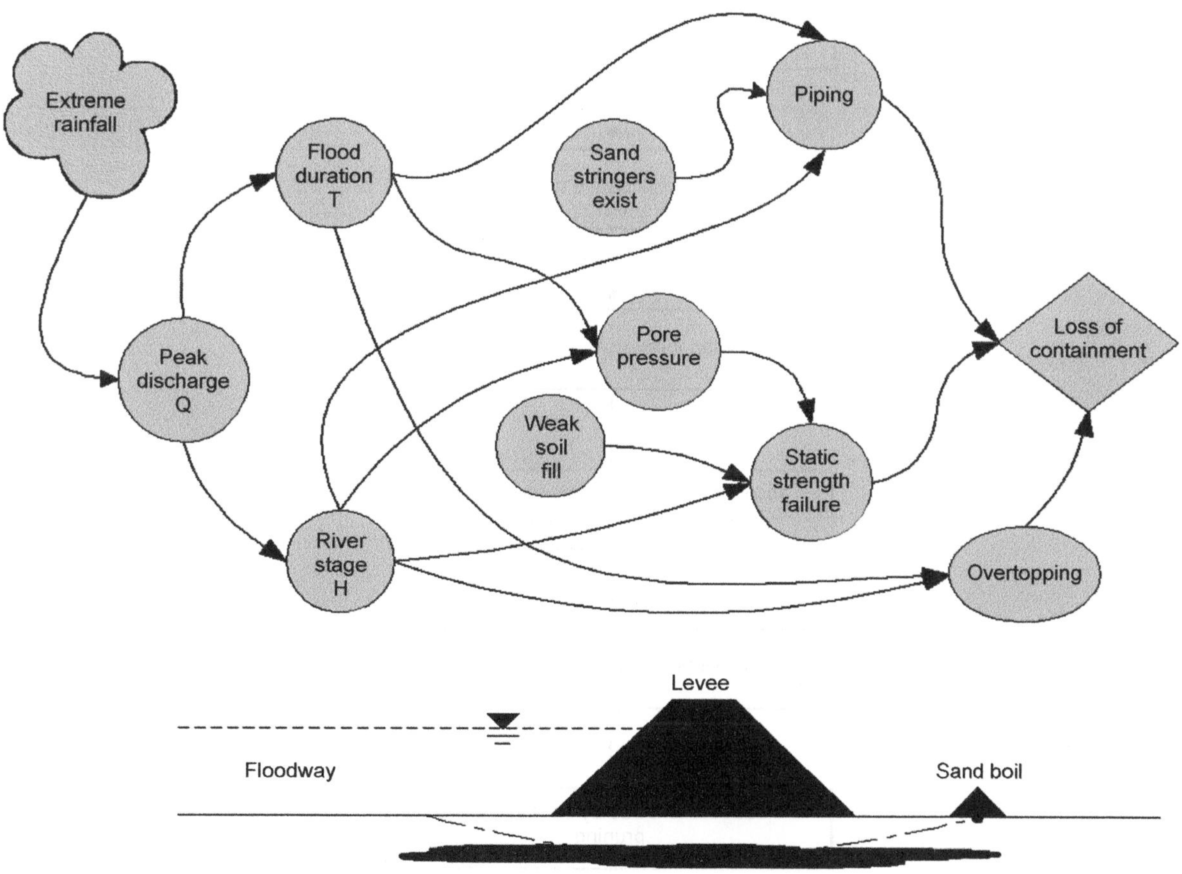

Figure 4-11. Influence diagram for levee risk analysis (from Chapter 8)

Steps in the event tree construction process

As a graphical representation of a model of a physical system, an event tree can be drawn directly from its influence diagram by ordering the nodes according to the way the directed paths of the influence diagram connect the events. The event tree can then be generated with the influence diagram nodes as vertices. In safety-system analysis, the most developed form of the ETA, construction process proceeds as illustrated in Figure 4-12.

As mentioned previously and as illustrated conceptually in Figure 4-9, once the various system states have been defined and enumerated, the states are then combined through branching logic to illustrate the various sequences of events that are associated with the given initiating event.

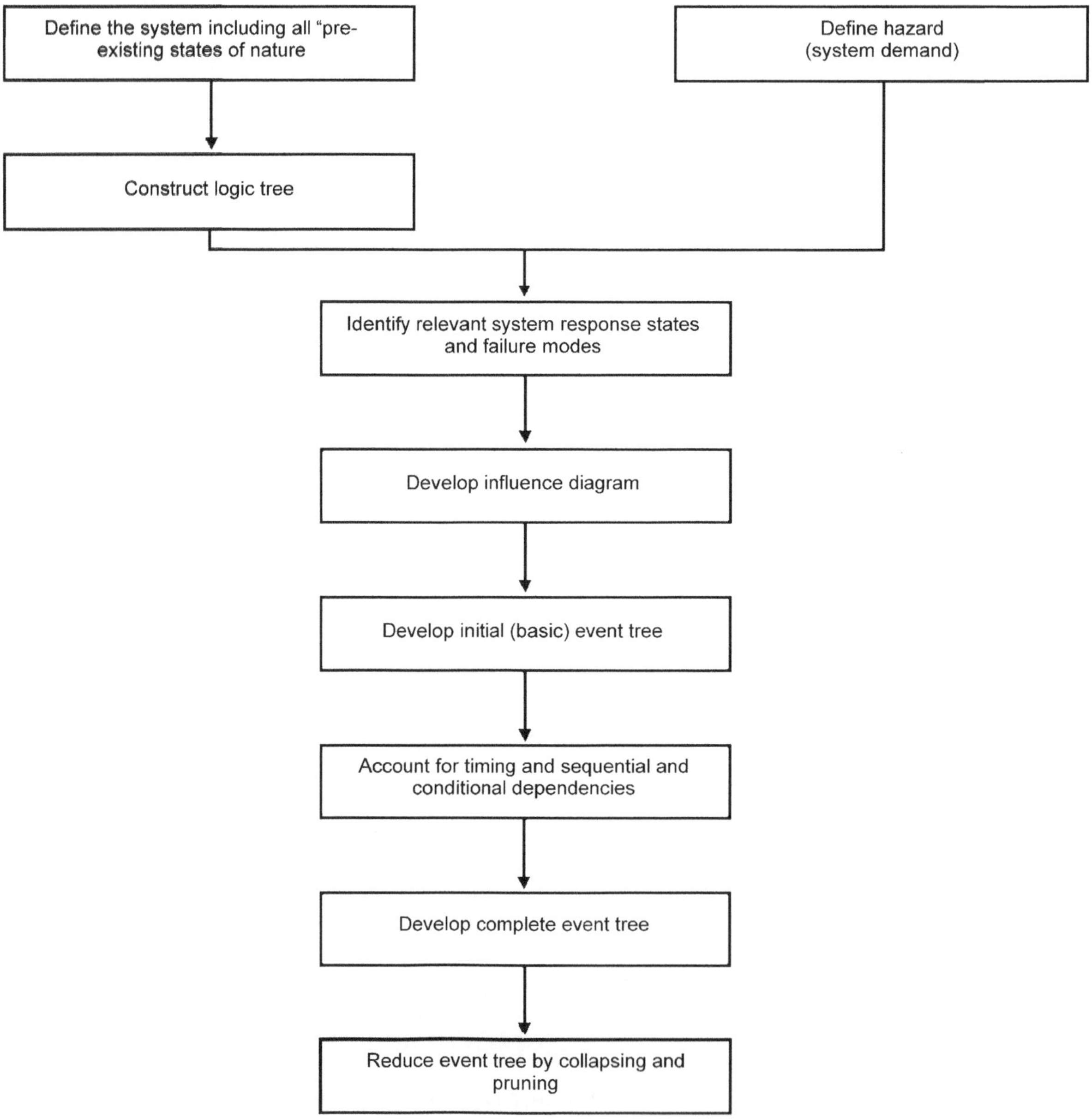

Figure 4-12. Steps in constructing an event tree

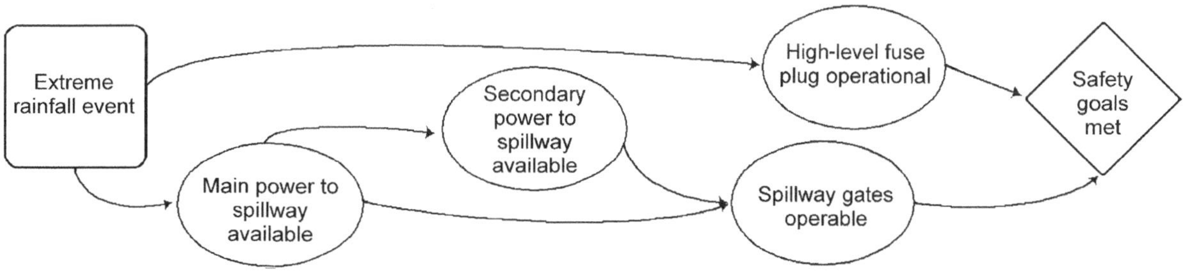

Figure 4-13. Influence diagram for extreme flood event

Event trees can rapidly become overwhelmingly large and complex. The size and complexity of the system and the complexity of system function determine the size and complexity of the event tree.

As in any modelling technique, it is important to simplify the description of reality without neglecting anything essential (Duesenberry, 1958). This capability is central to the craft of event tree construction. Pruning the tree is sometimes recommended, however, apart from simple cases where unimportant outcomes are known in advance; keeping the tree compact through pruning during construction is not generally feasible. In most cases, the tree must be essentially fully developed with probabilities assigned before pruning can take place.

Ground rules for event tree construction

Since the process of developing an event tree is based on decision tree branching logic, a brief description of decision tree construction follows. As mentioned previously, the essential difference between a decision tree and an event tree is that the decision tree branching points have two types of nodes, decision nodes and event occurrence nodes, whereas an event tree branching is done only in terms of event nodes. The reader is referred to the specialist literature with Raiffa (1968) being a useful introduction to the subject.

As mentioned previously, the event tree can be derived from its corresponding influence diagram (Figure 4-13). As a graphical representation of a model of a physical system all of the nodes in the tree, which are determined by the system states illustrated in the influence diagram, are combined through decision tree branching logic to obtain the various event sequences associated with the initiating event.

As mentioned above, the success and failure states of the system must be clearly defined. With reference to the upper part of Figure 4-12, the system state, the hazard, the system response states, and the conditions under which they cease to function successfully (failure modes) must all be defined before the influence diagram and initial (basic) event tree is constructed.

Once the systems and their operating states are defined, construction proceeds, beginning with the initiating event. Lines are drawn out towards the right to the first response state of the system. If binary or two-state modelling is employed then the tree will have a bifurcating structure at each subsequent node (Figure 4-14).

Again, with reference to the influence diagram, main power to the spillway activates the spillway that may or may not be operable (Figure 4-15).

One way for tree construction to proceed is from left to right. The tree is completed by extending all of the branches through all possible system states to yield the completed tree (Figure 4-16).

Logic and mathematical considerations

Independent of what the event tree is intended to represent (states of nature, sequences, system states or belief structures), an event tree is fundamentally a graphical statement

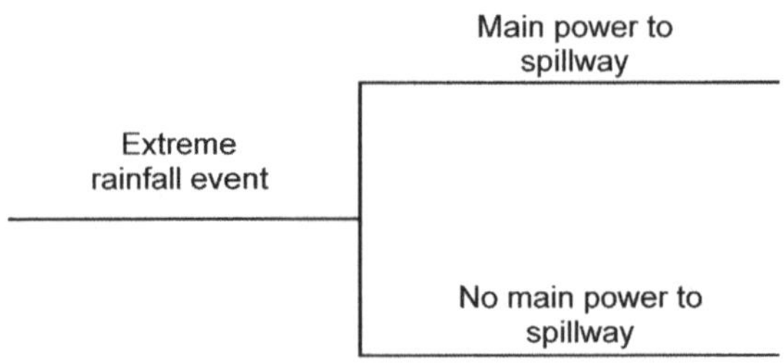

Figure 4-14. Step 1 in transforming the influence diagram to an event tree

in logic. This brings requirements for logical consistency in the event tree construction process. Thus, while event tree construction is a craft, it is a craft governed by the rules of logic.

Dealing with the "bushy mess" issue. Event trees rapidly become large and somewhat difficult to manage, especially if the system is complex. Therefore, it is desirable that only those branches that are essential for characterising the system behaviour be presented. In most real problems, it is necessary to seek a compromise between a complete representation of the system performance (possibly 10s of thousands of branches) and a pragmatic representation that simplifies the tree structure without neglecting anything that is essential. The following provides a brief summary of the Chapter 8 material that deals with how to keep event trees manageable.

Pruning. An event tree can be pruned by eliminating all the branches that have zero conditional probability for at least one event. Considering Figure 4-9, if the failure of system 1 caused the failure of system 2, then instead of considering the sequences I, F_1, S_2 and I, F_1, F_2, only the sequence IF_1 would be considered.

As discussed in Chapter 8, there are a limited number of ways that an event tree can be pruned. For the most part this pruning can only be done if the tree has been quantified as only then is it possible to estimate whether or not pruning a branch is likely to have a significant effect on the result. However, even this procedure is not as straightforward as it might seem because the net effect of pruning a large number of low probability events might actually be a significant proportion of the total failure probability.

Constraining the event tree by pruning during the construction effort poses certain difficulties, as the analyst must know the effects of the pruning effort in advance of knowing the result of the analysis. Further, constraining the tree in this way does not lead to a reduction in the analysis effort, as the effects of the constraints need to be checked by sensitivity analysis. The net result could well be that the overall amount of analysis is the same as if the full tree was constructed in the first place.

Partitioning a large tree into manageable sub-parts and then combining them at the end of the analysis is another way to keep the tree manageable. Typically, event trees are

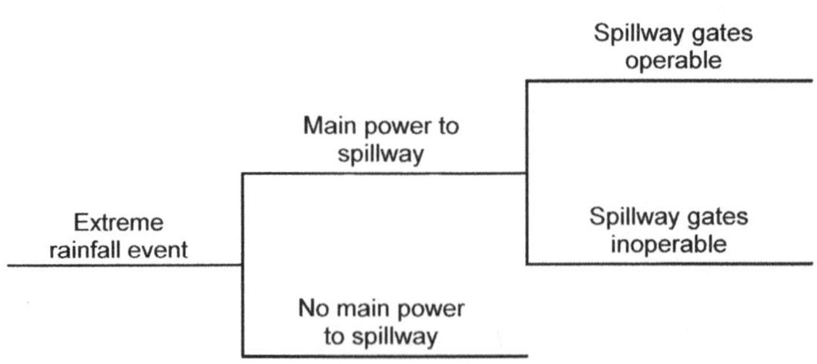

Figure 4-15. Step 2 in transforming the influence diagram to an event tree

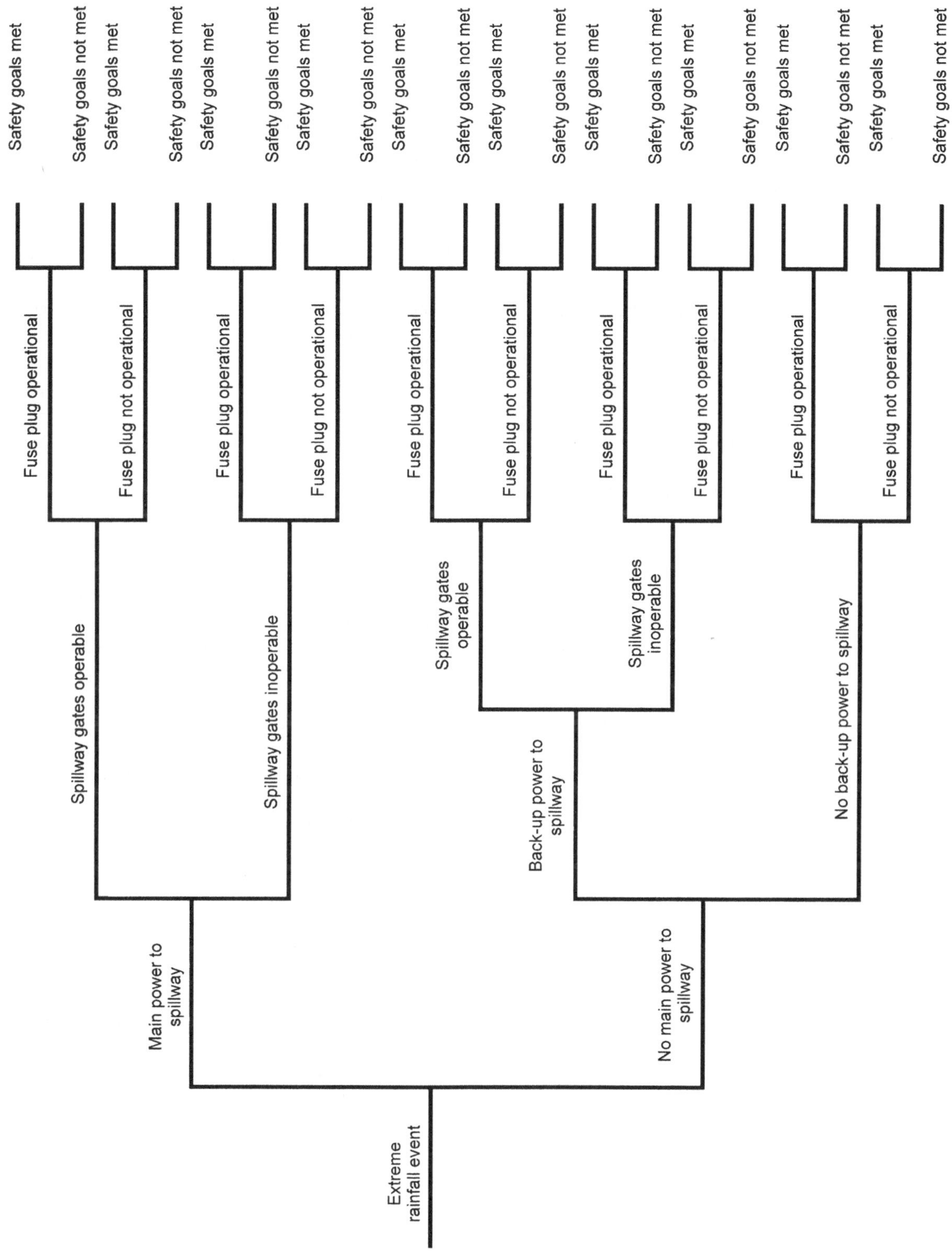

Figure 4-16. Fully developed tree

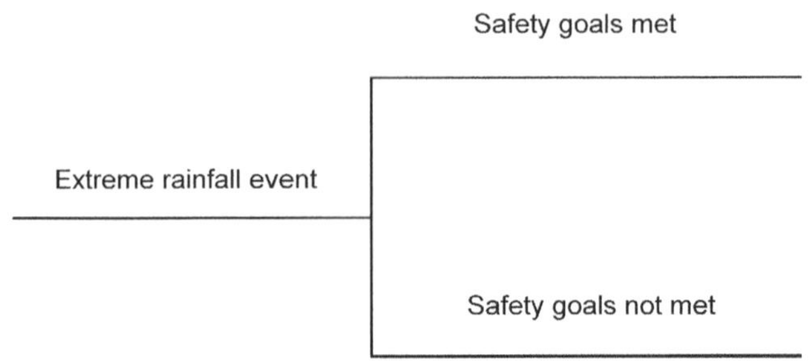

Figure 4-17. "Collapsed" version of Figure 4-16

constructed for the various hazards (floods, earthquakes human agency, etc.) that pose a threat to the safety of the dam.

For the purpose of presentation and communication, large event trees, once constructed and evaluated, can be transformed into simpler summary trees by "collapsing" them to a more a manageable size by combining several compatible nodes that lead to common outcomes. The tree illustrated in Figure 4-16, could be collapsed down to as little as two branches, or to any convenient intermediate size (Figure 4-17). Although this is not very informative technically, in its enumerated form it might be all that is required.

A more informative collapsed version is illustrated in Figure 4-18. Simplification of large and complex event trees by "collapsing" them into summary trees is always possible, thereby providing a readily understood and easily communicated way of presenting the results of an ETA. The extent to which a tree should be simplified for communication purposes can be tailored to suit the needs of the audience. In this way, large and complex trees, while cumbersome and often difficult to evaluate, are not necessarily an obstacle to communication as a little additional work can provide a very effective means of displaying the main features of the analysis.

Disadvantages of tree reduction. While reduction of the event tree for the purposes of communication and manageability has clear attractions, there is the danger that the complexity of the functioning of the system might be under-represented in the collapsed

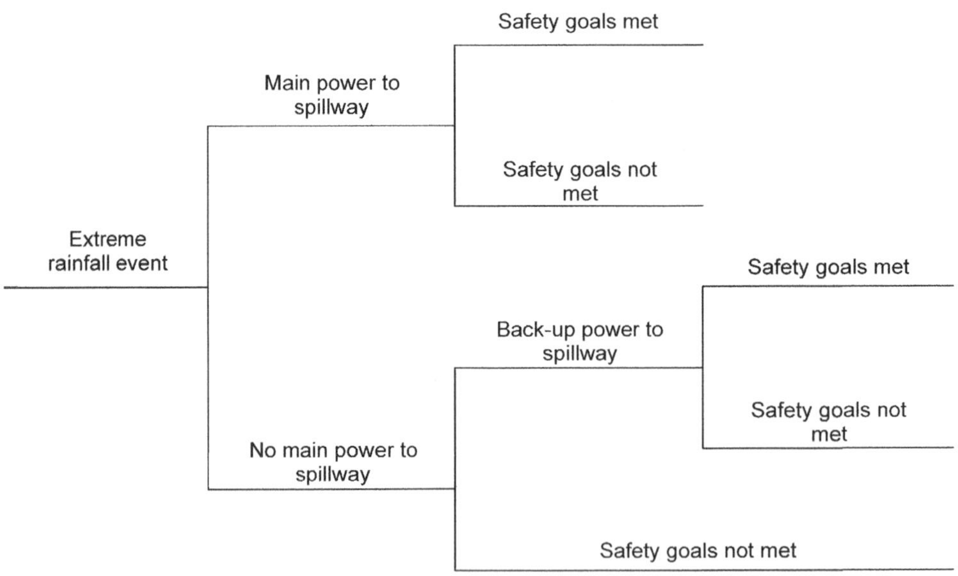

Figure 4-18. "Less collapsed" version of Figure 4-16

tree. Thus, while the collapsed tree is very useful as a means of conveying the result, only the complete tree can convey the complexity of the system and all of the functions necessary for the system to operate safely.

Excessive focus on "reduced trees" also has the potential to mask the complexity of the analysis and the difficulties faced by the analysts in determining the safety status of the dam under consideration. In this regard, only the complete tree can provide the insights into the reasons for the costs of the analysis effort and any resulting dam safety improvements.

Event tree evaluation

The process of evaluating the event tree is essentially one of interpreting what the tree represents and where necessary quantifying this representation. How an event tree is evaluated will depend on what it is intended to represent. Fundamentally, the evaluation process is one of explaining and interpreting cause and effect relationships for the system under consideration. In this regard, the exercise of event tree evaluation is fundamentally a matter of logic, be it a series of mathematical operations on statistical data or reasoned belief concerning the physical processes represented at the nodes of the event tree.

Qualitative evaluation

When used in a qualitative way in dam safety applications, an event tree provides the basis for inferring the performance of the dam given certain operating environments. The strengths of any inferences will obviously depend on the quality and extent of the data, knowledge and information embodied in the tree and the correctness of the logical and mathematical procedures used in constructing the event tree. Since an event tree is fundamentally a representation of a directed acyclic graph, event tree evaluation follows the general mathematical procedures for evaluating influence diagrams (Schachter, 1986) and Bayesian networks (Pearl, 2001). These concepts are outlined in Chapter 8.

Event trees provide a qualitative representation as to the various ways undesirable outcomes can be realised. The qualitative evaluation can be used to determine what safety management systems would need to be introduced to prevent various scenarios from developing. Used in this way, ETA provides a very useful risk management technique for identifying and designing risk mitigation measures. With reference to Figure 4-16, once the various ways all possible outcomes are realised, those undesirable outcomes associated with all initiating conditions can be isolated and measures to prevent progression to the undesirable outcomes implemented. Under such ideal conditions, the system can be made inherently safe, and the event tree for the inherently safe system would be as illustrated in Figure 4-19.

In terms of the above, the discharge systems are expected to work under all flood conditions up to the maximum capacity of the system. For floods up to a particular magnitude, only the spillway would be utilised until the discharge requirements exceed the spillway capacity thereby placing demand on the fuse plug. The backup power would only be required when the main power is deliberately taken out of service.

Conclusions that a dam is "safe" in terms of conventional deterministic approaches to dam safety assessment are amenable to representation in trees of this form, where judgements are made concerning the truncation of potential failure sequences. However, the usefulness of qualitative event trees diminishes rapidly once uncertainty is recognised or when it is not possible to make the system inherently safe, as illustrated in Figure 4-20, where qualitative likelihoods have been assigned to the branches of the event tree.

Figure 4-20 displays success and failure states for the two operating conditions described above, with success being defined as spillway discharge OR fuse plug

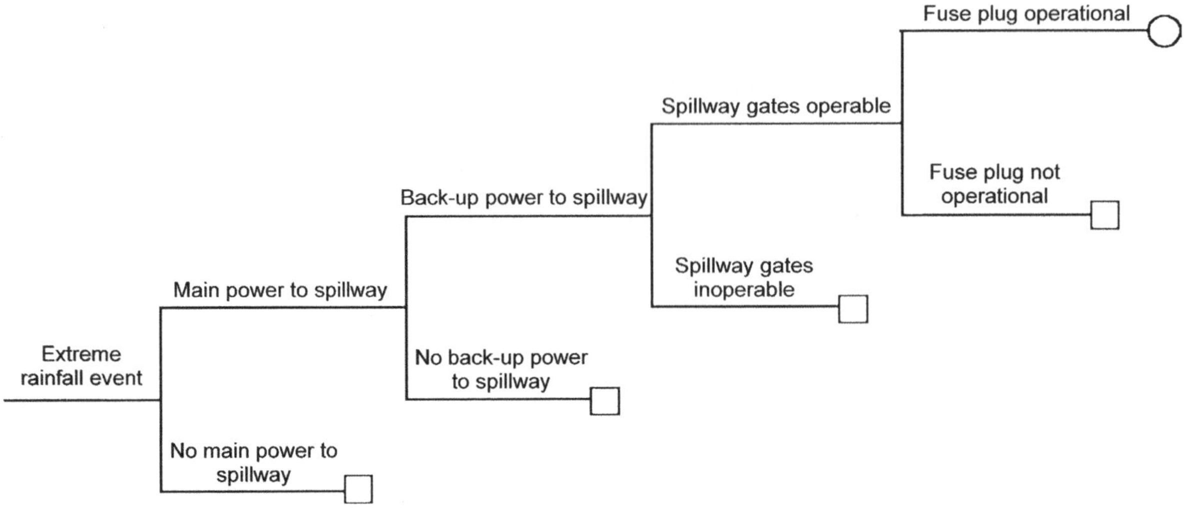

Figure 4-19. Event tree for inherently safe discharge system

Note: □ denotes termination of pathway

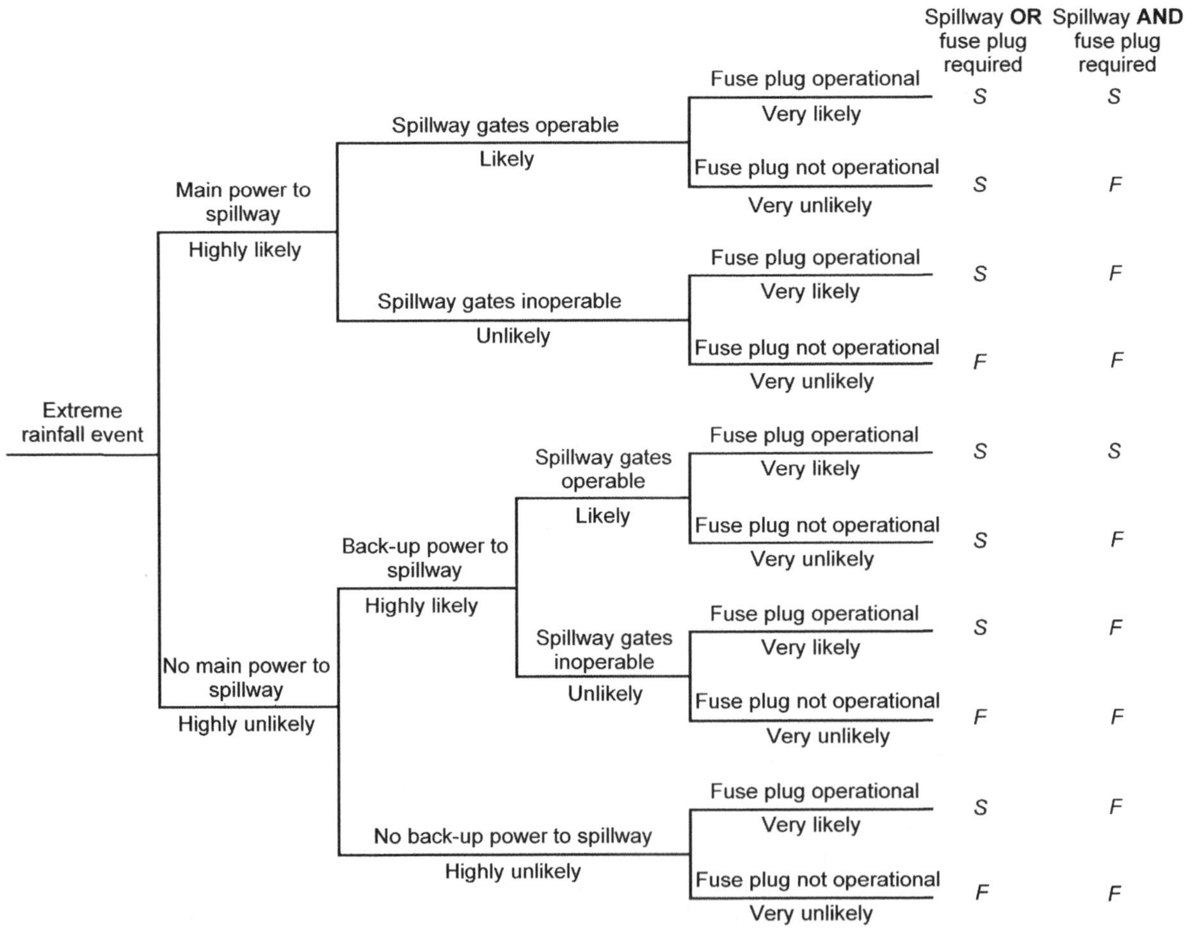

Figure 4-20. Qualitative event tree with uncertainty statements

Note: S denotes discharge system function success, *F* denotes failure

discharge being required for lower flows, and spillway discharge AND fuse plug discharge being required the higher flows.

Accepting that the verbal statements of likelihood – likely, highly likely and very likely – represent increasing degrees of likelihood above 50:50 can be assigned, the problem of how to interpret the tree arises as there is no rational way to combine the verbal statements of likelihood to obtain an overall likelihood of success. The difficulties in interpreting the tree now become immediately obvious, as do the problems associated with truncating failure paths (Figure 4-19) and with constraining the tree by pruning during construction. Further, while it might be possible to assign a likelihood to the tree as a whole by inspection of the terminal point success/failure states, there is no way to rationally relate the "assigned lumped likelihood" to the individual likelihood statements on the branches, even if there is a specified relationship between the statements of likelihood on the branches of the tree. The only way that the problem of event tree evaluation can be overcome is to proceed to quantitative evaluation.

Quantitative evaluation

Quantitative evaluation involves associating an appropriate value of probability for each branch. An individual branch probability is typically estimated in one of several ways, using:

- statistical (i.e. empirical) estimates;
- reliability models;
- fault tree analysis;
- expert opinions.

The second and third of these are normally grouped together under the name, "engineering reliability models". The use of statistical estimates and reliability models in event tree quantification are discussed in more detail in Chapter 8, where there is also some introductory discussion on the use of fault trees to estimate probabilities in event trees. Detailed treatment of FTA is presented below. Expert opinion is described in Chapter 7 and its use in event tree quantification in Chapters 6 and 11. The reader is also encouraged to refer to the established literature on probabilistic analysis of engineering problems.

Underlying principles. General principles for quantitative risk analysis have emerged over the past 10 to 15 years which provide useful guidance for applications of risk analysis to dam safety (Bedford and Cooke, 2001; Cooke, 1991; Howson and Urbach, 1991; Morgan *et al.*, 1990). A thumbnail sketch of the ways that event trees can be quantified is presented here.

The key principles of event tree quantification are that the quantification process should be based on the following principles:

- reproducibility;
- treatment of uncertainty;
- validation.

Reproducibility. It must be possible for scientific peers to review and, if necessary, reproduce all calculations. This entails that the calculation models must be fully specified and the ingredient data must be made available.

Uncertainty. Estimates of risk should be accompanied by an analysis of the uncertainty in the estimate. The analysis should be explicit about the sources, nature and magnitude of uncertainties.

Validation. Analyses should, in principle, be amenable to empirical control even though in practice this might be difficult and even impractical.

The idea behind these general principles is to provide guidance as to what should be striven for even if strict adherence to them is impractical. That they might not be achievable in a strict sense does not mean that they should be abandoned; rather the analyst is required to ensure that the analysis effort was guided with these principles as goals to be striven for.

Complementary nature of methods

Ideally, the four approaches to quantifying an event tree should complement each other, rather than be seen as equal alternatives. The use of a particular approach should be justified in the context of why an alternative technique was not utilised as the approaches are not equivalent and, depending on the situation, one can be more or less applicable than another.

Probability assignments are based on both data and phenomenological models. With extensive data but only a rudimental theoretical understanding of phenomena, statistical prediction is usually the best (or only) approach to probability assignment. In contrast, with good theoretical understanding and a representative model of the phenomena, but little data, a purely theoretical prediction is the common approach. Ideally, both good data and a good theoretical understanding with an associated robust model in which the parameters of the model are identified by extensive observation should be the goal to be striven for. In the less-than-ideal situation, where there is little in the way of good quality data nor a good theoretical understanding, at most a guess can be made when assigning probabilities. Judgement is overlain across the theoretical understanding, the model and the data, and the quality of judgement increases as both data and model completeness increase. The reader is referred to the companion section on probability assignment in Part III where a more detailed treatment of these considerations is presented.

Statistical estimates

Statistical estimates are characterisations or summaries of past observations. Statistical estimates reflect the frequency of occurrence of some condition in time or space. They deal with physical properties of the dam and its natural environment; that is, they deal with the frequency or rate at which certain conditions are observed. Often, probability (i.e. frequency) distribution models of regular mathematical form, such as normal, log-normal, or Poisson functions, are used to approximate historical data frequencies or physical properties of the dam such as material strength. This is mathematically convenient and usually allows stronger inferences to be drawn than does non-parametric (i.e. distribution-free analysis) inference.

Engineering reliability models

Engineering models are constructs based on reasoning from first principles of physics concerning the mechanical behaviour (response) of the dam under various influences expressed in terms of methods of applied mathematics. Models represent, more or less accurately, several technically important aspects of these mechanical behaviours. For example, dam stability might be analysed using limit-equilibrium models that involve analysis of stresses and strains. Strength parameters are estimated for the dam and then the stability of the dam under various loading conditions is calculated using the model. Uncertainties in parameter values and possibly in the model itself are propagated through the calculation to establish probabilities that the dam can resist certain loading conditions. Since engineering models are necessarily simplifications of reality, they represent both the degree of belief in the model and the model parameters as well the statistical properties of the dam.

Fault trees are a particular type of reliability model that model the logic of a system (e.g. how component malfunctions combine to cause failure of an overall system) rather than the physics of the system. Fault trees also reflect a combination of observed frequencies and degrees of belief in models and parameters. FTA is common in applications to mechanical and electrical systems, and is often associated with assigning probabilities to event tree branches related to the reliability of mechanical and electrical equipment.

Expert opinion

Expert judgement has always played a large role in science and engineering. Increasingly, expert judgement is recognised as just another type of scientific data, and methods have been developed for treating it as such. Expert opinion is most usually considered to be a statement of the reasoned degree of belief of the expert concerning a parameter, physical state or occurrence of an event. The alternative view is that expert judgement concerning parameters, states and events is intuitive, involving pattern recognition and even subconscious thought processes.

Structured expert judgement involves obtaining judgements and opinions of experts through formal processes of:

- selecting experts;
- definition of technical problems;
- choice of analysis methods by which the expert judgements are elicited and analysed.

Expert judgements can be elicited quantitatively or qualitatively. When expressed quantitatively, they can have several forms: probabilities; ratings; odds; weighting factors; and, possibly most importantly, probability distributions of the physical quantities of interest. Qualitative expression will involve a textual description of the expert's assumptions in reaching an estimate and natural language statements of probabilities of events such as "likely", or statements as to the expected performance such as "generally poor".

Expert judgements are necessarily conditioned on the data and information available to them at the time of the elicitation process. Therefore, their judgements should change as additional information is obtained. The challenge for the expert is to demonstrate that his/her revised judgement is consistent with all of the information available now as well as consistent with the previous judgement. The mathematical procedures to achieve this objective are presented in Chapter 11.

Uncertainty of assigned probabilities

Whether probabilities are assigned by statistical analysis, engineering modelling, expert opinions, or some combination of these approaches, they are almost never specified precisely. For fully engineered systems, such as spillway gates, given a robust model and quality data the quantification process might be expected to provide results within an order of magnitude or so of the long-run frequency or the actual future observation. Clearly, as uncertainty in the models and data increases, the uncertainty in the quantified risk also increases. However, while the result may not even be to within an order of magnitude, the process of quantification remains useful in that it permits an interpretation of the situation under consideration that cannot be achieved any other way.

4.3 Fault tree analysis (FTA)

This module presents some of the basic features of Fault Tree Analysis (FTA), its application and the availability of software for its implementation. There is extensive literature on FTA, including handbooks and application guides. Throughout this module specific terms are defined that relate specifically to FTA.

FTA is one of the tools available to the engineer conducting a reliability or safety analysis for a dam. It is a technique whose theoretical foundation is well-developed and that has been applied extensively in reliability and safety assessments for a wide range of engineered systems such as missile launch systems, chemical process facilities, nuclear power plants, dams, control systems and computers. In addition, the software and the databases available for conducting a FTA are sophisticated and add significantly to the efficiency of performing a risk analysis.

A FTA is an analysis process whose essential component is the fault tree. The fault tree is a graphical construct that shows the logical interaction among the elements of a system whose failure, individually or in combination, could contribute to the occurrence of a defined undesired event such as a system failure. Fault trees offer the analyst the capability to construct a logic model of a system that is visual and therefore is easy to view and read, and that provides a qualitative and quantitative insight to the system's operations and reliability.

It is important to note at the outset that FTA is one of many tools available to the risk analysis team. In a risk analysis for a dam system, various methods will generally be used to build a logic structure to analyse the expected future performance. As such, FTA will simply be one of the methods used. In the course of the risk assessment it is important to coordinate how a FTA for a system fits into the overall risk analysis model. This theme is critical to the risk analysis in general and to the FTA in particular and will be repeated throughout this chapter.

Origins

Fault tree analysis was developed in 1961 by H. A. Watson at Bell Telephone Laboratories. Watson and the team at Bell Labs were motivated by the need to evaluate the reliability of complex missile launch control systems. Following the initial work at Bell Labs, the Boeing Company, which modified the initial concept for computer evaluation, carried out further development in the 1960s.

Since its initial development in the early 1960s, the demand to assess the reliability and safety of complex systems contributed to the continued rapid development of FTA. In a little more than a decade, an extensive literature base and a number of computer programs were available.

By the 1970s, the range of FTA applications had expanded well beyond defence systems. During this period a number of organisations were developing computer software to perform qualitative and quantitative analysis. A keynote, non-military, application of FTA was the Reactor Safety Study (Rasmussen, 1975) published by the US Atomic Energy Commission. This assessment made extensive use of FTA to analyse the safety of commercial nuclear power plants. Since that time FTA has been an integral part of safety assessments for all commercial reactors.

With the advancement of desktop computer processing capabilities and the development of the graphic user interface, the construction, evaluation and assessment of fault tree models has made the use of FTA an efficient and effective means to construct, perform, document and manage system reliability assessments. Today, fault tree models are an integral part of facility and system risk management programmes.

Fundamental concepts

Before discussing the elements of FTA, it is instructive to consider the reasons for conducting the FTA in the first place. It is also worth considering where the FTA fits into a dam safety risk analysis, and how it differs from other systems analysis methods such as FMEA or an ETA.

A FTA is a systems analysis method that is used to obtain information about a defined undesired event. It is a deductive analysis, as opposed to an event tree that is an inductive analysis, in which the analyst reasons what can lead to the occurrence of a specified undesired event. In a top-down manner, a FTA works from the general to the specific. One of the early steps in a FTA is to specify a particular but general undesired event, such as failure of a system. The analysis then proceeds to determine what the specific causes or modes of system failure are.

There are three attributes that make FTA a valuable engineering and safety analysis tool: (1) it is graphic, (2) it provides qualitative insight to a system, and (3) it can be used to quantitatively assess a system's reliability.

Graphic tool. One of the attractive features of a FTA is the fact that a fault tree is a graphic display of a system and the events that lead to the occurrence of a specified undesired event; a system failure. As a result, the fault tree makes the logical interface of elements in a system tractable and understandable.

Qualitative analysis. A FTA is a qualitative technique developed to understand the potential modes of failure of a system. As the number of elements in a system and/or the complexity of its operations increases, it becomes difficult to determine the ways failure can occur. A primary result of a FTA is a listing of the system failure modes. It can be used to identify system vulnerabilities and to develop an order of magnitude sense of the failure modes that are the most significant for system failure to occur.

Quantitative assessment. Although a FTA is primarily a qualitative method, a fault tree is suited to quantify the probability of the undesired event and to determine the relative importance of events to the occurrence of the undesired event.

As part of a FTA, a model or fault tree is constructed that is designed to provide specific qualitative and quantitative information about a system, particularly its vulnerabilities and reliability. However, it has been suggested that constructing the model is not the point. Rather, the model is the analytical means to obtain the information that is required. Consistent with the Guiding principles of Part I, before a model is constructed, that is before the FTA begins, the analyst will have addressed certain questions:

- What is the system (what is it comprised of, what are its bounds)?
- What information do we want to obtain?
- What is the environment the system will operate in or be subjected to?

As is the case with ETA, these questions are vital to determining what will be analysed, the level of detail of the analysis, the modes of failure to be considered and the common cause effects that may exist.

On a broader scale, the analyst must establish how a particular fault tree fits into the scope of a risk analysis for a dam. For example, if a fault tree will be one of the inputs to an event tree or linked to other fault tree models there will be a variety of interface issues the analyst must address in the course of the risk analysis. It is in this context that the above questions are posed.

When a FTA is conducted for a "system" that is part of the overall dam system as defined in this volume, it is important that its bounds be clearly defined. Typically, there will be multiple systems, sub-systems, and sub-sub-systems that are considered in a dam risk analysis. From a practical perspective, each is part of the dam system. In a risk analysis, fault tree models may be developed for a number of different systems, many of which are functionally linked. Consequently, it is important that each system and its constituent components be clearly defined, avoiding overlap or oversight.

The environment a system operates in or is expected to encounter is a vital part of the analysis. The environment may vary from exposure to transient external forces such as those imposed by a seismic event to the internal conditions associated with changes in seasons. As different environmental conditions are considered, the scope and detail of the fault tree will vary. For example, a fault tree that is constructed to model the reliability of dam gate operations during normal conditions without flood or seismic events, will not be the same as a fault tree that models the performance of the same system during a flood or a seismic event. There are substantive issues involved in the development of a seismic fault tree that does not enter into the fault tree models for normal operations. In general, different fault tree models will be developed for different, major initiating events or hazards that are considered in the risk analysis.

Advantages and disadvantages

Like any engineering tool, FTA has its advantages and disadvantages. In many respects, fault tree construction and model evaluation is a craft that depends as much on the knowledge and depth of experience of the analyst as it does require sound engineering and scientific analysis techniques.

Building models that are too detailed or too coarse for a particular application are obvious downsides to any sophisticated tool. As a result, a premium is placed on experience, particularly when the systems to be analysed are large and complex.

Some of the recognised advantages of FTA include:

- It provides a logical and graphical means to model and analyse system failure modes, even for large systems.
- It is oriented to identifying system faults that have a bearing on the undesired event (e.g. system failure).
- As a modelling technique for assessing the reliability of systems it is well developed and accepted.
- It is an efficient tool when it comes to modelling the potentially large number of events and event combinations that can lead to failure.
- Sophisticated software tools make the job of fault tree construction, documentation and quantification an efficient and manageable task.

Disadvantages of fault trees include:

- A fault tree may not follow a system flow diagram and, as a result, it may not be easy to relate the system flow to the logic that leads to failure in the model.
- A large fault tree can be difficult to understand. From the graphic display of a fault tree, it can be difficult to "see" the system failure modes and to ascertain what are the important events in terms of those that cause system failure.
- Fault trees are not unique. Fault trees can be constructed in different ways and, therefore, have different appearances. However, the logic leading to the undesired event should be the same for any two "different" trees.
- Despite these disadvantages, a FTA often is the only alternative to rationally model a complex system, identify and understand its failure modes, and to quantify the likelihood of the undesired event.

General principles

This section presents a basic engineering introduction to FTA. The purpose here is to establish a framework for understanding the concepts. The elements of FTA are presented along with a description of the qualitative and quantitative results it provides.

A fault tree is the graphical display of the logical relationship of discrete elements in a system, as they relate to the occurrence of an undesired event, such as a system failure. Unlike a FMEA, a FTA does not consider all possible element failures. Rather, the

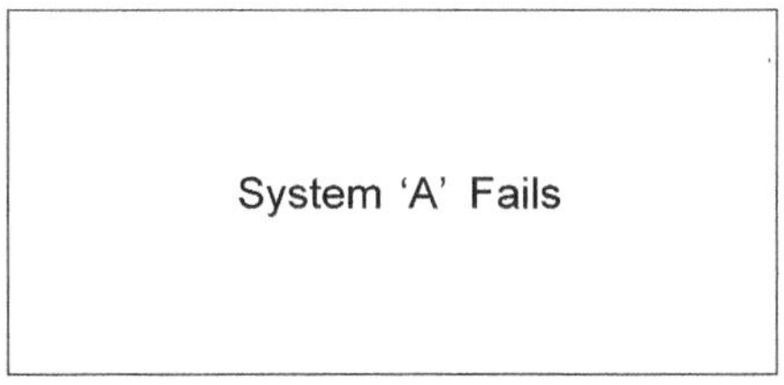

Figure 4-21. Top event

analysis focuses only on those failures that have a bearing on the occurrence of the undesired event. In this sense, the construction of a fault tree is a tailored analysis. Using a standardised system of event symbols and logic gates, a fault tree can easily be interpreted.

Elements of fault trees

In the following sub-sections, the symbiology of FTA is described. Standard event types and logic gates are described along with their graphic representations. Current fault tree software uses the standard symbols shown in Figures 4-21–4-25. In addition, they provide labels along with the symbols to identify event names and gates.

The top event

In a fault tree, the undesired event is located at the top of the tree and, as such, is referred to as the "top event" (see Figure 4-21). An important step in a FTA is to define the top event that will be analysed. To do this the analyst must consider:

- The condition or state that constitutes failure of a system to perform its function, whether it is to deliver power or to provide adequate drainage.
- The environment the system will operate in: seismic, flood, or normal conditions.
- Operating conditions and procedures.

When choosing a top event it is important that it not be too general, in which case the construction of the fault tree can become difficult to manage. Alternatively, if the top event is too narrowly focused, it may not provide a broad enough view of the system and its reliability.

Primary events

A primary event is an event, which for one reason or another, is not developed further in the analysis. These are events for which probabilities must be provided when the fault tree is quantified. There are four types of primary events. These events, their symbols and definitions are listed in Figure 4-22.

Logic gates

A logic gate defines the relationship between an output event and input events. It defines the logic flow and the relationship between higher events in the fault tree and lower, basic events. There are two general types of logic gates: the OR-gate and the AND-gate. Figure 4-23(a) shows an example of an OR-gate and the output and input events. The example in the figure states the top event – System A fails – occurs if one or both of the input events, "B" or "C", occur. Figure 4-23(b) shows a similar example when events "B" and "C" are in parallel. In this case an AND-gate is used to reflect the requirement for both events to occur for System "A" to fail. Events "B" and "C" in Figure 4-23 will typically refer to faults or failures of individual sub-systems or components.

The OR- and AND-gates are the primary logic gates used in FTA. There are, however, a number of other logic gates that are special cases of these. Figure 4-24 shows the various logic gates that are used in FTA. Each gate type is described in terms of its name, symbol, and definition.

Event	Symbol	Description
BASIC	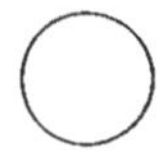	A basic initiating fault or failure that requires no further development
CONDITIONING	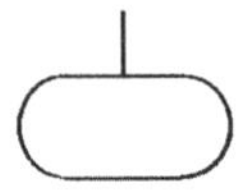	Specific conditions or restrictions that apply to any logic gate (used primarily with PRIORITY AND and INHIBIT gates, see Figure 4-28)
UNDEVELOPED	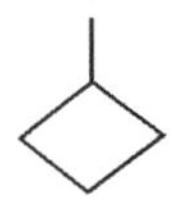	A fault that is not developed further because it is of limited consequence or relevant information is not available

Figure 4-22. Primary events and transfers

Transfer gates

Often a fault tree for a system will be linked to the fault tree for another system. For instance, this might occur in a fault tree for the operation of dam gates. The fault tree for gate operations might link to a fault tree for the electrical power system. When this occurs, transfer gates are used to indicate where the path of the fault tree logic goes. Two transfer gates are used to indicate TRANSFER OUT of, and TRANSFER IN to a fault tree. These symbols are shown in Figure 4-25.

Failures and faults

The distinction between failure and a fault can be important. As a result when faults are defined, the analyst should document what the undesired state of the component is and when its occurrence is unfavourable. The lesson here is that all failures are faults, but all faults are not failures.

Fault occurrence and fault existence

In a FTA a distinction is made between the existence of a fault and its occurrence. The term existence is used to denote that a fault has occurred and will continue to exist; the component is not repaired or replaced. The occurrence of a fault is used to denote the fact that a fault has occurred and can be repaired. The difference between occurrence and existence of a fault is not critical in the construction of a fault tree, but is important in the quantification of event probabilities.

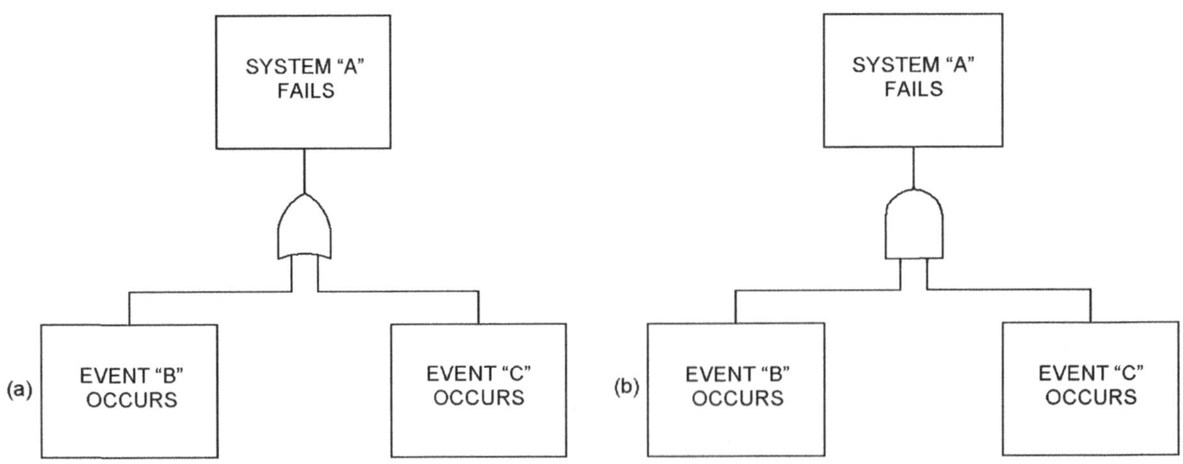

Figure 4-23. Illustration of the input and output events for: (a) an OR-gate, and (b) an AND-gate

Failures: mechanisms, modes and effects

Fault tree construction involves a search for the basic events, the root causes that contribute to the failure of a system. The fundamental elements of the process are illustrated in Figure 4-26.

Gate type	Symbol	Description
OR		Used to show that the output event occurs if one or more of the input events occur
AND		Shows that the output event occurs if all of the input events occur
EXCLUSIVE OR		The output event occurs if only one of the input events occur. This is a special case of the OR-gate
INHIBIT	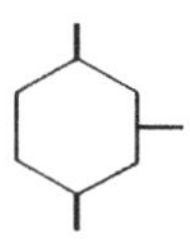	The output event is caused by a single event, given the occurrence of a qualifying condition (i.e. a conditioning event). This is a special case of the AND-gate
PRIORITY AND	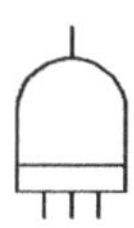	The output event occurs if all input events occur in a specified ordered sequence. This is a special case of the AND-gate
m out of n	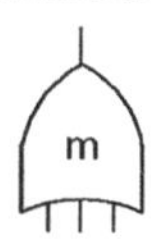	The output event occurs if m out of the n input events occur

Figure 4-24. Fault tree logic gate

Event transfer	Symbol	Description
TRANSFER IN		Indicates that the fault tree is developed further at the occurrence of the corresponding TRANSFER OUT (e.g. on another page)
TRANSFER OUT		Indicates that this portion of the fault tree must be attached to the corresponding TRANSFER IN (e.g. on another page)

Figure 4-25. Fault tree transfers

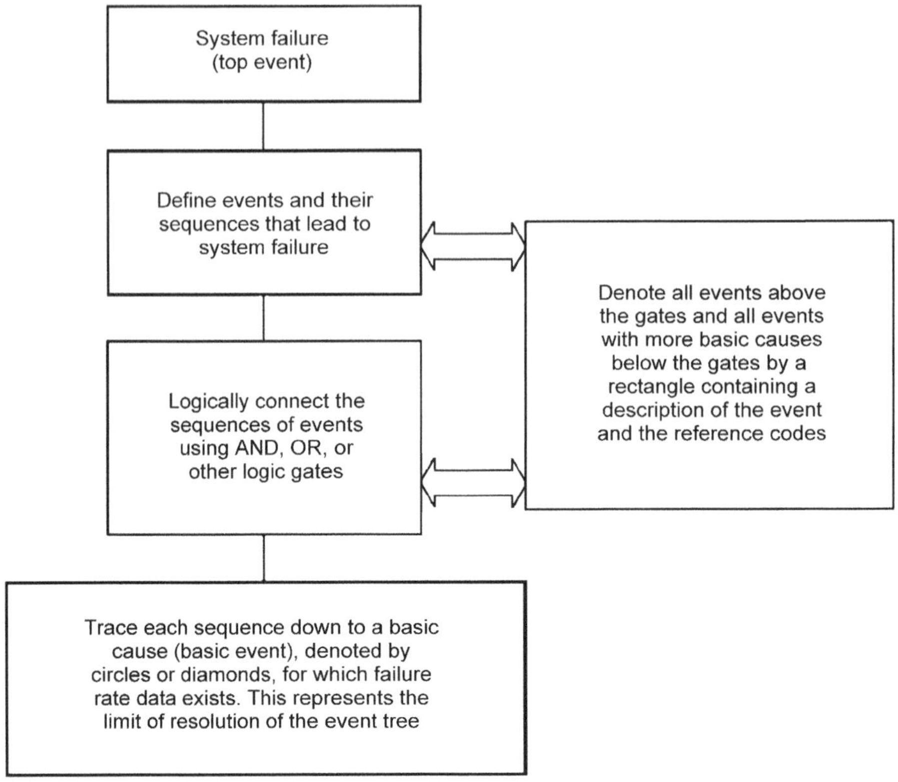

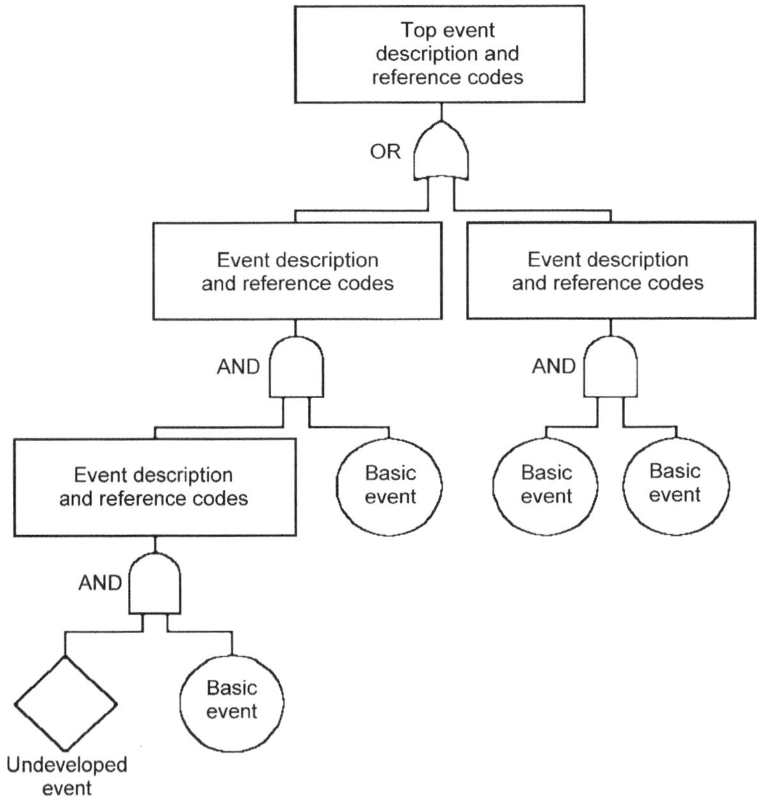

Figure 4-26. Elements of fault tree analysis

As a fault tree is constructed a top-down approach is taken. Beginning with a top event, the analyst proceeds to find the causes that will lead to its occurrence. The top-down building process begins with an effect such as the failure of a system. The next step is to determine what are the failure modes or event occurrences for this effect to be realised. A failure mechanism is a description of how a failure mode can occur. One characterisation of the process to determine root causes is in terms of the mechanisms, modes and effects of component failures. This is one reason why an FMEA can be a valuable input to a FTA.

Fault tree construction

The steps involved in constructing a fault tree define a process that includes the selection of the top event or a selection of logic gates and events, as well as the ancillary activities that are required to ensure completeness and accuracy of the fault tree model. The ground rules, on the other hand, provide specific fault tree building guidance.

Steps in the construction of a fault tree

As a starting point to a systems analysis, it is important that the analyst becomes familiar with the system in detail. The importance of this task is due in part to the fact that the fault tree analyst will generally not be the system designer or engineer who works with the system on a regular basis. Rather, they will be a systems analysis specialist who will work closely with those most familiar with the system, the utility engineers and operators to build a logic model for system failure. Therefore, the purpose of this step is to provide the analysts who will conduct the FTA the opportunity to become familiar with the system, its operation, success criteria, boundaries and interfaces. The FTA analyst should review design documents, operating procedures, maintenance procedures, the operating history, technical specifications, system descriptions that may be available, and as-built drawings. In addition, a site visit, including discussions with dam operations and maintenance personnel should be made.

The following steps are involved in the construction of a system fault tree:

- Define the system.
- Define the environment.
- Develop a system block diagram.
- Define the top event.
- Identify the immediate, necessary and sufficient causes of the top event and define the fault events.
- Determine the immediate, necessary and sufficient causes of the fault events.
- Repeat Step 6 until all fault events have been defined to the event level or level of resolution of the analysis.

The following describes each of these steps.

1. Define the system. When a risk analysis is performed for a dam system, there are a number of systems to be analysed which will have functional as well as physical interfaces. As an initial step in the risk analysis in general, and in the FTA in particular, each system must be defined in terms of its physical boundary and its functional interface with other systems. Experience from risk analyses for large systems, such as dams or power plants, suggests a clear definition of each system is important to the overall logic model development and the interface between system event and fault tree models.

In defining a system, the analyst must also establish its success and failure criteria. The fault tree will be built to consider only those causes that lead to the occurrence of the top event.

2. Define the environment. To develop a fault tree for a system, the conditions under which the system will operate must be defined. The environment will determine:

- The loads – temperature, lateral forces, pressures, etc. – that components in the system will be exposed to.
- The component failure modes that should be considered.
- The potential dependencies or common cause effects that may exist.

Depending on the nature of the various environmental conditions, the analyst may develop different logic models for a system. For example, the logic in a fault tree for a system that is part of an overall dam system during normal operating conditions would not be adequate to model the failure modes for the same system during earthquake conditions. As a result of the earthquake shaking, individual component failure modes will be different, common cause dependencies will exist that are not present during normal operations, thus dictating a different logic structure.

Experience suggests that different fault tree models will be developed for each major environmental condition. Modifying a well-designed fault tree to generate a tree that meets the needs for another environmental condition will often work.

3. Develop a system block diagram. A convenient, but not entirely necessary, stepping stone to building a fault tree is to construct a block diagram of the system. Once familiar with the system, the analyst needs to convert this information into a form that identifies the logical flow of the system operations. As a first step in this process, the analyst identifies the sub-systems and major components. The sub-systems should be defined and described in detail in terms of their function, physical elements and inter-relationships. As noted, this step is not necessary, but it does begin to put some order to the analysts understanding of the system. Done well, the block diagram makes the job of defining the top event and the construction of the fault tree much more efficient. It also provides a very useful communication tool, especially for managers and decision-makers.

4. Define the top event. The construction of the fault tree begins with selection of the top event. The top event must be carefully and specifically defined in order that the fault tree constructed satisfies the role it will play in the overall risk analysis, or the role it could play in another fault tree that links to it, or to an event tree. The top event should not be too broadly defined, or the fault tree may be too large or otherwise unsuited to the function it fulfils in the analysis.

5. Identify the immediate, necessary and sufficient causes of the top event. Once the top event is defined, building the fault tree will be an iterative process. In this step, the faults that will cause the top event to occur are identified. If a block diagram of the system was developed it will facilitate this process. To aid in the identification of the causes for the occurrence of the top event, the analyst should look to identify the immediate, necessary and sufficient failure modes or causes of the top event. The "immediate cause" concept is so called because it says that at this point in the analysis (at this level in the fault tree), the analyst need only look for the causes of the top event, recognising these are generally not basic causes of the top event. In fact, they are likely faults and failures of sub-systems. Once the list of causes of the top event is determined, this completes the top of the fault tree.

6 and 7. Determine the immediate, necessary and sufficient causes of the fault events. Applying the same approach as in Step 5, the analyst proceeds down the fault tree to determine for each cause of the top event or sub-system fault, their immediate cause. This process is repeated until the fault tree is fully developed in terms of basic events that define the limit of resolution of the analysis. With each iteration, the analyst attempts only to identify the immediate cause of an event or sub-system failure, recognising that with each iteration the fault tree gets closer to identifying the mechanisms that will lead to system failure.

Ground rules

In carrying out Steps 6 and 7, general ground rules have been developed that aid in the fault tree development process (USNRC, 1991). These ground rules are meant to bring a sense of order to the construction process as well as logical consistency to the fault tree. The ground rules for fault tree construction are:

- When a fault is entered into the tree, write in the event box or event database precisely what the fault is and when it occurs. This should be done at the time the fault is identified as an immediate cause of the event above it or the top event. The description should state what the condition of the component is and what makes it a fault event. The value of following this rule cannot be overstated. For the analyst and subsequent reviewers looking to verify the model logic, these notes are vital.
- Define each fault as either a component or system failure. If an event is classified as a component failure, add an OR-gate below it and look for the modes of failure of the component. If a fault is classified as a system failure, look for the minimum necessary and sufficient conditions to cause its occurrence.
- If the normal, operating or non-failed functioning of a component propagates a fault sequence, then the component is assumed to operate normally. This is to say that the analyst should not assume that a fault will occur and therefore defeat or halt the failure sequence. Alternatively, if the normal functioning of a component blocks a fault sequence, the propagation of the fault sequence continues only if additional faults occur.
- Define all the inputs to a logic gate before further development of any one of the inputs.
- Fully develop a branch in the fault tree to the basic event level before starting the development of another branch.
- Inputs to logic gates shall be defined fault events. A logic gate should not be an input to a gate.

Dependent events

To this point in the discussion of FTA, it has been assumed (at least implicitly) that events are independent or mutually exclusive. There are, as one might expect, dependencies that can exist between events. Dependence can be extremely important because of the impact it has on system reliability. Dependencies will tend to increase the likelihood of multiple, concurrent failures, thus defeating system redundancies. Different terms are used to describe the various sources of dependencies, such as common-mode failure (CMF), common-cause failure (CCF), and propagating failures. CMFs are multiple, concurrent and dependent failures of identical equipment that fail in the same mode. Propagating failures occur when failure of one item causes sufficient changes in conditions, leading to the failure of other components. Lastly, common-cause failures (CCF) are failures that occur simultaneously due to a single, shared cause. These commonly used terms do not fully capture all of the types or sources of dependency. As a result, the notion of dependent events or dependency that exists between events is used.

Two issues arise in a FTA when it comes to dealing with possible dependencies:

1. The analyst must identify the possible sources/causes of dependence.
2. Once identified, the dependencies must be modelled to account for their impact on event occurrences and the probability of the top event.

Table 4-4. Examples of causes of dependent failures (adapted from USNRC, 1981)

Cause	Example
Vibration	Machinery, earthquake
Moisture	Condensation, leakage/seepage
Chemical	Corrosion
Corrosion	Oxidation due to moisture or exposure to water
Freezing	Effects of water freezing that impacts operating equipment, imposes additional loads on structures and equipment
Temperature	Fire, lightning, electrical short circuits
Grit	Airborne dust, metal fragments from out of tolerance equipment
Maintenance, operations action	A maintenance crew makes the same error at similar equipment during maintenance activities

There is a wide range of factors or causes for dependence between events. These include ageing, actions of facility personnel, environmental conditions, and system or component physical or functional dependencies. Table 4-4 lists examples of causes that can lead to dependence between events. As shown in the table, the causes are varied and their impact on equipment potentially complex.

Table 4-5 attempts to put some order to the different sources of dependence by classifying them into groups. The table provides a general description of each group and some examples. Computer software tools are one way to identify system and component dependencies. These methods use algorithms to search the fault tree logic "minimal cuts sets"[1] for common susceptibilities to failure. However, these are not an exclusive solution to the evaluation of dependent events.

Methods for dealing with and modelling dependencies vary depending on the type of dependency and its importance to the risk analysis. Table 4-6 summarises the methods for modelling different types of dependencies. They include explicit modelling and parametric analysis. Explicit modelling involves the handling of dependencies in event-tree and fault-tree logic. This approach works well for external events that impose loads on all or part of a dam system. For example, event and fault-tree models can be developed to model the performance of a dam system during major external events such as floods, earthquakes, extreme wind, etc.

Parametric methods have been developed as an extension of reliability methods to model the occurrence of dependent failures. These methods estimate model parameters for the occurrence of one and two failures.

There may be cases where the most important dependencies are modelled and others are ignored or treated conservatively. This can be done in cases where the impact on the risk analysis results is not significant and the effort in data collection, analytical evaluation, and computational effort is extensive.

Fault tree evaluation

Once a fault tree model has been constructed, qualitative and quantitative evaluations can be performed. As part of a risk analysis for a dam in which the probability of dam failure will be determined, both evaluations will be performed. In general, however, depending on the purpose and scope of the FTA, only a qualitative evaluation may be conducted. Here, the term "fault tree evaluation" should not be confused with the term "risk evaluation", although there are some conceptual similarities. Furthermore, from a

Table 4-5. Types and causes of dependencies

Type	Description	Examples
Initiating event	An initiating event has the potential to change environmental conditions (e.g. loads on components) on multiple components or the entire dam system simultaneously, leading to an increase in the probability of occurrence of a basic event (e.g. failure)	Earthquakes, floods, explosions, wind, wave action, forest fire, loss of off-site power
System, sub-system, and component functional dependencies	Dependencies that exist due to the facility design that makes one system dependent on the status (failure or success) of another	A system that can only operate in conjunction with the successful performance of another system
Shared equipment	Equipment shared by multiple systems makes each of the systems vulnerable if the shared components fail	Shared control cabinet, common electrical bus
Physical interactions	Physical interactions occur when the failure of one component changes the environmental or load conditions on other components, thus increasing their probability of failure	Failure of a drain that leads to increased uplift pressures on a dam
Human interaction	Human actions can introduce dependencies as a result of errors of omission or commission in the course of different facility activities	Installation of equipment by the same crew; same crews following incorrect procedures during calibration, maintenance, or testing

engineering analysis perspective, the term fault tree evaluation relates to interpreting and understanding the fault tree and revealing the significance of the information embodied in the fault tree.

As constructed, a fault tree is a graphic of the deductive logic paths that lead to the occurrence of the top event. The first step in a fault tree evaluation is a qualitative analysis. The results of the qualitative analysis include:

Table 4-6. Methods for analysing dependent failures

Type	Method
Common-cause initiators (e.g. external initiating events)	Event-specific logic models (event and fault tree models) and computer-aided methods
System, sub-system, and component functional dependencies	Event tree analysis
Shared equipment	Event tree analysis, fault tree linking
Physical interactions	Event-specific logic models (event and fault tree models) and computer-aided methods
Human interaction	Event and fault tree analysis, human reliability analysis

1. Conversion of the fault tree from graphic to Boolean form.
2. Identification of the system failure modes or the combination of events that must occur for the top event to occur.
3. Measure of the qualitative importance of events.
4. Identification of the susceptibility to dependent failures such as common-cause and common-mode failures.

A quantitative evaluation is typically performed using the results of the qualitative analysis. The results of the quantitative analysis include:

1. Probability of the top event.
2. Quantitative measure of event importance and contribution to the top event.

Once the quantitative evaluation is set up it can, of course, be used to perform sensitivity analyses.

Fault tree evaluation is considered in more detail in Chapter 9.

Endnote

1. The failure modes for a system are referred to as the minimal cutset for the top event. Formally defined, a minimal cutset is the smallest combination of events (e.g. component failures) which, if they all occur, will cause the top event to occur. A fault tree will consist of a finite set of minimal cutsets which are unique for the top event and any one of which leads to the occurrence of the top event.

5 Consequence analysis

Understanding the consequences of dam failure is fundamental to the assessment of risk. The following outlines considerations in dam failure consequence estimation: an aspect of dam safety risk analysis that has received little attention in comparison to that afforded to dam failure analysis.

Like probability of failure analysis, dam failure consequence analysis is fundamentally an exercise in predictive modelling, through which the anticipated effects of dam breach inundation are described. Typically, dam failure consequences of interest will include:

- threats to public safety;
- environmental degradation;
- infrastructure and property damage and losses;
- socio-economic impacts, including political and public perception issues;
- owner's reputation and financial integrity.

Because of the broad range of considerations involved, consequence analysis is a multidisciplinary endeavour, which has many analytical components that are outside the realm of engineering. Typically, the engineering analysis pertains to modelling:

- reservoir operation;
- formation of the breach in the dam;
- characteristics of the dam breach flooding;
- damage state and magnitude of the loss in the affected areas downstream.

From an analysis perspective, dam breach consequences can be broadly considered to fit into two main categories:

1. Direct consequences attributable to contact with the floodwaters.
2. Indirect consequences that arise as a result of the direct consequences.

Typically, direct consequences, which are the focus of this chapter, are divided into three categories:

1. Life safety.
2. Economic and financial.
3. Environmental impact.

The following boundaries apply to this categorisation (model):

- Public safety including life safety considerations that apply to loss of life, physical injuries and emotional trauma caused by direct contact with flood waters.

- Economic losses pertain to all third party economic impacts whereas financial losses pertain solely to the dam owner.
- Environmental degradation refers to environmental losses that occur during the flood event, with collateral losses of habitat that impact migratory species being considered as indirect losses.

This chapter describes what has to be modelled in the endeavour of consequence analysis and how the various elements of the modelling effort are related. It does not present details of the actual models used, as some elements of the modelling endeavour, such as dam breach inundation modelling, are established disciplines outside the scope of this volume. The remaining elements are being dealt with elsewhere in the volume.

5.1 General considerations

In principle, the endeavour of dam failure consequence analysis involves developing a model of the reservoir and the region downstream of the dam (Figure 5-1) and then (as an alternative) perturbing the model by running the dam breach flood across it.

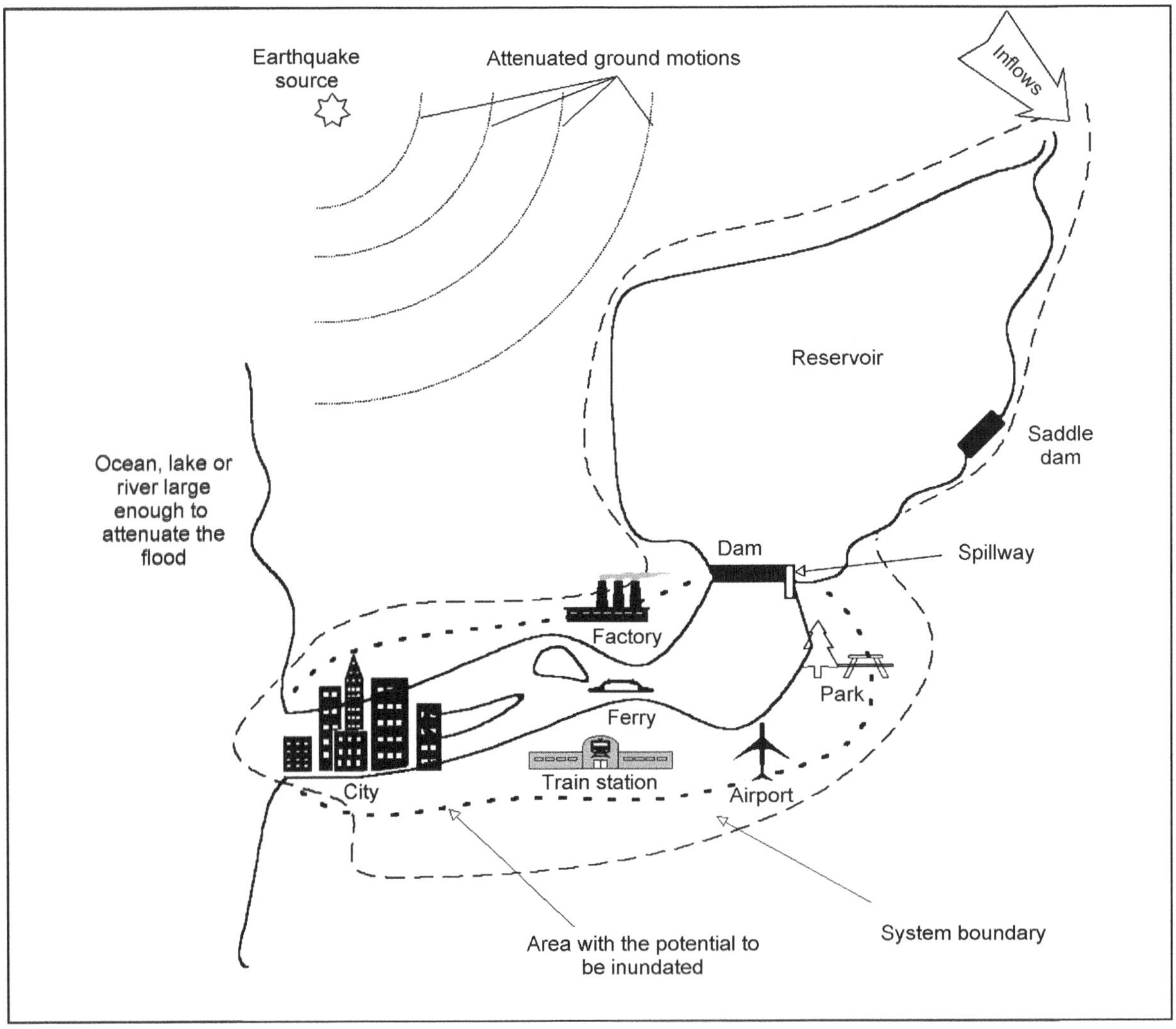

Figure 5-1. Spatial model of reservoir and downstream region

The complexity of the model and the extent of the modelling endeavour will vary from one situation to the next depending on the extent of the dam breach inundation, and the demographics and land use of the area affected by the flood. The system boundary may be limited to the extent of the inundation or it might be larger if wider environmental social and economic issues are considered.

As is the case with dam failure analysis, the level of the modelling effort will also depend on the degree of resolution required by the risk analysis. In keeping with the iterative nature of the risk analysis process, it is generally appropriate to begin with a relatively coarse representation of the downstream area, moving to more refined modelling techniques as the need arises.

From a modelling perspective, the region downstream of the dam can be considered to comprise "objects" or "groups of objects", where the term "object" is used in its most general sense. Some of these objects are interrelated, others are independent of each other.

Typically individual "objects" can comprise:

- people;
- buildings;
- structures and infrastructures;
- animals and wildlife species;
- habitat objects such as trees.

"Groups of objects" can comprise:

- families;
- social groups, such as schools, hospitals, car- or bus-loads of people, etc.;
- infrastructure networks;
- farmyards and concentrated wildlife populations;
- entire habitats.

For relatively straightforward situations, the downstream region can be modelled using maps, census data and one-dimensional dam breach modelling techniques. These modelling techniques have been used for many years in emergency planning activities. When appropriate, these techniques can be applied to situations ranging from those where there are few objects in the inundation region to very complex, high object density situations, where order of magnitude estimates of the consequences of failure are all that are required, initially at any rate.

However, for complex situations it may be appropriate that the modelling effort covers a great deal more than just estimation of potential loss of life and property losses. Additional reasons that more detailed analysis might be appropriate could include efforts to improve emergency planning procedures and risk mitigation measures, such as infrastructure changes (e.g. raising bridges, or changing road alignments).

Early stages of a consequence analysis might aim at estimating realistic bounds for the losses and damages. Further iterations can then be carried out to improve the resolution of the modelling endeavour to obtain the mean and confidence limits or to make a best estimate of these parameters. For example, an initial upper-bound estimate of the infrastructure damage could be the replacement costs of all structures impacted by dam breach flooding. While such an estimate process is obviously conservative, it may well be adequate for some decision purposes. The estimate can then be refined during additional iterations by modelling the resistance of the infrastructure to the forces of the floodwaters using structural reliability models.

5.2 General modelling considerations

The modelling approach to consequence estimation involves five steps:

1. Identification of consequences.
2. Characterisation of reservoir operations.
3. Modelling of dam breach behaviour.
4. Characterisation of the downstream geography and socio-economic conditions.
5. Modelling inundation.

Typically, several models are required to accomplish these tasks, which may be broadly divided into four categories:

1. Dam failure: reservoir operation, dam breach inundation.
2. Geographic and social scenarios: topography, environment, infrastructure.
3. Demographics and land use: buildings and structures, population characteristics.
4. Damage and loss functions: life safety, property infrastructure and transportation networks, environment, economic activity in the affected area.

These models are related to the hazard model (that characterises the perturbance that causes the system to respond), and the resulting dam response and system response models as illustrated in Figure 5-2.

Building on the influence diagram concept described in Chapters 4 and 8, Figure 5-3 depicts the relationships between the models in a manner consistent with the analytical models in Chapters 7, 8 and 11. The "pre-release models" element of Figure 5-3 relates to the logic tree, the hazard model and the dam response event tree model. The "inundation and damage models" element illustrates the interactions between the damage loss models and the dam breach inundation model and the interactions between the logic model, the event tree models and the inundation and damage model.

Consequence identification

The first step in consequence analysis is identifying the hierarchy of consequences of concern. A variety of techniques of consequence identification are used in routine practice, the most common of these are checklists, impact matrices, and relevance trees.

Consequence identification is essentially a surveying endeavour, and while there are established ways to conduct surveys once the objective of the survey has been defined, there are no defined procedures to comprehensively define what should be identified or how the identification procedure should be carried out. There is no deductive procedure by which one can assure that all-important consequences are identified and nothing of importance has been left out. In fact, there are no procedures to determine what features are "important" and what features are not. While importance is a matter of perspective, general groups, that encompass what appears to be important, include life safety, economic and environmental consequences.

Consequence identification is amenable to structuring in terms of a hierarchy, although there are no formal rules for establishing a hierarchical structure for such an endeavour.

A hierarchy can be relatively "broad and shallow" or relatively "narrow and deep". In both cases the considerations identified will be the same.

One approach is to categorise consequences by sector (e.g. economic, environmental, public safety, legal, political, etc.) and subsequently search for consequences in each sector. Another approach is to divide out consequences by the stakeholder group or economic interest affected.

Another way to categorise identification techniques is by the distinction between scanning and tracing approaches (Chapter 10). Scanning techniques use a single-step

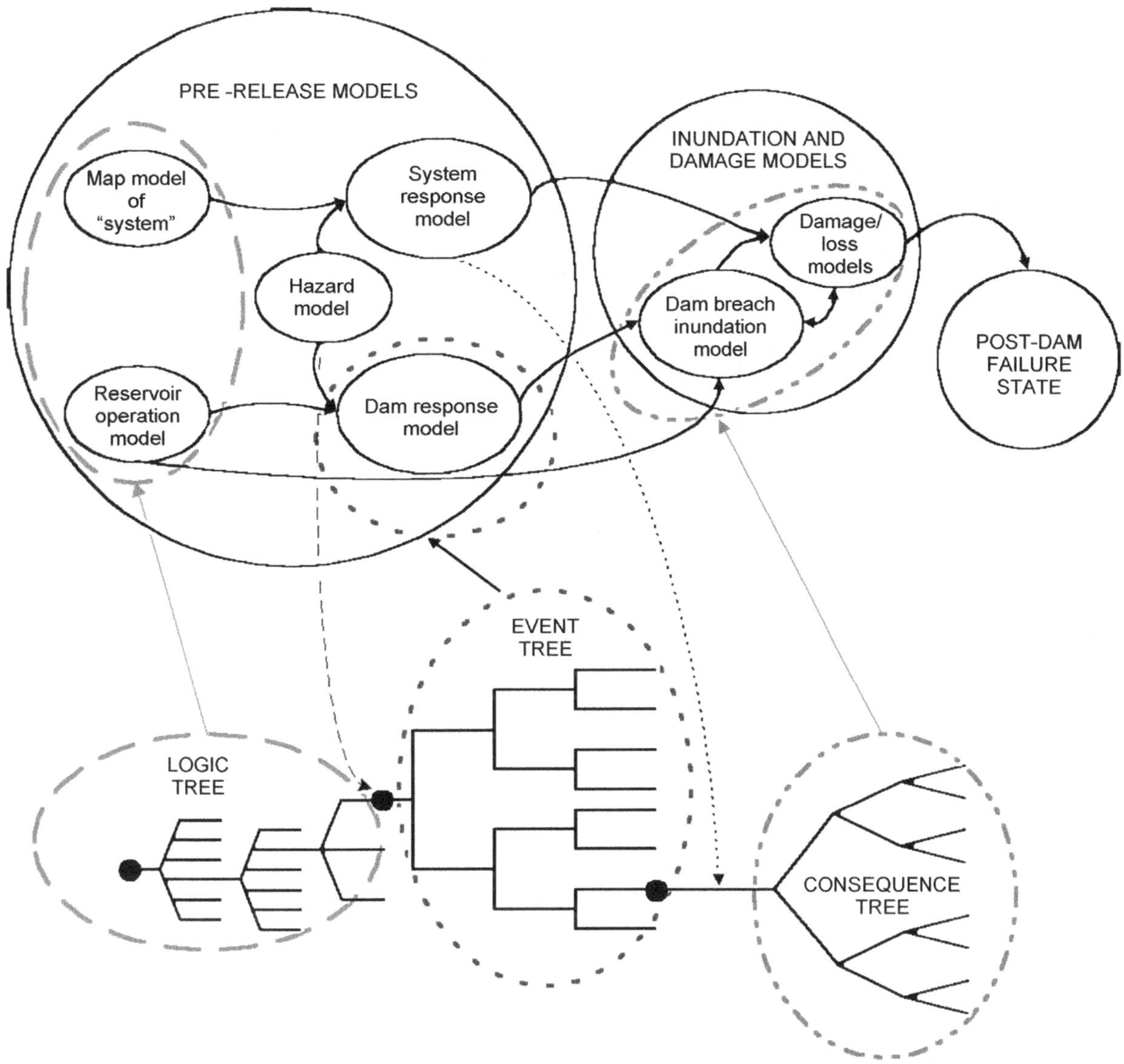

Figure 5-2. Modelling considerations

identification of consequences, relying on intuition and insight. Tracing techniques emphasise the structure within the consequence hierarchy as a road map for identifying consequences. In general, a small number of broadly encompassing significant and well-chosen consequences suffice to capture the important concerns of dam safety analyses.

The most common approach to consequence identification is a checklist, which is a simple form of scanning. Checklists attempt to guarantee, and often assure, some degree of comprehensiveness in identifying a complete list of consequences.

Matrix approaches are similar to checklists but add a second dimension to the taxonomy. They provide more flexibility in characterising interdependencies among consequences and the causal relationship to the initiating failure events.

Tracing techniques systematically start with high-level, broad consequences, and then work toward more detail to graphically represent the interrelationships among the different types of consequences of dam failure. Structuring consequences in a tree-like hierarchy facilitates determining whether the set possesses the desirable properties one

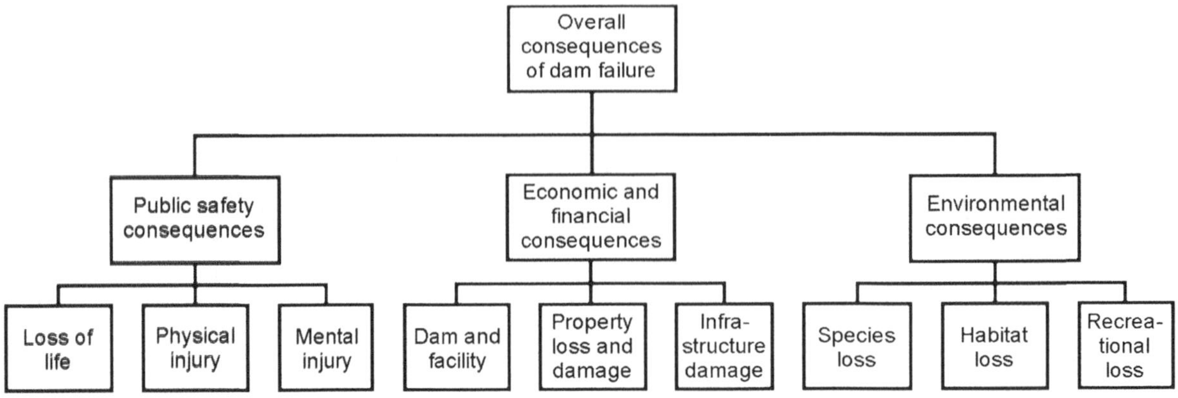

Figure 5-3. Example of hierarchical structure of failure consequences

seeks in a set of sub-consequences and attributes. The systematic characteristics of tracing techniques are derived from the inclusion of properties such as *completeness*, *operationality*, *decomposability*, *non-redundancy*, and *minimalness*, as discussed in Chapter 10.

Reservoir operation

The height and volume of water stored behind the dam are essential inputs to the dam breach modelling endeavour. In general, the height and volume of water is determined by the planned operational reservoir level and the naturally occurring variations in inflow (Figure 5-4).

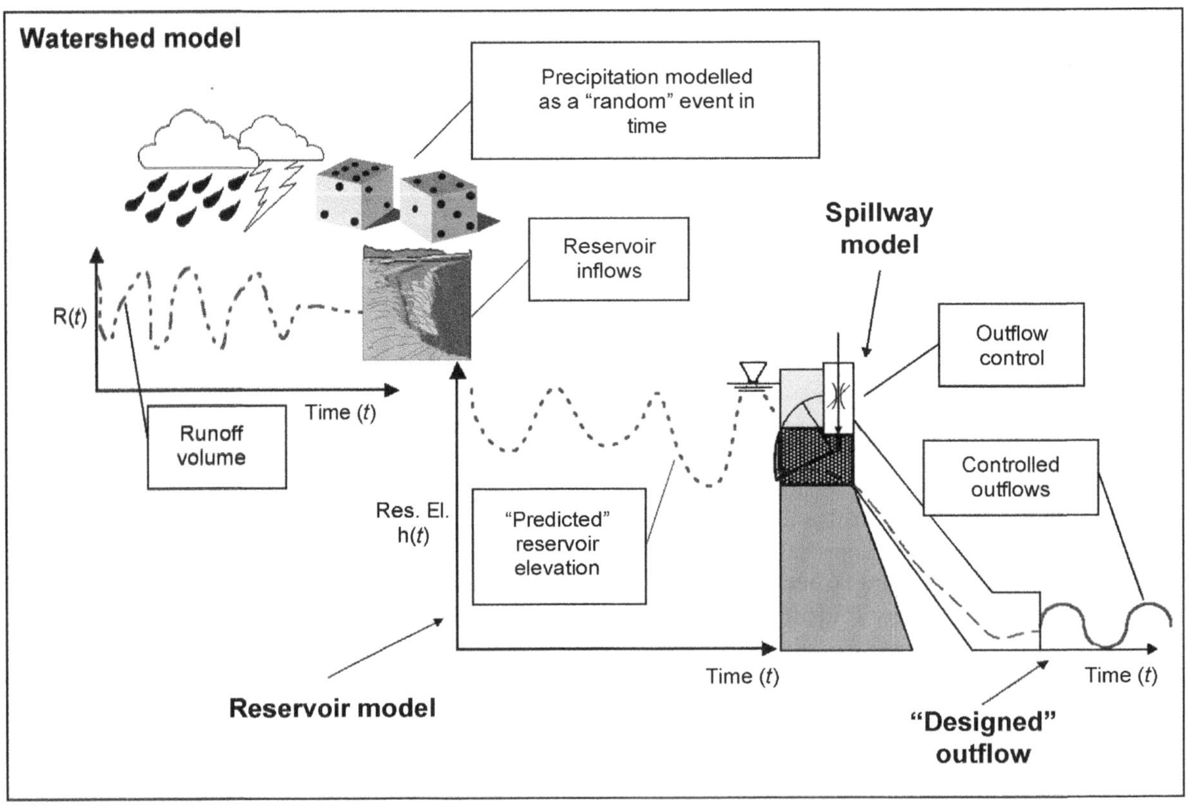

Figure 5-4. General form of reservoir operation model

The reservoir elevation and, therefore, the total volume that can be discharged is represented by a mathematical equation that essentially is the difference between the volume at time t plus the random inflows through the time period minus the controlled outflow.

The reservoir operation model is situation specific and can only be provided by those who plan the operational regime. As is often the case in risk analysis, the maximum and minimum possible reservoir elevations might be used to establish the bounds of dam breach flooding at the early stages of the risk analysis.

Dam breach modelling

Breach formation is usually modelled either: (1) semi-empirically, using data from field observations from experimental test dams, or (2) analytically, using engineering models based on hydraulics and the erosion properties of soils. Semi-empirical models structure the key parameters of an expanding breach into a predictive equation. Typically, these equations involve, at least, the average breach width, b, and the breach formation time, t (Figure 5-5).

Treatment of the details of the dam breach modelling endeavour is beyond the scope of the volume and accordingly the reader is referred to the expert literature on this subject.

Geographic and social characterisation

The modelling of the region around and downstream of the dam is a matter of geographical and socio-economic analysis. The endeavour is very well established with the basic principles set up several hundred years ago, with continual improvement in accuracy and resolution ever since. Today, geographic information systems (GISs)

Figure 5-5. Breach formation

provide a cost effective and efficient means of generating comprehensive high-resolution, multi-attribute maps that can be readily updated.

Map models, be they prepared using ancient techniques or digitally, are made up in layers with the basic topographic data serving as the platform on which all other information is superimposed. Figure 5-6 illustrates a Venn diagram representation of the elements of a comprehensive multi-attribute map of the region around and downstream of a dam (the system). The extent to which any or all of these elements are represented in the risk analysis depends on the purpose of the risk analysis.

A land-use map with relief contours is one common way to construct a model, which, depending on its purpose, would represent most of the attributes illustrated in Figure 5-6. Typically such a land-use model is constructed in "layers", beginning with the topographic survey data and adding the relevant detail on the basis of other surveys.

Typically the model need only include those layers necessary to meet the purpose of the risk analysis. If the analysis does not include an environmental component then the "environmental layer" might not be included in the map model. If there is no population at risk, then the "families and social groups" and "individuals" layers would not be required for the modelling endeavour.

Infrastructure data would normally be imposed on the topographic and environmental models (as is the case for maps used in environmental impact assessments for new infrastructure). Land use and demographics are then delineated using surveys, as are buildings and other structures. Finally, the locations of families, social groups and individuals are determined from census data and other surveys.

Of these, the topography, demographics and land use, and property and infrastructure models can, as a first approximation, be considered to be static in time with three

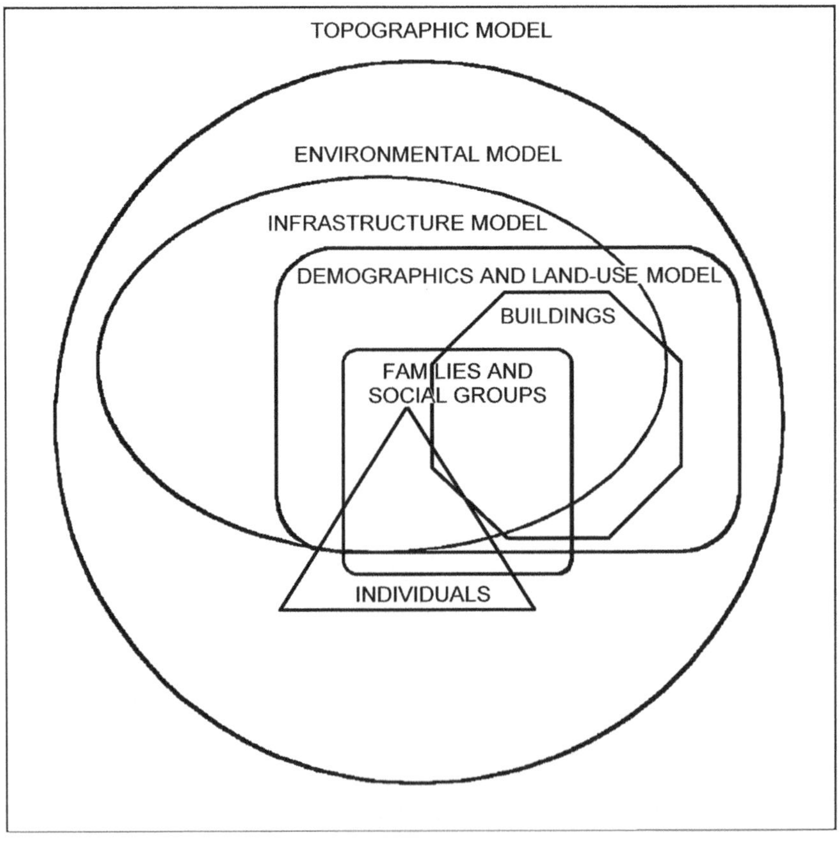

Figure 5-6. Schematic Venn model of "objects" and downstream region

dimensions (*x*-, *y*- and *z*-). Other modelling simplifications could include modelling the environment and economic activity as if they are static in time., Dam breach inundation and life safety are necessarily dynamic in time with up to four dimensions, *x*-, *y*-, *z*- and *t*-.

Topographic, demographic, land-use and population models are commercially available. These are not discussed further. Environmental and economic modelling for dam breach flooding has been carried out, although not extensively, and there are presently no agreed scientific procedures. These are areas where further research and development is required.

Inundation modelling

Once the model of the area of interest downstream of the dam is constructed, the dam breach inundation model can be applied as illustrated in Figure 5-7. The objective of the inundation-modelling endeavour is to determine the depth and velocity of the flood waters throughout the inundation area.

In general, the dam breach floodwaters can be expected to be laden with debris of various origins (silt, trees, damaged buildings and objects, etc.). Therefore, the mechanics of the fluid flow will, in general, be more complex than that of clear water. Other complications, such as river channel erosion and directional changes in the river bed as a result of this erosion, may also enter the inundation modelling endeavour.

Typical outputs of the inundation-modelling endeavour are the flood hydrograph (Figure 5-8) and flood depths and velocities, the resolution of which are a matter to be specified by the scope and requirements of the risk analysis.

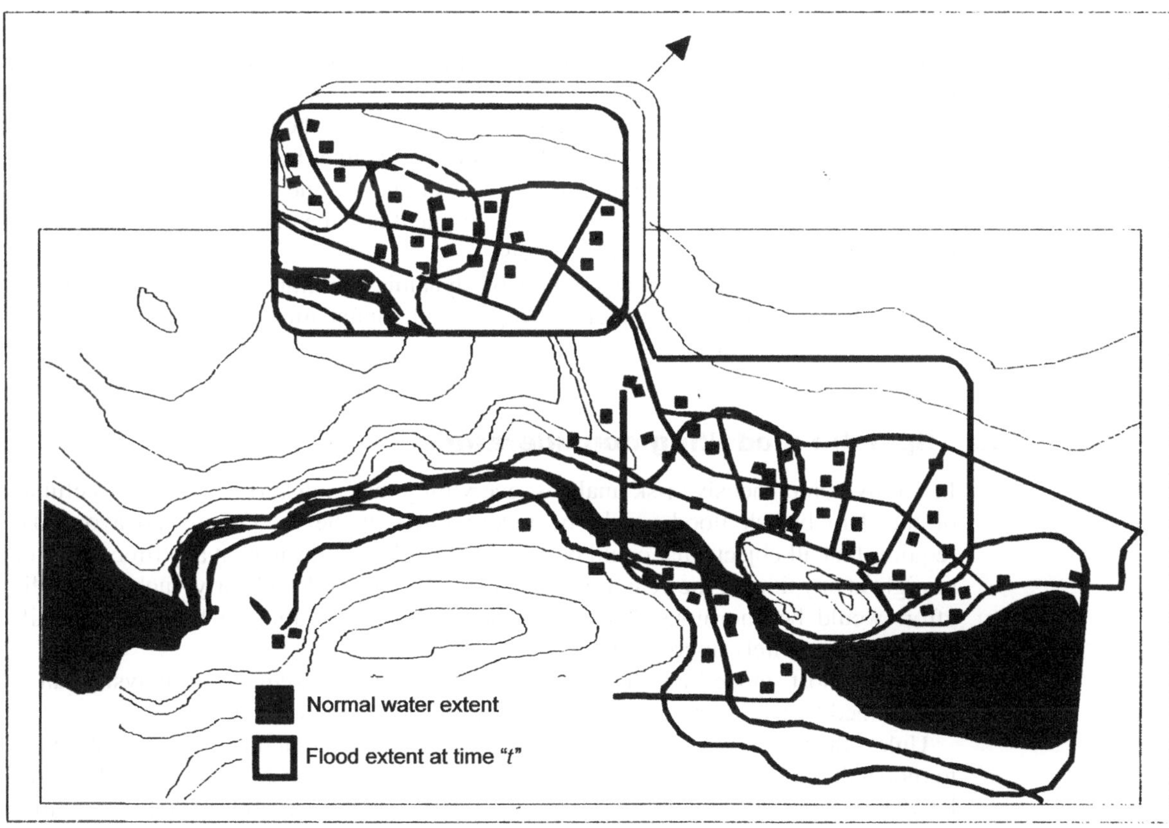

Figure 5-7. Dam breach flood model superimposed on land-use model (Assaf and Hartford, 2002)

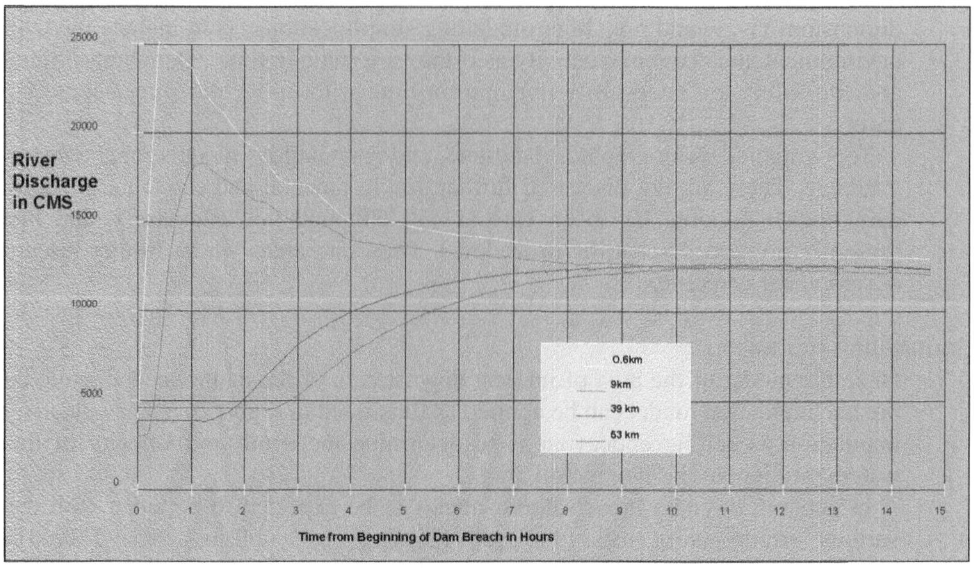

Figure 5-8. Dam breach flood hydrograph (Assaf and Hartford, 2002)

The depths and velocities will depend on several factors, including the stored volume, the headwater and tailwater levels at the time of breach, the rate of development and size of the dam breach, the downstream channel geometry, the morphology and erodability of the channel, and the land use and features of the landscape including buildings, structures and infrastructure.

The approximations in the hydrodynamic modelling should be specified, as they can be a significant contribution to the total uncertainty. In reality, the characteristics of the water-debris fluid change with distance from the dam as the debris load changes with destructive force of the flood waters. Therefore, modelling the inundation as clear water and laminar flow would represent a lower bound of the destructive force of the floodwater as, in general, the flow will be turbulent and the density and "angularity" of the flow waters will be elevated by debris.

In advanced modelling, there would be dynamic iteration between the inundation modelling and the damage modelling to permit the inundation model to "pick-up" and "deposit" debris during the flood event.

5.3 Specific modelling considerations

In a fully comprehensive risk analysis, everything that is disturbed by the forcefulness of the dam breach flood would be incorporated in the modelling endeavour. This requires that all objects that could be impacted by the floodwaters at any time of the day or night at any time of the year are accounted for in the model. Clearly, such a modelling effort would be enormous. This said, there may be situations where such detailed modelling is entirely appropriate.

As in most modelling endeavours, dam failure consequence analysis can be streamlined by modelling representative dam breach flood scenarios.

Unless the reservoir and downstream conditions are static in time, it may be necessary to model the full spectrum of inundation scenarios, as the consequences of failure could be highly dependent on the reservoir elevation and the variability of the conditions downstream of the dam. Such conditions are often modelled by selecting representative scenarios that cover the full range of the variations of reservoir elevation and downstream conditions.

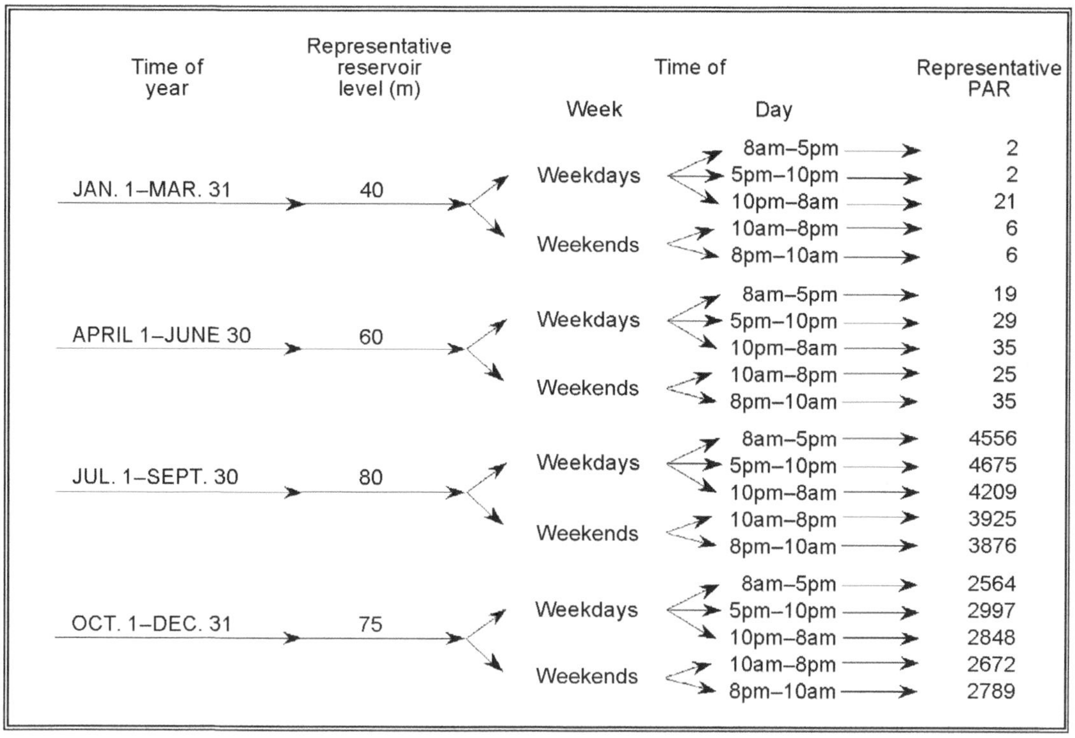

Figure 5-9. Temporal variation in representative population at risk (Assaf et al., *1997)*

In general, while nothing is absolutely static in time, many of the parameters can, as a first approximation, be considered to be "essentially static" in time, particularly over a period of a year or significant part of a year. However, an assumption that one or more parameters are "static in time" does not imply that there is no uncertainty around these parameters.

In addition to spatial and temporal variations in the conditions downstream, there may also be significant uncertainty associated with variability in the downstream conditions at any particular time. Thus, each of the parameters used to characterise the downstream conditions will, in general, be an uncertain variable represented by a probability distribution. For instance, the population at risk (PAR) can be expected to vary with different times of the day–week–month–year with additional uncertainty associated with each specific time frame (Figures 5-9 and 5-10).

This means that PAR is, in fact, a non-stationary stochastic process with highly complex cyclical structure. The largest cycle with the period of one year contains a number of looped smaller cycles. Since it cannot be handled effectively with the existing mathematical tools, the stochastic process $X_{(t,x)}$ is replaced by a series of random variables $Xt_{(x)}$ where each X_t is simply the value of the process at a pre-selected time t. If the pre-selection process is done right, a full range of consequences will be covered. When we select overall scenarios for risk analysis that aspect of the selection process should be kept in mind.

"Objects at risk"

For the purpose of this volume and concerning direct consequences of failure, the term "at risk" applies to everyone and everything that actually come into contact with floodwaters. The idea of treating people as "objects" does not denigrate the importance of human life in the modelling endeavour. It is simply a matter of modelling that permits the generalisation of the problem of dam failure consequence estimation.

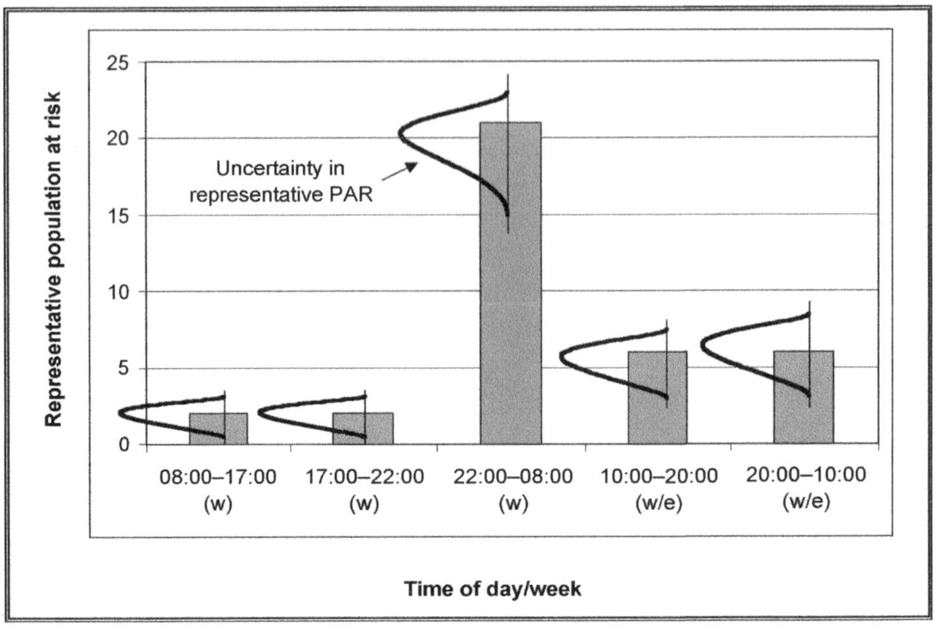

Figure 5-10. Uncertainty in representative population at risk

Damage and loss functions

Loss functions provide a mathematical formulation for dam failure consequence analysis. The imposition of the dam breach flood on the downstream sub-system represents a disturbing force on the "objects" in the inundation area. The responses of the "objects" are represented by damage and loss functions. These functions are then integrated with the land use and demographic models of the inundated area. A simplified object loss function might be in the form

$$Loss = \sum_{Categories} \sum_{(x,y,z,T)1}^{(x,y,z,T)n} Category\ Metric \qquad (5\text{-}1)$$

The determination of the responses of the "objects" disturbed by dam breach flooding, and thereby the damage and loss functions, is carried out in terms of the same general principles and techniques as the determination of the response of the dam system to the hazards. As described in Chapters 6 and 11 on probability assignment, these damage and loss functions can be determined in one of two ways, or in some combination thereof:

1. Statistical analysis of historical data pertaining to the effects of flooding.
2. Modelling from first principles of physics and engineering.

The two approaches provide complementary information.

The objective of the data analysis or physical modelling efforts is to obtain the probable value of magnitude of the loss or damage due to the forcefulness of the dam breach flood. Typically, the damage function will generate the probable values of "states" such as loss of life and will be described by probability distributions (Figure 5-11) that are dependent on time and space considerations.

Scenario modelling framework

The concepts presented above are intended to provide a generalised approach to dam failure consequence estimation, whereby the approaches to modelling are fundamentally the same for all consequence categories. Therefore, while the approaches to scenario modelling might be different for the different categories, the analytical procedures used

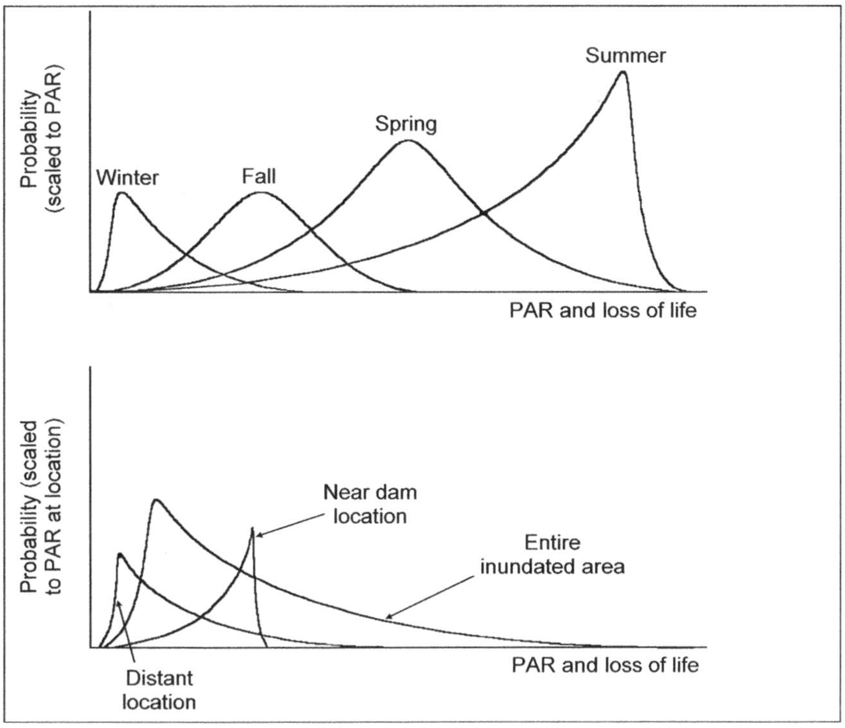

Figure 5-11. Hypothetical probability distributions of loss of life as determined by temporal and spatial conditions

to estimate the probable value of the loss or damage (the damage or loss function) given a particular scenario can be the same, or if not the same will be consistent mathematically.

Figure 5-12 illustrates a generalised response model for "objects" in an inundated area at the time of a dam failure. The fundamental characteristics of the "objects" are represented in a general way in that "objects" are either "animate" in the case of people and living species, or "inanimate" in the case of infrastructure, buildings and habitat.

The response scenario model includes consideration of emergency preparedness plans by the authorities and planned evacuation. In general, emergency plans are not always assured or, if established, they might not work. The formal removal of "objects" is accounted for, as is the non-existence or failure of formal removal procedures.

Up to this point, all "objects" are treated in the same way and when appropriate, the damage or loss functions can be applied directly at these nodes (a more detailed discussion of this concept follows). The process of scenario modelling illustrated here is dealt with in much greater detail in Chapters 4 and 8 on event tree analysis and Chapter 10 on consequence analysis.

At this stage in the modelling endeavour it becomes necessary to take account of the characteristics that are unique to "animate objects". Obviously, the considerations at this stage are extremely sensitive and must be handled appropriately. However, as mentioned previously, the modelling endeavour must account for all relevant aspects of the situation being analysed.

While Figure 5-12 attempts to bring out the level of detail that might be considered explicitly in the risk analysis, other variations of this model are possible and may be used as appropriate. One such model is illustrated in Figure 5-13, which combines a condensed form of Figure 5-12 with Equation 5-1. Another variant is to replace the

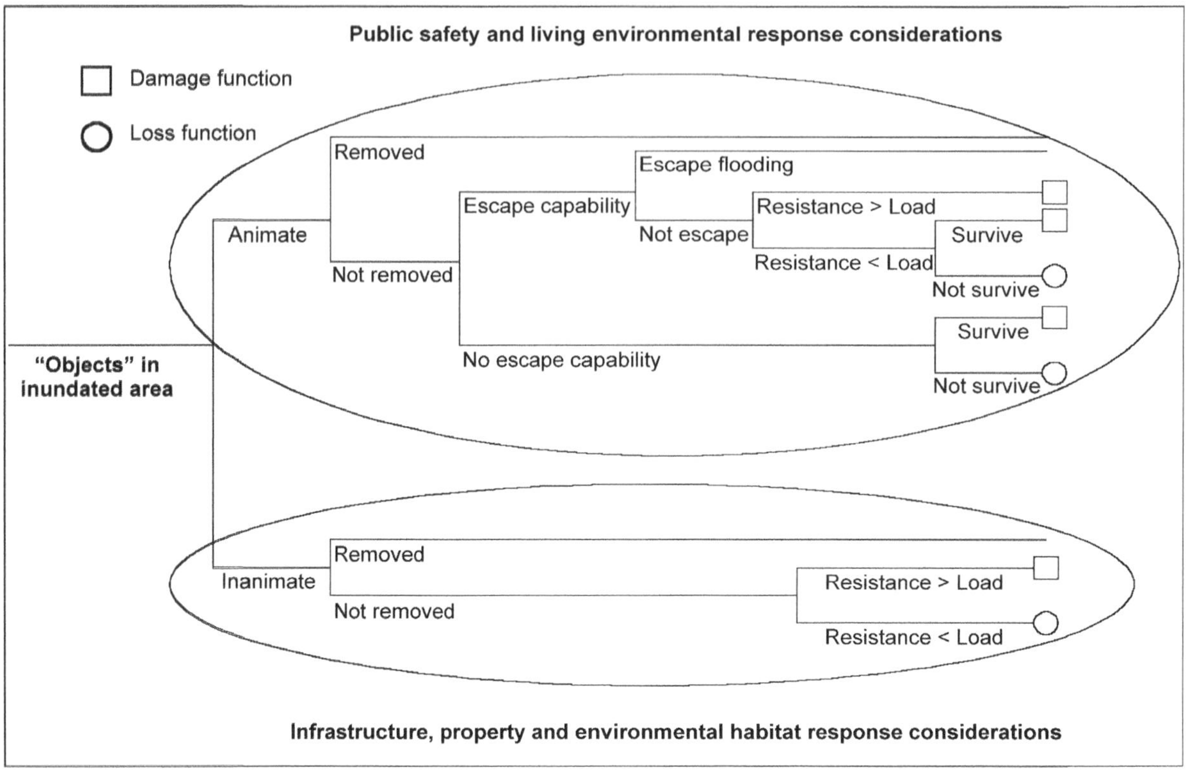

Figure 5-12. Consolidated response model

phenomenologically estimated object loss frequency (OLF) at the right-hand side of the event tree with a statistically estimated OLF.

Figure 5-13 is essentially a "lumped" model whereby the loss function consolidates all of the considerations of Figure 5-12 (which itself is a "lumped" model although to a lesser degree) into a single-loss function. In reality, for life safety analysis the situation is rather more complex than has been depicted here, however, consideration of these complexities is a matter for the analyst to deal with on a case by case basis.

5.4 Category considerations

The previous material suggests that there is considerable similarity of analysis methods between the three principal (from the perspective of direct consequences) consequence categories. This commonality is brought out through examination of experimental test data on buildings and on people, as illustrated in Figures 5-14 and 5-15.

Concerning Figure 5-14, the data of Black (1975) is theoretical, whereas the data of Clausen and Clark is interpreted from the available information on the failure of the Dale Dyke Dam in Sheffield, UK in 1864. Comparable data on tests on inanimate monoliths and human subjects is presented in Figure 5-15.

Figures 5-14 and 5-15 illustrate the complementary roles that phenomenological models and experimental and observational data have in the consequence estimation endeavour.

The general form of the relationships between the models of the various consequence categories and the dam breach flood model that comprise the relationships in the failure consequence modelling endeavour are illustrated in Figure 5-16.

Figure 5-16 together with Figure 5-12 provide a conceptual illustration of how the various models are combined to permit the estimation of the consequences of dam

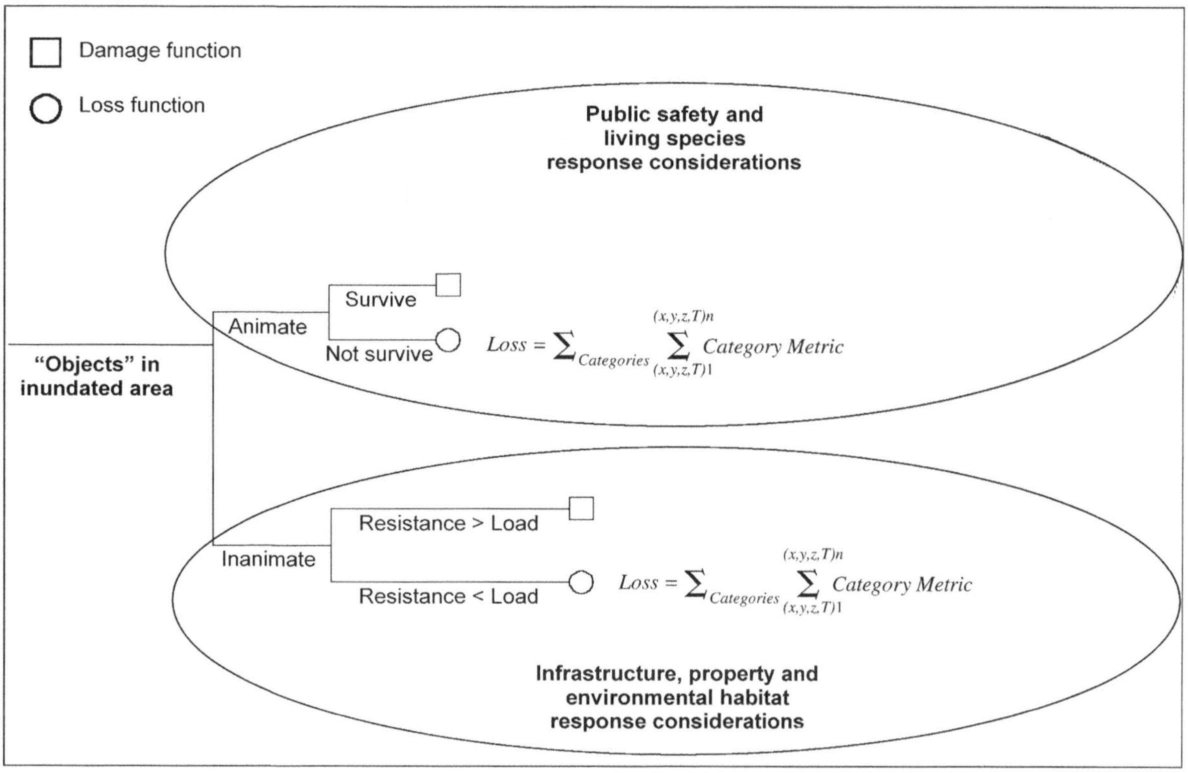

Figure 5-13. Simplified model of loss estimation process

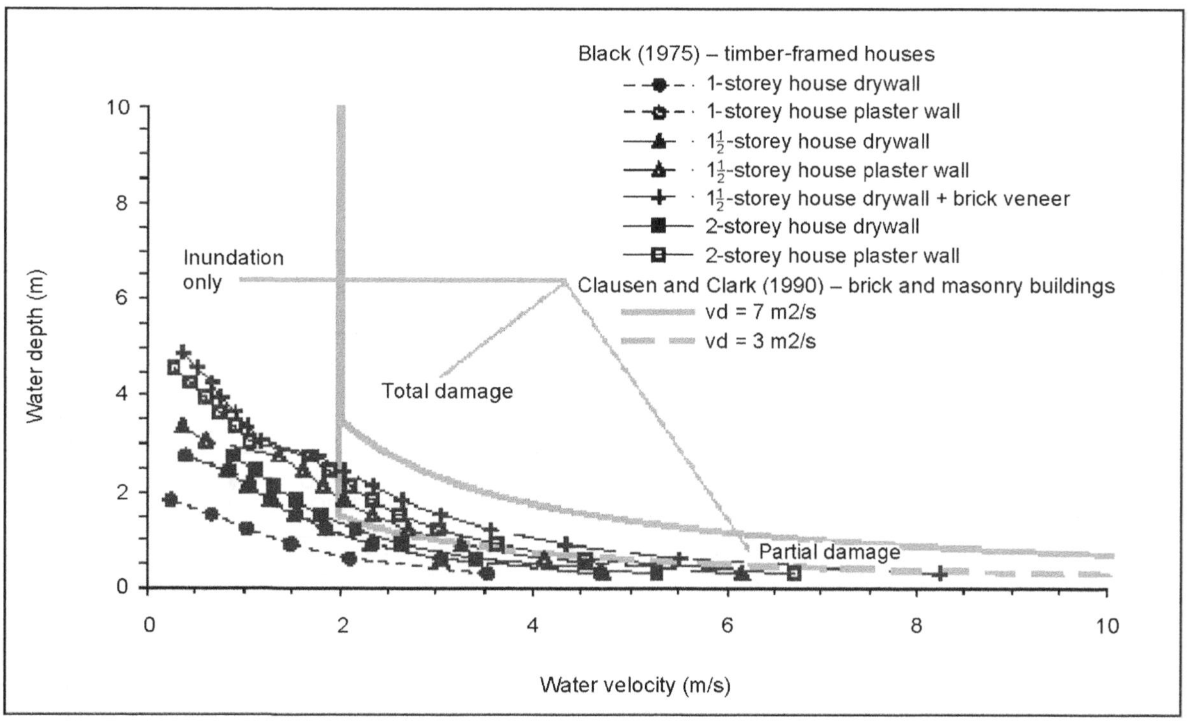

Figure 5-14. Theoretical and experimental data for building vulnerability (from Karvonen et al.*, 2000)*

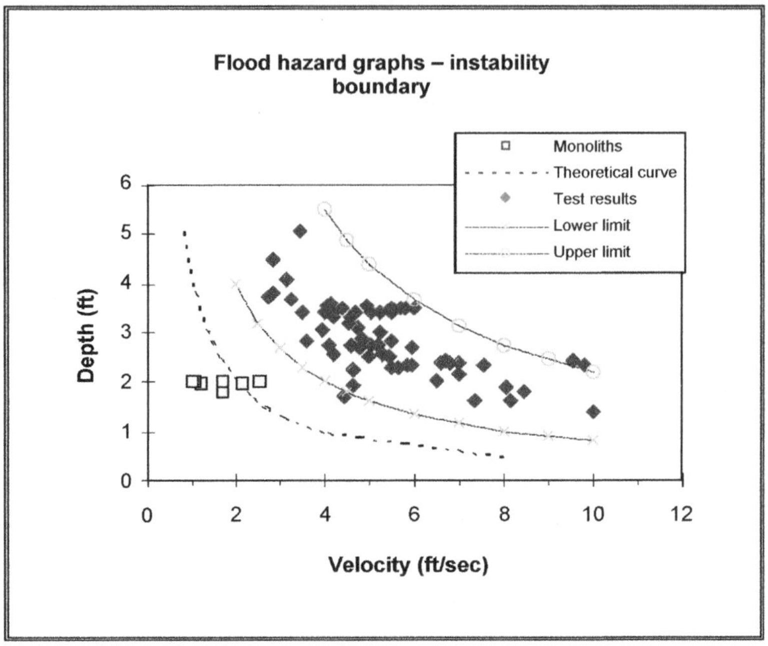

Figure 5-15. Theoretical and experimental data on humans and monoliths (after Lind and Hartford, 2000)

failure. In general, there will be complex interactions within and between the various models, which are intended to reflect the interactions that occur in reality. For example, the topographic changes (erosion) that occur as a result of dam failure have the potential to change the manner in which the other models respond to the effects of the floodwaters. Topographic changes that cause loss of infrastructure, such as roads and bridges, may change the escape potential for people fleeing the floodwaters

Social behaviour also influences the manner in which people respond to hazards. While at a fundamental level, the number of people in the inundated area at the onset

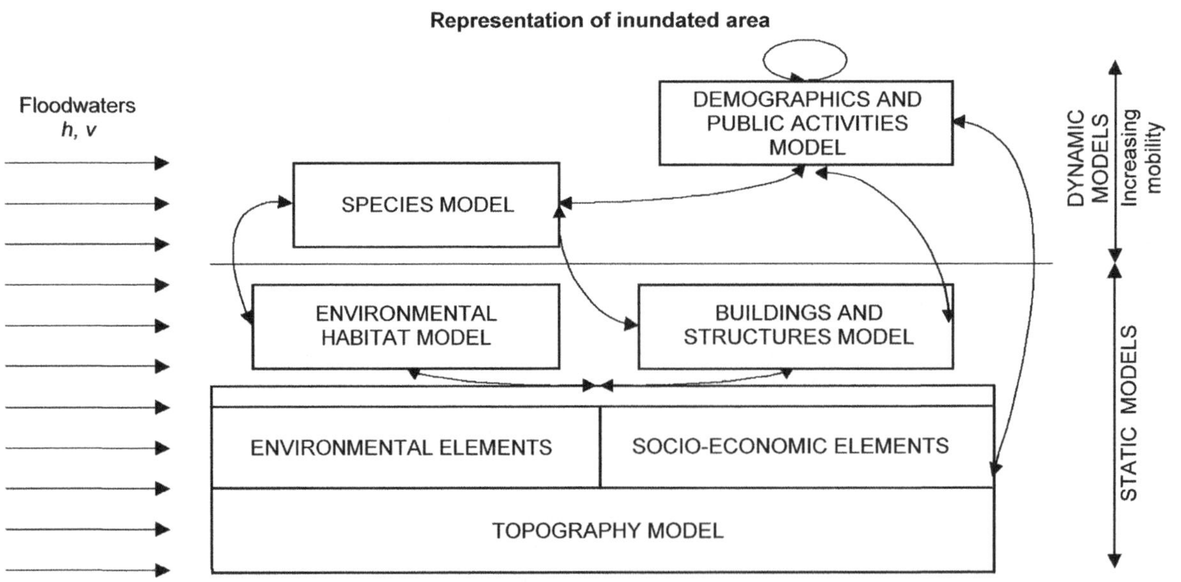

Figure 5-16. Relationships between consequence models

of flooding, the population at risk, is the most important parameter, all people do not respond as individuals. Families tend to try to stay together, and human nature being what it is often results in caring individuals further endangering themselves to assist people in apparently greater danger. There are many other considerations that occur in reality and which should be accounted for in some way in the modelling endeavour.

While there are clear similarities in the responses of the "objects" in the consequence categories, there are also distinct differences in the sequences of events that occur prior to the application of the hydraulic forces to the object.

The idea that some considerations can be modelled as essentially static, as first illustrated in Figure 5-12, is further developed in Figure 5-16. Here the term "static" applies to the ability of the "object" under consideration to react to the threat prior to coming into contact with the floodwaters. Thus, damage and loss functions for the infrastructure and structures and for the environmental habitat are modelled without any consideration of "reaction" to the threat posed by the flood. On the other hand, people and, to a degree, species of animals are more complex as these categories possess some degree of capability to detect and react to the threat prior to the arrival of the floodwaters. While animal species might react instinctively, the reactions of people are rather more complex.

Against this background, the modelling concepts of Figure 5-12 and Figure 5-16 combined with the procedures to determine damage/loss functions provide a relatively direct way of determining environmental habitat, infrastructure and property losses. Their application to matters of public safety, and to a lesser extent risks to living species, are rather more complex. The various ways that researchers have formulated approaches to apply the modelling concepts presented in this chapter are discussed at length in Chapter 10.

Owing to the importance of public safety in dam safety risk assessment, a more detailed treatment of the analysis of life safety considerations follows.

5.5 Life safety analysis

Human response to impending flooding is critical to loss of life (LOL) estimation and to emergency planning for risk management. However, human response is also highly variable and uncertain, be it an individual or a group. Human response has physical and psychological dimensions, which tend to be situation specific, making it difficult to characterise. Human response to impending flooding may not be rational and in general physiological factors will be involved in the response.

In general, to survive a dam breach flood, each individual in the affected area will have to do several things correctly. It may be a small number of things, it may be many things. This means that a range of choices will be available to an individual who receives a warning of impending flooding, with survival requiring not just the correct choices but the effective implementation of each choice.

This randomness and uncertainty mean that prediction of human behaviour, which is essential for predicting loss of life from dam failures, can only be done in a probabilistic way.

As there has been very little research into the psychology, physiology and physics of human responses to dam breach flooding, predictions of LOL will necessarily have very large uncertainty bounds. A draft review of some 178 case history accounts of flood disasters carried out at Utah State University (Bowles and McClelland, 1999, 2000) indicates LOL in dam breach flood events is determined by a very large number of factors. Bowles and McClelland have identified at least 100 variables that they consider to be determining factors in the probability of survival. While this work has not yet advanced to the stage of a formal predictive model based on historic records, it does help to demonstrate that analytical determination of human response to impending flooding is presently extremely difficult and in many cases impossible.

An alternative representation of the life safety analysis endeavour can be represented in a relatively straightforward way as illustrated in Figure 5-17.

The scenario model can be developed as outlined above resulting in specification of the people potentially inside or entering the danger area. Such specification should include characterisation of the uncertainty (in the demographic analysis) in the number of people exposed to the threat of being inundated at time "*T*".

Whether or not the people in or entering the inundated area will react to the threat of flooding and attempt to flee will depend on several factors. Given that individuals and/or groups attempt to flee, the success of these efforts will also be dependent on many factors. Thus, the "reaction" process to be modelled can generally be expected to be highly complex.

The conceptual reaction model of Figure 5-17 applies to individuals or, if aggregated, to groups. The overall objective of the reaction analysis endeavour is to estimate the number of ways that people can be exposed to the forcefulness of the floodwaters, and the number of people exposed for each way that (they can become exposed to) can become exposed. Once exposed to the floodwaters, the damage/loss function model can be applied to each end branch of the reaction model.

Conceptually, the relationship between a case history of LOL from dam failure and the generalised modelling concepts presented above for a single dam breach scenario are illustrated in Figure 5-18.

Figure 5-18 is intended to represent the general form of the modelling endeavour for the failure of an individual dam. The differences between the "actual" and "reported" distributions are intended to illustrate the uncertainty in the input data and in the output

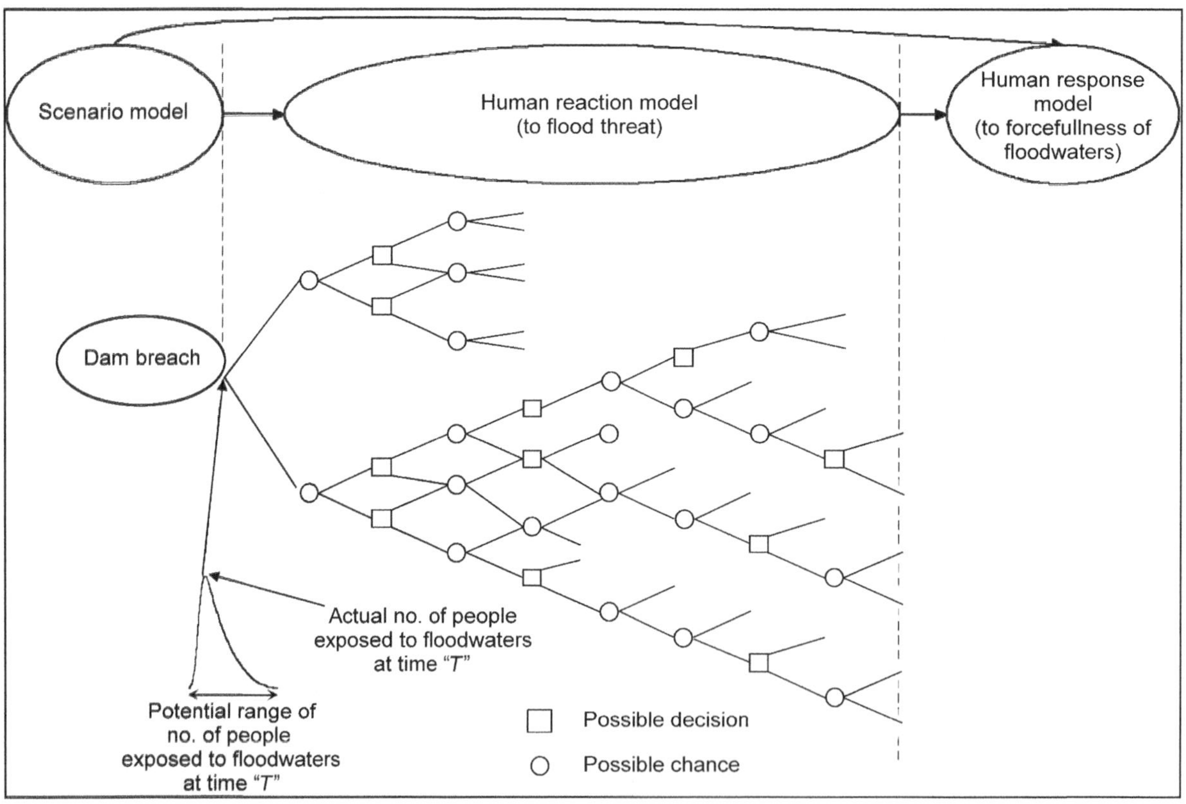

Figure 5-17. Simplified modelling concept for human reaction

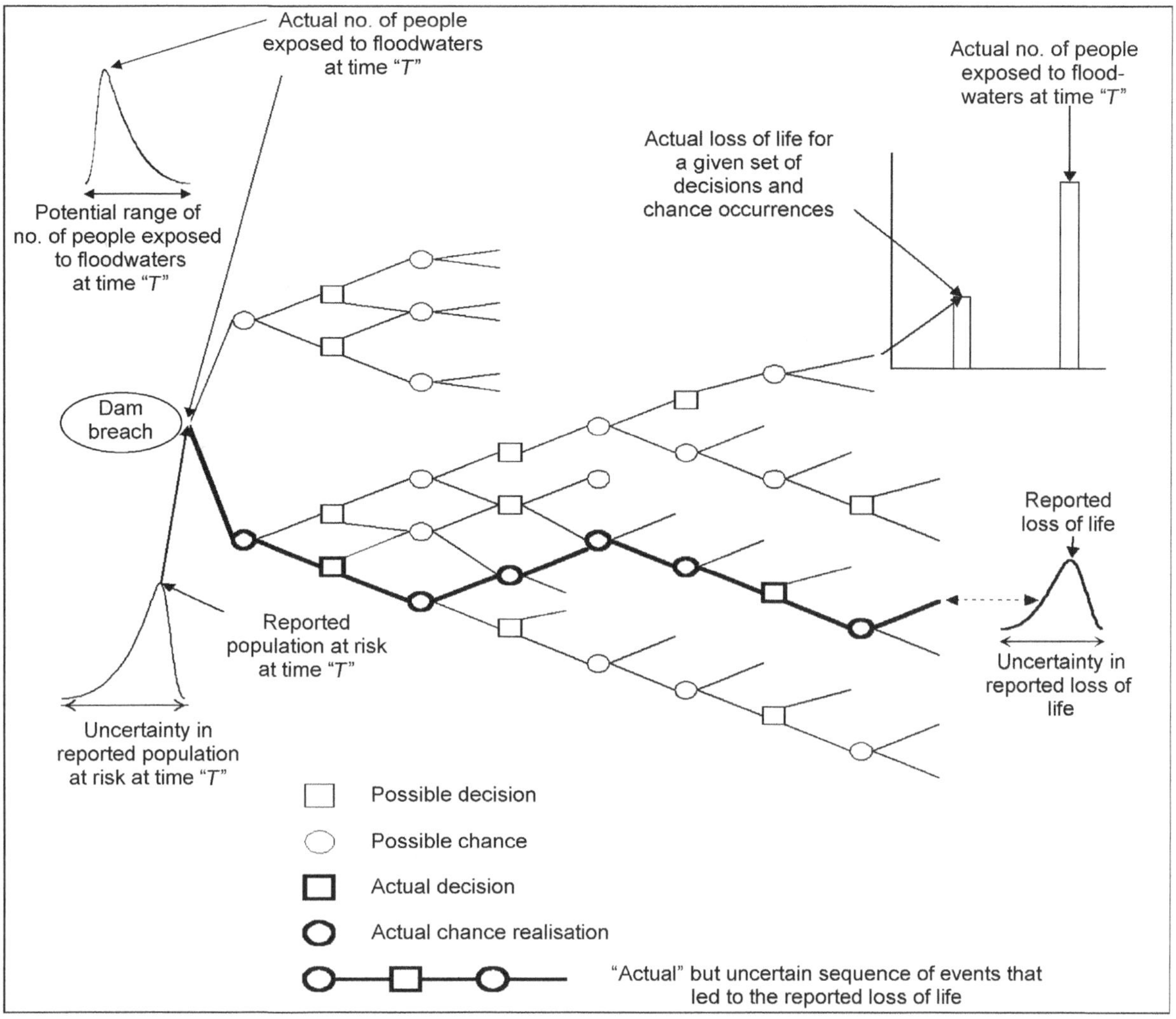

Figure 5-18. Relationship between case histories and generalised models

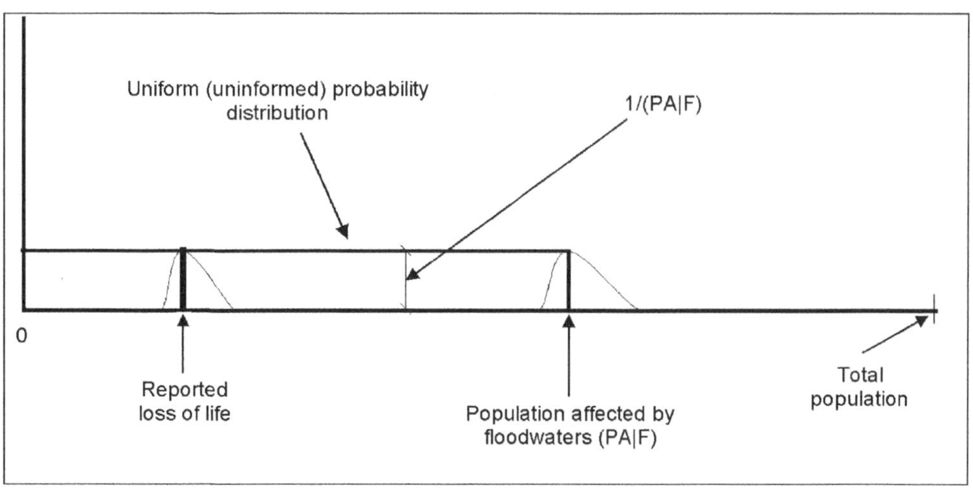

Figure 5-19. Possible initial probability distribution for loss of life from case history data

distributions at the ends of each reaction branch. A similar reaction tree structure would be generated for each dam breach scenario (x, y, z, T).

Thus, for all possible dam breach scenarios and all possible reactions, an initial (prior) probability distribution for the LOL from a dam failure could be as illustrated in Figure 5-19. Figure 5-19 embodies the assumption that all possible end branches, ranging from no LOL to loss of all life that came in contact with floodwaters, are equally likely.

Ultimately, the analysis through a process of iteration should result in a comprehensive set of probability distributions. These distributions can then be combined mathematically into a consolidated probability distribution of the LOL from dam failure (Figure 5-20).

Methods of applying the general modelling concepts outlined above have been proposed, all of which embody the ideas of modelling scenarios, human reactions and loss functions. These proposals include object loss frequency approach, statistical methods, and advanced phenomenological modelling and simulation.

The simulation approach revealed that the properties of the probability distribution for LOL from dam failures could not be obtained directly from data, the empirical formulae or the object loss frequency approach.

Regression approaches

Modelling the risk to human life using regression analysis from the historic database of failures have met with limited success. The number of persons at risk, P, and the warning time, WT, are the most important numbers correlated with the loss of life, L; both depend on other factors that are unknown.

The pioneering work by Brown and Graham (1988), proposed that for a given scenario, characterised by the population at risk (P), and warning time (WT) empirical equations fitted to the data. In terms of the *risk to life*, $R=L/P$ (dimensionless), the US Bureau of Reclamation (USBR) (1989) prescribed:

$$R=0.5 \text{ for } WT<0.25 \text{ hrs;}$$

$$R=P^{0.6} \text{ for } 0.25<WT<1.5 \text{ hrs; and} \qquad (5\text{-}2)$$

$$R=0.0002 \text{ for } WT>1.5 \text{ hrs}$$

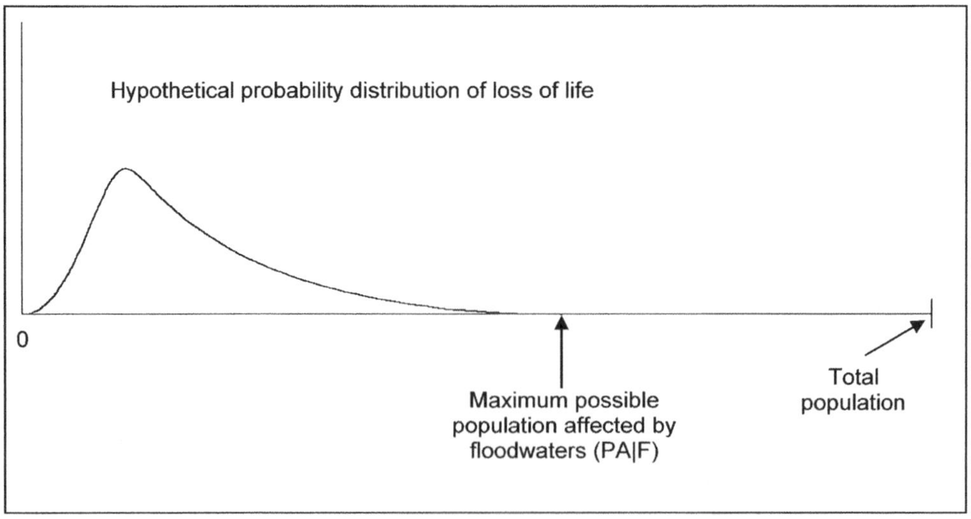

Figure 5-20. General form of output of a life safety analysis

The discontinuities in these formulae are severe. For example, for $P=10\,000$ the loss of life jumps from 251 to 5000 at $WT=0.25$ hrs, and then jumps from 2 to 251 at $WT=1.5$ hrs.

For a better fit, DeKay and McClelland (1993) introduced a distinction between "high lethality" (deep and swift floodwaters, e.g. in a canyon) and "low lethality" (e.g. on a plain). They proposed the continuous expressions

$$L=P(1.5207P^{0.513}e^{[3.83WT-4.012]})^{-1} \text{ for high lethality, and} \quad (5\text{-}3)$$

$$L=P(1+.5207P^{0.513}e^{[0.822WT]})^{-1} \text{ for low lethality} \quad (5\text{-}4)$$

For $P=10\,000$ at $WT=1/3$ hrs these two formulae give $L=255$ and 13 respectively, a 20-fold difference. Since the lethality is not defined quantitatively and is difficult to judge for a given topography, these formulae leave much to the analyst's judgement.

Object loss frequency (OLF) approach

The "object loss frequency" (OLF) concept coupled with scenario and human reaction models forms the basis of USBR's "flood severity based method" for estimating LOL from dam failure as described in Graham (1999). The LOL frequencies and the OLF, as described above, were calculated from case history accounts of dam failures. The DSO-99-06 "flood severity based method" attempts to match both the hydrodynamics (average of depth × velocity) of the case in hand and human reactions (in terms of warning time) and "flood severity understanding".

The fatality rate values (*OLF*) presented in Table 6 of DSO-99-06 are based on inspection (Figure 5-21). The data on flood severity (*DV*), warning time (*WT*) and flood severity understanding (*FSU*) are not always represented in the form of previously measured parameters in the case histories. Notwithstanding the efforts that have gone into collecting and analysing case histories of dam failures and flash floods, fatality rate data is extremely scarce and each case is unique.

In presenting 11 fatality rate models DSO-99-06 condenses all of these factors into the three parameters, *DV*, *WT*, and *FSU* by a process of inspection of available case history accounts (Figure 5-22). These models can be represented by the more generalised loss function (Reiter, 2001) of the form:

Flood severity	Warning time (minutes)	Flood severity understanding	Fatality rate (fraction of people at risk that died)	
			Average	Range
HIGH	No warning	not applicable	0.76	0.3 to 1.00
	15 to 60	vague	no case fit this category	
		precise	no case fit this category	
	more than 60	vague	no case fit this category	
		precise	no case fit this category	
MEDIUM	No warning	not applicable	0.14	0.02 to 0.43
	15 to 60	vague	0.014	only one case
		precise	0.01	only one case
	more than 60	vague	0.05	only one case
		precise	0.035	0.0 to 0.080
LOW	No warning	not applicable	0.007	0.0 to 0.025
	15 to 60	vague	0.0095	0.007 to 0.012
		precise	0.0	only one case
	more than 60	vague	no case fit this category	
		precise	0.0003	0.0 to 0.002

Figure 5-21. Fatality rate data (DSO-99-06)

Flood severity	Warning time (minutes)	Flood severity understanding	Fatality rate (fraction of people at risk expected to die)	
			Suggested	Suggested range
HIGH	No warning	not applicable	0.75	0.30 to 1.00
	15 to 60	vague	Use the values shown above and apply to the number of people who remain in the dam failure floodplain after warnings are issued. No guidance is provided on how many people will remain in the floodplain.	
		precise		
	more than 60	vague		
		precise		
MEDIUM	No warning	not applicable	0.15	0.03 to 0.35
	15 to 60	vague	0.04	0.01 to 0.08
		precise	0.02	0.005 to 0.04
	more than 60	vague	0.03	0.005 to 0.06
		precise	0.01	0.002 to 0.02
LOW	No warning	not applicable	0.01	0.0 to 0.02
	15 to 60	vague	0.007	0.0 to 0.015
		precise	0.002	0.0 to 0.004
	more than 60	vague	0.0003	0.0 to 0.0006
		precise	0.0002	0.0 to 0.0004

Figure 5-22. Proposed fatality rate models (DSO-99-06)

$$Loss\ of\ life = PAR \times FAT_{Base} \times IMPACT \times CORRFAT \qquad (5\text{-}5)$$

Where, PAR=population at risk, FAT_{Base}=base fatality rate of PAR, Table 7 of DSO-99-06, mean values, $IMPACT$=additional impact factor to account for flood severity impact (SEV), living environment impact (LOC) and vulnerability impact (VUL) derived in the RESCDAM LOL method using public population register, $CORRFAT$=correction factor to take account of warning efficiency and possible emergency/rescue action into account in each sub-area (re-arranged values of Table 7 of DSO-99-06).

Equation 5-6 is a more general and category specific version of Equation 5-5 with a correction factor $K_{Category}$ where $K = IMPACT \times CORRFAT$ for each category of object.

$$Loss = \sum_{Categories} K_{Category}\left(OLF_{Category} \times \sum_{(x,y,z,T)1}^{(x,y,z,T)n} Category\ Metric\right) \qquad (5\text{-}6)$$

The factor K takes account of the warning time and flood severity understanding in DSO-99-06 as well as other relevant factors. The factor $K_{Category}$ essentially comprises two components, one pertaining to the severity of the flooding (depth × velocity, DV) and the escape potential EP.

$$Loss = \sum_{Category} DV_{modifier} \times EP_{modifier} \times OLF \times \sum_{(x,y,z,T)1}^{(x,y,z,T)n} no.\ of\ Objects\ at\ Risk \qquad (5\text{-}7)$$

Although the ideas presented in DSO-99-06 and by Reiter (2001) as summarised above have been developed for LOL modelling, Equations 5-6 and 5-7 permit the application of loss frequency modelling to economic and environmental considerations. This wider application is achieved by setting the EP modifier to a default value of 1 to obtain the upper bound. Clearly, if there is ample intervention time, it may be possible to reduce economic and environmental damages by removing certain categories of "objects" such as personal possessions, and certain species of animals, from the inundated area.

Equation 5-7 permits the transformation of the average loss frequency over the entire inundation area as determined from case history data to location specific loss frequencies provided the "modifiers" are known. In instances where data is scarce, parameters such as *DV* can be obtained from experimental data and phenomenological modelling as described above.

The above demonstrates the complementary nature of data models (experimental and case history) and phenomenological models and the benefits of combining them to compensate for the lack of data and to use the available data to calibrate the phenomenological models.

Simulation approach

In recent years, scientific advances in dam breach modelling and simulation of emergencies have added a powerful new dimension to life safety analysis. These methods, while developed for the purpose of emergency management, also provide a means of estimating the probability distribution of LOL as one output of the modelling endeavour.

The simulation approach incorporates logical statements of the phenomena under consideration. Consideration of the hydrodynamic conditions which people might have to deal with during dam breach flood emergencies is clearly an important part of the modelling process. While the model of the scenarios considered by the analyst might be idealised, the logic is necessarily correct and therefore computable.

One way of modelling how people respond to flood warnings is illustrated in Figure 5-23. The logical limitations of this model pertain to the extent of the logic in the model, not to the process of logical inference, and can be resolved by expanding the logic to include more possibilities. The parametric uncertainties can be addressed through physical modelling, calibrated against historic data and scale-model tests.

The general approach to simulation modelling of life safety considerations is to view the downstream "objects" from two perspectives (Assaf and Hartford, 2002):

- a static view that describes long-term characteristics and relationships of people, and their world, which includes for example a person's age, societal, the construction type of a building, and the number of lanes of a road segment; and
- a dynamic view that describes a snapshot of the world at a given time of day, week and year, which includes for example locations of people, weather related driving conditions, and state of alertness at day or night.

These two views permit complete characterisation of the left side of Figure 5-17 and the interrelationships between the left-hand and right-hand sides of Figure 5-17. The procedure then combines this data with a reaction model (e.g. Figure 5-18) and a response model (Figure 5-12) that utilises stability models (Figure 5-15 and Figure 5-16).

The process proposed by Assaf and Hartford (2002) is a general emergency scenario-modelling environment that is intended to provide a basis for emergency planning for floods and other disasters. The probability distribution of LOL for the various scenarios is just one of the outputs of the simulation process.

This model, of people surviving floodwaters, has been invented to account for observed phenomena in a relevant way. In this way, the logic is firmly rooted in existing knowledge of LOL in flooding, knowledge of how people respond in emergencies in general and all of the location and environment specific characteristics of the downstream inundation area under consideration.

By accounting for warning time, severity of flooding (depth and velocity) and rate of escape, the logic captures the three fundamental elements of the procedure outlined in DSO-99-06, and extends them to include the dynamics (in time and space) of flood scenarios. In doing so, it provides a basis for complete characterisation of the scenarios outlined in Figure 5-18.

While simulation methods are at the early stages of development, they present an opportunity to provide improved procedures for LOL estimation. They also provide

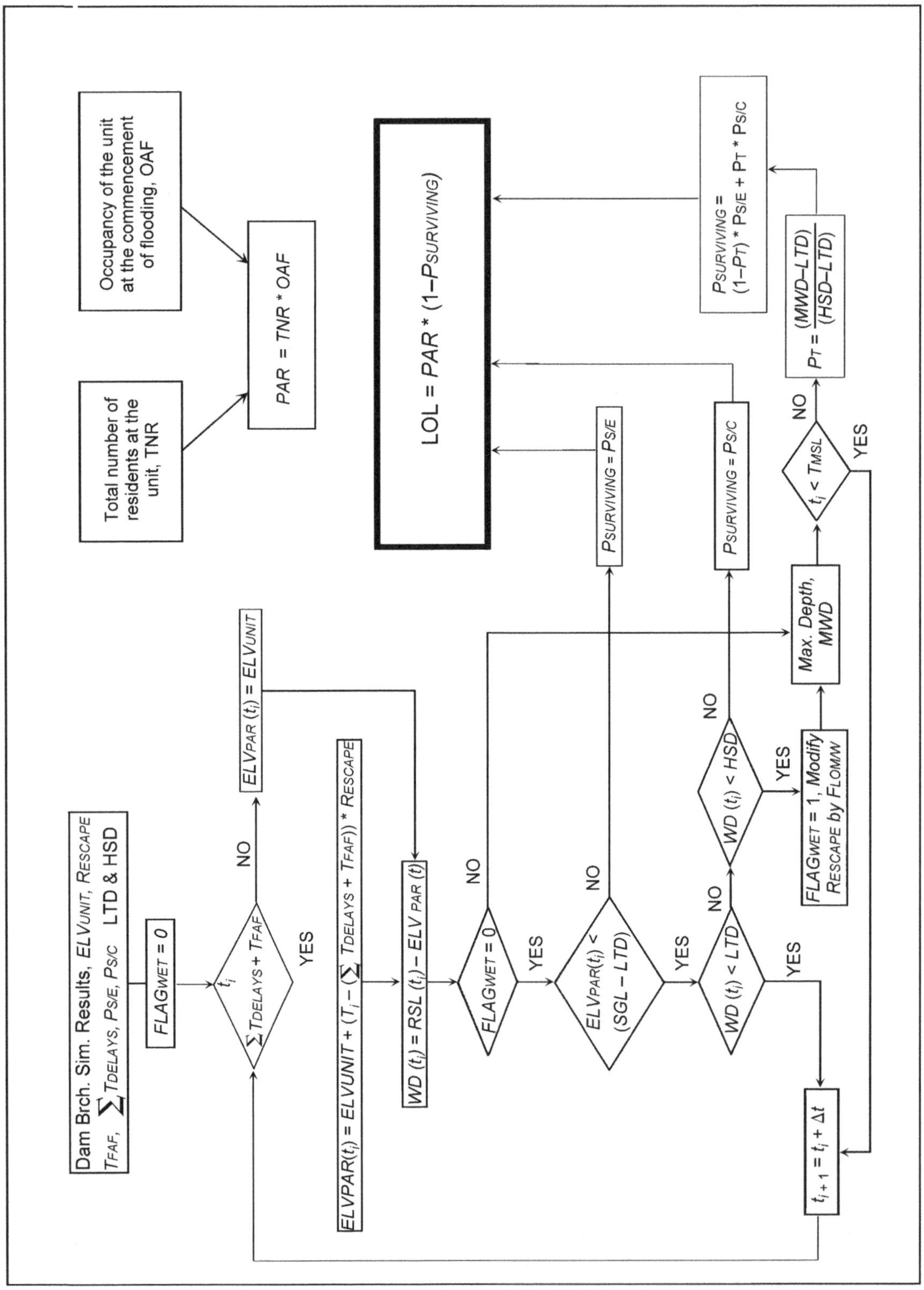

Figure 5-23. Proposed life safety analysis algorithm (Assaf et al., 1997)

opportunities to broaden the application of risk analysis in dam safety management and in emergency management in general.

Comment on uncertainty

In recent years, there has been emerging acceptance that estimates of LOL from dam failures using any of the above methodologies are highly uncertain.

Although based on the same data, the Brown and Graham formulae can give different results to those of the DeKay and McClelland formulae. The DSO-99-06 procedure in turn can lead to predictions that are significantly different from those of the empirical formulae.

The topic of uncertainty in LOL estimation is poorly understood and there has been little in the way of scientific investigation into the associated problems. Just how uncertain estimates of probability distributions of LOL actually are is unknown with the result that available methods cannot be construed as being "approximate" methods. The above discussion demonstrates that the proposed methodologies (Assaf and Hartford, 2002) are underpinned by concepts that are fundamentally sound. However, the lack of data to populate the databased models or to corroborate the phenomenological models means that a great deal of further research is required to provide analytically determined probability distributions of LOL from dam failures that could be demonstrated as being approximately correct.

6 Probability assignment

In general, probabilities are assigned to distinctly different variables in the risk analysis: (1) chance event variables associated with natural variability (also termed aleatory uncertainties) and; (2) state or condition variables associated with knowledge uncertainties (also termed epistemic uncertainties). Whether variables are chance variables or state variables is an artefact of the model within which the probabilities are assigned. Thus, the assignment of probabilities is inextricably linked to the model being used to characterise the performance of the dam, its loading conditions, and the consequences that follow dam failure.

The risk analysis model, typically in the form of an event tree or a fault tree, can be, a model of a physical system (e.g. a model of a particular dam), a statement about the joint probabilities of random variables, and an accounting scheme for information and beliefs, or any combination of these things. Typically, in dam safety risk analysis the complete risk model will comprise all of these forms of model.

6.1 Philosophical preliminaries

The frequency interpretation of probability is well established in engineering practice and the belief interpretation, which one suspects has been used implicitly and often unsuspectingly in engineering practice, is increasingly being used intentionally in engineering. Recent demands for explanation and transparency in dealing with uncertainty through risk analysis, means that, increasingly, the belief interpretation is being used explicitly.

In parallel to these alternative meanings for the concept of probability are alternative approaches to the way inferences are drawn from evidence, that is, from statistical data. Frequentist statistical inference is a widely applied body of doctrine comprising, among other things, estimator theory and the theory of significance tests.

Classical frequentist inference evolved in an effort to create an "objective" appraisal of scientific theories, and treats probability as frequencies of random variations in the physical world, for example, as the naturally occurring variations among experimental results. Howson and Urbach (1991) point out that the frequentist theory of estimation,

> (. . .) has two branches, known as point estimation and interval estimation. Point estimation aims to select a specific number as the so-called best estimate of a parameter; it is contrasted in the literature with interval estimation, a method of locating the parameter within a region and associating a certain degree of "confidence" with the conclusion that is drawn.

Inferences about degrees-of-belief, in contrast to frequentist inference, are built upon Bayes' Theorem. Bayes' Theorem describes the degree to which observed evidence should logically change a degree of belief (i.e. a probability) held before the evidence was observed, to a logically following degree of belief after.

Degree-of-belief inference has sometimes been called, "subjective probability", but Kaplan and Garrick (1981) put forward the view that the term is misleading, and that it has caused confusion and controversy. They put forward the view that the battle between the "frequentist school of thought" and the "Bayesian school of thought" has been due to a misunderstanding. The root cause being the desire for objective science on the part of the "frequentist school of thought", a view shared by scientists in general.

Kaplan and Garrick's views are not unique, similar arguments concerning the objective nature of the Bayesian approach have been put forward by other experts in probability and risk, the works of Howson and Urbach (1991) and Morgan and Henrion (1990) being just two examples.

Recently, Hacking (2001) presented the view that there are two types of belief probability: the interpersonal/evidential type and the personal type. This distinction permits the analyst to reveal the fundamental nature of each probability construct, the extent to which its probability constructs are founded in data and the extent to which the probability has been constructed in terms of established mathematical principles.

However, the belief interpretation of probability does not simply imply that, "a probability is what one believes it to be". As Orkin (1999) points out, "without essential mathematical form, anyone can say anything". Thus, while a loose interpretation might be attractive from the perspective of facilitating the easy generation of numbers, it has limitations and inconsistencies if the belief is not correctly constructed. The problems associated with such a limiting interpretation can be overcome by adding additional constraints that strengthen the interpretation of probability. The result of adding such constraints is an interpretation of the form "a probability is what the evidence, as correctly assembled in terms of the necessary mathematical and scientific principles, permits one to believe it to be as a basis for action". It is this combination of all of the information (ranging from objective data through rational judgement to entirely subjective senses) in a logically consistent way through mathematical procedures that provides a probability construct with its essential mathematical form.

Against this background, the interpretations of probability as belief and probability as frequency are considered to be complementary with both interpretations being necessary elements of probability theory as it applies to dam safety risk analysis. The interpretations are complementary, scientifically valid, and should be applied to the appropriate extent. There is no question of the two interpretations being mutually exclusive, or one being superior in its totality to the other, because the two approaches are complementary and, depending on the situation, the use of one interpretation may be more appropriate than the other. In this regard, the two interpretations are not necessarily alternatives.

Mathematical, logical and scientific considerations

As discussed in Chapter 7, probability theory is a branch of mathematics. It is logical and internally consistent in the sense that all the mathematics of probability theory can be derived from a small set of axioms. In essence, the axioms specify properties that "probability" must have, for example a value between zero and one.

Mathematical or logical proof is required to demonstrate something conclusively. Of course, in most real situations, mathematical or logical proof is unachievable and one is forced to accept that probabilities cannot be demonstrated as being correct in an absolute sense. While it would be very comforting to know that important decisions are based on "right" probabilities, we must accept that we can never be sure that they are.

In the absence of mathematical or logical proof, the process of scientific inference can be used to demonstrate the reasonableness of a decision. The following view concerning the difference between mathematical proof and scientific proof was put forward by Singh (1997).

> To appreciate the value of proofs they should be compared with their poor relation, the scientific proof. In science, a hypothesis[1] is put forward to explain a physical phenomenon. If observations of the phenomenon compare well with the hypothesis, this becomes evidence in favour of it. Furthermore, the hypothesis should not merely describe a known phenomenon but predict the results of other phenomena. Experiments may be performed to test the predictive power of the hypothesis, and if it continues to be successful then this is even more evidence to back the hypothesis. Eventually the amount of evidence may be overwhelming and the hypothesis becomes accepted as a scientific theory.

However, scientific proof can never be proved to the same absolute level of a mathematical theorem: it is merely considered highly likely based on the evidence available. So-called scientific proof relies on observation and perception, both of which are fallible and provide only approximations to the truth. Even the most widely accepted scientific "proofs" always have a small element of doubt in them. Sometimes this doubt diminishes, although it never disappears completely, while on other occasions the proof is ultimately shown to be wrong.

Scientific proof is inevitably fickle and shoddy. Mathematical proof is absolute and devoid of doubt. Science is operated according to the judicial system. A theory is assumed to be true if there is enough evidence to prove it "beyond all reasonable doubt". On the other hand, mathematics does not rely on evidence from fallible experimentation, but is built on infallible logic.

The process of scientific inference, whereby the observations and information are transformed into conclusions upon which actions can be based, is well established. Probability assignment is fundamentally a matter of reasoning under uncertainty whereby a mathematical representation of the state of knowledge is inferred by reasoning from incomplete and uncertain data and knowledge. Accordingly, the logic of scientific inference and probability are central to the assignment of probabilities.

Logic

A dictionary definition of logic, a cornerstone of scientific inference is: the science and art of reasoning correctly: the science of the necessary laws of thought: the principles of any branch of knowledge (Chambers, 1972). A process is logical if the line of reasoning involved is demonstrably correct. Logic permits the discrimination of correct reasoning from plausible but actually incorrect reasoning. Because probability assignment necessarily involves reasoning from incomplete data and knowledge, the probability construct is never completely defined. Concerning the frequency interpretation of probability, it is necessary to accommodate the fact that key information is not included in the mathematical construct, with the result that it is a partial representation of reality. Similarly, concerning the belief interpretation of probability, the logic involved is the logic of partial belief.

The belief interpretation of probability holds that $P(A)$ represents the degree of belief that an individual has in the occurrence of a future event A. The theory of subjective probability is normative in the sense that it is logical and, when used in this way, the laws of probability can be held as comprising the logic of partial belief. Probability is the logic of partial belief in that if one believes A and if one also believes that "if A then B" then logic prescribes that one should also believe in B. This said, there are numerous and frequent examples which demonstrate that the interpretations and actions of people do not conform to the laws of logic.

The issue of using correct logic in assigning probabilities can be illustrated as follows. While belief probability does not define the degree of belief that someone (preferably an expert) should have in the occurrence of an event A, it does require that the degree of belief in $A \cup B$ (A or B) should be at least as great as the degree of belief in A. (See Cooke, 1991, Chapter 3 for an explanation as to why probabilistic reasoning is much more difficult than deterministic reasoning.) According to Cooke (1991):

> Arguments that are valid when the premises are known with certainty are not "almost valid" when the premises are "almost certain." Premises that are equivalent when known with certainty are not "almost equivalent" when the premises are "almost certain." Rather, discontinuities arise and just the slightest bit of uncertainty can change a valid argument into an invalid one or can make equivalent propositions inequivalent (sic).

Mathematical form

Concerning "events" or "occurrences", a probability is represented by a number between zero and one, and as with the differential and integral calculus, Leibnitz is responsible for the following symbolism (Keynes, 1921),

If **P** is the relation of certainty, **P**$=1$
If **P** is the relation of impossibility, **P**$=0$
If **P** is a relation of probability, but not the relation of certainty, **P**<1
If **P** is a relation of probability, but not the relation of impossibility, **P**>0

However, these basic definitions do little to reveal the mathematical form and properties of a "probability". The form and properties are important because without them the assigned probability can be anything. Regardless of whether one uses the belief interpretation or the frequency interpretation of probability, the probability number has mathematical properties as determined by the information that is available and the logic used in combining this information and transforming it to a mathematical form. In the case of the frequency interpretation the data will be transformed into a probability using statistical procedures, whereas in the case of the belief interpretation the evidence is transformed into a probability using Bayes' theorem.

The mathematical definition of a probability is normally specified in terms of specific mathematical properties, which can be illustrated graphically by a probability distribution. In the case of the occurrence and non-occurrence of success states in an event tree the probability construct, the probability mass function (pmf), is of the form shown in Figure 6-1.

On the other hand, if the parameter of interest is continuous over an interval such as the shear strength of a material, the probability construct, the probability density function (PDF), will be of the form illustrated in Figure 6-2.

Although it is not absolutely necessary, it is easier to grasp the concept of probability form by considering a random variable defined on a sample space of events. Without going into detail (see Chapter 7), the random variable in question, $X(x)$, may have $x=1$ value associated with a success and $x=2$ value with a no-success. If the corresponding probabilities are known (say, 0.8 and 0.2, respectively), then the probability distribution takes the form of probability mass function (PMF) shown on Figure 6-3. The possible values of x are on the horizontal axis; the vertical axis represents probability. In this example, the distribution is of a discrete form since there are two possible outcomes for the event under consideration (corresponding random variables can have only two values).

If the uncertain quantity of interest is continuous over an interval, such as the shear strength of a material, the probability distribution can take the form of pdf as illustrated in Figure 6-4.

Since, in the continuous case, the pdf is simply a limiting form of a histogram, the ordinates of the pdf do not represent probability. Probability of any event (for example, the shear strength of material having values between x_a and x_b) is represented by the area under the curve.

Figure 6-1. Graphical representation of mathematical form of a "probability of an event"

Probability functions, be they discrete or continuous, have several mathematical properties. One of those properties is that the integral of the probability function is 1.0. In the case of the discrete probability function, the integral is the summation; whereas in the case of the continuous probability function it is simply the integral,

$$\sum_j P_j = 1 \tag{6-1}$$

$$\int_{-\infty}^{+\infty} f(x)dx = 1 \tag{6-2}$$

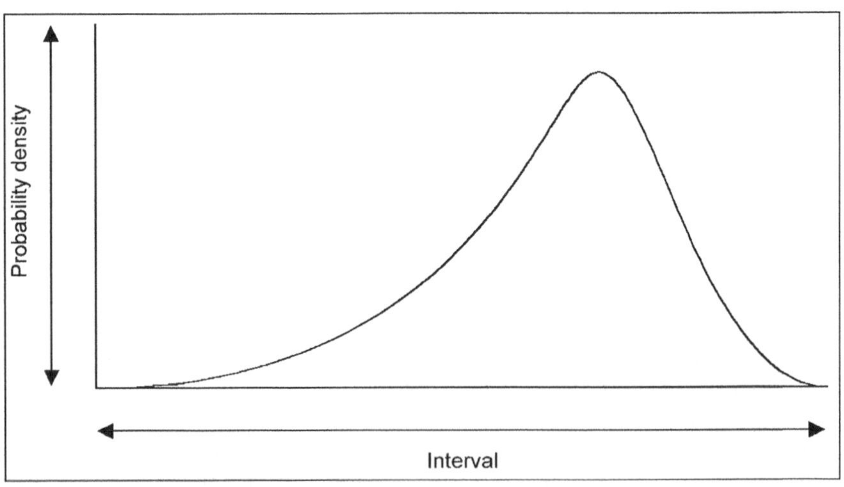

Figure 6-2. Graphical representation of the mathematical form of a "probable value of a parameter (or state)"

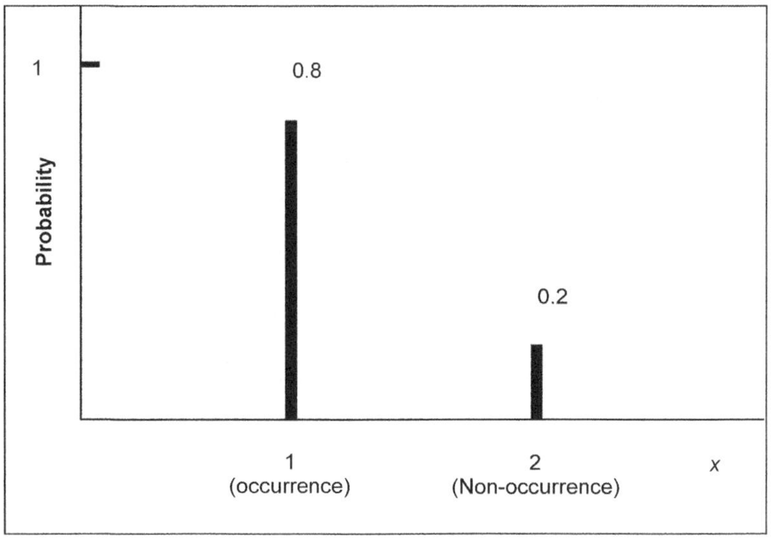

Figure 6-3. Alternative representation of the mathematical form "probability of an event"

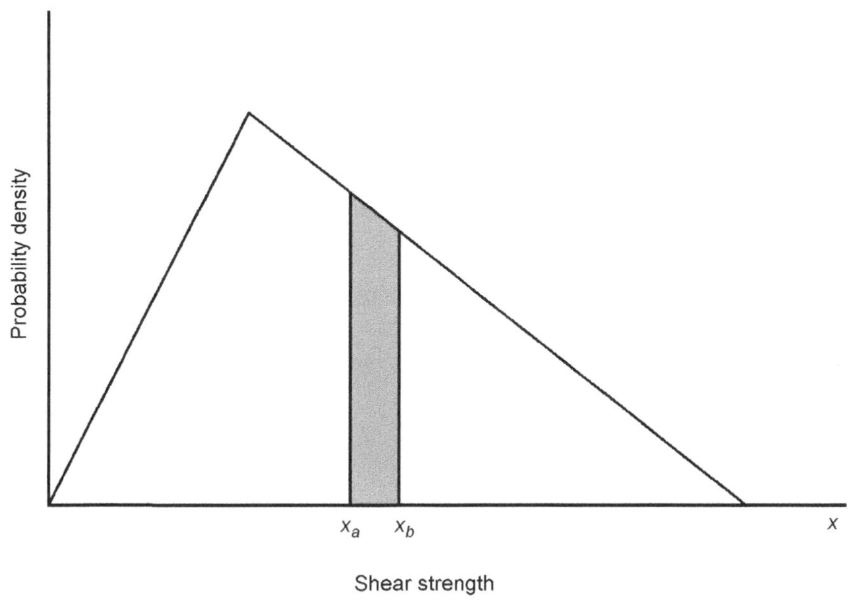

Figure 6-4. Alternative representation of the mathematical form of "probable values of a parameter"

In a general mathematical sense, the only difference between the general form of the properties of the discrete distribution and those of the continuous distribution, pertain to the differences between the summation and the integral. However, the actual form of these properties depends on the fundamental nature of the process that generates the raw data, which controls the form of the probability value: the key feature of the probability distribution.

Scientific considerations

The purpose of a probabilistic analysis is to bring an uncertain science/engineering problem to an acceptable science/engineering conclusion. The process requires data (including information and knowledge) in the broad sense of the term and involves the analysis of the data and the transformation of the individual elements of data into the mathematical form known as a probability.

For the purpose of this discussion, the term data is used in a general sense where the data to be analysed is obtained by observation or through phenomenological modelling. The observations can be of various forms, ranging from carefully controlled scientific measurements through approximate measurements to unverifiable observations. In this regard, the experts' knowledge can be treated as a source of data which can be analysed in the same way as other data types. The conventional approaches to data analysis are: Classical (frequentist) analysis and Bayesian statistical analysis.

Since these two approaches have the same initial two steps, the differences between them are due to philosophical matters associated with theoretical aspects of the processes of reasoning from the data to the conclusions. The steps involved in these processes are broadly similar, but there are important differences as illustrated in Figure 6-5.

The choice of model is normally based on knowledge and experience and under such circumstances will be subjective to a degree.

The classical analysis approach presupposes complete knowledge of the phenomenon that creates the events that generate the data. This is reasonable when it comes to situations where phenomenological models are reasonably well established through the normal process of scientific acceptance (by objective feedback, e.g. the way Newton's laws have become widely established in everyday life even though they are not accurate in every physical application) and also when extending the bounds of these phenomenological models utilising new data.

In terms of the Bayesian approach, the expectation generated by existing knowledge and experience, including knowledge of relevant phenomenological models and independent relevant data, is first stated and then modified by introducing the new data.

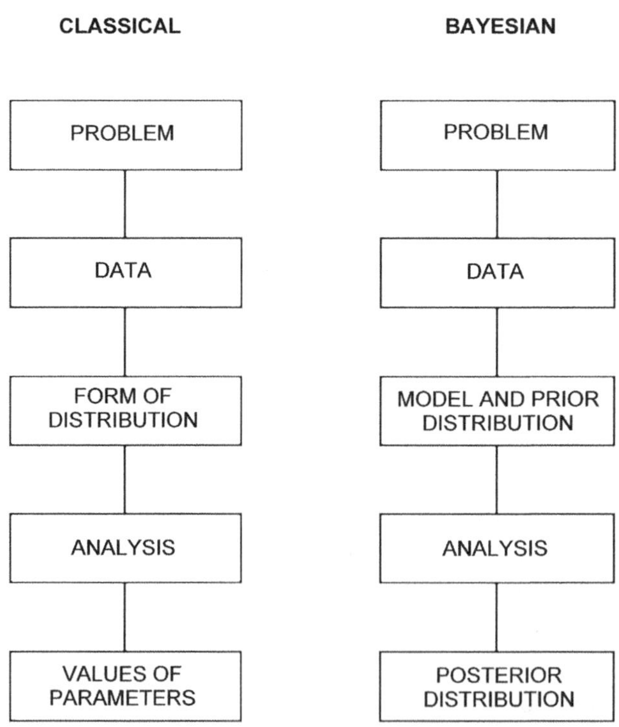

Figure 6-5. Steps in data analysis problems (adapted from NIST, 1998)

Preliminary data analysis by inspection which is not represented in flowchart form provides a means of forming opinions about data prior to formal analysis through statistical or Bayesian techniques. In many engineering applications, preliminary data analysis by inspection provides a stand-alone basis for data interpretation whereby the experience and judgement of the engineer interpreting the data is brought to bear in an entirely subjective way to reach conclusions as to the properties of the data. Material strength data or the results of geotechnical investigations are commonly only employed in terms of preliminary analysis by inspection whereby decisions are made by simply looking at "characteristics" of the data, such as the maximum or minimum values. Preliminary data analysis by inspection can be used to make approximate estimates of the properties of the data in cases where there are large amounts of relevant data available.

In reality, all three approaches are employed at the early stages of the process of interpreting data. Similarly, and fundamental to many engineering and scientific endeavours, it is common practice to adopt the inspection approach and plot the raw data to see what it "looks like", independent of the knowledge of the phenomenon that generates the data.

The Bayesian approach imposes probabilistic models, such as normality of the errors in the deterministic models, and Classical approaches impose deterministic models, such as regression models, on the data. On the other hand, the inspection approach requires the analyst to interpret the data directly and estimate admissible models that best fit the observed data.

Common to both approaches is the notion that the probability is a property of the data and the model selected to solve the solution to the problem at hand. In this regard, in Bayesian analysis, the engineering/scientific knowledge that is subjectively and possibly intuitively brought into the process of assigning the prior distribution, is essentially data and, once incorporated, it is treated as data in the remainder of the Bayesian analysis process. Engineering/scientific knowledge, through the exercise of judgement, is also brought into the process in the selection of the model that the subjective prior and the data are imposed on.

In terms of the Classical approach, the engineering/scientific knowledge brought into the process is in the selection, through judgement, of the statistical properties of phenomenological model used in the solution of the problem, with the remainder of the process being entirely deterministic. In the case of the inspection approach, the model is an artefact of the data and the analysts' insights into the data. These insights into the data come from the analysts' past experience pertinent to the phenomenon under consideration that permits pattern recognition and comparative analysis, which includes utilising engineering/scientific knowledge in the application of graphical techniques for revealing both the data and what is absent from the data.

Limitations of the methods

Although well established methods of engineering and scientific analysis, all three methods have limitations. The result is that in some instances, especially many relevant to dam safety, there is no way of knowing if the application of any or all of the techniques results in a good approximation of the "desired probabilities of parameters and states". Results arrived at through application of each of the three methods of data analysis are heavily dependent on the quality and the quantity of data used in the analysis.

A second limitation is that in practice, for some aspects of dam safety risk analysis, it is necessary to extrapolate beyond the bounds of all data and experience. There is no way of knowing if the properties of the known data are the same as the properties of the data that includes the extreme values. In view of these limitations, the seriousness of which often goes unappreciated, probability assignment is best carried out for phenomena that are amenable to empirical control.

6.2 Probability construct

Accepting the premises that a probability is a "mathematical construct of the degree of reasoned belief" that one can hold with respect to the proposition, there are a number of considerations that must be accommodated when constructing this degree of reasoned belief.

On this basis or probability as reasoned belief perhaps we can consider that the probability construct has three fundamental components:

1. Data analysis, as described above, where the term data is used in the broadest sense.
2. Phenomenological models.
3. Cognition (the subjective processes of the experts).

In principle, probabilities can be derived directly from data if sufficient data is available or from phenomenological models if there is sufficient knowledge of the process that generates the parameter or state of interest. Cognition (mental processes including imagination and intuition) alone does not provide a reliable means of estimating probable values or probabilities/states or probabilities of events without data and/or phenomenological models because cognition has to be applied to something.

The extent to which the probability construct comprises data analysis phenomenological modelling and the cognitive capability of the expert will vary from one situation to the next. However, one would normally expect the relative proportions to be specified.

The relative proportions of the three elements of the probability construct can be represented as illustrated in Figure 6-6.

Considering only data and phenomenological models, the domain of scientific probability can be illustrated as shown in Figure 6-7.

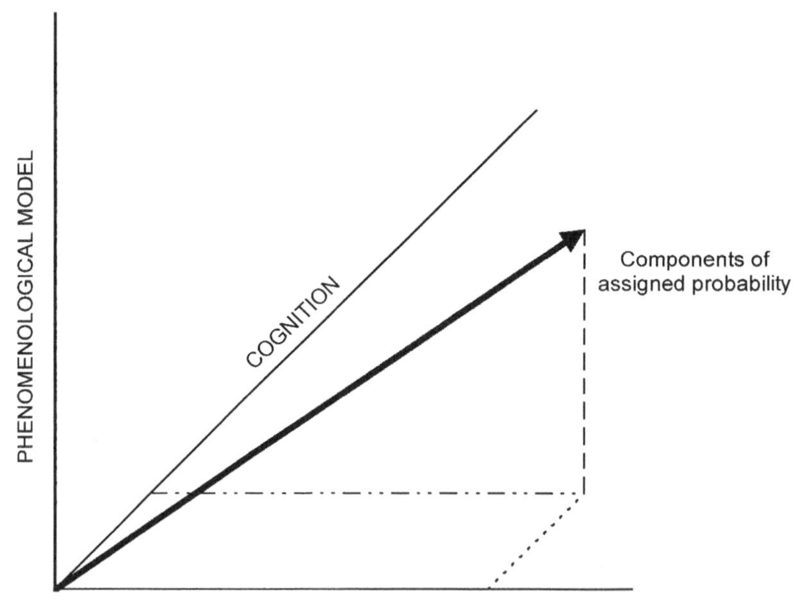

Figure 6-6. Components of assigned probability

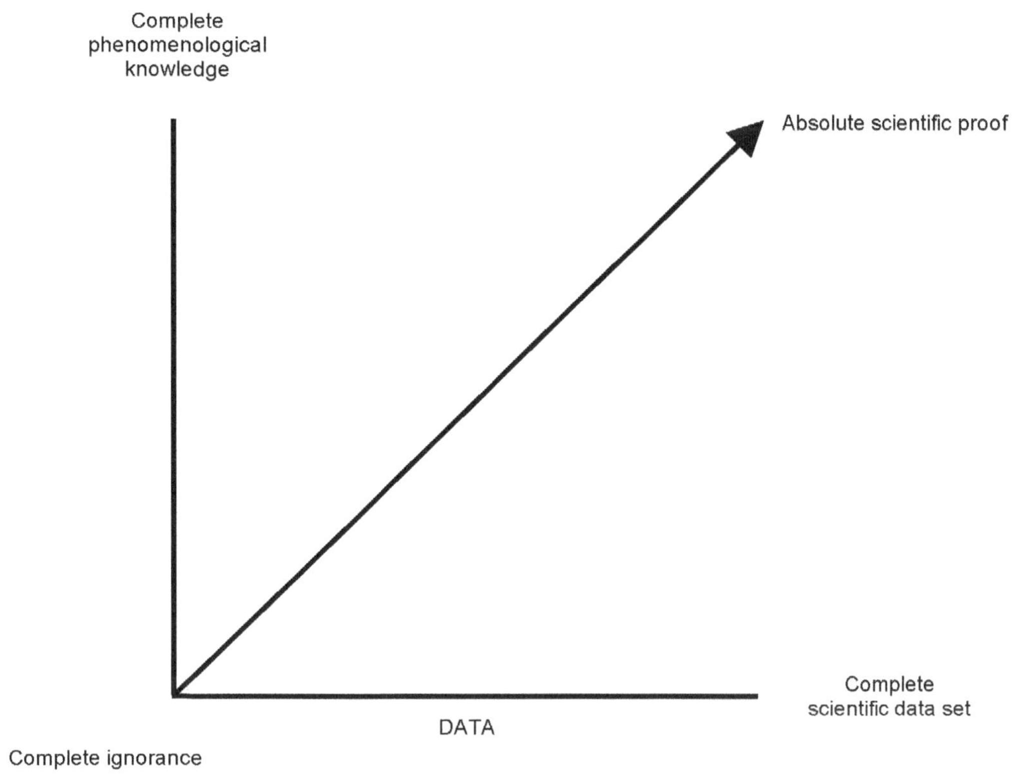

Figure 6-7. Domain of "scientific" probability

As mentioned previously, the scientific method is inevitably fickle, however, it is very useful and the closer one can get to complete data and complete phenomenological understanding, the closer one gets to overcoming the fickleness that results from having to extrapolate subjectively from limited data and understanding.

Regardless of how one goes about it, the process of constructing the probability requires a subjective step, which is an entirely cognitive process. The domain of cognition and data illustrated in Figure 6-8 is similar to that of Figure 6-7.

Combining Figures 6-7 and 6-8 yields the three-dimensional representation of the probability construct shown in Figure 6-9.

Considerations in constructing the probability as represented in these figures include:

1. The size of the subjective component can vary from almost 100% to almost zero with the credibility of the result varying inversely (although not necessarily linearly).
2. The size of the subjective component will be dependent on:
   - the amount of data available and the extent to which the available data is complete with respect to the complete data set (assuming that the size of the complete population is known – and ignoring the problem of it usually being unknowable);
   - the quality of data available;
   - the mathematical correctness of the process of updating the prior probability distribution when additional information becomes available;
   - the extent to which phenomenological models co-exist with data and the nature of these models (science models or predictive models).

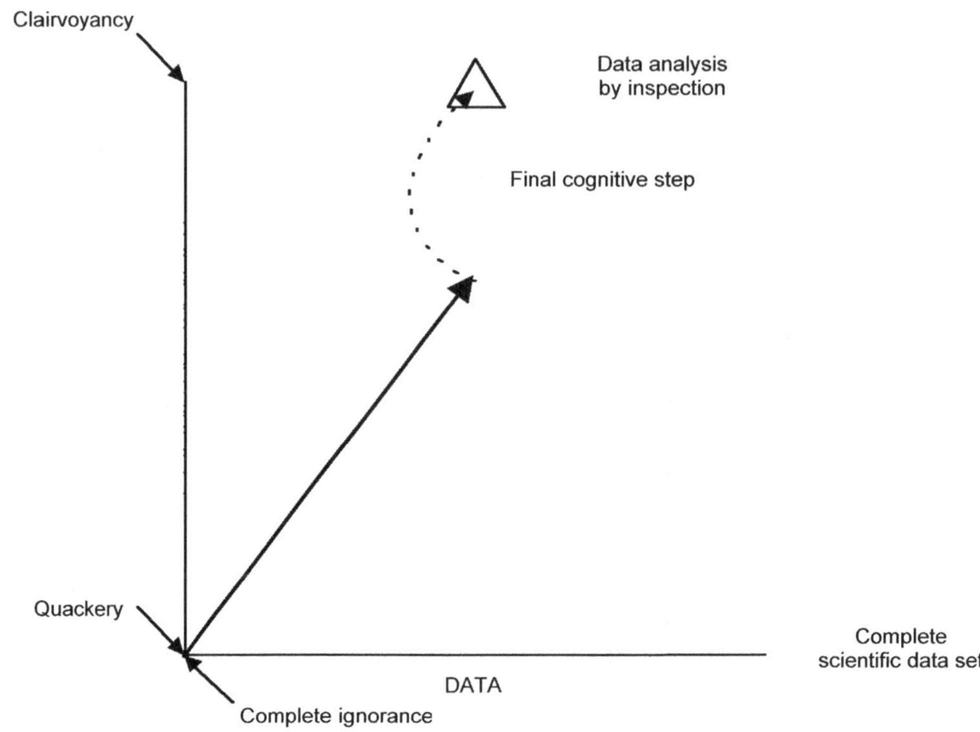

Figure 6-8. Domain of data and cognition

3. The validity of the subjective component will depend on:
   - the capability of the "expert";
   - the form (in the horse racing sense) of the expert at the time of the analysis;
   - the manner in which the expert interprets and uses the data and the phenomenological models in the final step of the process of finalising the form of the probability.

In terms of considerations 2 and 3, the subject matter knowledge of the expert is part of the data and modelling endeavour, and should be brought into the process as part of the data set as created in step 2.

4. The cognitive and subjective element, which is the final and deciding feature of the probability construct, is captured in step 3, the remainder can be handled through appropriate mathematical and scientific procedures.

In reality, the process can not be decomposed quite as simply as suggested here as, referring to Figure 6-9, the analysts bring subjective opinions and preferences into the processes of phenomenological modelling and data collection and interpretation. This difficulty can be handled in one of two ways; as in Figure 6-9.

- The domain of scientific proof is restricted to objective science whereby only that which is universally agreed is represented, and all of the subjectivity is captured in the cognition axis.
- The extent to which subjectivity exists in the data analysis and phenomenological modelling is represented in the domain of scientific proof. Under such circumstances it becomes necessary to identify the point at which the subjectivity in the scientific analysis ends and the subjectivity in finalising the statement of

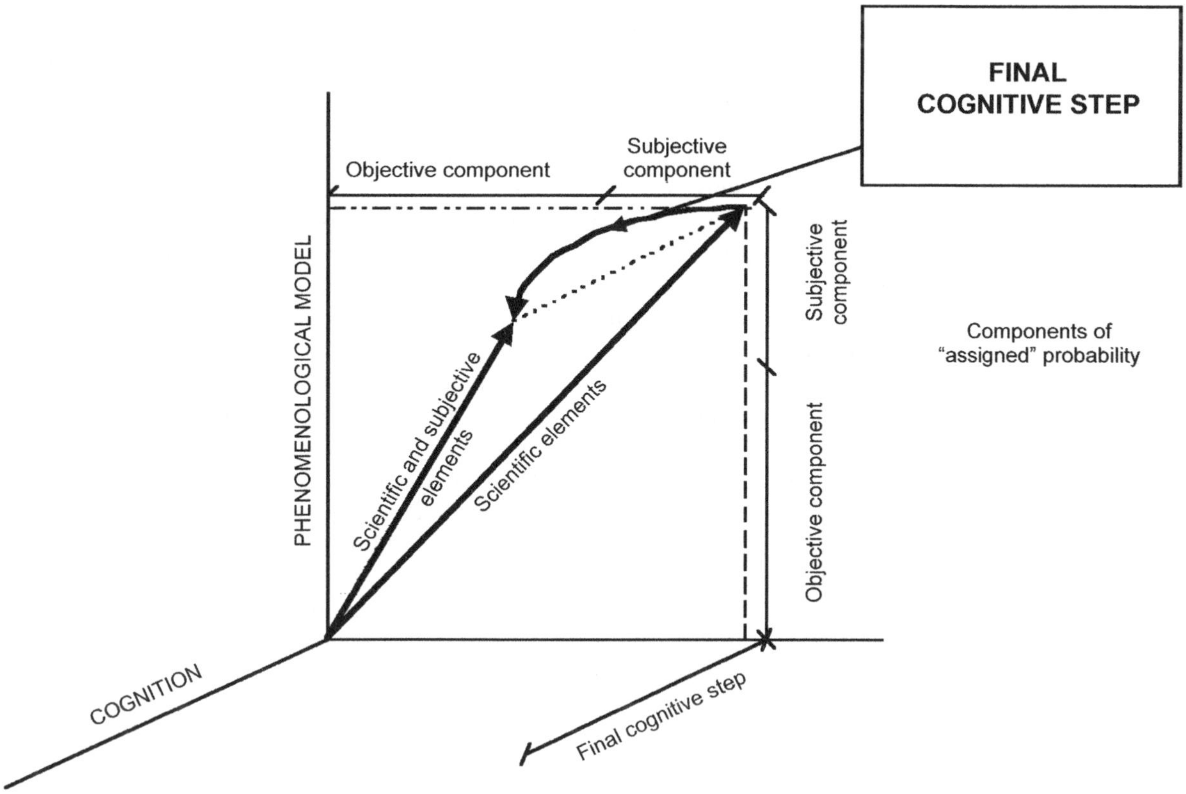

Figure 6-9. Relationships between elements of the probability construct

probability on the basis of the data and the phenomenological model begins (Figure 6-9 and Figure 6-6, with axes rearranged for clarity).

Because probability is a mathematical construct, it is necessary to follow the established mathematical and scientific rules in all stages of the process. It is for this reason that the practices that have emerged in recent years' elicitation of experts' belief probabilities is done in the presence of a normative analyst (experienced in probability issues) and a substantive analyst (experienced in the experts' field of interest).

6.3 Data

Data, expert knowledge and phenomenological models provide the basis for assigning probabilities. The matter of probability assignment can be approached from two perspectives, a "top-down" approach based fundamentally on available data and its interpretation, or a "bottom-up" approach of generating knowledge and then data based on fundamental physics. Judgement and opinion are interwoven in these two perspectives, whether applied individually or in combination, and arguably information obtained through one method is used to inform the judgements in the other and vice versa.

As will be explained below, neither perspective can provide a completely satisfactory approach to probability assignment, both in terms of fully characterising the "probability" and/or the "uncertainty in the probability". As explained in Chapter 7, because probability pertains to uncertainty it is never possible to know if the assigned "probability" and "uncertainty in the probability" are approximately correct in a measurable sense.

The ideal way to address these issues is to collect as much data as can be physically collected, and at the same time to find out as much as can be found out about the phenomena that create the observations.

As noted previously, probability assignment involves transforming data (ranging from carefully collected scientific measurements through various forms of objective data to the relevant internal knowledge bases of experts) into a mathematically valid probability construct. In this sense, one of the attributes of experts is that they serve the role of a database where their internal data has been objectively acquired through experience.

The problem of data is that, under the most ideal conditions in the physical sciences, it is rarely if ever possible to have the entire data set available to use in the analysis process. Thus, even under the most ideal conditions, one never knows the exact mathematical form of the probability distribution. To further complicate matters, one rarely if ever knows how large the complete data set would actually be with the result that it is never possible to know what percentage of the total data set is being brought to bear in the solution of the problem. There are countless examples in civil engineering and dam safety practice where just one extra piece of data (e.g. one more or a slightly deeper drill hole) would have led to a completely different conclusion. Thus, there is no way to demonstrate that sufficient data has been collected to form a mathematically valid approximation (or known degree of approximation) of the complete data set. Under such circumstances, the best defence against a false sense of security is the sceptical approach of the inquiring mind.

Aside from the problem of the quantity of data, the quality of the data is extremely important and it is generally advisable to qualify the data in some way prior to accepting it for use in the analysis process. Without such a process of qualifying data, additional uncertainties and even errors will be introduced into the analysis. There are established procedures to qualify data to protect the integrity of the analysis process. The procedures used to qualify the data and information used in the analysis will normally be specified.

Setting these problems of data aside, there is no doubt that data (of some form) is essential in the assignment of probabilities. In this regard, while in the absence of data, the "uninformed prior" might provide a means of getting started with a Bayesian analysis when there is no means of advancing the Classical approach, it does not provide a basis for ending the analysis.

There is also a hierarchy of data types that pertain to their reliability and appropriateness for use in risk analysis. Objective scientific data obtained under controlled conditions, with all uncertainties fully specified, is at the top of the hierarchy. Such data is typically stored and maintained by recognised institutions where maintaining the integrity of the data is a central function.

Data gathered in terms of established scientific methods, but which do not enjoy "official status" resides at the next level of the hierarchy. Further down the hierarchy are objective data sources that have not been qualified as scientific but which could be qualified if the necessary effort was expended to cleaning up the data.

Verifiable judgements and opinions of "calibrated" domain experts are at the next lower level of the hierarchy. When there are experts who have extensive and validated predictive experience of the phenomenon under consideration, it may be possible to use the experts' knowledge as a supplement for data when it is simply not possible to obtain objective data. Whether one uses the judgements and opinions of "calibrated" experts in lieu of collecting data will generally be determined by the nature and objectives of the risk analysis, and requirements for transparency, and reproducibility of the analysis. The uses of domain experts in this way is common in civil engineering practice, most usually at the early stages of the planning process but often throughout the design and construction process.

Unverifiable opinions of domain specialists are at the next lower level of the data hierarchy. The domain specialist might have a great deal of knowledge about the phenomena under consideration, but no direct knowledge or experience of the precise conditions pertinent to the problem at hand. The domain specialist might employ a process of reasoning from first principles to infer the form of the elusive data. An example of such a situation would be the requirement to elicit the probability distribution of the post liquefaction shear strength of a gravely material in the body of a dam. Under such circumstances, the line of reasoning from the premises (material grading, in situ shear strength, state of stress, effects of shaking, etc.) to the conclusion (the expected post liquefaction probability distribution of shear strength) can be drawn out and made available for inspection if required. The data, as such, is conjecture and would be best identified as such in the analysis.

The following levels of the data hierarchy arguably would not constitute "expert elicitation" processes, as there is no relevant scientific knowledge and therefore no experts in the science under consideration.

The second lowest level in the data hierarchy comprises opinions concerning matters where the phenomena concerned are beyond experience and domain specialisation, by individuals capable of extending the frontiers of domain knowledge. Data obtained in such a way is clearly a conjecture, and may be better represented as a guess.

Opinions of individuals who are not domain specialists, and have no background in constructing probability distributions, concerning the probability distribution of the parameter or state reside at the lowest level of the hierarchy.

6.4 Phenomenological models

Phenomenological models provide a means of generating synthetic data where no data exists. They also provide a means of modelling the process that is the subject matter of the risk analysis.

Phenomenological models are physics models of the processes of interest about which observations can be made. Such models are necessarily simplifications of reality and, as such, can never be complete or exact. However, phenomenological models may be desirable in the assignment of probabilities to complement the available data. They might be used, to determine the probability directly without resorting to data analysis, to verify the adequacy of probabilities derived from data analysis if there is sufficient data available, or to complement the data analysis process if there is insufficient data to define the probability directly.

As is the case for data analysis, phenomenological models necessarily embody the subjective beliefs of the individual that creates the model concerning what should be included in the model and what should be excluded.

There are many aspects to the appraisal of models used in risk assessment and only an outline can be given here (McQuaid, 2002). It is useful to distinguish between two types of model since different quality considerations apply. The two types may be called "science" models and "predictive" models. The predictive models may, and usually do, incorporate science models but go beyond them in having to deal with issues that cannot be subjected to the procedures of science.

A science model is a means of representing the state of knowledge or "science" concerning a phenomenon. It provides an interpretation, in mathematical terms, of what is currently known or accepted as physical descriptions of the phenomenon. The quality of a science model, or the "soundness" of the science, is assured by the normal processes of science. These operate on the way knowledge is acquired by experiment or practical observation, the robustness of the assumptions used in framing the mathematical description and the form of validation of the final result.

The domain of science models is the relatively easier one in which to perform quality assurance, with much reliance on the rigour of peer review in the open literature. Although often slow, this is a powerful means to root out lack of care, ignorance of relevant information or neglect of evidence running counter to the hypothesised model, and promotion of commercial interest by selective disclosure and biased examination of evidence as distinct from advancement of knowledge.

Predictive models represent a conjecture of what might happen under stated assumptions, for example if an earthquake were to occur at a particular location with associated population distribution, location of vulnerable structures, etc. Such a predictive model would incorporate science sub-models describing the progression to the defined failure state. The hypothesised progression will be identified by some systematic technique such as fault tree analysis.

A predictive model is a tool of risk assessment and incorporates assumptions and judgements about the effects of particular practical circumstances. Such assumptions and judgements will not be testable by the methods of science (otherwise the model would be a science model by the above definition). Where judgement has to be exercised, there is a need for conformity to some basic principles as outlined below.

A predictive model is not necessarily a mathematical representation of the conjectures underlying the model. The presentation of a predictive model needs to be clear about:

- What features of the practical situation are chosen to be represented and why?
- What features are judged not to need representation and why?
- What features cannot be represented and why?

Transparency is key to quality assessment to enable independent judgements to be made. Validation of models is the process of establishing the extent to which the application objectives of a model are attained, defined in terms of both the average or expected behaviour and the variability about the expected results. Validation of science models will obviously be provided by comparisons with relevant data. Validation of predictive models can rarely be completed in the same way, since experimental data will not be available or even not practically possible to acquire. Sometimes it is possible to obtain validatory evidence from observations following accidental events. Where validation of one kind or another is limited, it is essential to test the sensitivity of the model predictions to plausible variations in the input assumptions.

6.5 Experts and judgement

According to McQuaid (2002).

> The main predisposing factors for ensuring model quality are complete openness and diversity of expert inputs. These provide security against the operation, inadvertent or otherwise, of bias which can arise from effects such as "group think" and slavish conformity to scientific and risk assessment fashions.

Much has been written about the different types of uncertainty but the point is that the handling of uncertainty in all its forms requires expert assessment. The process of expert assessment covers a spectrum. At one end, the assessment can be made on the basis of formal and transparent analysis using models that are based on established physical laws and mathematical relationships. The models are supported by relevant and validated data. Moving across the spectrum, the role of formal analysis diminishes to the point where modelling may provide only a partial answer. There may be various ways of representing the features of the problem, each consistent with the available, usually sparse, data. Expert judgement figures increasingly in the conduct of the assessment. The quality of the judgement depends on depth of relevant experience and availability of evidence from similar circumstances accumulated over time or across different disciplines. Eventually the point is reached where the uncertainties and lack of understanding are such that the assessment must be based largely or wholly on expert opinion. The quality of the assessment then depends on such characteristics as credibility, standing and independence of the assessors.

The spectrum as described is in decreasing order of tractability, by which is meant the ease with which the reasoning behind the conclusions can be drawn out and explained. The process of drawing out or eliciting judgement has received much attention, particularly in nuclear safety studies. Formal methods of elicitation have been developed, covering the manner of numerically representing the judgements of individuals and of aggregating them into a combined view using standard statistical techniques. But much expert or professional judgement is elicited informally and the weighing-up of the consensus view can be largely subjective. The conclusion may simply emerge or appear to emerge.

The compression of what is often a complex mixture of analysis, judgement and opinion into a single numerical expression of risk can lead to an impression of certainty based on science as it is commonly understood.

This can raise a number of questions for the decision-makers and the affected public that point to a need for:

- procedural guidelines for decision-makers so that they can better organise the provision of the advice taking account of the nature of the decision and the extent to which they may have to defend it – a consideration particularly relevant to Government;
- recognised protocols for eliciting judgements from experts and for combining those judgements into a consensus view properly defined;
- the judgements to be made tractable in the sense defined above and; most importantly;
- the advice to be seen as a living entity, with the advice to be revisited in a preordained way as new information emerges which may challenge the basis of the original advice.

Structured expert judgment

A great deal has been written on the subject of structured expert judgement over many years since it was first introduced in the 1950s. The discipline is now well established outside the dams community and the reader is referred to the authoritative literature (Cooke, 1991; Cooke and Goossens, 1999; Hogarth, 1987; Morgan and Henrion, 1990) on the subject. Expert judgements are the expressions of informed opinions, based on knowledge and experience, that experts make in responding to technical problems. While expert judgement might often be incorporated implicitly, structured expert judgement involves obtaining judgements and opinions of experts through the formal processes of: (1) selecting experts; (2) definition of technical problems; and (3) choice of analysis methods by which the expert judgements are elicited and analysed.

Expert judgements can be elicited quantitatively or qualitatively. When expressed quantitatively they can have several forms: probabilities; ratings; odds; weighting factors; and, possibly most importantly, probability distributions of the physical quantities of interest. Qualitative expression will involve a textual description of the experts' assumptions in reaching an estimate, and natural language statements of probabilities of events such as "likely", or statements as to the expected performance such as "generally poor".

Expert judgements are necessarily conditioned on the data and information available to them at the time of the elicitation process. Therefore, their judgements should change as additional information is obtained. The challenge for the expert is to demonstrate that his/her revised judgement is consistent with all of the information available now as well as consistent with the previous judgement.

For most applications in dam safety risk analysis, the elicitation of expert opinions is concerned with uncertain quantities (parameters or states) that take values in some continuous range. As such the uncertainty is described by a belief probability distribution for uncertain quantities with values in a continuous range. When uncertainty is quantified in an uncertainty analysis it is always conditional on something. Thus, when expert judgments are cast in the form of distributions of uncertain quantities, the issues of conditionality and dependence are important.

It is essential to make clear the background information conditional on which the uncertainty is to be assessed. This information will generally be contained in the documents that are presented to familiarise the experts with the case under examination. In some practices this might be referred to as the "case structure" (Cooke and Goossens, 1999). The "case structure" document describes which areas of interest are to be assessed by the experts in greater detail. The questionnaire on the variables to be elicited is also derived from the "case structure" document. For that reason the "case structure" document should also describe which conditions the questions are based on, which issues should be taken into account in the uncertainty assessments, and which issues are omitted and why.

Protocols for expert elicitation

Protocols for expert elicitation are now well established with all accepted protocols having similar general structure. The following is a thumbnail sketch of the concepts and the reader is referred to the expert literature (Budnitz *et al.*, 1998; Cooke and Goossens, 1999; Morgan and Henrion, 1990) on the subject.

There are three principal elements to the process:

1. Preparation.
2. Elicitation.
3. Post-elicitation interpretation.

The preparation phase involves all of the necessary preparation by the analyst(s) to:

- frame the risk problem;
- identify the uncertainties of interest;
- establish the risk model and determine how best to focus the questions for the experts;
- identify and select the experts;
- structure the elicitation process;
- prepare the experts and refine the process (dry run).

In the formal expert elicitation session, the experts' individual judgements are discussed in the presence of a normative analyst (experienced in probability issues) and a substantive analyst (experienced in the experts' field of interest). The roles of the normative and substantive analysts can be combined if suitably qualified individuals are available.

The post elicitation process involves:

- Combining the opinions of the experts.
- Dealing with any logical inconsistencies.
- Reviewing the results with the experts.
- Integration of the expert opinions with the risk model developed in the preparation phase.
- Document the entire process.

Performance measures and rational consensus

The goal of applying structured expert judgement techniques is to enhance rational consensus. Necessary conditions for achieving this goal are laid down as methodological principles, as below.

Scrutability/accountability. All data, including experts' names and assessments, and all processing tools are open to peer review and results must be reproducible by competent reviewers.

Empirical control. Quantitative expert assessments are subjected to empirical quality controls.

Neutrality. The method for combining/evaluating expert opinions should encourage experts to state their true opinions, and must not bias results.

Fairness. Experts are not pre-judged, prior to processing the results of their assessments.

Cooke and Goossens (1999) claim that these are *necessary* conditions for rational consensus, but do not claim that they are sufficient as well. Hence, a rational subject could accept these conditions and yet reject a method which implements them. In such a case, however, he/she incurs a burden of proof to formulate additional conditions for rational consensus which the method putatively violates.

Empirical control

The notion of empirical control of expert subjective judgement appears to run counter to the notion of subjectivity. However, the idea of empirical control is reasonable if elicitation of expert opinion is restricted to situations where there is scientific expertise relevant to the problem at hand. This entails that there are theories *and* measurements relevant to the issues at hand, but the quantities of interest themselves cannot be measured in practice.

Cooke and Goosens (1999) provide the following examples. Toxicity of a substance for humans is measurable in principle, but is not measured for obvious reasons. However, there are toxicity measurements for other species, which might be relevant to the question of toxicity in humans. On the other hand, we may be interested in the dispersion of a toxic airborne release at 50 km from the source. Although it is practically impossible to measure the plume spread at 50 km, it is possible to measure this spread at 1 km. If a problem is an expert judgement problem then, necessarily, there will be relevant experiments which can in principle be used to enable empirical control.

Endnote

1. A hypothesis being a proposition made as a basis for reasoning without assumption of its truth.

Part III
Theoretical considerations

7 Essential elements of probability theory

The purpose of this chapter is to provide a conceptual foundation for the use of quantified probabilities in risk analysis for dam safety studies. The chapter begins by considering what it means to be uncertain, and why probabilities can be used to measure uncertainty. It continues on to discuss how the strength of evidence is measured and used to draw statistical inferences. Finally, it turns attention to the use of quantified expert judgement in risk analysis.

7.1 Uncertainty

The term, *uncertainty* means different things to different people and, therefore, a clear statement is needed of what it means to say that a natural process, performance prediction, or other aspect of dam engineering is "uncertain". This section discusses the meaning of uncertainty in relation to the performance of dams.

In the water resources literature, uncertainty is sometimes defined as "lack or absence of certainty", which is not very illuminating (Morgan and Henrion, 1990; National Research Council, 2000; US Army Corps of Engineers, 1992). Maass *et al.*, (1962) say uncertainty means that the "consequences of a decision cannot be foretold with confidence". In practice, one could argue that most people treat uncertainty as a primitive term, that is a term whose meaning is accepted by its usage but left formally undefined.

The US Water Resource Council's (1983) *Principles and Guidelines* for project evaluation attempts a more formal definition for uncertainty, saying that uncertainty describes situations wherein lack of certainty is not describable by numerical probabilities. This definition follows the distinction popularised by Knight (1921). In this view, risk refers to situations wherein one can assign probabilities to chance, whereas uncertainty refers to situations wherein one cannot. Keynes (1937) expressed the distinction thus:

> By uncertain [. . .] I do not mean merely to distinguish what is known for certain from what is only probable. The game of roulette is not subject, in this sense, to uncertainty [. . .]. The sense in which I am using the term is that in which the prospect of a European war is uncertain, or the price of copper and the rate of interest twenty years hence [. . .]. About these matters there is no scientific basis on which to form any calculable probability whatever. We simply do not know.

This narrow definition is no longer widely used in science and engineering, although it is still common in the popular literature on investing. A derivative of this view has re-emerged in the mechanical engineering literature, wherein risk is defined according to Knight, but uncertainty is divided into uncertainty and ambiguity (Smith, 2000). Smith defines risk as referring to situations wherein chance can be described by known probability distributions with specified parameters; uncertainty as situations wherein chance can be described by probability distribution, but the form or parameters of those distributions is unknown; and ambiguity as situations wherein "the functional form is completely unknown, and often [. . .] the relevant input and output variables are unknown".

Today, most decision theorists reject Knight's distinction. Current thinking is that, while a decision-maker may choose not to assign probabilities to vague uncertainties, he/she does do so implicitly in making practical decisions. Along this line, Keynes (1937) goes on to say:

> [under uncertainty] there is no scientific basis on which to form any calculable probability whatever. We simply do not know. Nevertheless, the necessity for action and for decision compels us as practical men to do our best to overlook this awkward fact and to behave exactly as we should if we had behind us a good Benthamite calculation of a series of prospective advantages and disadvantages, each multiplied by its appropriate probability waiting to be summed.

Any real decision involves quantified probability, whether explicitly or implicitly arrived at. Yet, despite the decision theoretic view of uncertainty as inevitably quantitative, there are issues about which it is nearly impossible to assess credible probabilities – except, of course, by back-calculating them from the decisions taken. An NRC panel (1996) includes among such issues, the economic impact of global climate change decades in the future, and long-term geological effects on nuclear waste repositories.

Necessity and chance

We begin with the notions of necessity and chance, notions that have occupied philosophers since antiquity. Is there a cause for everything? That is, do events in the world follow fundamental laws of nature that are, in principle, knowable? Or is there an irreducible level of randomness in the world? When we speak of uncertainties in dam safety, are we describing a lack of knowledge about the laws and properties of nature, or are we describing things that are, at some level, accidental and unpredictable?

Necessity

In the modern era, the notion of necessity is usually associated with Laplace. Laplace began the *Philosophical Essay on Probabilities* (1814) by saying:

> All events, even those which on account of their insignificance do not seem to follow the great laws of nature, are a result of it just as necessarily as the revolutions of the sun.

And the implications of this notion were profound:

> Given for one instant an intelligence which could comprehend all the forces by which nature is animated and the respective situation of the beings who compose it – an intelligence sufficiently vast to submit these data to analysis – it would embrace in the same formula the movements of the greatest bodies of the universe and those of the lightest atom; for it, nothing would be uncertain and the future, as the past, would be present to its eyes.

As the 19th century turned, great philosophers and scientists thought of the world as entirely determined by laws, a view that had been building since the great advances of Newton a century before (Porter, 1986). This became known as the doctrine of necessity.[1]

The problems raised by the doctrine of necessity were also profound, however. If the world is completely determined by natural laws, and therefore explainable and predictable, not only into the future but into the past (presuming these laws are reversible), what room is left for human agency, and is history itself predetermined? The mid-19th century saw a flood of popular intellectual discussion of this question following the publication of Buckle's *History of Civilization in England* (1858), which argued that climate, geography, and other environmental factors, rather than human will, affect the course of history.

Chance

James Clerk Maxwell was one of many contemporaries of Buckle who were deeply affected by the notion of necessity and its implications for history and free will (Tolstoy, 1981, p. 152). Maxwell's own work on statistical mechanics would profoundly affect the debate, and would lead in the latter decades of the 19th century to an emerging view of the world as driven by chance, not necessity. A view now held by many high-energy physicists.

The modern argument against necessity is usually attributed to Peirce (1998), who argued that the variety and increasing complexity of nature cannot be explained solely on the basis of "the rule of mechanical necessity", since mechanical law describes one determinate result, while "variety can spring only from spontaneity". The doctrine of necessity, according to Peirce, is "the common belief that every single fact in the universe is precisely determined by law", but this is incompatible with the observation that throughout nature "everywhere the main fact is growth and increasing complexity. [. . .] the essential of growth is that it takes place in one determinate direction, which is not reversed. Boys grow into men, but not men into boys." Thus, he concludes, the laws of nature must be probabilistic rather than deterministic. The world is not driven by necessity, but that "the apparently universal laws that are the glory of the natural sciences are a by-product of the workings of chance".[2]

In recent years, the notion of chaos has arisen, which is related to chance. The origin of chaos theory lies in the observation that, for complex, non-linear systems, small changes in initial conditions may lead to large changes in outcomes. Since changes in initial conditions may be too small to be modelled or even measured, outcomes become unpredictable, and for practical purposes, random (Gleick, 1988). The often cited example is meteorological modelling, wherein small disturbances in model input on one side of the world can generate large differences in model output for the other side of the world.

Probabilism and the presumption of randomness

Laplace's view, that the world is unpredictable only because our knowledge is incomplete, has come to be known as probabilism. It may be, of course, that our knowledge never can be complete. Heisenberg's insight was that there are fundamental limits to what can be measured and, therefore, to what can be known. Most engineering risk analysis is based on probabilism: the world is uncertain because knowledge is limited, not because nature is random. This is the view adopted throughout the present volume.

Despite this Laplacian view, risk analysis often models some natural processes as if they were random. For example, floods and earthquakes are treated as stochastic processes with probabilistic descriptions. How do we reconcile probabilism with this practical assumption that at least some of the world is random? The answer is that, it is simply convenient to model the world as if some uncertainties are due to inherent randomness, even if they are not. The predictions obtained by making this assumption may be as good or better than those obtained by more correctly modelling uncertainties as due to limited knowledge, since the latter may be much harder to do.

For example, flood flows (discharges) are often described by exceedance probabilities, the chance that certain flows are equalled or exceeded within some fixed interval of time, such as a year. The individual occurrences of flood flows are modelled as a random process. This assumption is made, not because rainfall and runoff are actually random in time and space, but because the hydrologist may know too little about the weather (and the basin) to predict flood flows deterministically. Using a random process model may yield a better prediction than would result from trying to quantify the uncertainties in our knowledge of the weather. Nonetheless, the assumption of randomness in flood flows is only a convenience of modelling. From the view of probabilism, the randomness is not inherent in nature. This has led Matheron (1989, p. 143) to say that "probability is in the model", not in reality.

Vocabulary of uncertainty

Having introduced the contrasting notions of necessity and chance, we return to the concept of uncertainty, and draw a distinction between randomness and uncertainty as those terms commonly appear in risk analysis. We suggest a taxonomy for the uncertainties accounted for in dam safety studies.

Randomness

Since ancient time, the notion of randomness has concerned natural processes that are inherently unpredictable.

> **Random** (adjective). Date: 1565. 1. a: lacking a definite plan, purpose, or pattern. b: made, done, or chosen at random. 2. a: relating to, having, or being elements or events with definite probability of occurrence; b: being or relating to a set or to an element of a set each of whose elements has equal probability of occurrence. (Merriam-Webster, 2000)

The role of dice, patterns of the weather, occurrence of an earthquake, and other such unpredictable occurrences have been called aleatory by Hacking (1975) and others (e.g. Cooke, 1991; Daston, 1988; Gigerenzer, 1989), after the Latin aleator, meaning "gambler" or "die caster". This term is now widely used in risk analysis, especially in applications dealing with seismic hazard, nuclear safety, and severe storms. The term probability, when applied to such random events, is taken to mean the frequency of occurrence in a long or infinite series of similar trials. This frequency is a property of nature, independent of anyone's knowledge of it. It is innate, and has a "true" value. Two observers, given the same evidence, and enough of it, should eventually converge to the same numerical value for this frequency.

To describe something as random is to presume that its occurrence can be described only probabilistically. By treating flood frequencies, for example, as random, we describe their occurrence by exceedance probabilities, and presume these occurrences to be generated by a celestial pair of dice. The annual peak discharge for a given year cannot be predicted. All that can be said is that in a long series of years some fraction will experience peak discharges larger than a chosen value.

Uncertainty

At least since the Enlightenment, the notion of uncertainty has concerned what we know. The truth of a proposition, the guilt of an accused, whether or not war will break out. These are not random, but unknown.

> **Uncertain** (adjective). Date: 14th century. 1: Indefinite, indeterminate. 2: not certain to occur: Problematical. 3: not reliable: Untrustworthy. 4 a: not known beyond doubt: Dubious; b: not having certain knowledge: Doubtful; c: not clearly identified or defined; 5: not constant: Variable, Fitful. (Merriam-Webster, 2000)

Such unknown things have been called epistemic, after the Greek, $\varepsilon\pi\iota\sigma\tau\eta\mu\eta$, meaning "knowledge". This term, too, is now widely used in risk analysis, to distinguish imperfect knowledge from randomness (Table 7-1).

The term probability, when applied to imperfect knowledge, is usually taken to mean the degree of belief in the occurrence of an event or the truth of a proposition. In this sense, probability is a property of the individual. We may or may not know what the

Table 7-1. Terms used in the literature to describe the duality of meaning for "uncertainty"

Terms pertaining to uncertainty due to naturally variable phenomena in time or space: "uncertainties of nature"	Terms pertaining to uncertainty due to lack of knowledge or understanding: "uncertainties of the mind"
Natural variability	Knowledge uncertainty
Aleatory uncertainty	Epistemic uncertainty
Random or stochastic variation	Functional uncertainty
Objective uncertainty	Subjective uncertainty
External uncertainty	Internal uncertainty
Statistical probability	Inductive probability
Chance [Fr]	Probabilité [Fr]

value of the probability is, but the probability in question can be learned by self-interrogation. There is, by definition, no "true" value of this probability. Probability is a mental state and, therefore, unique to the individual. Two observers given the same evidence may arrive at different probabilities and both be right.

The word uncertainty originates at least as early as the 14th century, having been used by Chaucer, and usually has synonyms such as, doubt, dubiety, scepticism, suspicion, or mistrust (Merriam-Webster, 2000). From its dictionary definition, uncertainty means a lack of sureness or a lack of confidence about someone or something, ranging from falling just short of complete sureness or confidence, to an almost complete lack of conviction about an outcome or result.

The synonyms amplify subtly differing facets of the notion of uncertainty. Doubt suggests an inability to make a decision. Dubiety suggests a wavering between conclusions. Scepticism suggests unwillingness to believe without conclusive evidence. Suspicion suggests a lack of faith in the truth, reality, fairness, or reliability of something. Mistrust suggests doubt based upon suspicion.

Three facets seem to appear: uncertainty with respect to the world means that an outcome or result is unknown or not established and therefore in question. Uncertainty with respect to a belief means that a conclusion is not proven or is supported by questionable information. Uncertainty with respect to a course of action means that a plan is not determined or is undecided. The term uncertainty has a variety of shades. Each of these expresses an aspect of uncertainty that comes into play somewhere in risk analyses.

A taxonomy of uncertainty

In modern practice, risk analysis usually incorporates uncertainties of both the aleatory and epistemic variety. That is, the term uncertainty is used as an over-arching term that includes randomness. The National Research Council (1996) describes different types of uncertainty, using the following terminology (Figure 7-1):

Aleatory uncertainty is attributed to inherent randomness, natural variation, or chance outcomes in the physical world; in principle, this uncertainty is irreducible. These uncertainties may include things such as stream flows, assumed to be random processes in time, and geotechnical properties of levees, assumed to be random processes in space. Aleatory uncertainty is sometimes called, random variability, stochastic variability, objective uncertainty, or external uncertainty. An NRC panel (2000) called this, natural variability.

Natural variability associated with the "inherent" randomness of natural processes, manifesting as variability over time for phenomena that take place at a single location

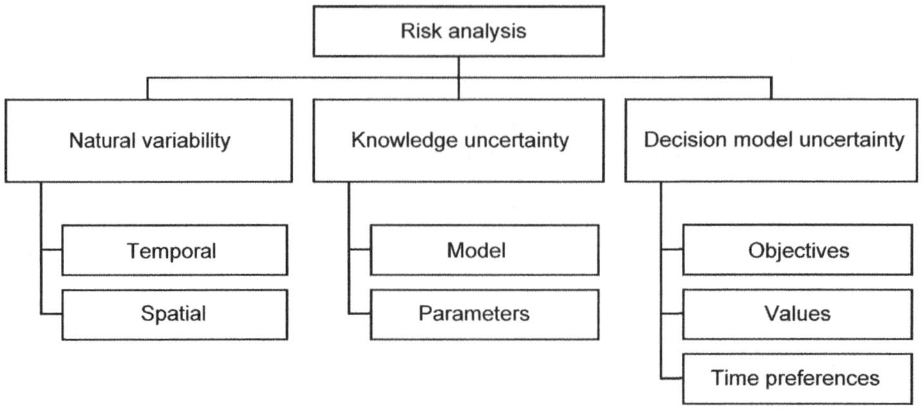

Figure 7-1. Categories of uncertainty entering risk analysis

(temporal variability), or as variability over space for phenomena that take place at different locations but at a single time (spatial variability), or as variability over both time and space. Such natural variability is approximated using mathematical simplifications, or models. These models may or may not provide a good fit to natural phenomena. In the best of cases, they are close but only approximate fits.

Epistemic uncertainty is attributed to lack of data, lack of knowledge about events and processes that limits our ability to model the real world. Epistemic uncertainty is sometimes called, subjective or internal uncertainty. Epistemic uncertainties divide into two major sub-categories: model uncertainty and parameter uncertainty. Model uncertainty has to do with the degree to which a chosen mathematical model accurately mimics reality; parameter uncertainty has to do with the precision with which model parameters can be estimated. The NRC panel called this *knowledge uncertainty.*

Model uncertainty reflects the inability of a model to precisely represent a system's true behaviour, or our inability to identify the best model, or a model that may be changing in time in poorly known ways (e.g. flood-frequency curve changing because of changing watershed). Parameter uncertainties, result from our inability to accurately assess parameter values from test or calibration data, from limited numbers of observations, and from the statistical imprecision attendant thereto. Parameter uncertainties may also arise from data uncertainties, including measurement errors, inconsistency of data, transcription errors, and inadequate representativeness.

An example of the interplay of aleatory and epistemic uncertainties in practice is, again, the flood-frequency curve. The flood-frequency curve describes natural variability or aleatory uncertainty of flood flows, while error bands about the curve describe epistemic uncertainty in the parameters of the flood-frequency model. The frequency curve reflects the irresolvable variation of nature, which we assign to chance. The error bands reflect our limited knowledge about the statistically estimated parameters of the frequency curve, which we assign to ignorance. Collecting more data would improve our estimates of the parameters, and thus reduce the error bands about the frequency curve, but no amount of data can reduce the underlying probability distribution represented by the exceedance curve.

Decision-model uncertainty is attributed to our inability to understand decision objectives, or at least, how alternative projects or designs should be evaluated. This is a third type of uncertainty, somewhat different in character from the first two. This includes, for example, uncertainty in discount rates and the appropriate lengths of planning horizons. In the current volume the influence of decision-model uncertainty is for the most part ignored.

Separating aleatory and epistemic uncertainties

As noted above, in any practical risk analysis some uncertainties are modelled as if they were random, even though they actually arise from limited knowledge. Consider soil properties. We approach data scatter in soil property measurements as if it arose from some random process. In fact, much of the scatter is due to spatial variation. It is a unique realisation. It can, in principle, be observed, given enough time, money and testing technology. Nonetheless, it is convenient to model soil variation as if it were random, because then we can use the powerful tools of random process theory to draw conclusions about the soil property variation and its implication for engineering performance.

Dividing the total uncertainty in this way into an aleatory part and an epistemic part is purely a modelling decision. It does not reflect a real division in the world, at least not if one adopts a Laplacian view of the world. This is presumably a simplifying assumption that transfers some of the uncertainty from the limited knowledge account to the randomness account, where it may be easier to handle. Moving the boundary between the two types of uncertainty does not lessen the total, but it does have implications other than just simplifying the calculations.

The way a risk analysis separates aleatory from epistemic uncertainty affects both the conclusions that can be drawn and the meanings of those conclusions. Think of drawing a curve through a set of data. A high-order curve may fit the data closely, but the uncertainty in the parameters of the curve will be large because there are only a fixed number of data and many parameters to be estimated from them. In contrast, a line may not fit all the data as closely as the curve does, but the uncertainty in the slope and intercept of the line will be small, because there are many data from which to estimate only the two parameters. The uncertainty in the parameters of the curve is epistemic. It comes from limited knowledge and can be reduced by collecting more data. The data scatter about the curve is aleatory. It is natural variation about the given curve and cannot be reduced by collecting more data. The decision about what order curve to fit and, thus, about what part of the data scatter is irreducible, has little to do with the natural world, it is simply a modelling decision.

This modelling decision on whether and how to separate aleatory and epistemic uncertainty has important implications in a dam safety risk analysis. This is because aleatory and epistemic uncertainties are fundamentally different notions of uncertainty and, as a result, they behave differently in risk analysis calculations. In the example above, the data to which a curve is fit might be soil strength data in an embankment. In a stability calculation, the aleatory data scatter about a spatial trend averages out from one section of the embankment to the next. In some places the strengths are higher, some lower, but the average is unaffected. In contrast, epistemic uncertainty about the average strength, in this case the trend line or curve, introduces a systematic effect into a stability calculation. If mean strength is over-estimated at one location, it is over-estimated at every location, thus the whole embankment would be more likely to fail.

A second implication is on what is meant by a predictive probability. Consider the "probability of excessive settlement" of a long levee, or by the same token, the "probability of excessive lateral deformation" in a long excavation. What does it mean that this probability equals, say, 10%? Does it mean that 10% of the levee should be expected to fail? Does it mean that there is a 10% chance that the entire levee will fail? Does it mean something between these two? The answer depends on how the total uncertainty is divided between aleatory and epistemic. To the extent that all the uncertainty is assumed randomness, the probability refers to a temporal or spatial fraction. To the extent that all the uncertainty is assumed due to limited knowledge, the probability refers to a chance of complete failure. Almost always, the uncertainty is apportioned between randomness and limited knowledge, so the probability itself is a mixture.

A third implication is on scale effects. To the extent that uncertainty is presumed to be due to randomness, the uncertainty averages over space and perhaps time. The variability of measured performance among long test sections will be less than the variability among short test sections. This was alluded to above. The variability of soil properties among large specimens will be less than among small specimens. The variability among in situ tests that mobilise large soil volumes will be less than the variability among in situ tests that mobilise small soil volumes. The converse is true of behaviours that rest on extreme soil or formation properties. Seepage conditions and piping that depend on the most transmissive element of a formation become both more variable with scale and also more extreme. Rock slope failures that depend on the least favourably inclined joint become more variable and also more probable as the volume of rock mass increases.

Finally, separating out some fraction of the uncertainty into an aleatory part creates an irreducible level of uncertainty in the analysis and corresponding predictions, by presuming a fraction of the total uncertainty to be unknowable. The trade off against this irreducible level of uncertainty is that one hopes the modelling assumption will allow powerful methods of mathematics to be applied to the problem of inference and estimation and thus, in the end, a more precise outcome to be achieved.

7.2 Probability

This section discusses the meaning of probability and the use of probability as a measure of uncertainty in risk analysis. Today, probability and statistics are the default way to measure uncertainties in science and engineering, and increasingly in public policy as well. Yet, the historical development of the theory and meaning of probability over the past four centuries has led to rich and subtle discussions about the role of probability in human affairs, and has led to a duality in the way probability is interpreted by mathematicians, scientists, and philosophers. This duality parallels the distinction between the concept of uncertainty as reflecting natural frequencies vs reflecting lack of knowledge. This parallel distinction between the nature of uncertainty in risk analysis and the nature of probability as a philosophical concept is exploited in practical applications. The section ends with a discussion of issues surrounding vagueness in the language we use to describe risks and uncertainties, and of attempts over the past few decades to grapple with vagueness and imprecision.

We take for granted today that uncertainties and risks should be denominated in the language of probability, and that calculations of risk should be based on inferences made by statistical reasoning. Underlying this simple supposition is a rich history of philosophical inquiry and debate over the human enterprise(s) to which probability theory applies, and to the meaning of the word, probability. This debate is sometimes simplified to contrasting the meaning of probability as either frequency of occurrence or degree of belief – often referred to as the controversy between frequency and belief – but the intellectual issue is more subtle, extending to epistemology and the social sciences. For engineering purposes, the concepts of probability as frequency and as belief co-exist comfortably in risk analysis. There is no competition between these meanings, rather they apply to different and equally important things. As seen in the previous section, the nature of uncertainty itself comprises both natural variation and lack of knowledge.

The above is a practical engineer's point of view. There are those who argue that only objective frequencies of the world can be logically treated as probabilities (Von Mises *et al.*, 1939), and there are those who argue that even the frequencies we observe in the world can only be interpreted as the subjective belief of the observer (De Finetti, 1990). These immoderate positions are philosophically engaging, but often difficult to implement in a practical risk assessment. As Gigerenzer (1991) points out, many if not all people perceive uncertain events in the world through the lens of frequency, no matter the philosopher's view. Thus, although by adopting the Laplacian view, all the uncertainties of risk analysis in a philosophical sense have to do with limited knowledge, nonetheless, a modelling approach that combines frequencies pertaining to natural variations with degrees of belief pertaining to knowledge uncertainties is practical.

What is *probability* and why use it?

Ideas about human reasoning have long shaped the meaning of the term probability, and remain central to understanding contemporary interpretations. Indeed, the topic has even been the subject of a popular recent book (Bernstein, 1996). The thing many practitioners overlook, however, is that, while all other terms in the probability calculus have well-defined meanings, probability itself does not; probability is a privative term, the meaning of which is not part of the formal system of probability theory (Salmon, 1998).

Probability theory is a logical construct based on a small number of axioms. The mathematical aspects of these axioms and their implications are discussed in more detail in Chapter 11. As long as one accepts the axioms, all the results of mathematical probability theory follow necessarily. But, within the axioms only the properties of probability as a mathematical measure are defined. Nowhere do these axioms, and therefore nowhere does all of mathematical probability theory, address the meaning of the concept of probability. The meaning we give the concept in risk analysis is a philosophical question independent of the mathematics. The meaning we assign, however, does affect what can be done with the mathematical theory, and consequently the meanings of the numbers that result. So, it is pertinent to review how probability theory evolved, and the philosophical interpretations imputed to probability over the course of time.

Historical development of probability theory

The intellectual development of probability theory is a rich and fascinating story (Daston, 1988; David, 1962; Gigerenzer, 1989; Hacking, 1975, 1990; Pearson and Pearson, 1978; Porter, 1986; Stigler, 1986). Early modern references to a theory of probability as applied to games of chance date to Cardano in the 16th century, but today the origin of probability theory is usually traced to a correspondence between Fermat and Pascal, the latter of whom in 1654 developed the combinatorial mathematics which underlie probability as applied to games of chance. The Chevalier de Méré has become a footnote in history by having suggested to Pascal the famous "problem of points", which became a topic of the correspondence with Fermat, and in turn led to the arithmetic triangle of the binomial coefficients (David, 1962). The problem of points has to do with the question of how to divide the stakes in a gambling game that is interrupted before one side wins. Note, that in this early development, the meaning of probability lies in the game of points, and reflects frequencies of the roll of dice in such games.

In 1662, shortly after the Fermat–Pascal correspondence, the Englishman John Graunt compiled a Bill of Mortality for the City of London, charting births and deaths from 1604 to 1661, and published it with commentary on the meaning of the data. Thus, for the first time, statistical data were collected on a social process and published with an intent to inform trade and government.[3] The popularity of the endeavour quickly expanded to the continent, where the French undertook a similar compilation for Paris in 1667 (Bernstein, 1996). Graunt even foreshadows modern concerns with risk, in commenting that while many people fear "notorious diseases, [he] will set down quantitatively how many actually die of them, so that those persons may the better understand the hazard they are in".

The development of probability theory beyond questions of games and population statistics to issues of physics – principally astronomy and geodesy in the earlier years and to a broad range of applications by today – flowered in the 18th and 19th century. This was the work of many people, whom we now think of as among the great scientists, mathematicians and philosophers of the modern age. Among these were James and Daniel Bernoulli, Laplace, Poisson, Legendre, Gauss, Maxwell, Boltzmann and Gibbs.

So, too, during this age of enlightenment, did the nature of uncertainty became part of the emerging inquiry into human knowledge, both among proponents of logical deduction from first principles, and among proponents of inductive reasoning from empiricism. The former were concerned with fundamental truths, the latter with observed frequencies (Daston, 1988; David, 1962; Hacking, 1975). Between 1710 and 1750, Jacob Bernoulli, John Locke, David Hartley, and David Hume put forward views on knowledge and reason that accommodated both objective and subjective concepts of uncertainty (Gigerenzer, 1989).

A seminal work of this period was Jacob Bernoulli's *Ars Conjectandi* (published posthumously in 1713), which set the stage for the use of probability as logic or, in his terms, the art of conjecturing. The probability of Bernoulli was to play a central role in moral philosophy, and its implementation was to use the model of law (jurisprudence) rather than that of empirical science. This view followed closely upon that of Leibniz who had many times in his life recommended that a serious effort be made toward developing a theory of probabilistic reasoning. Sivia (1996) notes, that within this practical discipline, mathematics was to provide a rigorous method of reasoning and a way of ensuring consistency when the complexity of the situation might otherwise overwhelm normal reason.

By about 1850, growing misgivings about the rationality of the common man gave way to the rise of statistical inference and its stature in the increasingly influential fields of physics and social science (Porter, 1986, 1995). Throughout most of the 19th and early 20th centuries, probability became defined by frequencies of occurrence in long series of similar trials, largely to the exclusion of degree of belief. In engineering, this view was most associated with the writing of Richard von Mises (Von Mises *et al.* 1939).

Beginning with the pioneering work of Ramsey *et al.* (1931) and De Finetti (1937), followed by Savage (1954) after World War II, a degree-of-belief view gradually regained acceptance. The rapid advance of operations research and decision analysis during the war had been an important factor. This duality of frequency and belief remains today, with frequency dominating the experimental sciences, and belief prevailing in economics, management and public policy. A clear exposition of the practical differences is given by Barnett (1999). The advent of modern concepts of inferential statistics in the 20th century has led to renewed interest in the question of frequencies of natural variation vs degrees of inductive confirmation.

The meaning of probability

The original axiomisation of probability theory is due to Kolmogorov. If one accepts the axioms, then all the results of mathematical probability theory hold necessarily; they can be deducted from the axioms. Thus, there is no disagreement about the properties that probability displays. These properties can be found in any textbook. The issue has always been, and remains, what does the word *probability* mean?

Since the beginning of modern probability theory in the 1600s, there has been a dual meaning: (a) relative frequency in a long or infinite number of trials; and (b) objective or subjective degree of belief. In this light, it is interesting to note that Poisson and Cournot used the French words *chance* and *probabilitè* to refer to the two distinct concepts of frequency and belief and, by so doing, suggested that the two concepts were not completing but simply different. The terms have more distinct meaning in French than their translations in English suggest. *Chance* denotes an objective propensity of an event, the facility with which it can occur. An event will have, by its very nature, a larger or smaller *chance*, known or unknown. *Probabilitè* denotes credibility or degree of reasonable belief. The probability of an event is the reason for thinking that the event did or will occur.

How can a theory describe two quite different concepts? As Pólya (1954) notes, this is not at all unusual in science. Laplace's equations for differential fields are used to describe the flow of incompressible non-viscous fluids, as well as the distribution of forces in an electrostatic field. There is no reason that a single mathematical theory cannot describe any number of physically different processes. Thus, these meanings for probability are not necessarily in opposition to one another, although arguments in the literature might lead one to think they were. These meanings describe distinct concepts, each of which is part of almost any complex risk analysis. The concepts co-exist in most practical applications, because the two types of uncertainty, natural variability and knowledge uncertainty, co-exist in most practical applications. The real issue is how to convolve the two distinct types of probability and thus the two types of uncertainty to which they refer?

Probability as frequency

In much of engineering, probability is interpreted to be the frequency of occurrence of some event in a long series of similar trials. The principal early proponent of this view was John Venn (1834–1923); a more modern one, Richard von Mises (1883–1953). Venn (1896) described probability as:

> The fundamental conception is that of a series which combines individual irregularity with aggregate regularity. Probability has no meaning except in connection with such a series. Any probability must be referred to a series. The probability of an event is its relative frequency in the series.

A trial is an individual occurrence producing an outcome of some sort. For example, each individual lift of soil placed in a compacted embankment might be considered a trial. The frequency of soils having low moisture content among these lifts (i.e. among the trials) would be the probability of soil with low moisture content. Similarly, each maximum annual discharge in a stream might be considered a trial. The frequency of peak annual flows above a certain magnitude would be the probability of floods of at least a certain size.

Frequency definitions of probability are the ones that non-statisticians usually think of first. They are based on the relative number of times a certain outcome occurs in a long series of similar trials. There are three traditional definitions of frequency: the, so-called, classic, Venn limit and infinite trials. The classic definition defines probability *a priori*. If there are n possible alternatives, for m of which q is true, then the probability of q is m/n. The Venn limit says that if an event occurs a large number of times, the probability of q is the limit of the ratio of the number of trials on which q is true to the whole number of trials, as the number of trials tends to infinity. The infinite trials definition assumes an infinite number of trials and defines the probability of q as the ratio of cases in which q is true to the whole number. The classic definition is associated with the work of DeMoivre and Cramér, the Venn limit with that of Venn and Von Mises, and the infinite trials with that of Gibbs and Fisher. The implicit assumption underlying all these definitions is that one cannot place probabilities on states of nature or on unique events. States of nature are constants (perhaps unknown), not random variables. To ask the probability of a flood stage higher than some value occurring within one specific year is undefined.

The frequency approach requires at least the potential for large numbers of similar trials, for example, many realisations of a process, or a sample of many measurements

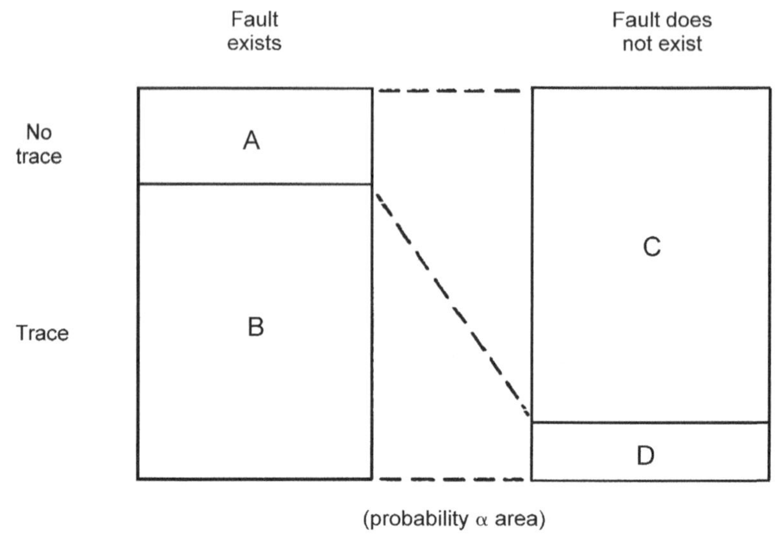

Figure 7-2. Conditional probabilities of observing evidence of a geological fault. Probability is taken proportional to area. States of nature (a fault exists or it does not) are shown vertically. Statistical evidence (trace appears in air photos) is shown horizontally (Baecher, 1972)

from a population. The frequency interpretation pertains to variability among collections of things. Proponents say that the frequentist interpretation is objective, because it deals only with observable patterns of data for groups of similar things. Inference from these patterns to states of nature does not follow necessarily from probability theory, but within the frequentist school that is acceptable. Inference within the frequentist school is based on refutation or falsification along the lines of Popper (Barnett, 1999).

Consider the following statement about the geology of a dam foundation. If a fault existed at this site, then aerial photographs would show its trace 90% of the time. This is not the same as saying: if a trace does not appear in the aerial photographs, then the probability of a fault is only 10%. The probability of a trace showing the fault in air photos given that a fault exists at the site is proportional to the areas B/(A+B) in Figure 7-2, which is permitted by frequentist theory. The probability of a fault existing given a trace is observed in the air photos is proportional to B/(B+D), which is undefined in frequentist theory.

Probability as belief

The belief interpretation of probability, quite common in civil engineering, holds that probability is a rational degree of belief. The probability that a solution cavity exists in a limestone dam abutment is typical of a dam-related problem that cannot easily be approached using a frequency definition of probability; the uncertainty is fundamentally epistemic because there is only one site in question. Such probabilities have to do with one-time events, past experience and amounts of information. They are personal and subjective, and not easily related to frequencies: actual or conceptual. Regarding such situations, the mathematician Augustus de Morgan said:

> Probability is a feeling of the mind, not the inherent property of a set of circumstances. [. . .] the amount of our belief of one proposition varies with the amount of our belief of other propositions with which it is connected.

A degree of belief is the propensity to take action when faced with a situation of uncertainty. According to Ramsey and Braithwaite (1931), to say that someone has degree of belief p that a proposition is true, is to say that:

> If the individual is offered an opportunity to bet an equal amount on the truth of the proposition or on "red in a single draw" from an urn containing r red and b black balls – in which for the integers r and b with $r/(r+b)<p$ – he will prefer the first bet; if $r/(r+b)>p$, he will prefer the second bet.

Degree of belief is manifest in action, not in introspection. Which bet an individual favours is a direct reflection of his/her degree of belief, and thus is his/her probability.

The belief approach does not require a number of trials. It provides that probabilities can be assigned to unknown outcomes of particular realisations of a process, or to the very properties (states) of nature itself. The belief interpretation pertains to the weight of evidence leading to a quantitative statement of uncertainty about nature. Thus, a belief interpretation can be applied to the unique occurrence of an event. On the other hand, the belief interpretation necessarily implies that probability is in the eye of the beholder.

The notion of degrees of belief as subjective probability was much debated in the early half of the 20th century. While, today, the notion is widely accepted, in earlier times significant efforts were devoted to developing an objective degree of belief theory. This theory holds that two observers of the same data necessarily arrive at the same degree of belief (Jeffreys, 1948). Subjective degree of belief theory, in contrast, says that two observers of the same data may arrive at different degrees of belief and both be right! The objective degree of belief theory is most notably associated with the work of Sir Harold Jeffreys, the Cambridge geophysicist, but has fallen into disfavour in recent times.

Jeffreys postulates an objective degree of belief by presuming that every observer starts from a condition of complete ignorance about the quantity to be inferred, and in this state should have the same prior probability distribution. Jeffreys' *principle of indifference* states that, in a condition of ignorance, one should assign equal probabilities to all possible outcomes of an experiment or values of a parameter. Thus, if a coin may land head-up or head-down, and if one has no information about the possible bias of the coin, then the rational assignment of probability *a priori* is to treat each possible outcome as equi-probable, that is, to assign each the value 1/2. This uniform assignment of prior probability is called "non-informative". In the same way, if one had no information about the value of, say, the mean strength of a soil, then the prior probability distribution over its possible values should be taken as uniform within the region of the possible outcomes. The principal difficulty of this approach is that the resulting prior probability distributions expressing total ignorance are not invariant to transformations of variable.[4] As a result, objective degree of belief theory has fallen from favour.

A subjective approach considers information of all kinds, statistical and otherwise, as admissible in developing a probability estimate. People use various types of information in formulating subjective probabilities, adopting statistical and knowledge-based reasoning or a mixture of the two depending on the circumstances. A familiar case is a weather forecaster's prediction of rain. While climatological records for the day in question provide a base-rate frequency, information from satellite photographs, atmospheric pressures, and even a glance out the window are also incorporated in a probability estimate. In fact, weather forecasters have served as subjects in behavioural studies for just this reason, and they have been found to forecast probabilities which correspond closely to long-run frequencies (Murphy and Winkler, 1974).

Likewise in geotechnical applications, all the information at hand, together with personal experience and judgement, are brought to bear in estimating the subjective probability of an event or condition. Because subjective probability inherently depends on one's state of knowledge at the time it is formulated, any such value can be expected to vary, both from one person to another and with time as information and knowledge are gained, in much the same way as engineering judgement itself.

Non-standard logic

The study of decision-making under uncertainty has a history interwoven with probability theory, regardless of whether one adheres to a frequentist or belief point of view. Yet, in practical situations it is sometimes difficult to specify probabilities with precision, whether in frequency or belief. The result is that in recent years a cottage industry of academics has arisen with the aim of developing less restrictive measures of uncertainty and more flexible algebra of inference than those offered by mathematical probability.

During the 1970s and 80s, researchers in cognitive psychology observed that human subjects frequently deviate from normative (i.e. probabilistic) models of rational decision-making, especially when faced with poorly specified conditions; for example, outcomes that are not clearly circumscribed, probabilities that are not well known, or utilities that are inadequately characterised. From this, a number of behavioural paradoxes arose, some of which have become canonised in the judgement and decision-making literature, for example Allais' paradox (1953) and Ellsberg's paradox (1961). Traditional decision theorists dismiss these paradoxes as the very reason one needs formal decision analysis tools, to wit, to protect vulnerable human beings from falling prey to intuitive fallacies (Raiffa, 1968). On the other hand, other workers began to question whether such deviations from normative behaviour should be judged irrational, as the deviations were common to large numbers of subjects. Thus, attempts were made, and continue to be made, to develop a theory of "behavioural decision-making" under uncertainty that aligns with the way people actually behave, as opposed to the way people "should" behave.

In recent years there has been increasing interest among certain risk analysis practitioners in mathematical approaches that are *non-bivalent*, that is, approaches that allow the truth of a proposition to be something other than simply true or false. Perhaps the most popular of these alternative approaches is fuzzy logic (Zadeh, 1977). It should be noted that this approach is controversial. In general, probability theorists (Lindley, 1987) and philosophers of science (Haack, 1996) dismiss fuzzy logic as being unnecessary and addressing concerns other than those traditionally dealt with by probability. The approach has had little impact on practical risk assessment.

Fuzzy logic deals with objects as sets with boundaries that are imprecise. The significance of that development lies in creation of a multi-valued logic, as opposed to the two-valued logic that is at the foundation of probability theory. In two-valued logic, if **A** is a set, an entity x either does or does not belong to **A**. In fuzzy logic, by contrast, if **A** is a fuzzy set, a statement "x belongs to **A**" is not necessarily true or false but it may be true (or false) only to some degree. This capability of fuzzy logic sets to express gradual transition from membership to non-membership can be utilised as a tool to express ambiguity about the meaning of a set definition, and is thus said to handle the unavoidable linguistic ambiguity of natural language better than probability does. It is important to note that fuzzy logic addresses linguistic ambiguity, not vagueness of probability estimates. The standard engineering response to ambiguity is to define terms more precisely, and that seems the best approach for dam safety studies.

7.3 The strength of evidence

This section discusses how observations and other evidence affect probabilities: how do we draw probabilistic inferences from data (Figure 7-3)? The tool for making inferences

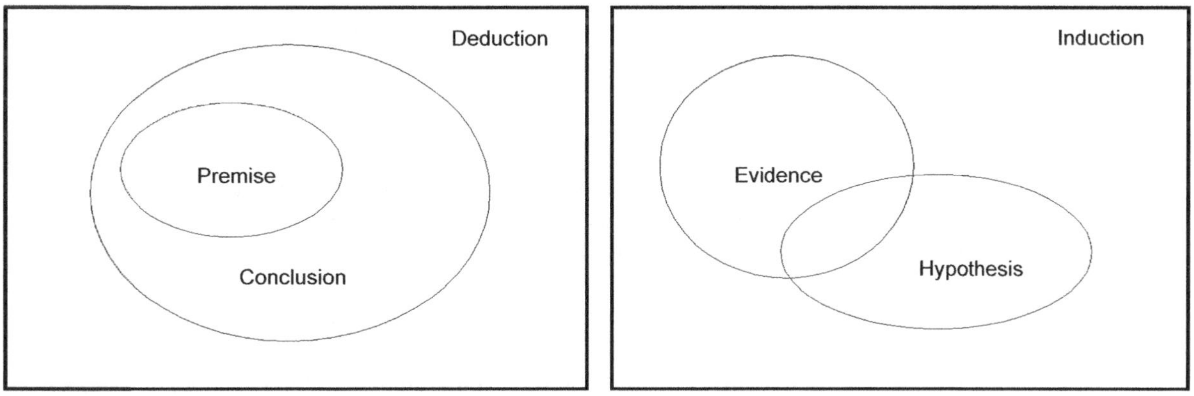

Figure 7-3. Schematic distinction between deduction and induction

about the world in the face of uncertainty is the set of statistical methods that derive from probability theory. It should not be surprising, however, that just as there are different schools of thought on probability, there are also different schools of thought on statistical inference.

Inductive reasoning

Deductive logic in the Aristotelian sense resolves to the repeated application of two syllogisms:

If A is true, then B is true
A is true
Therefore, B is true

If A is true, then B is true
B is false
Therefore, A is false

Starting from a set of premises, the syllogisms lead to a set of necessary conclusions.

It would be nice if we could use this strong deductive logic in all our daily affairs, but more typically, the information available is insufficient. Thus, a weaker form of *inductive* reasoning is required:

If A is true, then B is true
B is true
Therefore, A is more plausible

In this case, the evidence does not prove that A is true, but suggests that confidence in the truth of A should increase. Pólya (1954) called this *plausible reasoning*. Note, however, that while two observers should conclude that A is more likely, they would not necessarily agree on how much more likely. That is, qualitatively they should agree, but perhaps not quantitatively. Thus, in a practical decision (e.g. given the evidence, should we invest in yet more data collection?), the two observers might choose different options.

The "problem" of induction

Most geologists believe that glaciers once covered large parts of North America, although this was not observed by modern scientists. Most people believe the sun will rise tomorrow, although it is yet to happen. Science develops general theories that we use to make predictions in risk analysis, and these theories are presumably more than just the sum of past observations: we use them as if they had predictive power.

There is, however, a flaw underlying this presumption. Theories are based on inductive reasoning; yet, Hume famously demonstrated that inductive reasoning is not based on Aristotelian logic (Hume and Smith, 1963). Inferences drawn from the first syllogisms, above, are *demonstrative*: the premises necessitate the conclusions.

Inferences drawn from the last syllogism are *non-demonstrative*. Knowledge, unlike mere belief, is presumably based on rational argument, but Hume shows that it is not.

Ampliative inferences, like those of the last syllogism, expand knowledge and lead to theories. The conclusions contain more than the premises. Demonstrative inferences enjoy deductive validity, and may rearrange content in useful ways, but at the expense of being ampliative. What Hume said is that there are no ampliative inferences that are also demonstrative. We have observed the sun rise each morning since the beginning of time, and tomorrow is another morning. Yet, we cannot conclude necessarily that the sun will rise tomorrow, because something might change. The solar system could be sucked into a black hole (which, of course would also change the meaning of "tomorrow").

A common justification of induction is that nature is uniform. What has happened in the past will happen in the future, and laws that apply in one place apply everywhere. But, how do we know this? We know this because we have observed uniformity in the past. But that means uniformity itself is a concept derived inductively. The argument is circular. The uniformity of nature is a metaphysical faith, yet without such faith any sort of engineering is inconceivable. Hence the dilemma, known as *Hume's Paradox*. All of us have strong beliefs about things we have not observed: the future, the past, and places we have not been. Are these all without rational justification? Despite Hume's Paradox, the scientific method based on induction has been wildly successful. Yet, this, too, is a circular argument.

Hypothetico-deductive method

In response, an explanation has arisen of how we acquire scientific knowledge that has gained broad acceptance in the experimental sciences. The hypothetico-deductive method works as follows. From experience and some process of intuitive discovery, a hypothesis is developed. A set of specific predictive conclusions is deduced from this hypothesis. Experiments are performed to test the predictive conclusions against observations. If the observations are consistent with the predictions, the hypothesis is to some extent confirmed. If the observations are inconsistent with the prediction, the hypothesis is disconfirmed. As more and more such experimental confirmation accumulates the hypothesis becomes ever more generally accepted, but if at any time disconfirming evidence occurs, the hypothesis is immediately disconfirmed. The only non-deductive part of this method is the initial discovery of the hypothesis, which is said not to be a matter of logic and therefore not to require logical justification.

The standard criticism of the hypothetico-deductive method is that it presumes the hypothesis to be a premise when, in fact, the hypothesis itself is really a conclusion. Deduction is a major part of the hypothetico-deductive method but it is not the only part, and the other part is not logically sound (Salmon, 1967). The inference from observations to hypothesis is, by presumption, not logical. This can be easily seen in that the same set of observations may lead to multiple and incompatible hypotheses.

Falsifiability

Among the most influential modern approaches to the justification of scientific logic is Popper's *deductivism* (Popper, 1968). Popper sides with Hume in saying that there is no such thing as correct ampliative inference, but goes on to say that induction has no role in scientific logic whatsoever. Popper (1968) argues that deduction is the only correct model of inference in science:

> Now in my view there is no such thing as induction. Thus inference to theories, from singular statements which are "verified by experience" (whatever that is), is logically inadmissible. Theories are, therefore, never empirically verifiable. [. . .]
>
> But I shall certainly admit a system as empirical or scientific only if it is capable of being tested by experience. These considerations suggest that not the verifiability but the

> falsifiability of a system is to be taken as a criterion of demarcation. [. . .] It must be possible for an empirical scientific system to be refuted by experience.

Popper's basic tenet is *falsifiability*. Generalised scientific statements based on empirical observation differ from those of metaphysics in that they are falsifiable by as little as one negative instance, but they are verifiable by no finite number of positive instances. If a statement is tested and the outcome is negative, it can be rejected; if the outcome is positive, all that can be said is that the experiment failed to falsify the statement. Popper rejects the hypothetico-deductive method because the method maintains that confirming observations enhance the probability of the hypothesis being true, which Popper rejects.

There is nothing unique about a statement that has not been falsified. Many such statements exist. When a statement is falsified, it can be replaced by any of these. However, clearly not all unfalsified hypotheses are equally acceptable. For practical purposes, there must be a mechanism for selecting among unfalsified statements and, for Popper, falsifiability is again the criterion. Statements differ in the degree to which they can be falsified. The more highly falsifiable the better. The more specific a statement, the greater number of possible states of the world that are incompatible with it, the greater its content, and the more likely it is to be falsified. A highly falsifiable hypothesis which is severely tested becomes highly corroborated.

Popper (1968) dismisses the use of probability to express degrees of confirmation:

> The attempt has often been made to describe theories as being neither *true* nor *false*, but instead more or less *probable*. [. . .] Yet in my view, the whole problem of the probability of hypotheses is misconceived. Instead of discussing "probability" of a hypothesis we should try to assess what tests, what trials, it has withstood; that is, we should try to assess how far it has been able to prove its fitness to survive by standing up to tests [italics in original].

Criticisms of Popper's theory rest on at least two issues. The first, is that it does not in fact get around the problem of induction. The second, is that, when faced with practical decision (e.g. in risk analysis), the method gives no quantitative guidance on how to compare competing non-falsified theories.

Probabilistic induction

Peirce (1998) drew a distinction between two types of non-deductive reasoning: inference under uncertainty, and theorising.[5] Peirce said that inference under uncertainty is specifically probabilistic. That is, this form of inductive reasoning involves not proof but degrees of confirmation. A theory or set of theories is at hand, and thus a space of hypotheses can be determined within which probability calculations can be made. Theorising, on the other hand, according to Peirce, involves the abstract creation of new theory to explain observations, what might be called speculative reasoning. New theories are compared to existing observations for consistency, and used to forecast new observations or discoveries.

The logic of probability was thought by Laplace (Laplace and Dale, 1814), Bernoulli (1713) and others as an appropriate basis for inductive reasoning, that is for calculating degrees of confirmation based on observations. Inductive arguments intend to support their conclusions only to some degree. This is consistent with Maxwell's (Tolstoy, 1981) view:

> The actual science of logic is conversant at present only with things either certain, impossible, or entirely doubtful, none of which (fortunately) we have to reason on. Therefore the true logic for this world is the calculus of Probabilities, which takes account of the magnitude of the probability which is, or ought to be, in a reasonable man's mind.

Using probability to measure degree of confirmation implies a degree of belief perspective. The third syllogism can be formulated as a relationship among conditional

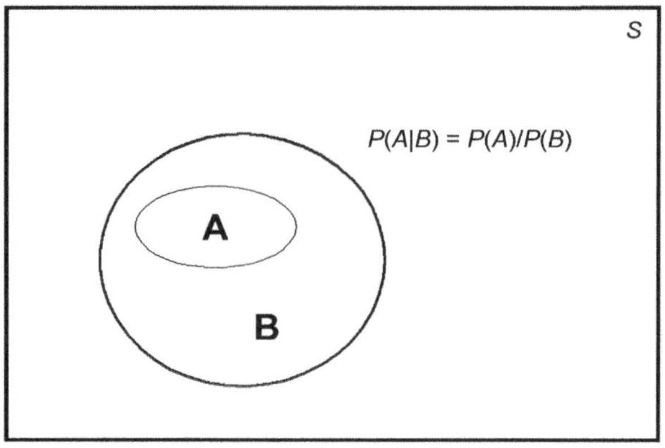

Figure 7-4. Degree of confirmation in syllogism III

probabilities. Since A implies B, the conditional probability $P(B|A)=1$. Thus, as shown in Figure 7-4:

$$P(A)=P(B)P(A|B) \tag{7-1}$$

$$P(A|B)=\frac{P(A)}{P(B)} \tag{7-2}$$

First observe, that since $P(A)\leq P(A|B)$ is necessarily the case, if B obtains, the credibility of A must increase. Then from equation (7-2), the less likely B is, the greater the confirmation afforded A. If A implies some wholly probable event B, such as the sun rising tomorrow, when that likely event occurs, little increased confirmation is afforded A. In contrast, were B some highly improbable event, such as the sun not rising tomorrow, and B turned out to obtain, it would afford considerably increased confirmation to A. This is consistent with Popper's concept of corroboration: "In appraising the degree of corroboration of a theory we take into account its degree of falsifiability. A theory can be the better corroborated the better testable it is" (Popper, 1968). This can also be seen by rearranging Bayes' Theorem (Section 7.5) to obtain, after noting that $P(B|A)=1$:

$$P(A|B)=\frac{P(A)}{P(A)+P(A^c)P(B|A^c)} \tag{7-3}$$

in which A^c is the complement of A (i.e. "*not-A*"). B is implied by A, but the greater the probability of B also in the case of A^c, the less the confirmation of A increases when B obtains.

We have dealt with the case in which A implies B, and then B obtains and, in so doing, increases the confirmation of A. What happens in the case that A implies B, but A does not obtain? Does this change the degree of confirmation of B? The answer is yes. Again rearranging Bayes' Theorem and noting $P(B|A)=1$:

$$P(B|A^c)=\frac{P(B)-P(A)}{P(A^c)} \tag{7-4}$$

from which, $P(B|A_c)\leq P(B)$. Thus, the degree of confirmation of B can only diminish if a premise leading to it does not obtain. Furthermore, the greater the degree of confirmation in the premise itself (i.e. $P(A)$), the greater the decrease in confirmation of B when that premise fails to obtain.

Statistical inference

Statistical inference is the enterprise of drawing probabilistic conclusions about the world from specific observations. These conclusions might be in the form of probabilities associated with states of nature, probability distributions over properties in nature, tests of hypotheses, or many other things.

The basic question of statistics is, given a set of observations, $\mathbf{z}=\{z_1, \ldots, z_n\}$, that arise from an experiment on an uncertain process, where the uncertain process is described by the probability distribution, $f_Z(z|\theta)$, which in turn depends on the parameter(s) θ, one would like to be able to say something about the uncertainty in θ that exists after the observations are made. That is, one would like to make estimates of θ from the data $\mathbf{z}$.

Modern statistical practice is dominated by two schools of thought, each with sub-groups. These might be called the "traditional" and "Bayesian" schools.[6] The division between these follows the meaning given by each to the term probability. The traditional school takes probability to mean frequency, while the Bayesian school takes *probability* to mean belief (Figure 7-5).

The "traditional" school of statistics

The traditional school defines probability as frequency. It rejects direct probabilistic estimates of θ from the data $\mathbf{z}$ as undefined. The state of nature θ is not a random variable, it is a fixed constant. It is meaningless to specify probabilities over θ because probabilities are frequencies and θ is not a frequency.

The approach of the traditional school is one of refutation (falsification). Probabilities are not placed directly on states of nature, but confidences are calculated in a form of reverse reasoning by excluding states of nature for which the chance (i.e. frequency) of observing the set of data actually observed would be small. While the state of nature θ is not a frequency, the outcomes of the observations $\mathbf{z}$ over repeated experiments can be described by frequencies. Therefore, it is permissible to speak of the conditional probabilities of certain observations occurring in repeated experiments, given an assumed value for θ.

The traditional-school approach is built upon the concept of *estimators*. Estimators are mathematical functions of the observed data that can be used to infer the value of an unknown but constant state of nature θ. Consider a set of sample observations $\mathbf{z}=\{z_1, \ldots, z_n\}$ from which an estimate of some population parameter θ is desired. Any mathematical function $g(\mathbf{z})$ of the sample observations is said to be a *statistic* of the data, T, or simply a sample statistic:

$$T=g(\mathbf{z}) \tag{7-5}$$

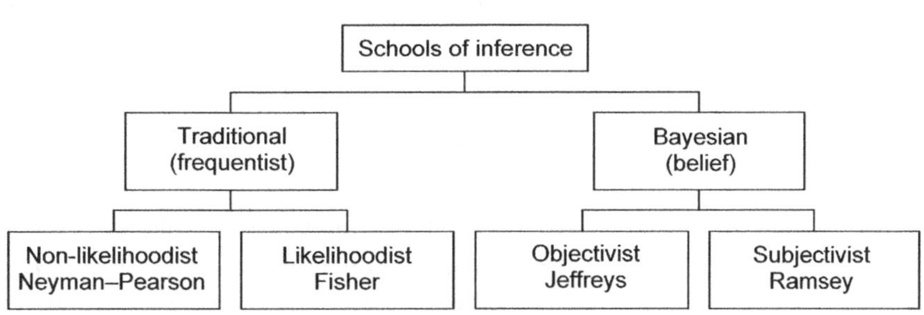

Figure 7-5. Taxonomy of statistical thought

For example, commonly used sample statistics are the arithmetical average, median, standard deviation, range, and so on. Appropriately chosen sample statistics are used to estimate a population parameter, as:

$$\hat{\theta} = T = g(\mathbf{z}) \tag{7-6}$$

where "appropriately chosen" reflects any of a set of commonly assumed good properties of such estimators.

The sampling distribution of an estimator is its probability distribution in repeated sampling from the same uncertain process. The sampling distribution describes the frequencies of the estimator over all possible ways a set of observations might be made from the sampled process. Certain properties of the sampling distribution are thought to provide good estimators of the unknown state of nature. The more common of these good estimator properties are given in Table 7-2. A good estimator is one whose sampling distribution is concentrated in a narrow region near the true value of θ.

Consider the common case of sampling from a Normal process. The sampling distribution of $\mathbf{z} = \{z_1, \ldots, z_n\}$ is:

$$f_z(\mathbf{z} \mid \mu, \sigma) \propto \exp\left(-\sum_{i=1}^{n} \frac{z_i - \mu}{2\sigma^2} \right) \tag{7-7}$$

in which μ is the mean and σ the standard deviation. If we consider the sample average $\bar{z} = \frac{1}{n} \sum z_i$ as an estimator, its sampling distribution is:

$$f_{\bar{z}}(\bar{z} \mid \mu, \sigma) \propto \exp\left(-\frac{n(\bar{z} - \mu)^2}{2\sigma^2} \right) \tag{7-8}$$

which has the property that $E[\bar{Z}] = \mu$, so $\bar{Z}$ is an unbiased estimator of the parameter μ.

Within the traditional school, tests of hypotheses are addressed through refutation. An hypothesis is considered to be true unless the results $\mathbf{z} = \{z_1, \ldots, z_n\}$ would be so improbable were the hypothesis true that there is no choice but to reject it. Otherwise, if the results are not so improbable conditioned on the hypothesis, the hypothesis is taken to be true, in the sense that there is inadequate negative evidence to refute it.

This leads to the concept of *confidence intervals* in estimating the value of θ. The confidence interval of an estimate of θ is that interval of the θ-axis for which the conditional probability, given θ, of observing $\mathbf{z} = \{z_1, \ldots, z_n\}$ is larger than some arbitrarily small value, p. Typically, a confidence interval might be set at $p = 0.05$ or

Table 7-2. Common "good" properties for statistical estimators

Estimator property	**Definition**
Unbiasedness	The expected value of T over all ways the sample might have been realised from the sampled population equals the parameter to be estimated ($E[T] = \theta$)
Consistency	T converges to θ, as n becomes large ($T \underset{n\to\infty}{\longrightarrow} \theta$)
Efficiency	The variance of the sampling distribution of T is minimum
Sufficiency	The estimator T makes maximal use of the information contained in the sample observations
Robustness	The statistical properties of T in relation to θ are insensitive to deviations from the assumed underlying pdf of $\mathbf{z}$

$p=0.10$. Values for θ outside this interval have conditional probabilities $P(\mathbf{z}|\theta)$ less than p. Thus, those values of θ are deemed to be refuted, whereas those within the confidence interval are deemed to be not refuted and thus accepted.

Presuming, for sake of example, that the standard deviation of equation (7-7) is known, then the variance of the sampling distribution of $\bar{Z}$ is

$$Var[\bar{Z}]=(\sigma/\sqrt{n})^2 \tag{7-9}$$

Thus, for any value of μ within the interval $\bar{z}-2(\sigma/\sqrt{n})\leq\mu\leq\bar{z}+2(\sigma/\sqrt{n})$ the probability of observing a difference not greater than $|\bar{z}-\mu|$ is approximately 95%, and this is said to be the $p=0.05$ confidence interval for μ.

The traditional school of statistical inference dominated experimental science from the 1920s to the 1970s. Fisher's (1925b) *Statistical Methods for Research Workers* went through 13 editions, the last in 1960. The approach became so dominant that research papers in the fields of biology, medicine, psychology, and others could not be published if they did not present statistical analyses conforming in detail to Fisher's methods (Jaynes, 2003).

A difficulty of the traditional school is that the collection of methods is *ad hoc*. There is no unifying theory through which the methods are connected to a larger principle of inference. In every new problem, one must guess on the basis of intuitive judgement what a good function $T(\mathbf{z})$ might be, and then test the various suggestions by calculating their respective sampling distributions, for the sampling distribution is the only criterion for deciding whether an estimator is good or not.

Principle of maximum likelihood

The likelihood principle due to Fisher (1921), says that all relevant information about a set of parameters θ arising out of an experiment with outcomes $\mathbf{z}=\{z_1, z_2, \ldots, z_n\}$ is contained in the likelihood function:

$$L(\theta|\mathbf{z})=L(\theta|z_1, z_2, \ldots, z_n)=Pr[z_1, z_2, \ldots, z_n|\theta] \tag{7-10}$$

which is the conditional probability of the observations $\mathbf{z}$ given θ. Note, the likelihood function is defined over the variable θ for known realisations of the observations $\mathbf{z}$. For independent observations arising from an uncertainty described by the probability density function (PDF), $f_Z(z|\theta)$, the likelihood becomes,

$$\begin{aligned} L(\theta|\mathbf{Z}) &= f_{\mathbf{z}}[z_1, z_2, \ldots, z_n|\theta] \\ &= \prod_{i=1}^{n} f_{\mathbf{Z}}[z_1|z_2, \ldots, z_n|\theta] f_{\mathbf{z}}[z_2|z_3, \ldots, z_n|\theta] \ldots f_{\mathbf{z}}[z_n|\theta] \qquad (7\text{-}11) \\ &= \prod_{i=1}^{n} f_Z[z_i|\theta] \end{aligned}$$

Fisher recommends using as an estimator the value of θ that maximises the likelihood of the observations (MLE). The MLE has several desirable properties. It uses all the information contained in the sample, a property called sufficiency, and it is asymptotically ($n\rightarrow\infty$) unbiased.

A principle division within the traditional school is that between statisticians who believe that only the likelihood function is important for inference, and those who believe that the entire experimental frame of reference, including outcomes that would have been possible but that were not actually observed, is important. For example, continuing a series of experiments until one has collected enough evidence to support a particular hypothesis and then stopping is perfectly acceptable to a likelihoodist, but not to a non-likelihoodist. The latter would argue that by stopping an experimental series when there was just enough evidence one could prove any hypothesis. They would argue that the experimental frame is needed to supply the conditional probability assessment for other possible sample outcomes, which, in fact, have not occurred.

The Neyman–Pearson school of statistics

The Neyman–Pearson school of inference, in contrast to Fisher, is not based on the likelihood principle, but rather is associated with unbiased estimators, confidence intervals, and hypothesis testing. Within the traditional school it is a rival approach to Fisher. According to Neyman, probability enters the problem of inference as a way of characterising the experimental process itself: to express how reliably the testing process discriminates between alternative hypotheses, and how well it facilitates learning from error. One cannot automatically assess the credibility of any particular estimate, one can only assess the long-run reliability of a system of estimation. We can draw inferences in a way that is usually right, but we cannot measure the credibility of our doing the right thing on any individual occasion.

Neyman and Pearson championed a view contrasting with Fisher's which, while still within the frequentist tradition, focuses on balancing the chance of rejecting a correct hypothesis with the chance of accepting a wrong hypothesis (Neyman and Pearson, 1967). The Neyman–Pearson school holds that experimentation and accepting or rejecting the null hypothesis are decision problems. When considering alternate rules for accepting or rejecting hypotheses, each having the same probability of rejecting a correct hypothesis (type-I error), one should favour those rules with the lowest probability of accepting an incorrect hypothesis (type-II error). The tool with which this comparison is made is called the power function of a rule (or error characteristic), which describes the conditional probability a type-II error, given θ, resulting from that rule. The best rule is the one which has the universally lowest power function.

The difficulty with such a criterion is that one can usually obtain better error control over the entire parameter space only by increasing the cost of experimentation. Wald (1971) expanded the solution to the decision problem by adding the concepts of cost and worth of consequences into his formulation. As in the Neyman–Pearson and Fisher schools, Wald did not place probabilities on the state of nature, but in arguing the worth of consequences he reduces the choices among which a decision must be made.

Bayesian school of statistics

The Bayesian school of statistical inference is based on Bayes' theorem, and thus is consistent with Fisher's likelihood principle. The central difference between the Bayesian approach and the traditional approach, is that the former admits of probability statements directly on states of nature, whereas the latter does not. The Bayesian approach treats uncertainty as degree of belief and, therefore, does not view states of nature as unknown constants but rather as variables for which probabilistic description is appropriate. Thus, starting from some prior probability distribution over the state of nature, $f_\Theta(\theta)$, the likelihood function of the data is used to update these probabilities into a posterior probability distribution conditioned on the observed data by means of Bayes' theorem. A longer discussion of the statistical methods that derive from the Bayesian school is deferred to Chapter 11.

Bayes' theorem is a relationship among conditional probabilities:

$$Pr(A \mid B) = \frac{Pr(A)Pr(B \mid A)}{Pr(A)Pr(B \mid A) + Pr(\bar{A})Pr(B \mid \bar{A})} \tag{7-12}$$

in which A is some proposition, and $\bar{A}$ is its complement (i.e. "not-A"), and B is some other proposition possibly related to A. The conditional probability of B given A is the likelihood. Note that the denominator of the right-hand side (RHS) of equation 7-12 is simply a normalising constant ensuring that the sum of $Pr(A|B)$ and its complement is 1.0. In its inferential use, Bayes' theorem is often simplified to,

$$P'(A|B) \propto P^{o}(A)L(B|A) \qquad (7\text{-}13)$$

in which $P'(A|B)$ is called the "posterior" probability of A, $P^{o}(A)$ is the "prior" probability, and $L(B|A)$ is the "likelihood" function. The terms posterior and prior (sometimes *a posteriori* and *a priori*) imply the notions of after and before the data B are known. The proportional symbol is used to avoid the necessity of writing out the sometimes long integral expression for the normalising constant.

Equation 7-12 can, in principle, be used to "update" the probability of A from $Pr(A)$ to $Pr(A|B)$, based on the observation of B. For example, given some background rate on the probability of liquefaction in the foundation materials of dams, and given that low standard penetration test (SPT) blow count materials are more likely to liquefy than are high blow count materials, then the information that SPT blow counts in a particular foundation are low can be propagated through equation 7-12 to calculate an increase in the probability that this particular foundation will liquefy. In this way, Bayes' theorem is used as a vehicle for drawing inferential conclusions from observations.

There are two things worthy of note. First, in the words of De Finetti, "data do not speak for themselves" (De Finetti, 1990). Data only tell us how to modify the prior probabilities assigned to propositions to arrive at probabilities one should rationally assign after the data are observed. Second, the only relevant information for evaluating the magnitude of this change (i.e. establishing the strength of evidence in the data) is the likelihood function.

7.4 Expert opinion

Many important uncertainties in risk analysis are not amenable to quantitative estimation from data. In some cases there are no data at all, only the judgement of experts. These uncertainties have traditionally been treated using expert opinion. The tacit knowledge of experts is based on intuition, unenumerated past experience, subjective theory and other important but qualitative beliefs. This section surveys the growing field of expert elicitation of subjective probabilities, and summarises the emerging understanding of the psychology of probability assessment.

In almost all risk analyses there are some uncertainties that are simply not amenable to quantitative estimation based on data and models. These may reflect unique situations that are not found in the historical record of experience with dams, they may reflect uncertainties associated with poorly understood physical phenomena, they may reflect conditions for which data could, in principle, be collected but only at a prohibitive price, and so forth. Formerly incorporating such uncertainties in a risk analysis relies on professional judgement. In most cases, this judgement has to do with tacit rather than explicit knowledge. It is based on intuition, qualitative theory, anecdotal experience, and other sources that are not easily amenable to mathematical representation. Yet this judgement of experts is important information in analysing risk.

Within the degree-of-belief school of probability, personal opinion is an admissible basis for estimating probability, as long as that opinion is quantified in a way that maintains consistency. That is, given certain assumptions, degrees of belief have been shown by Ramsey and Braithwaite (1931) and others to satisfy the axioms of probability theory and, thus, the mathematics of probability theory can be used to operate on them, and they can be formally incorporated in a quantitative risk analysis. Degree-of-belief probabilities, which mostly have to do with epistemic uncertainties, may or may not be appropriately convolved with probabilities associated with natural variability. Whether they are or not depends on the decisions at hand and not on matters of principle. In principle, from the probability-as-logic point of view, quantitative probabilities inferred from data and associated with natural variation may be mixed with degree-of-belief probabilities inferred by eliciting expert opinion.

How do people estimate subjective probabilities?

While it is evident that subjective probability requires integrating information of various kinds within a consistent framework, it is less clear how people do this. Subjective probabilities should be concordant with probability theory (e.g. they should sum to 1.0), and one would prefer they be calibrated to observed frequencies in the physical world (i.e. they should have predictive value). These properties are called *coherence* and *calibration*. A considerable body of behavioural research indicates that people are not well equipped for mentally processing uncertainty according to probability theory, or in harmony with observed frequencies (Hogarth, 1975). The mathematics of probability theory describes how people ought to quantify uncertainties, not how they do.

In practice, it appears that people use simple mental strategies or rules of thumb to simplify the task of quantifying subjective probabilities. In the literature of cognitive psychology, these are called *heuristics*. A large literature has grown up around heuristics and, as importantly, around the systematic errors to which heuristics lead. The literature calls the latter *cognitive biases*. The conclusion is, people use rules-of-thumb to simplify judgements about probability, and to the extent these fall short of normative standards, so do the assessed probabilities. A surprising number of engineers take subjective probability assessments for granted, treating them as though they were coherent and calibrated, when, in fact, they often are not. This *heuristics-and-biases* school of thought is most associated with the work of Edwards and Tversky (1967), Kahneman *et al.* (1982) and their colleagues. Criticism of the approach can be found in Gigerenzer (1991, 1999). Applications to geotechnical engineering have been reported by Folayan *et al.* (1970), Baecher (1972), Hynes and Vanmarke (1976) and Vick (2002), among others. The work has been used extensively in nuclear risk assessment (Meyer and Booker, 1990; Mosleh *et al.*, 1987) and in seismic hazard analysis (Budnitz *et al.*, 1998). Common heuristics revealed by behavioural research and some of the biases they lead to are described in the sub-sections of this and the next section. These have mostly been identified in controlled settings. In practical settings, heuristics are more difficult to distinguish.

Representativeness

Representativeness is more often illustrated than precisely defined (Kahneman and Tversky, 1982). In general, it reflects subjective probability judgements based on the resemblance of particular conditions in one circumstance to those in another. In a classic experiment, subjects are provided with a detailed profile of the behaviour and personality characteristics of a hypothetical person, then asked to estimate the probability that the person is a lawyer vs an engineer. Subjects told that the individual was drawn from a group of 70 lawyers and 30 engineers produced the same estimates as those told that the group contained 30 lawyers and 70 engineers. The subjects' judgements are based on matching the description to stereotypes of lawyers and engineers, ignoring other information – in this case prior frequencies.

In geotechnical practice, representativeness is often encountered in over-reliance on complex models, while discounting simple observations. Uncertainties remain disguised by embedded approximations, simplifications, and assumptions so that analysis results are taken as uniquely representative of field conditions with near-certainty, because "it's the best analysis we have".

Anchoring and adjustment

Anchoring and adjustment is easily illustrated: when asked to estimate a quantity or an uncertainty, people often start with a "best estimate", and adjust up or down.

Unfortunately, people tend to stick too close to the initial value, not adjusting sufficiently to reflect uncertainty.

Asked to estimate the undrained shear strength of a clay and the uncertainty in that value, one's natural reaction is first to think about a typical value. What's the average shear strength for this type of soil across the many sites I have dealt with? How does the present site differ? Do I think the soil here is stronger or weaker, stiffer or looser? How much should I adjust up or down? Might this soil be 10% stronger? How different could this site be to suggest upper and lower bounds? These are the questions one asks oneself, but this chain of reasoning is exactly that which has been shown to lead to significant over-confidence in resulting estimates.

A quite different result is obtained if one first states the largest value the strength might have, then the lowest, and only afterwards hones in on a central value. The latter yields a broader range of assessed uncertainty and a better calibration to the physical world. Folayan *et al.* (1970) present an example of this effect as observed in geotechnical practice.

How well do people estimate subjective probabilities?

Heuristics deal with *how* people quantify probabilities, but they also influence *how well* people do so. The many fallacies that people – even technically trained people – exhibit at the gambling table should dissuade us from thinking that one's natural tendencies concerning probability are well-calibrated to the physical world. People behave as if games of chance even out, or as if pulling the slot machine handle oneself improves the chance of winning, or as if small numbers of observations are highly representative of a random process. These things are all false and, left to one's own devices, the probabilities we estimate are usually neither coherent nor consistent. In particular, people tend to over-confidence in their assessments, and mis-calibration seems to vary systematically with the difficulty of the assessment.

Over-confidence

Over-confidence is the most pervasive bias in assessing subjective probability (Lichtenstein *et al.*, 1982). It manifests in probability estimates that are too extreme at both ends of the probability scale and estimated distributions having insufficient dispersion about the mean. People – even experts – rarely assess their uncertainty to be as large as it usually turns out to be, sometimes to a shocking extent, as reported in a well-known study by Alpert and Raiffa (1982).

Figure 7-6 plots the results of experiments reported by Fischhoff *et al.* (1997) in which three groups of subjects provided answers to general-knowledge questions as well as estimated probabilities that their answers were correct. The estimated error probabilities were found to be reasonably well calibrated relative to the actual error frequencies only within a limited range for probabilities no less than about 0.1. Their over-confidence, expressed as the difference between actual and judged error probabilities, increased dramatically at more extreme values with estimated error probability of 1:1,000,000 corresponding to actual error frequency of about 1 : 10 or a ratio some five orders of magnitude. Moreover, the subjects showed little ability to distinguish among varying degrees of extreme likelihood, with judged probabilities ranging from 10^{-2} to 10^{-6} despite actual error frequencies hovering about 10^{-1}. A related and surprising finding is that the harder the probability estimation task, the greater the associated over-confidence, and for quite easy tasks sometimes under-confidence is displayed, although this effect is poorly understood.

Neither experts in general nor geotechnical experts in particular seem immune from over-confidence. Hynes and Vanmarke (1976) give figures showing best estimate and 50% confidence level predictions of embankment failure height made by seven internationally known geotechnical engineers for a test embankment on soft clay at the

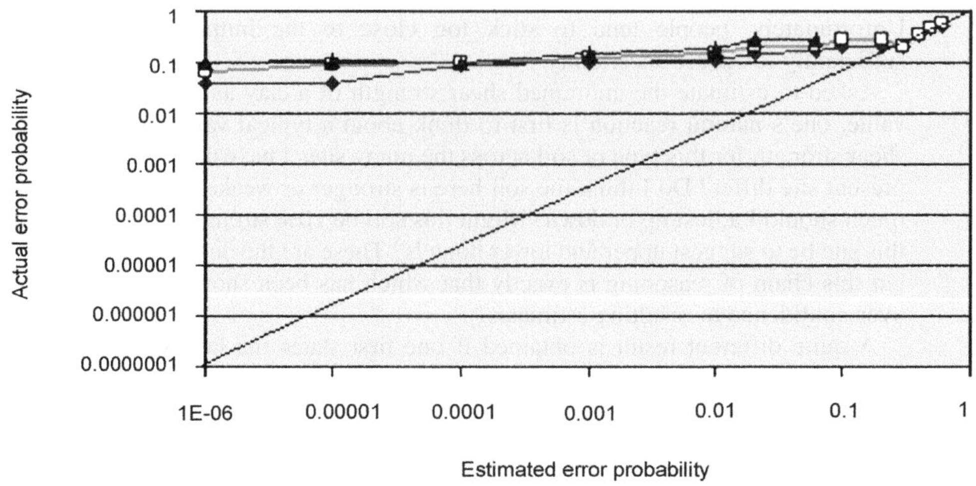

Figure 7-6. Subjectively estimated vs actual probabilities (Fischhoff et al.*, after Vick, 1997)*

Massachusetts Institute of Technology (MIT) I-95 test site. While the average of the seven best estimates is reasonably close to the outcome, no individual estimate had 50% error bounds large enough to encompass the actual outcome. Had the estimates been unbiased, half would have encompassed the actual failure height at the 50% confidence level, but none did so. Slovic *et al.* (1982) and others have cited these and similar findings in suggesting that *substantive expertise*, or capability within one's specialised knowledge domain, has no necessary relationship to *normative expertise* or ability to provide coherent and unbiased probability judgements. On the other hand, at least some evidence with experts assessing probabilities about professional subjects with which they are familiar suggests that they may be more calibrated than are non-experts (e.g. in the case of weather forecasters (Winkler and Murphy, 1968) and auditors (Smith and Kida, 1991)).

Similar over-confidence effects have been shown in subjective estimates of probability distributions provided by geotechnical engineers, Folayan *et al.* (1970) obtained estimated distributions for compressibility parameters of San Francisco Bay mud from engineers with up to 17 years of experience. Baecher (1972) further analysed these prior distributions in comparison to that obtained from subsequent laboratory tests. As shown on Figure 7-7, the estimated means were lower than that measured, but more significantly, over-confidence produced distributions too narrow to encompass most of the measured data.

Why over-confidence? There are many suggestions in the literature. Keren (1994) suggests that anchoring and adjustment may be to blame. People may anchor on a probability estimate reflecting intermediate difficulty, say 75%, and adjust up or down – but not sufficiently – depending on the perceived difficulty of the estimation task. Ferrell and McGoey (1980) present a similar argument, but the literature contains other attempted explanations as well (McClelland and Bolger, 1994).

Neglect of base rates

Neglect-of-base-rate bias reflects people's tendency to judge probability by the similarity of one circumstance to another. Typically, this manifests in ignoring the average frequency or *a priori* probability of an event in light of new information, even when that new information may have limited diagnostic strength. A typical result is that an event, known to be rare in the physical world, will be assigned a reasonably high

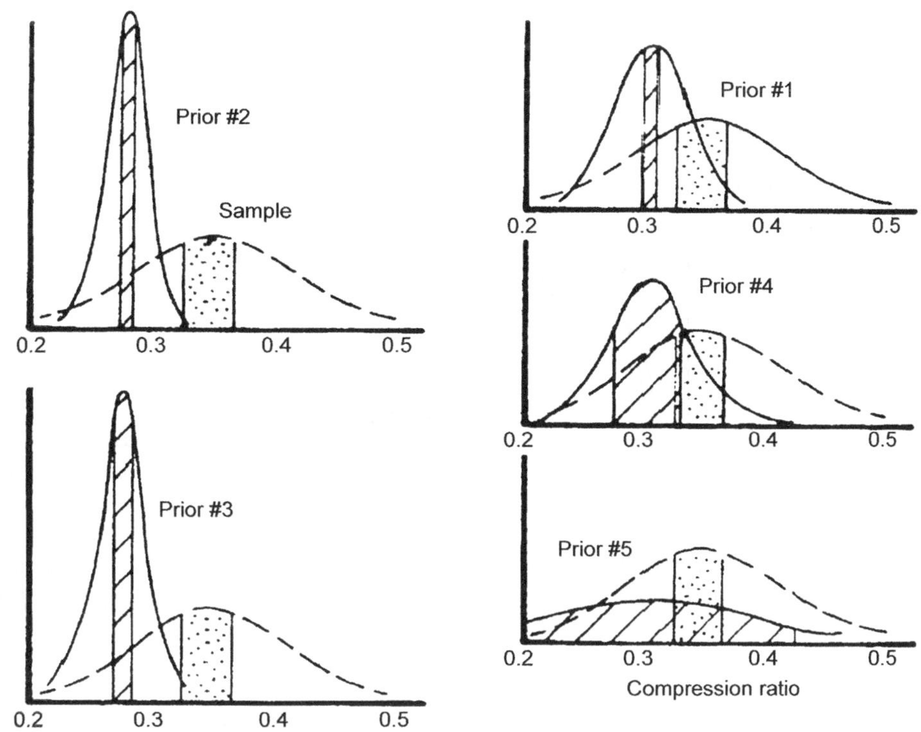

Figure 7-7. Subjective estimates of the compressibility of San Francisco Bay mud compared to test results for five experts (after Folayan et al.*, 1970)*

probability based on fragmentary evidence. Forensic engineering (failure–cause analysis) is a rich breeding ground for examples of neglect-of-base-rate bias.

The presumption is that people, even people with expertise in an area, focus on recent identifiable results as if they in themselves were wholly typical of some uncertain process or quantity. In so doing, people forget about – or at least under-rate – prior information or probabilities.

Results of a seismic hazard project are reported by Baecher and Christian (2003) in which liquefaction potential index (LPI) (Yegian and Whitman, 1978) was calculated for an industrial site in Tokyo Bay. The LPI was used in the analysis to predict ground failure. The result, expressed as a PDF, is shown in Figure 7-8. About the time the analysis was done, a Richter magnitude 6.5 earthquake occurred in the Tokyo area, causing 65 gal peak ground acceleration at the site, and evidence of liquefaction within the facility perimeter. The question was, how should this affect the calculated probability distribution of LPI?

Subjective estimates by several engineers suggested considerable adjustment, yet a calculation of the posterior PDF of LPI using Bayes' theorem (Figure 7-8) suggest only minor adjustment. The statistical calculation shows less updating and more influence of the prior PDF than do the subjective estimates. Because many uncertainties are involved in calculating LPI (corrected below count, water level, effect of vertical stress, cyclic shear strength, etc.), the combined effect of small changes in each of the many uncertain quantities easily accounts for the observed liquefaction. Our intuition suggests that the updating should be greater than statistical theory suggests.

A particularly malicious manifestation of base-rate neglect appears in what Kahneman *et al.* (1982) have called, the "law of small numbers". Recall that the law of large numbers states that as the number of observations (i.e. sample size) becomes large, statistical patterns of the sample asymptotically resemble those of the sampled

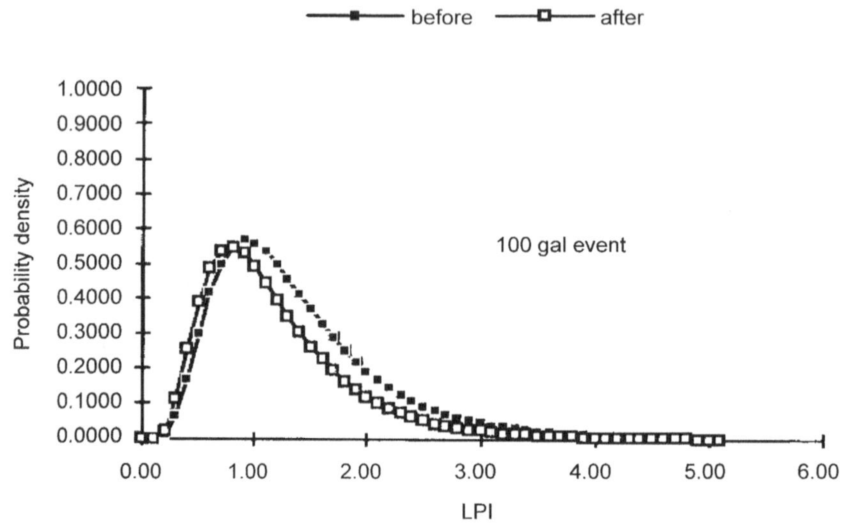

Figure 7-8. Liquefaction potential index (LPI) probability distribution for 100 gal event, before and after updating by observed earthquake performance

population. Kahneman and Tversky's law states that people intuitively believe that the same resemblance manifests in small samples. The enormous variability in statistical properties of small samples is largely ignored by engineers. Thus, we should not be surprised when subsequent measurements vary considerably from those taken initially but we usually are. Interestingly, Gigerenzer (1991) argues that the neglect of base-rate effect can be reduced or eliminated by describing priors in frequency terms rather than as probability functions but, while interesting, this is beyond the present scope.

Misperceptions of independence

Over-confidence and neglect of base rates deal with single variables. In risk and reliability analysis a critical factor is often the probabilistic independence – or lack thereof – among variables. How good are people at quantifying conditional probabilities or correlations? Somewhat less work has been done on this topic.

Some evidence exists that people judge the probability of the combined occurrence of events as being higher than the probabilities of the constituent events, when the events are representative of a pattern. This has been called the conjunction fallacy by Kahneman and Tversky (1982). Consider the following question: "A small dam was built across a stream with similar geology at both abutments. The regional geology is flat-lying sedimentary rocks. A spring flowing muddy water appears at the downstream toe of the dam, creating a small volcano of silty sand. Which of these two alternatives is more probable?: (a) There is a geologic fault beneath the dam, or (b) there is a geologic fault beneath the dam allowing internal erosion of the embankment". While many experienced people judge the second alternative as more probable, clearly it cannot be, because the joint occurrence of two events is necessarily less likely than either event alone.

Another place where misperceptions of independence arise is in combining statistical data with judgement. This is especially apparent in quantifying the uncertainties in soil engineering parameters, which may be statistically correlated. For example, soil cohesion and friction angle are in essence regression parameters. Larger values of c imply smaller values of ϕ, and vice versa. Thus, judgemental estimates of uncertainties in c and ϕ cannot be made separately, although common practice is to do so.

Can people learn to be well-calibrated?

A great deal of work has gone into finding out whether people can be trained to be better calibrated in their probability assessments. The common conclusion is yes, but the results of training seem not to generalise well to new tasks (Ferrell and McGoey, 1980; Keren, 1994; McClelland and Bolger, 1994). People seem to provide better calibrated probabilities when asked questions within their realm of expertise, although this is not without controversy.

An encouraging result comes from the realm of weather forecasting. In a series of studies, Murphy and Winkler established that National Weather Service forecasters are perhaps the world's best-calibrated probability assessors (Murphy and Winkler, 1974, 1977a, 1977b; Winkler and Murphy, 1973). Why should this be? The authors attribute the result to: (i) practice (forecasts are made every day); (ii) immediate feedback (outcome on the following day); (iii) quantitative scoring of performance; and (iv) promotion and pay incentives for accuracy. The important suggestion for geotechnical practice is that the very act of risk assessment and quantifying subjective probabilities may improve practitioners' ability to assess their own personal uncertainties, whether or not a project calls for formal risk analysis.

Protocol for assessing subjective probabilities

The preceding discussion has summarised empirical findings suggesting that people are not inherently adept at quantifying intuitive probability values, at least in the sense of providing numbers that are consistent, coherent and well calibrated. This should not surprise us. People are not adept at all sorts of intuitive mathematical tasks, nor for that matter are they adept at consistently judging the loudness of sounds or other psychophysical phenomena (Stevens, 1951). That is why we calculate probabilities and perform risk analysis in the first place. The question is: what do we learn from the preceding evidence that helps us structure a protocol for assessing subjective probabilities that supports useful outcomes?

People, even geotechnical engineers, do not enter a situation with a well-structured, mathematical conception of the probabilities of events in their minds. The protocol of assessment must evoke such a structure. Current usage calls this process *elicitation*. The protocol cannot simply ask a subject to guess a number for the probability for an event, and expect that numbers so generated will be consistent, coherent and well calibrated.

Even if the cognitive influences of heuristics and biases would seem to paint a bleak picture, people are nevertheless remarkably well adapted to dealing with uncertainty in everyday life, and successfully accommodating uncertainty has always been a hallmark of geotechnical practice. The problem is less one of expressing uncertainty judgements than in doing so in ways congruent with probability theory. The first requirement is internal validity, or *coherence*, with imposed requirements for satisfying basic probability axioms: a set of subjectively estimated probabilities for mutually exclusive and collectively exhaustive events must sum to 1.0 just like any other.

Beyond this, internal consistency depends on the sensitivity to how information may be presented or synthesised, and here is where heuristics and biases come into play. Defeating or at least reducing their effects is one important factor but, unfortunately for practitioners, the heuristics-and-biases school is mostly silent on this topic. Elicitation procedures from related fields, however, provide a framework for subjective probability assessment that acknowledges behavioural effects.

Conducting a process to elicit quantified judgement

A common misconception in eliciting expert judgement is that people carry fully formed probabilistic opinions around in their heads on almost any subject of interest, and that the focus of an elicitation process is merely to access these pre-existing opinions. Actually, people do not carry fully formed constructs around in their heads but develop them during the process of elicitation. Thus, the elicitation process needs to help experts think about uncertainty, needs to instruct and clarify common errors in how people quantify uncertainty, and needs to lay out checks and balances to improve the consistency with which probabilities are assessed. A successful process of elicitation is one that helps experts construct carefully reasoned judgements. The process should never be approached as a "cookbook" procedure, for the results will be unsatisfactory, as will the risk assessment in which the results are used.

The steps in using expert elicitation to quantify judgemental probabilities are the following:

1. Decide on the general uncertainties the probabilities of which need to be assessed.
2. Select a panel of experts displaying a balanced spectrum of expertise about the identified uncertainties.
3. Refine issues in discussions with the panel, and decide on the specific uncertainties the probabilities of which need to be assessed.
4. Expose the experts to a short training programme on concepts, objectives, and methods of eliciting judgemental probability, and on common errors that people make when trying to quantify probability.
5. Elicit the judgemental probabilities of individual experts on issues pertinent to their individual expertise.
6. Allow the group of experts to interact, supported by a facilitator, to explore hypotheses, points of view, and quantified estimates of probability, toward the goal of aggregating probabilities and resolving the breadth of opinion.
7. Document the specific process used to elicit judgemental probabilities and communicate the results back to the panel of experts.
8. It is important for credibility – as well as defensibility – that the process be well documented, open to inspection, and methodologically transparent to peer review.

Choosing experts: who should be on the panel?

The matter of choosing experts who should be on the panel is dealt with in Chapter 11.

Systematic process of elicitation

The process of eliciting expert opinion in judgemental probabilities has five phases; (1) motivating; (2) training; (3) structuring (deterministic); (4) assessing (probabilistic); and (5) documenting. The first two, motivating and training, set the stage for the whole process. The next two, structuring and assessing, comprise the central effort of the process, the former establishing a structure for the uncertainties to be assessed, and the latter performing the actual quantification. The process is complete with documentation, the purpose of which is to create a defensible set of results.

To obtain consistent and defensible results, it is important that a systematic process like this be used, although individual organisations may tailor the details of the protocol to their own special needs.

Motivating stage

The motivating stage intends to develop rapport with the experts, to explain why and how judgemental probabilities will be elicited, and how the results will be used in risk assessment. Experts are justifiably reluctant to participate in probability elicitation unless assured about how resulting assessments will be used in practice. During this phase, the basic philosophy of judgemental probability is reviewed, and an attempt is made to bring motivational biases within the expert panel out into the open. Motivational biases include those factors that might lead experts to provide assessments that do not accurately or completely reflect actual beliefs, whether generated consciously or unconsciously. Examples of motivational bias include the desire to appear knowledgeable and, thus under report uncertainty; the desire to influence a decision, and thus prejudice answers; and the desire to advance a special cause, and thus refuse to credit alternate points of view.

Training phase

The training stage has the purpose of making the experts aware of the processes and aids people typically use in quantifying judgemental uncertainties, and how well-calibrated judgemental probabilities are with respect to observed frequencies of assessed events in the world. The goal of this training is to encourage the experts to think critically about how they quantify judgement, and to avoid the common biases encountered in quantifying judgemental probability.

A typical training phase involves having the experts explain how they proceed to make probabilistic judgements, and how they use data and other information in arriving at expressions of uncertainty about the technical issues to be faced in the elicitation. Common cognitive biases are discussed, and warm-up exercises involving probabilistic predictions are used to illustrate these biases. It is helpful if these exercises involve uncertain quantities of an engineering nature, such that the experts feel professionally engaged. A useful exercise is to have the experts explain retrospectively how unanticipated outcomes of an engineering project might have occurred; that is, "if you know that an earth embankment settled excessively on what would otherwise appear to be a stable and homogeneous sand foundation, how might you explain that observation?" The benefit of such thought experiments is that they open up the range of considerations the experts might consider, and illustrate how easily one can become over-confident in assessing probabilities.

Deterministic phase

The structuring or deterministic phase has the goal of defining the specific uncertainties to be addressed, and the relationships among those uncertainties that allow individual probability assessments to be combined. The outcome of the deterministic phase should be clear; specific definitions of the uncertainties to be assessed, such that were a new expert added to the panel, the descriptions of the uncertainties to be assessed would stand on their own without further clarification. The goal is to make unstated assumptions explicit, and to disaggregate the technical problem into components that experts can readily deal with.

The deterministic phase begins with a review of the technical issues to be addressed in the elicitation. If time beforehand permits, a literature review and collection of documents should be made. A summary of this material should be prepared, along with graphs and tables comparing principal results or issues. Copies of the summary book should be distributed to the experts before the meeting, and time should be allocated at the beginning of the elicitation sessions to review the information. This helps the experts understand why the elicitation is approached the way that it is, and establishes a sense that the elicitation process is being attempted in a serious, complete and professional manner.

Two procedural issues affect the outcomes of a risk assessment: the way failure sequences are conceptualised, and the way probabilities are assessed. The deterministic phase considers the way failure sequences are conceptualised, and then decomposed into tractable parts.

One reason for decomposing failure sequences is that research suggests that people tend to be better at working with decomposed problems, and with estimating the probabilities associated with components, than they are at working with entire systems, and assessing probabilities for holistic processes (Gettys *et al.*, 1973). One reason for this is that, by design, engineered systems have very low probabilities of adverse performance or failure. From empirical experience, people are not well able to accurately estimate these low probabilities. Thus, by decomposing a system into a series of events, the conditional probabilities associated with individual components are made larger and people can more readily deal with them.

Asking questions about problem structure. The most common way of decomposing a dam safety problem is by building an event tree or a fault tree. An event tree starts with some initiating event, and then considers all possible chains of events which could lead from that first event to various performances by the system. Some of these chains of events lead to adverse outcomes or failures, some do not. For each event in the tree, a probability is assessed presuming the occurrence of all the events preceding it in the tree, that is, a conditional probability. The total probability for a particular chain of events or path through the tree is found by multiplying the sequences of conditional probabilities. Bury and Kreuzer (1985, 1986) describe how an event tree might be structured for a gravity dam.

A fault tree decomposes the problem in the reverse direction. The fault tree starts with some failure condition, and then considers all possible chains of faults that could lead to that failure. Conditional probabilities for each fault are assessed in the same way as with an event tree, but the total probability is calculated by starting at the failure and moving backwards, rather than starting from an initiating event and moving forward, as in an event tree. The advantage of the event tree is that it comprehensively uncovers combinations of events leading to failures; the disadvantage is that event trees can become bushy messes. The advantage of a fault tree is that it focuses only on chains of events leading to failures; the disadvantage is that it may fail to uncover important combinations of events.

An event tree can be used to decompose a problem at different levels of detail. Usually, judgement is more easily applied to smaller components, and research suggests that more detailed decomposition, within reason, enhances the accuracy of calculated failure probabilities. One reason, presumably, is that the more detailed the event tree is, the less extreme the conditional probabilities which need to be estimated.

Both event trees and fault trees require a strict structuring of a problem into sequences. This is what allows probabilities to be decomposed into manageable pieces, and provides the accounting scheme by which those probabilities are put together. In the process of decomposing a problem, however, it is sometimes convenient to start not with a highly structured event or fault trees, but with an influence diagram. An influence diagram is a graphical device for exploring the interrelationships of events, processes and uncertainties. Once the influence diagram has been constructed, it can be readily transformed into event or fault trees. Stedinger *et al.* (1996) show how influence diagrams can be used in assessing hydrologic risks of dam projects.

Finding an appropriate level of detail. Decomposition of a probability estimation problem relies on disaggregating failure sequences into component parts. Usually, these are the smallest-sized pieces that can be defined realistically and analysed using available models and procedures. Decomposition can be used for any failure mode that is reasonably well understood. Clearly, decomposition cannot be used for failure modes for which mechanistic understanding is lacking. Internal erosion leading to piping is arguably one such poorly understood failure mode.

In most cases, the extent of decomposition, that is the size of the individual events into which a failure sequence is divided, is a decision left to the panel of experts. Most real problems can be analysed at different levels of disaggregation. Considerations in arriving at an appropriate level of disaggregation include the availability of data pertinent to the components, the availability of models or analytical techniques for the components, the extent of intuitive familiarity experts have for the components, and the magnitude of probabilities associated with the components. Typically, best practice dictates disaggregating a failure sequence to the greatest degree possible, subject to the constraint of being able to assign probabilities to the individual components.

Usually, it is a good practice to disaggregate a problem such that the component probabilities that need to be assessed fall with the range 0.01 to 0.99 (see, e.g., Vick, 1997). If this range can be limited to 0.1 to 0.9, all the better. As will be discussed below, people have great difficulty accurately estimating judgemental probabilities outside these ranges.

Assessment (probabilistic) phase

Research in psychology has shown that people have limitations in their ability to consistently assess numerical probabilities representing their beliefs. Therefore, procedures which aid probability assessment, and which help minimise common forms of assessment bias, should be routinely used in expert elicitation. The elicitation of judgemental probability borders on experimental psychology, and as such is strongly influenced by procedural details.

Associating numbers with descriptive statements. Probability theory is one way of measuring uncertainty. Probability theory expresses uncertainty as a number between 0 and 1. People also use descriptive phrases to express the notion of uncertainty. For example, people say that something is "likely", or "probable". People might say, "there is a very good chance of that happening". These verbal descriptions carry meaning but they are less precise than numbers. Thus, it is tempting to attach verbal descriptions to the numerical scale of probability, and then to use verbal descriptions as a means of measuring judgemental probability. Common experience suggests that, at least in the early stages of expert elicitation, people find verbal descriptions more intuitive than they do numbers. Such descriptions are sought for components within the event or fault tree. Then, using approximate transformations between verbal descriptions and quantitative judgemental probabilities approximations can be assigned to component events. The warning about using verbal descriptions is that the range of implied numerical probability different people associate with verbal descriptions can be wide, and the number even an individual associates may change with the semantic context in which the verbal description is used.

Table 7-3, taken from Lichtenstein and Newman (1967) and Vick (1997), shows a simple list of verbal descriptions of uncertainty that have been used in the course of dam safety studies. More extensive studies have also been performed in the psychology literature to attempt to quantify the relationship between verbal descriptions of uncertainty and probability. The results show encouraging consistency from one person

Table 7-3. Empirical translations of verbal descriptions of uncertainty (sources: Vick (1997) and Lichtenstein and Newman (1967))

Verbal description	Probability equivalent	Low	High
Virtually impossible	0.01	0.00	0.05
Very unlikely	0.10	0.02	0.15
Unlikely	0.15	0.04	0.45
Fairly unlikely, rather unlikely	0.25	0.02	0.75
Fair chance, toss-up	0.50	0.25	0.85
Usually, good chance, probable, likely	0.75	0.25	0.95
Quite likely	0.80	0.30	0.99
Very likely, very probably	0.90	0.75	0.99
Virtually certain	0.99	0.90	1.00

to another, but the ranges of responses are large; also, mirror-image pairs sometimes give asymmetric results. Thus, verbal descriptions of uncertainty are a useful tool but are only a starting point.

Avoiding intuitive or direct assignment. It is common for experts who have become comfortable using verbal descriptions to describe probability, and to want to begin directly assigning numerical values to those probabilities. This should be discouraged, at least initially. The opportunity for systematic error or bias in directly assigning numerical probabilities is too great (e.g. Luce and Suppes, 1965; Phillips and Edwards, 1966). More experience with the process on the part of the experts should be allowed to occur before directly assigning numbers. At this initial point, no more than order of magnitude bounds on the elicited numerical degrees belief are a realistic goal.

The direct numerical values generated by experts unfamiliar with judgemental probability elicitation tend to be unstable, and often violate basic axioms of probability. Because the judgemental probabilities elicited for component events are almost always conditional on the occurrence or non-occurrence of other events in the event schema, it is useful to use these numerical values to calculate total probabilities and then to review implications with the panel.

Intuitive approaches which simply ask people what they consider a probability to be, tend to make no attempt to foster careful weighing of judgemental opinions or to disassociate events from their consequences. One result, for example, is that people tend to over-estimate the probability of events that have a favourable consequence and to under-estimate the probability of those with an unfavourable consequence. Intuitive approaches make use of specific numerical scales, and these may be suggestive of things not intended.

Action approach to elicitation. The degree-of-belief theory is based on the notion that judgemental probability and actions are inseparable: judgemental probabilities may be inferred from behaviour, but they are not necessarily intuitive. As a consequence, simply asking an expert what he judges a probability to be will not necessarily result in an accurate assessment of his judgemental probability (i.e. as might be reflected in a situation requiring him to make a decision). Thus, to accurately assess judgemental probability one needs some form of "action" approach in which probabilities are inferred from behaviour in a controlled situation.

The action approach to assessment presents subjects with comparisons and decisions. The approach places gambles involving the probabilities to be assessed. In the literature, the hypothetical gamble is called a "reference lottery". Judgemental probabilities are inferred from the resulting decisions.

The expert is given the choice between two lotteries: one presents a probability p of winning a significant cash prize, C, with a complementary probability $(1-p)$ of winning nothing; the other presents the same cash prize C if a discrete event A occurs, and winning nothing if A does not occur. The expert is asked to adjust the value of p until he is indifferent between the two lotteries. The resulting p is assumed to be the same as the judgemental probability of A.

Consider a dam site at which one potential mode of troublesome performance involves the unlikely but possible existence of a fault or shear zone in the rock stream formation under the dam. If such a zone were there, problems of uplifting and potentially with internal erosion would be of increased concern. Some amount of site characterisation has been carried out, but the results are not definitive, and the resulting uncertainty is sought as part of a risk assessment. That uncertainty depends in part on the prior experience of the expert making the judgement, and in part on the inconclusive information from the site characterisation programme. How might this judgemental uncertainty of the fault existing be quantified?

Were no other information available (e.g., regional frequencies of shear features), the expert's uncertainty about the fault could be approximately measured by comparison to familiar events. For example, if the expert prefers to bet on the toss of a fair coin rather than on the existence of the fault, by implication his/her judgemental probability of the fault existing must be less than $\frac{1}{2}$. Should he/she prefer to bet on the existence of the fault over the roll of a six-sided dice, then his/her judgemental probability of the fault existing must be greater than $\frac{1}{6}$, and so forth, leading to a bounding of the uncertainty. Changing the payoff odds on the gambles is another means for bounding the assessment.

Research on expert elicitation has addressed a number of methodological issues of how probability questions should be formulated. For example, should questions ask for probabilities, percentages, odds ratios, or log-odds ratios? In dealing with relatively probable events, probabilities or percentages are often intuitively convenient to experts, but in dealing with rare events, odds ratios (such as, "100 to 1") may be easier because they avoid very small numbers. Also, do devices such as probability wheels – which spin like a carnival game and represent probability as a slice of the circle – help experts visualise probabilities? Definitive conclusions from this research are lacking, and in the end, facilitators and experts must pick and choose a protocol that is comfortable to the individuals involved.

Quantifying judgemental probability distributions. Not all uncertain quantities involve simple probabilities of discrete events. Many of the parameters needed for engineering analysis have a realised value along a scale, and the issue facing expert elicitation is to assess the uncertainty about the parameter's value over that scale.

For example, the base friction between a concrete mass and its foundation, measured as a drained friction angle, could, in principle, have a value anywhere between zero and ninety degrees. Uncertainty about the interval of this scale in which the actual value lies is described by a probability distribution. A probability distribution summarises the relative uncertainty about the parameter's value lying within specific intervals of the scale. In expert elicitation it is often convenient to represent probability distributions as cumulative functions, which graph the scale of the parameter along the horizontal axis, and the (judgemental) probability that the realised value of the parameter is less than specific values along the vertical axis.

The process starts by asking the expert to suggest extreme values for the uncertainty quantity. It is useful to have the expert describe ways that values outside these extremes might occur. Then, the expert is asked to assess probabilities that values outside the extremes occur. Starting with extreme values rather than best estimates is important in guarding against over-confidence and anchoring. Asking the expert to conceive extreme scenarios makes those scenarios "available" to him/her, and allows him/her to think about the extremes more readily.

As numerical values are elicited, the facilitator should begin plotting these on graph paper; however, at this point the plot should not be shown to the expert because it might bias future responses to conform to the previous ones. As ever more assessments are made, these are plotted on the graph to begin establishing bounds and to point out inconsistencies.

After establishing extreme values and their exceedance probabilities, the next step is to fill in values and their corresponding probabilities between the extremes. As in the beginning, it is good practice to avoid early assessment of the mean or mode, and rather to focus on values without particular significance to the expert. The judgemental probability that the realised value of the uncertain quantity is less than a specific number can be elicited using comparative gambles, a probability wheel or other aid, as discussed above. The plot of assessed values is used to identify gaps where more assessments are needed, and inconsistencies in the expert's responses, where further assessments or clarification is called for. Toward the end of the process, the interval technique might be used to elicit the median and quartiles of the distribution.

In checking for consistency, it is useful to compare numerical results elicited as values with those elicited as probabilities. In the *fixed probability* approach, the expert is given a probability and asked for a corresponding value of the uncertain quantity, or given a probability interval, and asked for corresponding ranges of the uncertain quantity. For example, "what value of friction angle do you think has a $\frac{1}{3}$ chance of being exceeded?" "What values of friction angle do you think have a 50:50 chance of bounding the true value?" In the *fixed value* approach, the expert is given a value of the uncertain quantity and asked the probability that the true value is less than that value, or the expert is given a range of values and asked the probability that the true value lies within that range. In the *interval* approach, the expert is asked for the median of the uncertain quantity, then the half points (quartiles) of each resulting interval, then the next half points (octiles), and finally the 0.01 and 0.99 points.

Limited research suggests that, in general, fixed value procedures produce probability distributions that are more diffuse and usually better calibrated than do fixed probability or interval procedures (Seaver *et al.*, 1978; Winkler, 1967). Furthermore, for reasons of avoiding over-confidence, the interval approach is discouraged.

Normalised frequency approach to estimating probabilities. The normalised frequency approach to assessing a judgemental probability starts with an observed, empirical frequency of similar events in dam inventories and adjusts those background rates either up or down to reflect local conditions.

A number of issues arise in using the normalised frequency approach. The first of these is identifying the relevant sub-category of events or processes in the dam inventory to which the present project relates. The collection of dams in the inventory is heterogeneous and often of insufficient size from which to draw firm statistical conclusions. Unique aspects of the project at hand may make it difficult to select a sufficiently similar subset within the inventory from which to infer base rate frequencies. Second, failures or dam incidents within the inventory are seldom simple cause and effect. Therefore, it is often difficult to isolate the empirical frequency of narrowly defined events and processes. Third, the calculation procedures for adjusting base-rate frequencies to the peculiarities of the current project are themselves not without difficulties. On the other hand, the normalised frequency approach has intuitive appeal in that it begins with empirical frequencies.

Two approaches have been used for adjusting base-rate frequencies either up or down to account for local conditions. Each of these starts with a list of factors to differentiate the current project from the aggregate of projects in the dam inventory, that is the factors that make a particular dam either better or worse than average. The question is how to combine these unique factors with the base-rate frequencies. The less formal approach is to make adjustments directly, through intuition and discussions among the expert panel. A significant hazard to this informal approach is that it exacerbates the anchoring bias discussed below. Nonetheless, experts often find direct, intuitive adjustments reasonable.

The more formal procedure is to use Bayes' theorem to update the base frequency. Bayes' theorem provides a calculation procedure for formally combining prior probabilities (that is, base rates) with other information that can be captured in likelihood functions (that is, conditional probability statements). To apply Bayes' theorem, first features are identified which distinguish the current project from those in the dam catalogue. Then, estimates are made of the conditional probability of these features being associated with a project performing adversely and, correspondingly, the conditional probability of those features being associated with a project not performing adversely. The ratio of these two conditional probabilities, the likelihood ratio, is multiplied by the ratio of the base-rate frequency and its complement (that is, one minus the base rate) to obtain an updated odds ratio. While, in principle, Bayes' theorem provides a vehicle for quantitatively updating a base rate for project specific factors, in application it requires that the reliability of indicator factors are known in both a false-positive and false-negative way, which is often difficult to assess.

Consider again the uncertainty surrounding the possible existence of a fault of modest size in the foundation of a proposed dam site. Presume that in the experience of the expert panel, the incidence of such faulting is about 0.2 with projects in similar geology. That is, the base rate, as a first approximation, is $p(fault)=0.2$. In the present project, however, a reasonable number of borings has failed to detect faulting. How should the base rate be adjusted? The base rate can be updated from Bayes' theorem, expressed as odds (i.e. the ratio of the probability in favour to the probability against, $p/(1-p)$):

$$\frac{p(faulting \mid data)}{1-p(faulting \mid data)}=\frac{p(faulting)}{1-p(faulting)}\,\frac{p(failed\ to\ find \mid faulting)}{p(failed\ to\ find \mid no\ faulting)} \tag{7-14}$$

The expression reads, the "odds ratio that faulting exists given the observed data, equals the product of the odds ratio before observing the data, times the ratio of the likelihoods (the likelihood ratio) of the data presuming that faulting were present compared to were it not present. Let us say that the expert panel, using geometric arguments, estimates that the probability that the boring grid would have successfully detected existing faulting is $\frac{2}{3}$. Conversely, if no faulting existed the probability is by definition, zero. Thus:

$$\frac{p(faulting \mid data)}{1-p(faulting \mid data)}=\frac{0.2}{1-0.2}\,\frac{1-0.67}{1.0}=0.08 \tag{7-15}$$

So, the site characterisation data reduces the odds from 1:4 to 1:12, and the corresponding probability from 0.2 to 0.08.

Reliability modelling for assessing probabilities. For some component events, engineering models are available for predicting behaviour. In these cases, reliability analysis can be used to assess probabilities associated with the components. Reliability analysis propagates uncertainty in input parameters to uncertainties in predictions of performance. The assessment problem is changed from estimating probabilities of adverse performance directly to estimating probabilities for the input parameters. Once probabilities for the input parameters are assessed, any of a variety of simple mathematical techniques can be used to calculate probabilities associated with performance. Among these are: first-order second-moment approximations; advance second-moment techniques; point-estimate calculations; or Monte Carlo simulation. Sometimes, experts elect to assess an additional component of uncertainty in the reliability analysis to account for model error. While there are many ways to do this, the most common is to assign a simple, unit-mean multiplier to the model output, having a standard deviation estimated by the experts to reflect model uncertainty.

Experience with panels of experts suggests that model uncertainty is among the least tractable issues dealt with. The difficult questions about model uncertainty have to do with underlying assumptions, with conceptualisations of physical processes, and with phenomenological issues. Experts tend to have strongly held beliefs on such matters, so that discussions can become intense. Nonetheless, model uncertainty is a critical aspect of risk assessment. Most models engineers deal with in their daily work were developed for design purposes. They deal with incipient failure conditions and with assuring that loads in resistances remain within working ranges. Risk assessment deals with adverse performance and failures. Thus, models which were developed to prevent the precursors of failure are now used to model failure processes themselves. Failure processes involve strongly non-linear behaviours in considerations of time rates and sequences.

Traditional engineering models may require a good deal of manipulation to draw conclusions about failure processes, and typically require a good deal of qualitative reasoning by experts.

Correlations among uncertainties. Most experts find the notion of judgemental probability intuitively reasonable and, with practice, develop proficiency at assessing their own uncertainty about individual events or parameters. On the other hand, most people, whether experts or not, have difficulty thinking about the correlations among the uncertainties pertaining to different events or parameters. Correlation means that a person's uncertainty about one event or parameter is affected by knowing whether another event occurred or by knowing the value of another parameter.

People usually require significant analytical assistance when grappling with correlations, so it is good practice not to be overly aggressive in trying to assess correlations. The easiest way to assess probabilistic dependence between two uncertain quantities, x_1 and x_2, is first to assess the conditional probabilities for x_2, assuming various values for x_1, then to assess the marginal probabilities for x_1 (i.e. irrespective of the value of x_2). The joint probabilities are found from the relationship:

$$P(x_1, x_2) = P(x_2 | x_1) P(x_1) \tag{7-16}$$

in which:

$P(x_1, x_2)$ = the probability of x_1 and x_2 occurring together,

$P(x_2 | x_1)$ = the probability of x_2, given the value of x_1, and

$P(x_1)$ = the probability of x_1 irrespective of the value of x_2.

This approach requires that multiple assessments of the conditional probabilities $P(x_2 | x_1)$ be made for various values of x_1, but the advantage is that the expert does not have to grapple explicitly with the concept of correlation coefficients, which tend not to be intuitive. The reverse conditional probabilities, $P(x_1 | x_2)$, which are often needed for the risk assessment, can be calculated using Bayes' theorem.

In practice, it is better to attempt to restructure a problem rather than to assess correlations among uncertainties. This can be done, for example, when two uncertainties are correlated because they each depend on some third uncertainty, as in the case of downstream costs of flooding. Emergency mobilisation costs and property damage costs in the future may be correlated because each depends on inflation. It would be more effective to assess the uncertainties in each cost conditioned on inflation, and then combine the two independent assessments, rather than attempt to assess the correlated behaviour of the two uncertainties. Of course, not all correlated uncertainties can be handled in this convenient way.

Verifying assessed probabilities. Once a set of probabilities has been elicited, it is important to check for internal consistency, which statisticians sometimes call "coherence". Coherence means that the numerical probabilities obtained are consistent with probability theory. This can be done, first, by making sure that simple things, such as the probabilities of mutually exclusive and collectively exhaustive events adding up to 1.0, are confirmed. Second, it is also good practice to reword or restructure questions in logically equivalent ways to see if the resulting answers change. Also, redundant questions can be asked about events or parameters whose probabilities could be calculated from previous answers. Inconsistencies are resolved by discussion and reconsideration.

Beyond coherence, the implications of the elicited probabilities for risk estimates and for the ordering of one set of risks against other sets is also useful feedback to the experts. Sometimes, seeing unexpected or unanticipated implications of a set of assessments causes experts to reflect again upon the answers that have been given and the probabilities that have been elicited.

Practical suggestions and techniques

Elicitation of subjective probabilities is best conducted with the assistance of probability analysts or facilitators versed in probability and decision theory, behavioural effects, and ideally geotechnical engineering as well. Their job is principally one of ensuring external validity and promoting internal consistency of the probabilities produced, much of which involves a process of prompting and querying the assessor. For example, probabilities of 0.4 and 0.8 for the presence of either soil or rock at a particular location in the ground would fail the requirements of external validity. Further questioning of the underlying reasoning would be necessary either to reconcile the discrepancy, or to reveal hidden assumptions (perhaps a perceived potential for residual weathering products at the soil/rock interface) which could then be made explicit by restructuring the problem with a revised set of possible outcomes.

Formal elicitation techniques customarily rely on education and training of assessors for reducing bias, under the precept that an assessor conscious of bias can and will compensate for its effects. General knowledge questionnaires are sometimes used to demonstrate over-confidence on a personal level. However, research findings on the effectiveness of such *debiasing* techniques are mixed, with some (Alpert and Raiffa, 1982) showing subjects to be nearly impervious to training even when offered rewards for unbiased results. Another perhaps more consistently effective debiasing technique for over-confidence is to request that the assessor specify and list the reasons why an elicited probability might be wrong (Koriat *et al.*, 1980; Slovic and Fischhoff, 1977), for instance by imagining reasons that might explain the occurrence of a less-likely outcome after the fact. Notably, listing both the supporting and conflicting evidence did not produce the desired effect. The authors are unaware of such specific devices for reducing other biases, but encouraging the assessor to consider all types of available information can reduce the effects of the representativeness heuristic, as the previous example of base-rate neglect showed.

Encoding refers to the actual specification of numerical probability values, and various aids are useful in geotechnical applications. One such technique adopted for risk analysis in tunnelling applications (Einstein *et al.*, 1976) uses a "probability wheel" device that separates a circle into adjustable sectors of different colour. A visual reference is obtained by adjusting the sector size until the assessor is indifferent between its relative proportion of the circle and the likelihood for the outcome in question.

Like people in general, most assessors express uncertainties, more readily, verbally than they do numerically, and another encoding technique used extensively in dam safety risk analysis exploits this trait using transformations from verbal to numerical designations. Table 7-3 shows conventions defined for mapping of verbal expressions of uncertainty to probability values. Also provided are median values and ranges for these and similar terms from a considerable body of behavioural research summarised by Reagan *et al.* (1989), who show that not only do people adopt these transformations in fairly consistent ways but that they also use them commutatively in converting both words to numbers and numbers to words.

The conventions help promote consistency among likelihood judgements expressed by a particular assessor, and from one assessor to another. In practice, the verbal expressions allow encoding to proceed along the lines of a multiple-choice exercise in first establishing a bounding range on verbal expressions by eliminating those which do not apply, then converging on the most appropriate one(s), followed by refinement of an interpolated probability value as needed. The defined conventions are purposefully truncated at probabilities of 0.01 and 0.99 since cognitive discrimination limitations usually produce little meaning for probabilities outside this range. At the same time, this maintains encoded values within or at least reasonably near the well-calibrated range and thereby provide some structural limitation on the effects of over-confidence bias. These constraints are relaxed in some circumstances, such as when indicated by further decomposition or by applicable base-rate frequency information.

7.5 Appendix: mathematics of probability theory

This section discusses the mathematical theory of probability and summarises a selection of results of importance to risk analysis for dam safety. We use probabilities to quantify frequencies and degrees of belief because probabilities provide consistency, and also because probabilities allow complex calculations to be made about uncertainties. Probability functions are introduced for describing uncertainties with respect to events or propositions, and discrete and continuous variables. In risk analysis one desires not only to quantify basic uncertainties, but also to derive the implication of those basic uncertainties on calculated results. This is accomplished using derived distributions which result from propagating uncertainties through mathematical models.

Mathematical theory of probability

Probability theory is a branch of mathematics. It is logically consistent in the sense that the mathematics of probability theory can all be derived from a set of axioms. These axioms specify properties that *probability* must have, but do not say what *probability* is. As a result many interpretations of probability are possible.

There are more than one set of axioms that probability theory can be based on. One such set is:

Axiom 1. The probability $P[A]$ of event A has a value between 0 and 1: $0 \leq P[A] \leq 1$.
Axiom 2. The sum of the respective probabilities of each of a set of mutually exclusive and collectively exhaustive events $\{A_i\}$ is 1.0: $\Sigma_i P[A_i] = 1$.
Axiom 3. The probability that two independent (defined below) events A_i and A_j both occur equals the product of their respective probabilities: $P[A_i \text{ and } A_j] = P[A_i]P[A_j]$.

All the mathematical relationships of probability theory derive from these simple axioms.

Random events

The mathematical theory of probability deals with experiments and their outcomes. An *experiment* is a random process generating specific and *a priori* unknown results or *outcomes*. The set of all possible outcomes of an experiment is called the sample space, S. Individual outcomes (points) within that space are called sample points.

The sample space of Figure 7-9 contains discrete sample points; the sample points may also be continuous. An event, A, is a collection of sample points within S. A simple event consists of a single sample point; a compound event consists of more than one sample point. The complement of A, denoted A^c, comprises all sample points in S not included in A.

The union of two events A and B (denoted A-or-B, or symbolically $A \cup B$) is the collection of all sample points contained in A or B or both. The intersection of two events A and B (denoted A-and-B, or symbolically $A \cap B$) is the collection of all sample points contained in both A and B. These relations are conventionally illustrated by the well-known Venn diagram (Figure 7-10). The term *mutually exclusive* means that two (or more) events cannot occur together, that is, the events share no sample points in common, while *collectively exhaustive* means that at least one event within such a group of events must occur, that is together the events comprise all the sample points in S.

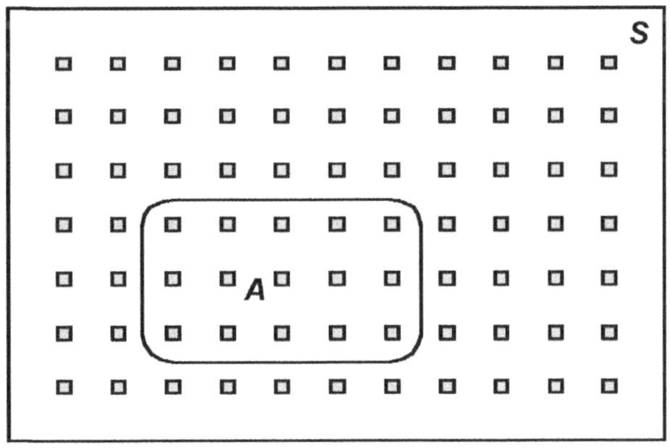

Figure 7-9. Sample space showing sample points and an event A

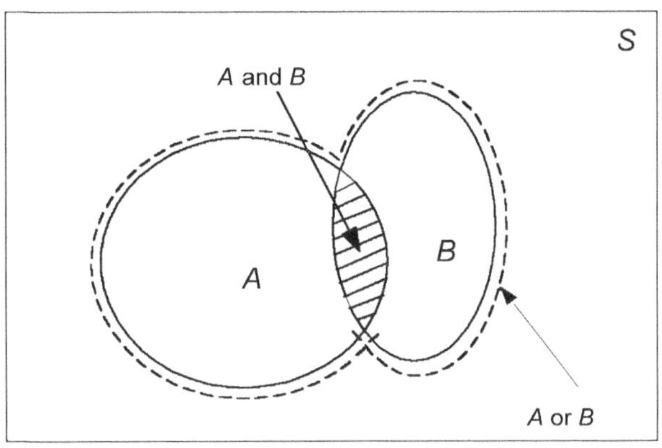

Figure 7-10. Venn diagram of events A *and* B *in sample space* S

Conditional probability, dependence and independence

One of the most important concepts in probability theory (and of paramount importance to risk analysis) is that of conditional probability, with its correlates, *dependence* and *independence*. Expressed in the formality of experiments and sample spaces, two events A and B are independent if they contain no sample points in common. That is, A and B cannot both be the outcome of a single experiment: an outcome can be A or B, or neither A nor B; but it cannot be both. Two events are dependent if they share sample points.

Conditional probability has to do with the notion that the probability of an event A may be influenced if another event B is known to obtain. For example, the probability assigned to high values for the undrained strength of a clay will presumably be larger than it would be otherwise, if we know the clay to have high preconsolidation pressure. From empirical evidence and from soil mechanics theory, high values of strength are associated with high preconsolidation pressures, and conversely for low values. Thus, the probability of high undrained strength given high preconsolidation pressure differs from the probability of high undrained strength in the case that nothing was known of preconsolidation pressure. This latter probability is said to be the *marginal* probability. The conditional probability of event A, given event B, is written, $P[A|B]$, while the marginal probability is written, as above, $P[A]$.

For the case in which the conditional probability is the same as the marginal, $P[A|B]=P[A]$, A and B are *independent*. For the case in which the conditional probability differs from the marginal, $P[A|B]\neq P[A]$, A and B are *dependent*. The intuitive meaning of independence is that the probability of A is not changed by knowing that B has occurred or is true. The definition is symmetric in A and B.

Table 7-4. Elementary relationship among the probabilities of events

Description	Equation
Event and its complement	$P[A^c]=1-P[A]$
Intersection of two events	$P[A \text{ and } B]=P[A]+P[B]-P[A \text{ or } B]$ $P[A \text{ and } B]=P[A]P[B]\vert A]$
Intersection of mutually exclusive and collectively exhaustive events	$P[A_1 \text{ and} \ldots A_n]=\prod_{i=1}^{n} P[A_i]$
Union of two events	$P[A \text{ or } B]=P[A]+P[B]-P[A \text{ and } B]$
Union of mutually exclusive and collectively exhaustive events	$P[A_1 \text{ or} \ldots A_n]=\sum_{i=1}^{n} P[A_i]$
Total probability	$P[A]=\sum_{i=1}^{n} P[A \text{ and } B_i]$, s.t. B_i me & ce
Conditional probability	$P[B\vert A]=\frac{P[A \text{ and } B]}{P[A]}$

From an informal consideration of Figure 7-10:

$$P[A \text{ and } B]=P[A]P[B|A]=P[B]P[A|B] \tag{7-17}$$

which can be rearranged and combined with the second axiom to give the relationship among conditional probabilities known as Bayes' theorem,

$$P[A|B]=\frac{P[A \text{ and } B]}{P[B]}=\frac{P[A]P[B|A]}{P[A]P[B|A]+P[A^c]P[B|A^c]} \tag{7-18}$$

In summary, a number of mathematical relationships derive immediately from the axioms. These are presented without proof in Table 7-4. The derivations of these and other elementary relationships of probability theory can be found in Benjamin and Cornell (1970), or Ang and Tang (1975) or in any introductory text on probability.

Describing empirical data

The most common way to graphically represent scattered data is in a histogram. A histogram graphs the number of measurements falling within specific intervals of value as a vertical bar. Thus, a histogram is sometimes called a bar chart. The height of the bar above each interval shows the number of measured values within the interval and the sum of the heights of the bars equals the total number of measurements. The histogram divides the data into fixed intervals. The choice of intervals is arbitrary, but the intervals should be of uniform width and have convenient end points. If too many intervals are chosen the general picture of relative frequencies will not be obtained and, conversely, if too few intervals are chosen the general picture will be blurred. A frequency distribution is constructed from a histogram by dividing each vertical bar by the total number of measurements. This gives the relative frequency of observed value in each interval. The sum of the heights of the bars in a frequency distribution is 1.0.

A cumulative distribution is constructed by summing relative frequencies starting at the lower-value end of the data and proceeding toward the upper-value end. The cumulative distribution, denoted $F(x)$, gives the fraction of measurements less than or equal to a particular value, $f_x(x_o)$=*fraction of measurements*$\leq x_o$. The cumulative distribution has the properties:

$$F_x(x_o) = \int_{-\infty}^{x_o} f_x(x)dx \tag{7-19}$$

$$F_x(+\infty) = \int_{-\infty}^{+\infty} f_x(x)dx = 1.0 \tag{7-20}$$

For discrete data the cumulative distribution is a step function increasing to the right. For continuous data the cumulative distribution is typically, although not necessarily, a smooth S-shaped curve.

Frequency distributions give a summary view of the variation in a set of data. The shape of the distribution suggests whether the data have any central tendency and, if so, where along the value axis the data are concentrated. The width of the distribution indicates the dispersion or scale of variation of the data. Some frequency distributions have a single point of concentration and are thus called unimodal. Others have more than one and are called multimodal. Most data sets encountered in dam safety studies are unimodal. Multimodal distributions may indicate a mixture of data of different types of inhomogeneous data. The frequency distribution also shows whether the variation in data is symmetric or asymmetric, that is whether high and low variations are evenly balanced. For data that are asymmetrically distributed, large variation from the central tendency of the data set are more frequent on one side of the centre than on the other.

Random variables

In engineering modelling it is generally convenient to fit mathematical models of variation to data distributions, and thus to represent the uncertainty in material properties, parameter values, or other quantities as probability distributions.

Discrete random variables

A random variable X having a sample space of discrete values is described by a *probability mass function* (PMF):

$$p_X(x) = P[X = x] \tag{7-21}$$

The generic random variable is denoted in upper case; a realised value in lower case. The values of the PMF are such that:

$$0 \leq p_X(x) \leq 1.0 \tag{7-22}$$

$$\sum_i p_X(x_i) = 1.0 \tag{7-23}$$

$$P[a \leq X \leq b] = \sum_{x_i \geq a}^{x_i \leq b} p_X(x_i) \tag{7-24}$$

The cumulative distribution function (CDF), sometimes called the cumulative mass function, for discrete variables is the sum of the $p_X(x)$ up to the value in question,

$$F_X(x) = P[X \leq x] = \sum_{x_i \leq x} p_X(x_i) \tag{7-25}$$

Thus, the probability $P[X = x_i]$ can be found from differences in the CDF:

$$P_X(x) = F(x_i) - F(x_{i-1}) \tag{7-26}$$

in which x_{i-1} is taken, for notational convenience, to be the next smaller point relative to x_i.

Continuous variables

For continuous random variables, the cumulative distribution function similarly describes the probability that a realisation is less than or equal to a specific value $F_X(x)=P[X\leq x]$, and the probability of the realisation lying within an interval $[a, b]$ is the difference of the CDF evaluated at a and b. As the interval $[a, b]$ approaches zero, the probability of X lying within the interval also approaches zero, the differential form of equation 7-26 expresses the density of the distribution of probability at a point rather than probability itself. This differential form is thus said to be the *probability density function* (PDF) of the random variable, which describes the density of probability at points along the x axis:

$$f_X(x)=\frac{d}{dx}F_X(x) \tag{7-27}$$

from which:

$$Pr(x_1\leq x\leq x_2)=\int_{x_1}^{x_2} f_x(x)dx \tag{7-28}$$

$$F_x(x_o)=\int_{-\infty}^{x_o} f_x(x)dx \tag{7-29}$$

$$F_x(+\infty)=\int_{-\infty}^{+\infty} f_x(x)dx=1.0 \tag{7-30}$$

$$F_x(-\infty)=0 \tag{7-31}$$

Moments

Probability distributions can be characterised by their moments. These are mathematically equivalent to physical moments in mechanics. The nth moment of $f_X(x)$ is:

$$\mu^n=E[x^n]=\int_{-\infty}^{+\infty} x^n f_x(x)dx \tag{7-32}$$

in which, $E[\cdot]$ is the *expectation*, or the weighted integral over $f_X(x)$. The first moment of $f_X(x)$ is the *mean* or arithmetical average, μ. This is the centre of gravity of the PDF. Moments taken around μ, rather than about 0, are said to be central moments. The nth central moment is:

$$\mu_n=\int_{-\infty}^{+\infty} (x-\mu)^n f_x(x)dx \tag{7-33}$$

The first central moment is zero. The second central moment, or average squared variability about the mean, is the *variance*, $Var(x)$. The square root of the variance is the *standard deviation*, σ. The ratio of standard deviation to mean is the *coefficient of variation*, $\Omega_x=\sigma/\mu$.

Another useful dimensionless ratio of moments is the coefficient of skewness, the ratio of third and second central moments,

$$\gamma_1 = \mu_3 / \sigma^3 \tag{7-34}$$

The coefficient of skewness is zero for symmetric distributions, positive for distributions with a long upper tail, and negative for distributions with a long lower tail.

Mathematical models for probability distributions

For many problems it is convenient to approximate the probability distribution by a mathematical function. Surprisingly, a comparatively small set of mathematical functions can be used to fit a broad range of frequency distributions encountered in practice.

Baecher and Christian (2003) tablulate the mathematical equations, ranges, and moments for a variety of commonly used probability density functions. By far the most important of these for continuous random variables (PDFs) is the Normal distribution (also called Gaussian and bell-shaped). Among other useful distributions are the log Normal, Exponential, and Beta distribution, although a great many other models exist (Johnson and Kotz, 1969, 1970, 1972, 1992). Baecher and Christian also tabulate equivalent information for commonly used probability mass functions. In the late 19th century, Karl Pearson attempted to systematise families of probability distributions. This system, using a convenient graph to plot the relationship of families of distributions to functions of their moments, is widely used in hydrologic and geotechnical risk analysis (see, e.g. Lumb, 1974). Detailed explanation of Pearson's families of distributions is given by Ord (1972).

Multiple variables

When more than one variable is of concern, the concept of probability mass function is directly extended to the joint PMF and the joint CDF, for the bivariate case:

$$p_{X,Y}(x, y) = P[(X=x) \textit{ and } (Y=y)] \tag{7-35}$$

$$P_{X,Y}(x, y) P[(X \leq x) \textit{ and } (Y \leq y)] \tag{7-36}$$

The joint probability density function is the continuous analog of the joint PMF:

$$f_{X,Y}(x, y) = \frac{\partial^2}{\partial x \partial y} F_{X,Y}(x, y) \tag{7-37}$$

The marginal distribution of one variable irrespective of the other(s) is found by integrating (summing in the discrete case) over the distribution of probability in the other variable(s), for example:

$$f_X(x) = \int_Y f_{X,Y}(x, y) dy \tag{7-38}$$

The conditional distribution of one variable given a particular value of the other variable(s) is found from the relative probability density with the other variable(s) fixed by renormalising the probability distribution such that its integral is unity:

$$f_{X|Y}(x|y_o) = \frac{f_{X,Y}(x, y)}{f_Y(y_0)} \tag{7-39}$$

For the special case in which x and y are independent, the following relations hold:

$$f_{X|Y}(x|y)=f_X(x) \tag{7-40}$$

$$f_{Y|X}(y|x)=f_Y(y) \tag{7-41}$$

$$f_{X,Y}(x, y)=f_X(x)f_Y(y) \tag{7-42}$$

$$F_{X,Y}(x, y)=F_X(x)F_Y(y) \tag{7-43}$$

$$F_{X|Y}(x|y)=F_X(x) \tag{7-44}$$

$$F_{Y|X}(y|x)=F_Y(y) \tag{7-45}$$

In the case where the variables are dependent, these simple multiplicative relations do not hold.

Derived distributions

In modelling one is often interested in functions of random variables. Since the arguments of the functions are random so, too, are the outputs of engineering performance random variables. The probability distributions of the output variables can be derived from those of the input variables and the form of the functional relations within the model. For relatively simple models, analytical results for derived distributions can be formulated. On the other hand, for the more complicated models typical in practice, approximations are required.

Analytical solution for derived distributions

Let y be a function of x through the relation:

$$y=g(x) \tag{7-46}$$

Presume y is a monotonically increasing function of x; then, for each value of x there is a unique value of y, and vice versa. In this case, the probability that $Y \leq y_0$ equals the probability that $X \leq x_0$, where $y_0=g(x_0)$. Thus:

$$F_Y(y_0)=F_X(x_0)=F_X(g^{-1}(y_0)) \tag{7-47}$$

and taking the derivative of $F_Y(y)$ and rearranging terms,

$$f_Y(y)=\frac{dx}{dy}f_X(x) \tag{7-48}$$

For multiple variables (i.e. vector values $\mathbf{x}$ and $\mathbf{y}$) the general relationship is:

$$f_{\mathbf{Y}}(\mathbf{y})=\frac{f_{\mathbf{X}}(\mathbf{x})}{J(\mathbf{x}, \mathbf{y})} \tag{7-49}$$

in which $J(\mathbf{x}, \mathbf{y})$ is the Jacobian of the transformation from $\mathbf{x}$ to $\mathbf{y}$,

$$J(\mathbf{x}, \mathbf{y})=\begin{vmatrix} \frac{\partial y_1}{\partial x_1} & \cdots & \frac{\partial y_1}{\partial x_n} \\ \vdots & \ddots & \vdots \\ \frac{\partial y_m}{\partial x_1} & \cdots & \frac{\partial y_m}{\partial x_n} \end{vmatrix} \tag{7-50}$$

First-order second-moment approximation

In many cases the analytical solution of equation 7-49 is not possible, and an approximation is needed. The most common is the first-order second-moment approximation (FOSM) using a Taylor's series expansion truncated to linear terms.

Consider the function $y=g(\mathbf{x})$ of $\mathbf{x}=\{x_1, x_2, \ldots, x_n\}$. If there is only one independent variable x and the value of g is known for some value of x, say $\bar{X}$, then $y=g(\mathbf{x})$ can be found for any other value of x using the Taylor series:

$$g(x)=g(\bar{X})+\frac{1}{1!}(x-\bar{X})\frac{dg}{dx}+\frac{1}{2!}(x-\bar{X})^2\frac{d^2g}{dx^2}+\frac{1}{3!}(x-\bar{X})^3\frac{d^3g}{dx^3}+\ldots \qquad (7\text{-}51)$$

This equation is exact, provided all terms to infinity are used. In many practical applications x is near $\bar{X}$, so higher order terms become small and the series can be truncated after only a few terms.

In practice, more than one independent variable usually appears, so a generalisation is needed. There are several equivalent forms of the Taylor series for multiple variables, but a convenient one is:

$$\begin{aligned} g(x_1, x_2, \ldots, x_n)={}&g(\mu_{x_1}, \mu_{x_2}, \ldots, \mu_{x_n})+\frac{1}{1!}\sum_{i=1}^{n}(x_i-\mu_{x_1})\frac{\partial g}{\partial x_i}\\ &+\frac{1}{2!}\sum_{i=1}^{n}\sum_{j=1}^{n}(x_i-\mu_{x_i})(x_j-\mu_{x_j})\frac{\partial^2 g}{\partial x_i \partial x_j}\\ &+\frac{1}{3!}\sum_{i=1}^{n}\sum_{j=1}^{n}\sum_{k=1}^{n}(x_i-\mu_{x_i})(x_j-\mu_{x_j})(x_k-\mu_{x_k})\frac{\partial^3 g}{\partial x_i \partial x_j \partial x_k}+\ldots \end{aligned} \qquad (7.52)$$

The partial derivatives are taken at μ_{x_1}, μ_{x_2}, . . . , etc. Ignoring higher order terms:

$$g(x_1, x_2, \ldots, x_n)\approx g(\mu_{x_1}, \mu_{x_2}, \ldots, \mu_{x_n})+\sum_{i=1}^{n}(x_i-\mu_{x_1})\frac{\partial g}{\partial x_i} \qquad (7\text{-}53)$$

To find the expected value of g, it is necessary to integrate g multiplied by the joint probability distribution function of the variables x_1 through x_n from $-\infty$ to ∞. Equation (7.53) is the sum of $n+1$ terms, so each term can be integrated in turn and the results added. The term $g(\mu_{x_1}, \mu_{x_2}, \ldots, \mu_{x_n})$ is a constant. So are all the partial derivatives, for they have been evaluated at the means of the individual variables. Also, the integral from $-\infty$ to $+\infty$ of a probability density function multiplied by a constant is simply that constant. This all leads to:

$$\mu_g\approx g(\mu_{x_1}, \mu_{x_2}, \ldots, \mu_{x_n})+\sum_{i=1}^{n}\int_{-\infty}^{\infty}(x_i-\mu_{x_1})f_{X_i}(x_i)dx_i \qquad (7\text{-}54)$$

but each of the terms after the summation sign must be identically zero, so:

$$\mu_g\approx g(\mu_{x_1}, \mu_{x_2}, \ldots, \mu_{x_n}) \qquad (7\text{-}55)$$

and

$$\sigma_g^2\approx\sum_{i=1}^{n}\sigma_{x_i}^2\left(\frac{\partial g}{\partial x_i}\right)^2+\sum_{i=1}^{n}\sum_{j\neq i}^{n}Cov(x_i, x_j)\frac{\partial g}{\partial x_i}\frac{\partial g}{\partial x_j} \qquad (7\text{-}56)$$

Simulation (Monte Carlo method)

A wide range of engineering and scientific disciplines use simulation methods based on randomised input, often called Monte Carlo methods. They have been employed both to simulate the behaviour of stochastic systems and to evaluate deterministic functions.

Because developments are often bound intimately to the disciplines in which the techniques have found application, notation and nomenclature sometimes reflect the preferences of a particular field.

To study the behaviour of random processes, a random value is generated for each uncertain variable and calculations are performed to yield a solution for that set of values. This gives one sample of the process. The trials are repeated many times, giving many samples of the process. Once a large number of runs have been completed, it is possible to study the output statistically and to obtain values of means, variances, probabilities of other fractiles, and other statistical parameters. Two important points should be noted. First, regardless of the number of stochastic variables, each run gives one sample of the process. Hence, increasing the number of stochastic input variables does not increase the number of runs for the same level of accuracy. Second, the technique is a repeated sampling of the stochastic process, and statistical methods can be used to examine the accuracy and precision of the results.

The unbiased estimate of the mean of a random process is simply the mean of the values returned by the process: $E[\mu_X]=\frac{1}{n}\sum_{i=1}^{n} x_i$; and the standard deviation of this estimate is related to the standard deviation of all the values by, $\sigma_{\bar{X}}=\frac{\sigma_X}{\sqrt{n}}$. If the process is Normally distributed, the variance is Chi-squared (χ^2) distributed. It can be shown (Benjamin and Cornell, 1970) that the upper confidence limit at the $(1-\alpha)$ level is:

$$(\sigma_2)_{1-\alpha}=\frac{(n-1)s^2}{\chi^2_{\alpha,n-1}} \tag{7.57}$$

in which s^2 is the variance computed from the sample, and the denominator is the cumulative χ^2 distribution with the parameters α and $(n-1)$. These equations can be used to obtain estimates of the number of samples necessary to achieve a desired level of accuracy.

The above shows what might be called the brute-force Monte Carlo method. A set of randomly distributed points is generated; they are used to compute a sample of values of the desired function or integral; the statistical properties of the result are calculated from the sample. Obtaining satisfactory accuracy requires that a large number of random points be used. Is there any way to reduce the computational burden or, equivalently, to increase the accuracy for the same number of points? It turns out that this is indeed possible, and the techniques usually described as *variance reduction methods*. Rubinstein (1981) summarises the goals and conditions of these approaches.

Monte Carlo simulation uses randomly generated points to cover the range of values that enter into a calculation. In most cases the problem involves integration, but it can arise in deterministic or probabilistic circumstances. Some of the computations in probabilistic applications are in fact evaluations of integrals rather than simulations of stochastic processes. The technique has the advantage that it is relatively easy to implement on a computer and can deal with a wide range of functions, including those that cannot be expressed conveniently in explicit form. The major disadvantage is that it can converge slowly.

Several techniques are available to accelerate convergence or, equivalently, to reduce variance. Each can be effective for some problems and counter-productive for others. The choice of which method to employ depends on knowledge of the problem and its components. Sometimes a preliminary calculation with a limited number of simulation points will indicate which variance reduction method can be expected to work best. In many cases two or more techniques can be combined effectively.

Endnotes

1. The argument for necessity, like so many things in history, was not new to Newton and Laplace. Leucippus (ca. 5th century BC) said that nothing happens by chance, that everything happens out of reason and by necessity; and that to say that something happens by chance means not that there is no cause for the thing, but only that the cause is unknown (Bennett, 1998).
2. The argument for chance, too, was not new to the 19th century. In opposition to the Atomist view of Leucippus came the Epicurian (3rd century BC) view of free will, the notion that people's actions are unconstrained. One can choose to do what one wants, and is responsible for ones actions. This view conflicted with necessity because if all things follow necessarily from causes, so, too, do human actions, and thus there could be no free will.
3. The quote from Graunt, J. "Natural and Political Observations Made Upon the Bills Of Mortality", is "[. . .] to know how many people there be of each sex, state, age, religion, trade, rank, or degree &c. by the knowing whereof trade and government may be made more certain [. . .]." (quoted in Newman, 1966).
4. For example, one may take the *a priori* uncertainty as uniform over the parameter itself, or over its square, or over its cube, about any of which one is equally ignorant, and yet the different *a priori* probability distributions lead to different *a posteriori* probability distributions.
5. Peirce called the former *induction* and the latter *abduction*, but the usage has not caught on.
6. What is here called the "traditional" school has been called "orthodox" by Jaynes (2003) and "classical" by Raiffa (1968), but neither of these names is wholly satisfactory because each conflicts with other uses of the term in the historical probability literature. While "traditional" itself is also not wholly satisfactory, it seems less objectionable than the other two. This school, deriving most notably from the work of Fisher in England, and that of Neyman and Pearson in England and America, might also be called the Anglo-American school.

8 Event tree analysis

While a variety of methods are available for analysing engineering risks, event trees have become the common approach in dam safety studies. There are a number of reasons for this. Event trees provide an intuitive structure within which engineers can organise the variety of considerations of concern about a particular dam. As event trees typically progress from start to finish in chronological order, they allow one to think through and record how a chain of events might unfold. They are also highly versatile in adapting to unique conditions at a particular dam. All of this is comforting to practical people facing practical problems: Problems are decomposed into tractably small pieces and then brought back together. At the same time, alternative approaches to risk analysis, most specifically fault tree methods, have proven awkward or at least difficult to apply in practical dam safety studies. Unlike a piece of mechanical equipment, a dam is not easily broken down into a fully enumerated set of components, and it is not easy to unequivocally link failures among a subset of those components to subsequent failures of others. It may turn out in future research that fault tree approaches shed new light on dam safety assessments, but for now event trees are the standard approach.

Ultimately, event tree analysis is used to inform a decision process by explaining how a dam might be expected to perform. It adds considerations that in the past had not been considered formally: namely the likelihoods of various performance modes and the consequences to the dam and to downstream uses should such performance modes occur. Approached from this perspective, event trees are a diagnostic tool; they are not intended to generate numbers alone but to draw inferences about how a dam might perform when subject to service conditions.

Part II of this volume provides a description of how event trees are constructed and analysed in practical problems. The present section supplements that discussion with theory. The creative act of developing event trees and performing analyses upon them is determined by the purpose to which the risk analysis is set. Consideration of purpose needs to be made from the start because the end result is determined by it. This chapter intends not to describe the mechanics of creating event trees, but to focus attention on background considerations. It considers human biases that may enter the development of event trees, relates the specification of probabilities in event tree analysis to the discussions in Chapter 7, and addresses computational issues.

This chapter is organised in three general parts. The first sections focus on conceptual issues. They start with historical background and develop alternative concepts of what an event tree attempts to model. The middle sections focus on events and their interrelationships. They start with a simple example and develop the concepts of influence diagrams, causality and the structure of event trees. The final section considers quantitative issues. It looks first at the assignment of probabilities to event tree branches, then discusses computation.

8.1 Background

In the context of risk analysis, *system failure* means the cessation of proper functioning or performance of what is expected of the dam as a whole. Since the function of a dam as a whole is to retain water, with some allowable seepage, functional failure occurs when the system ceases to retain water. By system is meant the group of interacting elements that form the complex whole of a modern dam. The utlimate goal of event tree analysis is to provide insight into the functioning of a dam, and into the associated uncertainties about the way the dam functions. Along the way, this leads to a quantification of the probability that the system, the dam, may cease to provide its essential function. This is the probability of system failure. This probability reflects the aggregate uncertainty in knowledge about the functional performance of the dam and in the environmental loads and service conditions that the dam will face.

In a simple way, an earth dam might be represented as in Figure 8-1. System failure of the dam occurs if more than a nominal amount of water escapes from upstream to downstream. Event tree analysis uses the standard engineering approach of decomposing the system into a tractible number of "parts" – rip rap, upstream and downstream shells, filters, core and so on – for which probabilities of component performance can be estimated. These estimates are recombined to assess the performance of the dam as a whole system.

As in any modelling activity, assumptions and simplifications are made at each step in conceptualising the dam as a system within the context of an event tree. Sometimes, particular failure modes are screened out early in an analysis and later turn out to be important. Sometimes interrelationships or interactions among events are eliminated that may turn out to have non-negligible influence on probabilities of system failure. Fundamentally, creating an event tree model of a dam is a knowledge-based endeavour. Different analysts will have different ways of defining events, different ways of linking events together, and different ways of estimating parameters and assigning probabilities to events. All these things, combined with inadequate data and poorly understood models, mean that event trees and their numerical results are never unique. In theory, there ought to be a way of tracking all the uncertainties that are generated by these limitations, but in practice there seldom is.

An event tree reflects a belief structure about a dam, about the natural environment within which the dam resides, and about the natural and human processes that affect dam performance. The uncertainties that enter an event tree analysis – both in the way

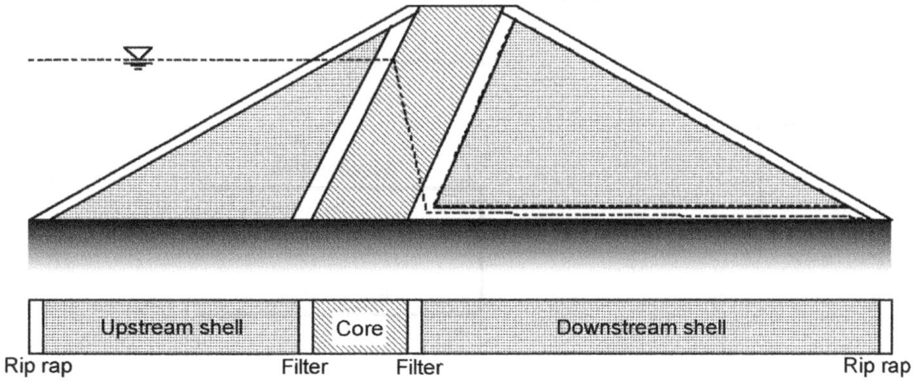

Figure 8-1. Simple representation of an earth dam system

events are structured in the tree and in the way numerical values of probability are assigned to branches – have mostly to do with limitations in knowledge not with random processes, although for modelling convenience they may be represented either or both as natural variations (aleatory uncertainties) and knowledge uncertainties (epistemic uncertainties). This is true about the external environmental forces acting on a dam, about internal response of the dam to those forces, and the estimation of properties and parameter values that enter the calculations.

Event trees

An event tree is a graphical representation of the many chains of events that might result from some initiating event; a few of which, should they occur, would lead to system failure. As the number of events increases, the diagram fans out like the branches of a tree, suggesting the name (Figure 8-2). An event tree is an expanding structure that starts from one or a few initiating events and widens combinatorially as the many subsequent events that might occur are added on. A fundamental presumption of event tree analysis is that engineering judgement is most powerful at detailed levels of dam behaviour, involving specific models, parameters and assumptions; it is less powerful at the aggregate level of an entire dam.

Probabilities can be associated with the occurrence of each event in a chain of events leading to failure, and a joint probability calculated for all the events occurring together and thus leading to failure. This yields a probability that failure occurs in this one particular way as a result of this one particular chain of events. Summing the probabilities of all the chains of events that lead to failure yields the total probability of failure due to this one initiating event. Summing the probabilities over all initiating events yields the total probability of system failure. Bury and Kreuzer (1985, 1986) and Kreuzer and Bury (1984) give examples for a gravity dam.

An event tree begins with some accident-initiating event (Figure 8-3). This might be a large flood, an earthquake, another natural hazard, human agency (e.g. operator error or sabotage), an internal flaw in the dam, or something else. From this point, the analysis attempts to generate all possible subsequent events, and correspondingly, events that might follow the subsequent events, and so on. These event outcomes are represented as branches issuing from the chance node representing a particular event. This process

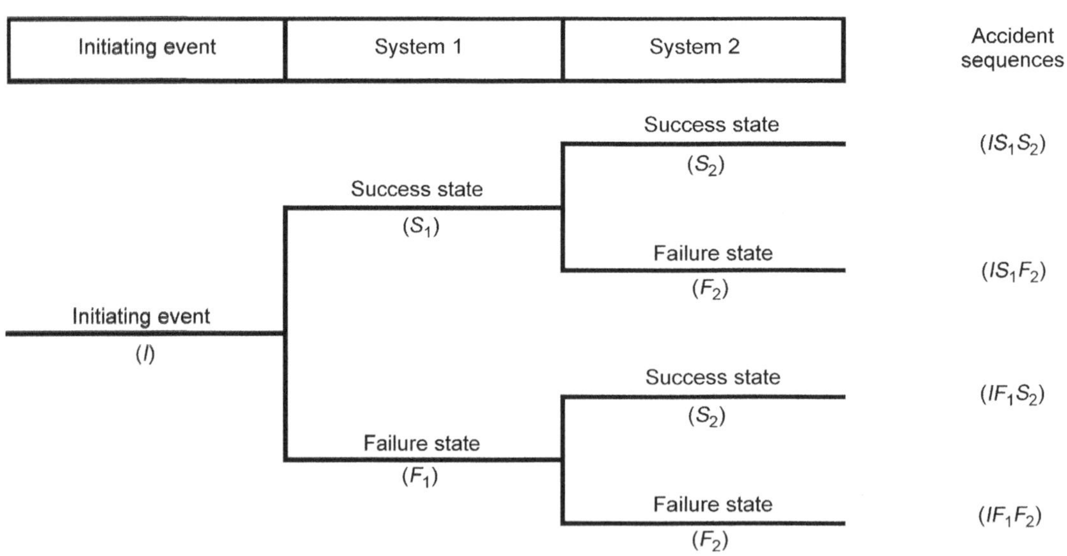

Figure 8-2. Illustration of a simple, generic event tree (US Nuclear Regulatory Commission, 1975)

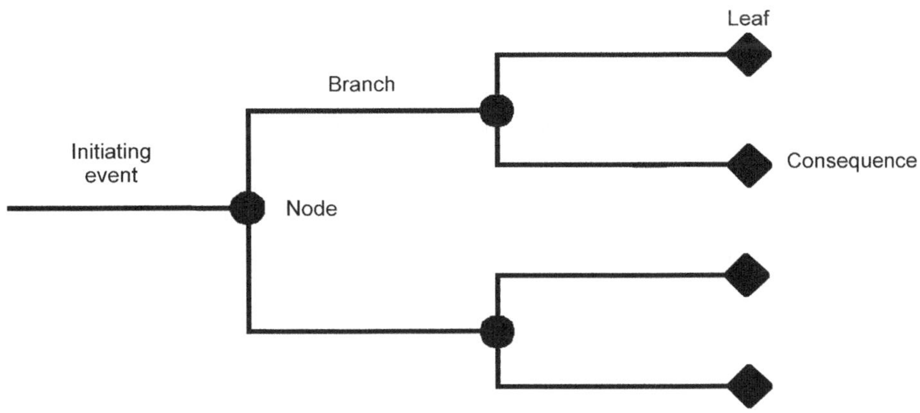

Figure 8-3. Event tree terminology

follows until many chains of events are generated, some of which lead to adverse consequences or failure, but most of which do not.

In the risk analysis literature for power plants, aircraft and other mechanical equipment, events within event trees are often limited to dichotomous outcomes (Leveson, 1995; McCormick, 1981). This is overly restrictive for most situations in dam safety. It is reasonable for events to have many possible discrete outcomes or even to have continuous outcomes. For computational purposes, however, continuous outcomes are usually replaced by a discrete approximation. The only theoretical requirement on the outcome space for events within an event tree is that the outcomes be mutually exclusive and collectively exhaustive. This implies that the outcome of a chance node follows exactly one branch.

The terminal node at the end of any chain of events through an event tree is referred to as a *leaf*. Each leaf in the event tree has associated consequences. The consequences are the costs or benefits accruing should the particular chain of events leading to that leaf obtain. In most cases, these consequences themselves may be complex and are analysed by constructing a *consequence tree*. A consequence tree structures the set of considerations involved in estimating consequences of failures in the same way that an event tree structures chains of events leading to possible systems failure.

Each branching path through an event tree represents a particular chain of events, or *accident sequence*, leading from an initiating event to a consequence (possibly benign). In Figure 8-4 these accident sequences are denoted at the right-hand side (RHS) of the

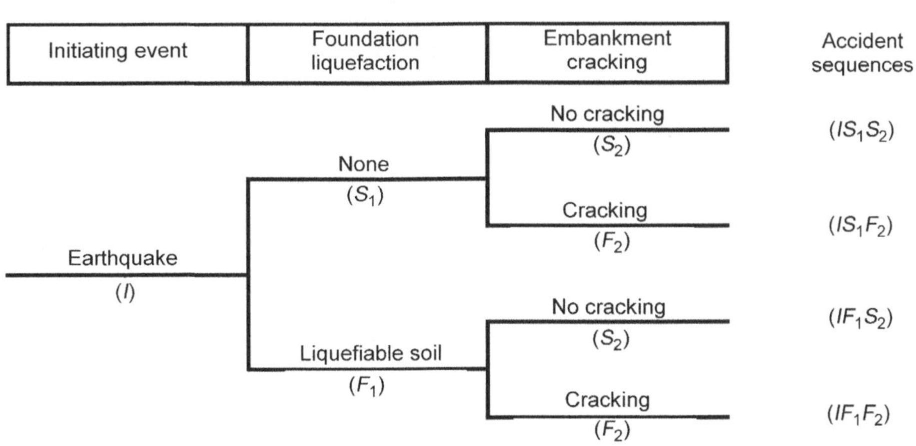

Figure 8-4. Illustration of a simple event tree for an embankment dam

tree by the concatenating symbols representing the individual events leading to that consequence. This has emerged as a standard way of representing accident sequences in the systems reliability literature. For example, the top-most sequence of branches starts with the initiating event (I), following which System One performs successfully (S_1) in inhibiting adverse consequences, and then System Two performs successfully (S_2). This sequence of branches is denoted, ($I\ S_1\ S_2$). Note the implication that System One is called first, followed by System Two at a later time. This chronological ordering is not necessary to the logic of an event tree, although it is often convenient.

An event tree is associated with the modes of behaviour of a system subject to a particular initiating event, and a separate event tree is usually created for each category of initiating event. The categories of initiating event could be external loads, internal changes of state, human activity, or any number of other things. Typically, however, event tree analyses for dam safety focus on three distinct initiating events: (1) severe storms; (2) earthquakes; and (3) design or construction flaws. Thus, an event tree analysis for a dam safety study might be built around three trees, with more limited risk analyses addressing such things as the mechanical performance of spillway gates or operator errors.

Generating an event tree is an inductive process, so, except in simple cases, the end result will depend on the skill and insight of the people doing the work.[1] Once an event tree is constructed, it can be used in a deductive way to calculate necessarily following conclusions about joint probabilities, risks, and the consequences of interventions.

There is no unique way to represent dam failure processes in an event tree. Different people may separate events and uncertainties in different ways. They may use different levels of decomposition. They may use different degrees of discretisation. Practical cases have shown that different groups of people with fundamentally different geological models for a dam site can arrive at quite different risk estimates, but this should hardly be surprising. Event tree analysis is simply a logical scheme for structuring concepts of how a system works and accounting for the uncertainties in that scheme. Differing concepts of geology or differing interpretations of systems behaviour may lead to different predictions. If such differences are critical, the event tree analysis can be used as a vehicle for focusing attention on how and why opinions differ, to see if further analysis lends credibility to one or another interpretation.

Modes of failure

Most published risk studies for dam safety focus on three broad categories or modes of failure (Lafitte, 1993):

- *Hydraulic failures* due to abnormally high pool. These include, among other things, overtopping and subsequent erosion of embankment dams, overturning of gravity dams, and downstream sliding on a foundation. Hydraulic failures may also include damage to spillway gates or operator errors associated with gates and spillways.
- *Mass movements* due to extraordinary loads, inadequate material properties, or undetected geological features. These include, among other things, limiting equilibrium instability of embankment dams, settlement leading to overtopping, liquefaction of foundation soils, abutment or foundation instabilities, rapid drawdown failure of upstream face and reservoir landslides leading to overtopping.
- *Deterioration and internal erosion*. These include, among other things, development of sinkholes in the dam embankment, piping within the dam core and erosion of foundation soils or joints.

Retrospective studies of historical dam failures seem to identify these three categories or modes as the more common ways that dams fail in practice (Baecher *et al.*, 1979; Foster *et al.*, 2000; Gruetter and Schneiller, 1982; Gruner, 1967; Jansen, 1983; Serafim,

1984). Important failure modes are identified ahead of time in constructing event trees, and thus are normally suspected from the outset, although event tree analysis may also lead to the identification of modes of failure that were not obvious *a priori*. The event tree reconstructs how initiating events and subsequent system responses unfold to lead to failure modes.

Initiating events

Initiating events are the first nodes in an event tree, the events that precede or initiate subsequent chains of the events leading to failure. The identification of external initiating events, especially those due to natural hazards, is relatively straightforward. External initiating events have to do with extreme loads, extreme hydrological events, seismic ground shaking, and other occurrences outside the envelope of the system being analysed (i.e. the dam). Usually, initiating events and their consequences are considered in isolation from one another. That is, the joint probability of such initiating events occurring simultaneously is considered negligible.

While a number of initiating events, in principle, could lead to each of the failure modes listed in the preceding section, most risk studies for dam safety focus on a small number of initiating events. Typically, these include:

- extreme storms leading to large inflows of water into upstream reservoirs;
- earthquakes causing seismic ground shaking;
- design or construction "flaws", typically associated with the impervious core of the dam or with filters installed to protect the core from erosion;
- equipment failures, for example of spillway gates or gauges; and
- human agency including operator error.

Initiating events are not necessarily one-to-one associated with the three failure modes of common concern. Each initiating event may contribute separately and in combination to the probabilities of each failure mode. Retrospective studies of historical failures appear to support the assumption that storms, earthquakes, flaws and equipment failures or operator errors are the common initiators of dam failure (Jansen, 1983).

Hydraulic failure, such as overtopping, results from an external initiating event, usually an extreme storm leading to a large inflow of water into the reservoir behind the dam, but may also be influenced by settlement of the dam crest, or by the performance of a gate or operator. Mass movement due to strength instability, too, is usually associated with an external initiating event, such as ground shaking, although it may be caused by gradual deterioration or changes in pore water pressures within the embankment or its foundation. Internal erosion has sometimes been modelled as caused by a flaw in either the design or construction of the dam, but it, too, is driven by external initiating events, specifically prolonged high pool levels or the first incidence of high pool level.

Conceptually, the analysis of initiating events and failure modes is similar across the many ways a dam behaves. Initiating events are treated as naturally varying phenomenon occurring randomly in time. Even though the uncertainties associated with external initiating events may be attributable to limited knowledge, in practice they are normally modelled as due to natural or random variability (i.e. as aleatory uncertainty). This implies annual probabilities of events of given size occurring or being exceeded as, for example, in flood frequency relations or earthquake recurrence functions. These initiating events are brought into an event tree which, in essence, can be thought of as a fragility curve representation of the system's response to the loading. The term *fragility curve* arises primarily in structural engineering, where it is used to mean a summary of structural response, expressed as the probability of failure as a function of applied load of one form or another. The fragility curve is treated as primarily reflecting limited knowledge of system behaviour, modelled as epistemic uncertainty.

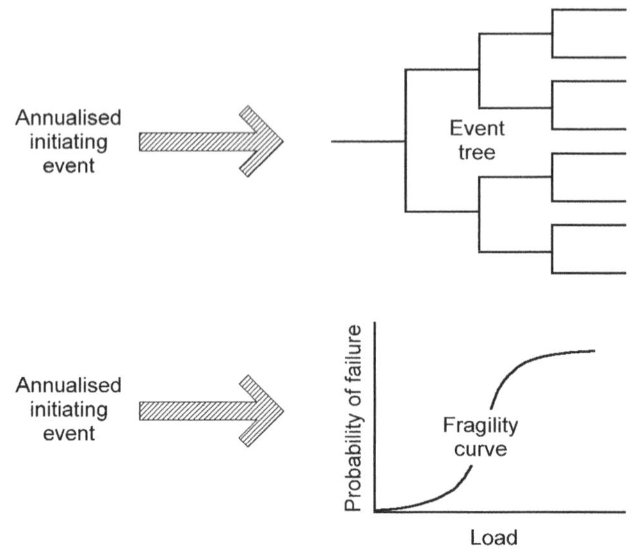

Figure 8-5. Annualised initiating event generates input to a system failure model that can be represented as an event tree, a fragility curve or some other description

The use of event trees in modelling any of these failures, for example, in generating a fragility curve, faces a number of difficult questions, including the absence of scientifically verified models for soil liquefaction, the start and continuation of internal erosion, the gradual development and redistribution of pore pressures, and other physical behaviours of dams. It also suffers a lack of good information about the as-constructed internal condition of most dams.

A significant fraction of dam failures occur upon first filling, when a dam is subject to operating loads for the first time. Sometimes, as in the case of flood control dams, this first filling to design capacity may occur only late in the life of the structure, when a low probability storm event finally occurs. In these cases, design and construction flaws – broadly defined to include things such as undiscovered site conditions, inappropriate design assumptions, poorly executed plans and so on – come to light. The initiating event is first filling of the reservoir, which is not necessarily a random event, and thus is not necessarily modelled as an annualised, random (aleatory) variable. The probability of failure is primarily related to limited knowledge as represented in the fragility curve (Figure 8-5). This is a place where event tree analysis is especially useful in systematising belief structures, setting out a framework for uncertainties based on limited knowledge and analysing the implications of those uncertainties.

Origins of event trees

Event tree analysis – whether by accident or design – is the most common approach to dam safety studies involving probabilistic risk analysis. Yet, the discipline is not supported by the same extensive literature found in the discipline of fault tree analysis or engineering reliability modelling. Why should this be? First, fault tree analysis has been applied to a large number of problems in mechanical and electrical systems reliability, many more than the number to which event tree analysis has been applied in dam safety and related areas of civil engineering. Second, fault tree analysis, with its sophisticated Boolean mathematics, makes a better topic for scholarly journals than does the practical know-how associated with event tree analysis. Third, the theoretical underpinning of event tree analysis appears straightforward, at least superficially, and not worthy of extended scholarly discussion.

Event trees most likely originated in efforts to enumerate (also to visualise) multivariate outcomes in probability sample spaces. Event trees appear in the work of early writers on probability theory (Daston, 1988; Hacking, 1975). In recent times, event trees have formed the basis for statistical (Bayesian) decision analysis (Raiffa, 1968) with its corresponding decision trees. Some writers have linked the origin of event trees to decision trees (Leveson, 1995), but event trees seem to have an earlier provenance.

The NRC *Reactor Safety Study* (Rasmussen, 1975) (colloquially known as the Rasmussen report) used event trees in combination with fault tree analysis to conduct probabilistic risk analyses of nuclear reactor safety (US Nuclear Regulatory Commission, 1975). The CIA and other government agencies have used event trees to forecast the outbreak of international hostilities and other such security issues as industrial espionage. Today, event trees have come to be the principal way that most risk analyses of civil works are conducted (Stewart, 2000; Whitman, 1984, 2000), as well as such applications as fire safety, and facilities operations.

Fault trees are the principal way mechanical systems such as power plants are analysed. Yet, even in the large probabilistic risk analyses (PRAs) of nuclear power plants, the problem is often first structured as a rudimentary event tree, and then filled in with detailed fault trees representing individual event sequences. Strictly speaking, there is no reason that dams cannot be modelled by fault trees. The top level event could be something along the lines of "failure to function adequately", or "failure to retain pool". Vick and Bromwell (1989) and Bury and Kruezer (1985, 1986) both report risk analyses of dam safety using fault trees.

Event trees and fault trees

Event trees, at least in concept, start with initiating events or causes to the left-hand side (LHS) of the drawing and progress toward ever more detailed consequences to the right-hand side. The ordering of events in an event tree can be rearranged so long as the relations among conditional probabilities are adjusted (more below), but in concept, the logical progression from cause to effect in a tree is an important, if sometimes concealed, principle of event tree analysis. In contrast, fault trees start with consequences (i.e. failures), typically at the top of the diagram, and progress backwards toward the ever more detailed causes, typically at the bottom of the diagram (see Chapter 9). Thus, the logical structure of fault trees is reverse to that of event trees, in that the logic moves from consequence to cause (Figure 8-6). A system failure mode is considered the 'top event' and a fault tree is developed in branches below this event showing causes. As suggested in Part II, event tree analysts ask, "what might happen if an initiating event occurs?"; while fault tree analysts ask, "how can a particular outcome come to pass?" In concept, fault trees are photographs showing conditions at an instant in time. They may also show conditions at the transition between two events in an event tree, or the top event may be used as the initiating event in an event tree.

It is sometimes suggested that event trees and fault trees can be used interchangeably, but this is true only with difficulty. In fact, event trees and fault trees evolved out of different needs and are adapted to different tasks (Bier, 1997). A fault tree presents all possible ways of producing a single event, using binary logic from Boolean algebra and resembling a "root system rising to a main stem". It is most suited to well-defined systems, such as mechanical equipment, for which all the components can be enumerated and their interrelationships specified. It is less well-suited to poorly defined situations, such as earth structures or the geological subsurface, for which discrete components may not be differentiable or interrelationships may not be unequivocal. It is also poorly suited to non-Boolean conditions in which component performance may have multiple states.

Fault trees identify specific outcomes and reason backward to uncover the possible ways that outcome might come to pass. A set of associated event trees may accomplish

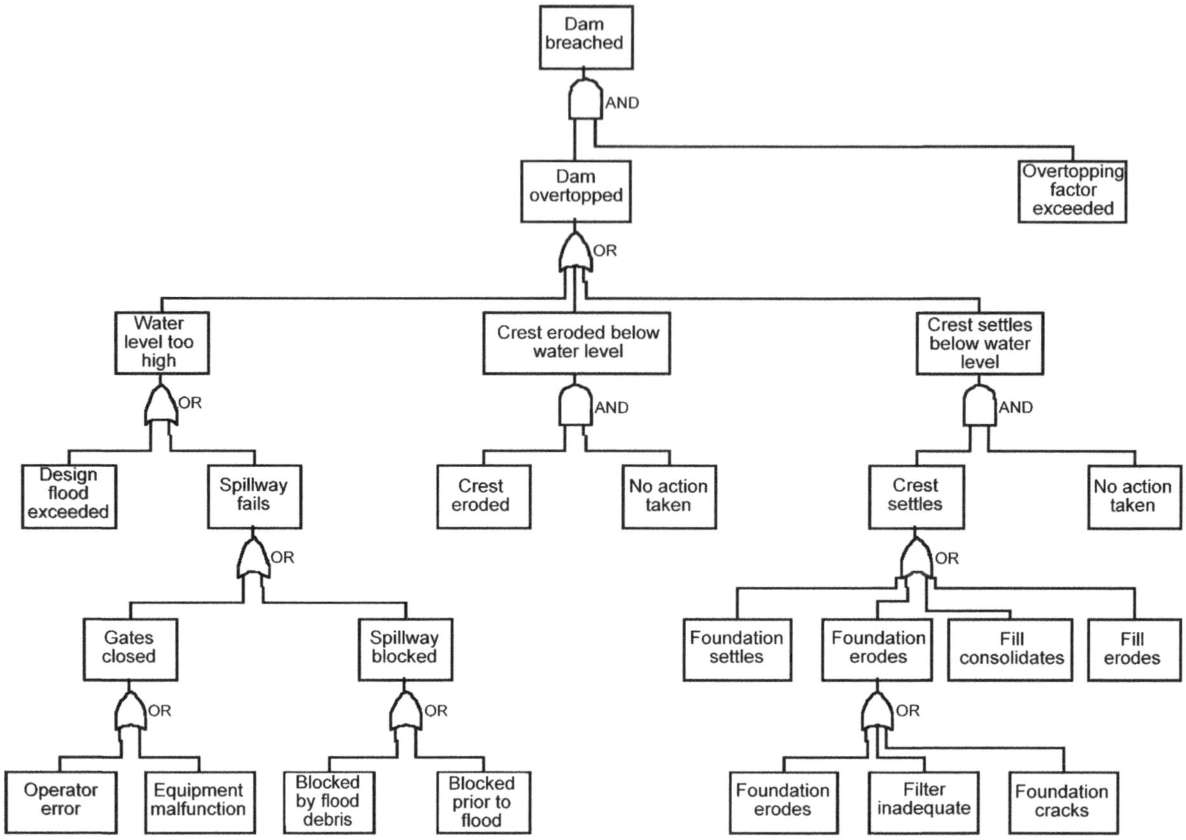

Figure 8-6. Example of a fault tree applied to the problem of dam failure (after Parr and Cullen, 1988)

this same result when their respective critical paths are considered as a group. To this end, fault trees can be more powerful and parsimonious ways of representing systems performance than an event tree, but only if the system is such that fault trees are well accommodated.

Sometimes fault trees can be used as the reliability analyses that underlie the estimation of probabilities for event branches (Figure 8-7). For example, spillway gate structures are well-modelled using fault trees because they are mechanical systems and share all the characteristics discussed above as amenable to fault tree characterisation. The probabilities associated with gate malfunction can be calculated using a fault tree just as they might be using some other structural reliability approach, and then input as the probability associated with a simple event branch, “spillway gate fails to function”, in an event tree.

8.2 Concepts

What does an event tree intend to be? The answer to this question – and no one answer is correct – dictates how an event tree is constructed and what conclusions can be drawn from it. Perhaps the most common concepts of event trees are the following three:

1. A model of a physical system (e.g. a model of a particular dam).
2. A statement about the joint probabilities of random variables.
3. An accounting scheme for information and beliefs.

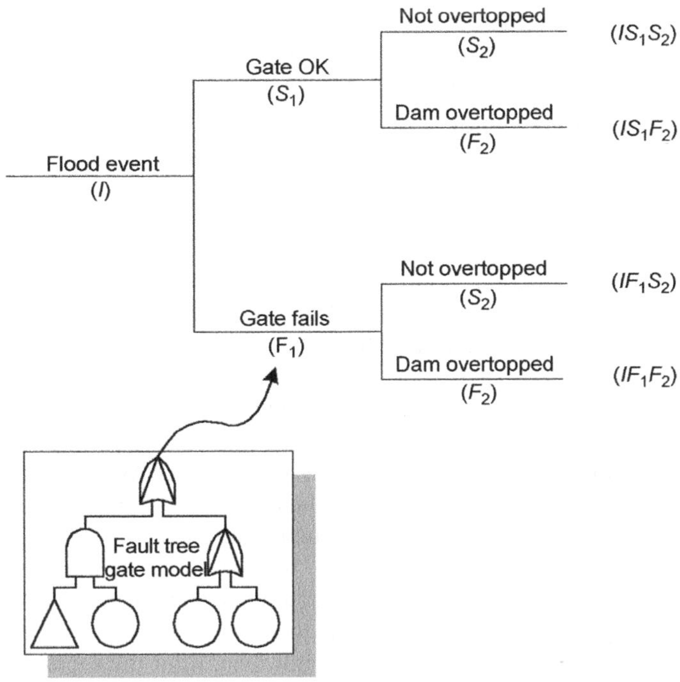

Figure 8-7. Event tree showing event probabilities calculated with associated fault trees for individual events (after McCormick, 1981)

If an event tree is viewed as a model of a physical system, then only events happening in time or space, and happening to or within the system, should be represented as nodes in the tree. Condition variables that describe the natural environment or parameters of engineering models should enter the tree only through reliability models used to assess branch probabilities.

If, on the other hand, an event tree is viewed merely as a visualisation of the possible joint outcomes of a set of random variables, then both events and condition variables can co-exist in the tree, if one adopts a degree-of-belief (i.e. Bayesian) view of probability, or, only events are represented in the tree, if a frequentist view is adopted. In either case, the ordering of the events and variables can be arbitrarily rearranged, and the tree can be used principally as a calculation engine.

If an event tree is viewed as a logical statement of information and beliefs then, again, both events and condition variables can co-exist in the tree. The uncertainties associated with events and condition variables can be convolved, or they may be separated into two distinct parts: an event tree to summarise occurrences in time or space, along with a logic tree to summarise the state of knowledge about the dam and its environment.

Decision trees (Raiffa, 1968) contain two other types of nodes – decision nodes and information nodes – that one typically does not find in event trees. Decision nodes represent an intervention or action point. Information nodes describe outcomes of experiments undertaken to shed light on uncertainties about condition variables.

Events and states of nature

This section addresses the nature of the events and uncertainties represented in an event tree, and how these different uncertainties and events relate to one another.

The dictionary definition of event as used in common speech has to do with outcomes and occurrences: with things that come to pass or come into being, with "things that happen" (Merriam-Webster, 2000).

> **Event.** 1 a: outcome, something that follows as a result or consequence; b: the final outcome or determination of an action; c: a postulated outcome, condition, or eventuality; 2 a: something that happens: occurrence, a noteworthy happening.

This definition is less restrictive than, although similar to, those normally found in textbooks on probability, which are more along the line

> **Event.** 1 a: outcome, or a collection of outcomes, of a defining experiment; b: a set of sample points in a sample space.

In probability theory, the notion of an event has to do with definitions of random experiments, sample spaces, and sets of points within sample spaces. This has led workers in applied risk theory to think of event trees merely as the graphical expression of sample spaces of experiments. When viewed this way, notions of causality, time ordering, and other logical structures within an event tree are moot. Events can be arbitrarily rearranged as long as the conditional probabilities are adjusted accordingly, and there is no notion of one event leading to another. While such rearrangements are correct in probability theory, it may be helpful from an engineering view to retain the chronological ordering or the logical structure of the set of events within an event tree.

Chapter 7 discusses the distinction between uncertainties arising from natural variations in space or time, aleatory uncertainties, and uncertainties arising from limited knowledge, epistemic uncertainties. The uncertainties of concern to event trees for dam safety are of both types. Some workers, however, limit the events represented as nodes in the event tree to only those having to do with natural variation: things that happen in space or time. In this approach, limited knowledge is represented in the event tree as manifesting in states of nature. A state of nature is a fixed condition of the world, the value (or state) of which may be unknown. For example, a "constant" in the sense of an engineering model, is a fixed parameter, the value of which is known only imprecisely.[2]

It may be convenient to characterise soil properties or extreme storms as random processes because in that way the powerful tools of probability and statistics can be brought to bear on predicting their occurrence. The same is true of earthquake epicentres, which in common engineering practice are assumed to vary randomly in both space and time. The way an event tree is constructed is inseparable from the way uncertainties are divided between natural variations and limited knowledge, that is, between aleatory and epistemic uncertainties. Two analysts could separate uncertainties in different ways and, as a consequence, end up with different events, and event structures, in their respective event trees.[3]

Event trees as models of physical systems

To the extent that an event tree is a model of the physical dam system, in the same way that a fault tree might be a model of a gate structure, each node should represent an identifiable behaviour of the dam or its physical components. Each event should be something that happens in time or space.

A stricter interpretation of event-trees-as-systems-models is that only nodes that have to do with system states of the dam should be included in the event tree. Using this interpretation, a node such as "liquefaction slumping of crest greater than 3 m" would be included in the event tree, because it describes a physical state of the dam. An event such as, "heroic effort made to protect eroding toe" would not be included because, even though it is an event that occurs in time, it is not the description of a physical state of the dam. Similarly, the event "slow detection of failure" does not describe a system state and thus would not be included in the event tree.

In contrast, existing but unknown conditions may be important uncertainties determining probabilities associated with event outcomes, but these uncertainties correctly belong in the reliability analysis leading to branch probabilities, not as event nodes in the tree. These conditions would be viewed as states of nature. Thus, for example, the suspected but uncertain existence of a low density fill in a dam foundation is important to the probability that the fill liquefies under seismic ground shaking, but it would not be an event node itself in the tree. It would, of course, influence the probabilities associated with events in the tree.

Approaching an event tree as a model of a physical system greatly simplifies the tree structure, because condition variables that might appear as nodes are now subsumed within the reliability models leading to branch probabilities.

Event trees as statements about the joint probabilities of random variables

At a simple level, event trees can be used to identify and describe the probabilistic outcomes of random experiments. The random experiment is a basic paradigm of probability theory. It is an experiment whose outcome cannot be predicted with certainty before the experiment is conducted. Consider rolling dice as a random experiment. The paired occurrence of the six faces of each die form a sample space comprising the potential outcomes of the experiment (Figure 8-8). The *sample space* of a random experiment is the set that includes all possible outcomes. For a simple experiment, the sample space may be just the set of possible outcomes. For more complex experiments, the sample space may be a mathematically convenient set that includes the possible outcomes and perhaps other elements as well. Subsets of this sample space are called events. For example, the 36 individual pairs {(1,1), (1,2), . . . , (6,6)} are the simple events in this sample space.

These paired outcomes could also be represented with a tree diagram (Figure 8-9), which provides as convenient a mechanism for enumerating all possible outcomes of the experiment as does the sample space diagram itself. One might describe event trees as nothing more than the visual representation of the joint probability distribution of a possible large number of random variables. Thus, as in Figure 8-9, any path through the event tree leads to one joint realisation of a set of random variables, in this case the numbers on each of two dice. The ordering of the random variables is unimportant, and indeed, can be arbitrarily changed.

Consider the simple event tree of Figure 8-10, showing flood discharge in volume of water per unit time, and flood stage in height of water surface in some suitable unit. The probabilities associated with the first branching are the marginal probabilities of flood volume taken from the flood frequency curve. The probabilities associated with the second branching are the conditional probabilities of water height (stage) for a given discharge. Changing the order, to put water height as the first branching event and

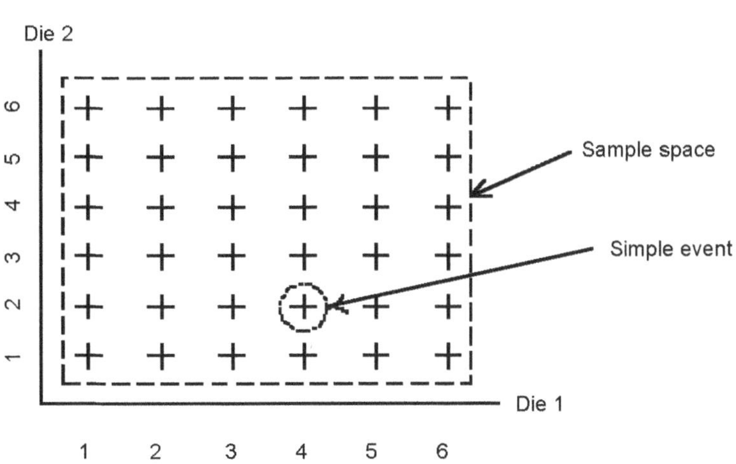

Figure 8-8. Sample space for the roll of two dice

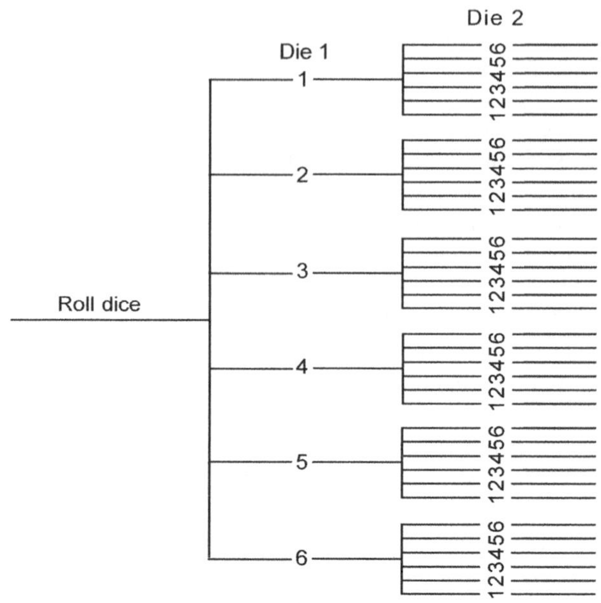

Figure 8-9. Event tree for an experiment consisting of rolling dice

discharge as the second, the only thing that changes are the branch probabilities. The first set now becomes the marginal probabilities of stage, and the second set now becomes the conditional probabilities of discharge for a given stage. The two sets of marginal and conditional probabilities can be transformed from one to the other by Bayes' theorem. The joint probabilities of the RHS outcomes are the same in both cases. Thus, the event tree is merely a convenient way of consistently representing the relationships among events and conditional probabilities.

Viewing event trees as statements of joint probability presumes no causality from parent node to child node, and no temporal ordering. Events can logically appear in the tree in arbitrary (e.g. reverse) chronology. Causality in such a tree, if that term has meaning at all, would simply mean that the realisation of an event *A* influences the probability of an event *B*, and from Bayes' theorem such causality of influence implies causality in the reverse direction, too.

Is there a correct ordering of events in such an event tree? In principle, the order of events can be rearranged arbitrarily to suit any pragmatic purpose. From probability theory, only the probabilities within a tree are conditional; the events need not be cause and effect. This applies even though events may, in the standard meaning of the term, be causally related, in that the occurrence of an event *A* leads with some probability to the occurrence of an event *B*.

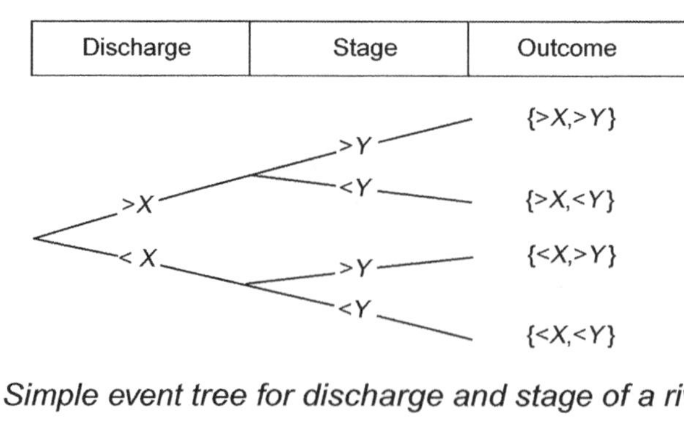

Figure 8-10. Simple event tree for discharge and stage of a river reach

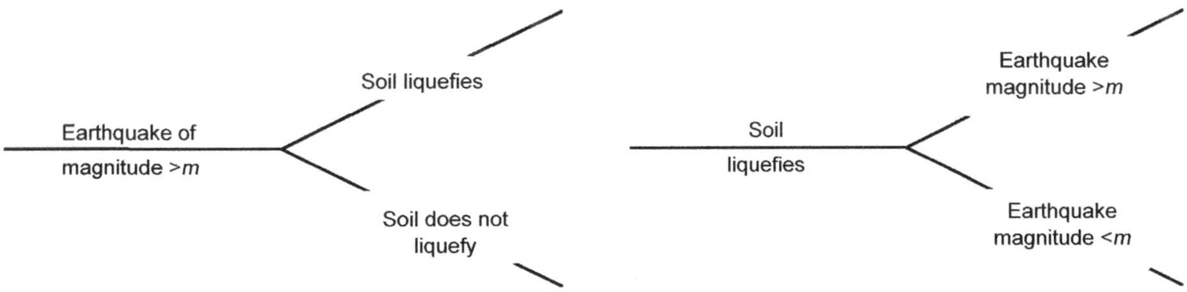

Figure 8-11. Alternate orderings of events in an event tree

Another example is an earthquake of given magnitude that may cause a particular soil deposit to liquefy. An event tree might be structured as the LHS of Figure 8-11. An earthquake of magnitude greater than m occurs, and then either the soil liquefies with probability $Pr(liquefaction \mid earthquake>m)$, or not. This would be the causal ordering of events as a geologist might understand the world. On the other hand, the events could be reordered as the RHS of Figure 8-11. For instance, this might be a convenient ordering if the task at hand involved inferential work about a prehistoric earthquake. In this case, the corresponding probability of interest would be, $Pr(earthquake>m \mid liquefaction)$, which is related to the former conditional probability by Bayes' theorem. This would be the causal order of events as an archaeologist might understand the world.

In practice, the order of events in an event tree may matter for purely operational reasons, because of the way people think about cause and effect, or because of the way they think about temporal orderings of occurrences (e.g. predictive vs forensic reasoning). If it does not matter what order the events are placed in within an event tree, then why not place them in a logically satisfying order?

Event trees as accounting schemes for information and beliefs

In contrast to the event-tree-as-system-model view, an event tree can also be a representation of a logic of belief. An event tree would thus be a representation of what we know, not a model of a particular dam. In this case, the tree may contain not only performance measures (aleatory events) but also condition variables (states of nature). The advantage of the event-tree-as-belief structure is that the full breadth of uncertain quantities affecting dam safety manifests in the tree. The principal disadvantages are increased complexity and loss of clarity about what the tree represents.

Evolving practice in seismic hazard, nuclear safety and some other disciplines is to separate events modelled as variations in time or space (i.e. aleatory random variables) from those presumed to represent knowledge limitations (i.e. epistemic uncertainties about states of nature). Two trees are thus created: an event tree and a logic tree. The event tree contains only aleatory events; the logic tree contains only epistemic uncertainties. The logic tree is a representation of the various possible realisations of the multivariate state of nature controlling probabilities within the event tree.

The event tree structure appends to individual "leaves" of the logic tree, as suggested by Figure 8-12, so that all calculations within the event tree are conditioned on the realised states of nature in the logic tree. For example, the estimation of branch probabilities for failure states F_1 and F_2 in the figure would be conditioned on the outcomes of the states of nature α,β,γ leading to the leaf at which the event tree begins, $Pr(F_1 \mid \alpha,\beta,\gamma)$ and $Pr(F_2 \mid F_1,\alpha,\beta,\gamma)$. This provides the notable advantage of simultaneously accounting for correlations anywhere in the event tree due to the common realisation of uncertainties about states of nature appearing in the logic tree. The final combined tree need be no more complex than before, since the number of combinations of all uncertain events, both aleatory and epistemic, remains the same.

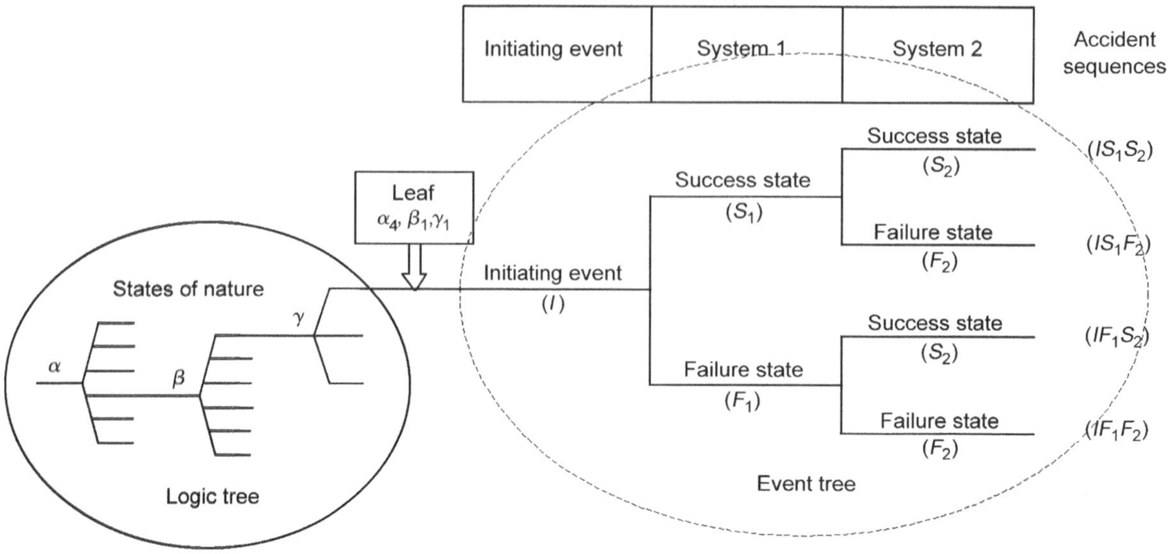

Figure 8-12. Logic tree describing fixed-but-unknown conditions (states) of nature, as conditioning point for event tree of system

Levee safety example

An homogeneous, compacted soil flood levee is shown in Figure 8-13. At certain times of the year the levee may be subject to high water in the floodway it serves. The functional purpose of the levee is to retain floodwaters within the floodway during high-water events and, thus, to keep the flood plain dry outside the levee.

As a first approximation, this levee might "fail" – and thus allow flood waters to escape into the surrounding river valley – in one of three ways: (1) high flood waters (stage or water height) might overtop the levee, causing erosion of the embankment, and consequent loss of containment; (2) the levee itself might fail due to inadequate soil strength and high internal pore pressures; or (3) seepage in the natural soils beneath the levee might lead to piping (evidenced by sand boils) and a deep sliding foundation failure.

Some of the considerations in analysing the risk that the levee will fail (at least in this one reach) are suggested in Table 8-1. First, at least five system states are relevant to a potential failure: the flood height relative to the top of the levee, internal pore pressures in the levee and its foundation, the geotechnical strength stability of the levee proper, and the presence of piping (internal erosion) in the foundation. For purposes of this example, piping in the levee itself is ignored as it is assumed that the levee has been

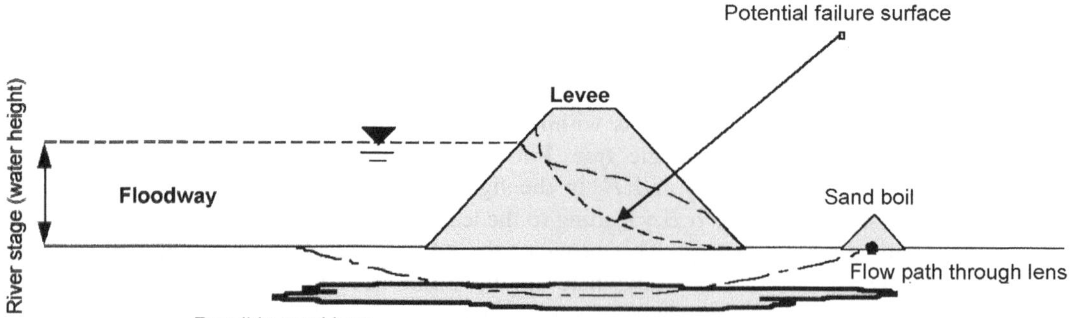

Figure 8-13. Example levee subject to flood flow

Table 8-1. Important system states for levee example

Failure modes	System states	State variable
Overtopping of levee	Overtopping of levee	Relative water height
Structural or geotechnical instability in fill	Internal pore pressure in levee	Pressure on pertinent failure surface
	Subgrade water pressure difference	Differential water pressure
	Levee strength stability	Levee deformations
Erosion of the foundation	Subsurface piping	Rate of seepage in subsurface

Table 8-2. Important uncertainties for levee example

Naturally varying (aleatory) conditions	Knowledge (epistemic) uncertainties
Extreme storm in upstream basin	Average strength of the levee fill
Peak flood discharge (flood frequency)	Presence of "flaw" zones in the levee
Peak river stage for a given discharge	Existence of permeable lenses in foundation
Flood duration given peak flood	
Spatial variation of soil properties	Critical duration of overtopping to cause failure

designed and constructed to perform satisfactorily, even though this can be an important contributor to risk.

The chance that the levee system occupies a specific set of states depends both on naturally varying conditions and on knowledge uncertainties (Table 8-2). Indeed, the probabilities associated with each system state may depend on the same aleatory and epistemic uncertainties and, thus, be probabilistically or statistically dependent (National Research Council, 2000).

The separation of natural variations from knowledge uncertainties as suggested by Table 8-2 is, in fact, a modelling decision that could be made in different ways by different people, or made in different ways by the same person if the purpose of the modelling changes. The separation here might be considered typical of many studies, but it is not necessarily the only way that natural variation and knowledge uncertainties can be divided. Event tree analysis, like all of engineering, combines art with science and probability with the "art" of judgement.

8.3 Influence diagrams

An influence diagram graphs relationships among initiating events, states of nature, conditions of the system, and consequences. Creating an influence diagram is, in essence, creating the basic model of the risk analysis from which event trees, reliability calculations and other models arise. There is not a rigid set of steps or a recipe for creating an influence diagram, but only a structured procedure to help maintain logical consistency.

As in all modelling, the influence diagram should represent the logic of the system, influences upon the system and uncertainties affecting system performance. To the extent possible, it should do so parsimoniously; it should seek an efficient representation. Obviously, this, too, is an art requiring interpretation and judgement. The enterprise of creating an influence diagram seeks a relatively simple representation with comparatively few parameters which, despite its simplicity, closely represents the behaviour of the dam.

An influence diagram provides a visual model of the structure of a risk analysis problem. This includes the timing of events, relationships among risky outcomes, and uncertain events, quantities, or variables. The intent of an influence diagram is to help visualise how system components, uncertainties and outcomes are interrelated; and especially to support the development of a systems risk model of the components, uncertainties and outcomes. The influence diagram involves no mathematical model; it deals only with relationships among entities.

Two important advantages of influence diagrams over event or decision trees make them useful in the early stages of risk analysis. First, even a complicated problem can sometimes be represented in a simple influence diagram. An influence diagram represents a single uncertain quantity by a single node, and does not lead to the combinatory explosion of branches associated with event trees and fault trees. Second, influence diagrams explicitly show interrelationships among events. In an event tree there is usually no immediately obvious way to track the causal dependence of one uncertain quantity on another.

Flow charts and fishbone diagrams

Engineers and analysts in different fields have developed a variety of related approaches to charting the relationships among systems states, uncertainties and outcomes as a precursor to constructing event trees or fault trees. The more common of these in engineering risk analysis is the influence diagram, but flowcharts and fishbone diagrams (sometimes called Ishikawa diagrams) are common in quality assurance, business management and other fields (Figure 8-14).

Flowcharts, familiar to anyone who has studied computer programming, are similar in concept to influence diagrams, but are less free-wheeling in the way they structure a sequence of events or a chain of cause and effect. The American National Standards Institute (ANSI) has created a standard set of symbols for flowcharting, and these may

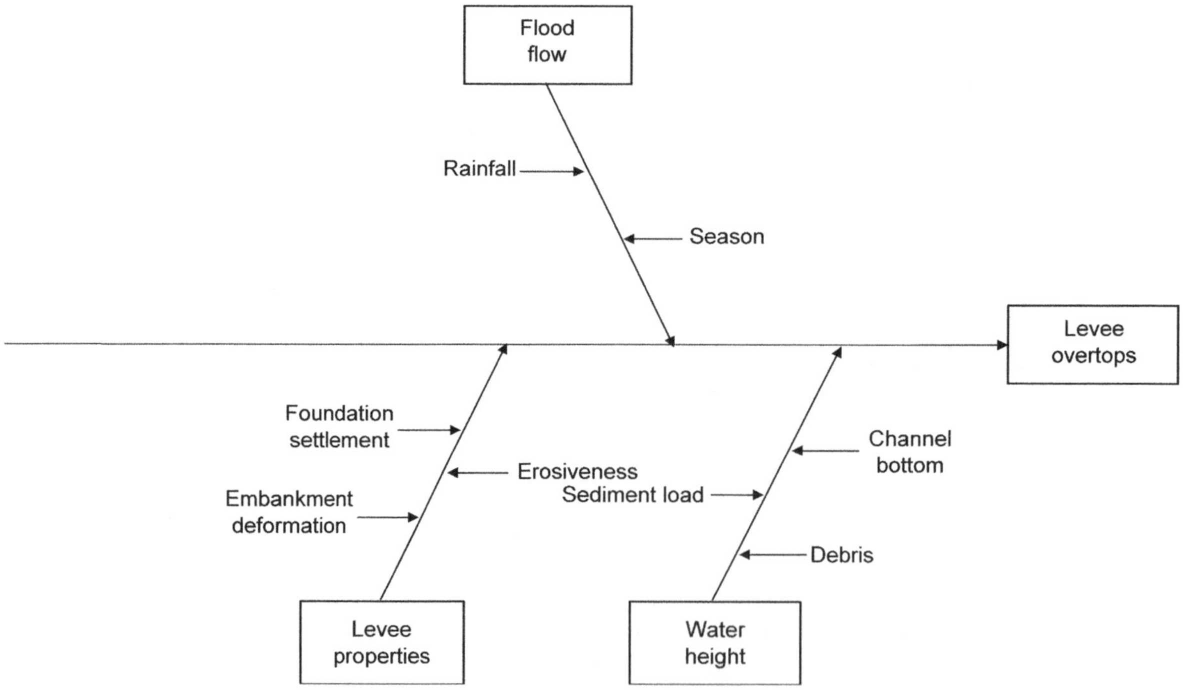

Figure 8-14. Fishbone diagram of levee failure shown schematically in Figure 8-13

be found in common software applications for creating flowcharts. The most common uses of flow charts in risk analysis are tracking logical processes, and tracking strictly chronological processes.

Fishbone diagrams are common in the business process re-engineering and quality assurance literatures (Harrington, 1991). To construct a fishbone diagram, one starts with an outcome at the RHS, which in risk analysis is usually an adverse consequence or failure. Influences on this outcome are placed as diagonal arrows into the spine or backbone leading to the outcome, much as in a fault tree but without the mathematical rigor. Sometimes these primary influences are ordered in reverse chronology and sometimes in order of importance, but not consistently across different practitioners.

Root causes are placed directly along the backbone, while secondary or lower causes are placed along (with arrows into) the primary causes. Sometimes, individual segments or legs of the chart are separated off for more detailed analysis.

Influence diagrams and event trees

Influence diagrams and event trees are alternate representations for the same systems. It is often convenient to first structure a systems model as an influence diagram, and then to use the insight gained from the influence diagram to structure an event tree for the same system. Each tool provides a means of graphically representing events and the interrelationships among events, although these means differ between influence diagrams and event trees. Each tool can be used to quantify the effects of external risks on process activities, or on engineered or natural systems (Marshall and Oliver, 1995).

An influence diagram uses event nodes representing the uncertain conditions surrounding a process, as well as activity nodes representing discrete process steps. Relationships among nodes are shown by directed paths, which sometimes are referred to as arcs or influence lines. The graphical display of events, process steps and consequences, and their relationships to components of the system being analysed is used to visualise the impacts of external events or uncertainties. An influence diagram is often used as an exploratory tool leading to the development of more formal event trees and, as such, system states and uncertainties of various types may be mixed within the same diagram.

The quantification of uncertainty in an influence diagram is accomplished by building a compatible event tree in conjunction with the diagram. Such a tree consists of chance nodes corresponding to the risk and activity nodes in the influence diagram, connected by branches delineating their paths of influence (Clemen and Reilly, 2001; Rowe, 1977).

Relationships among events in event trees follow a hierarchy from independence, through causality, to networked causality and to temporal orderings. These may be shown in a number of ways, as suggested in Figure 8-15, although these formats are not used in the present examples. Many analysts make no distinction between statistical correlations and causal effects, in that the mathematics of probability theory can be made to account for inverted ordering of events, even if the events are causally related.

Elements of influence diagrams

Standard practice in constructing influence diagrams (Marshall and Oliver, 1995), similar to standard practice for decision trees (Pratt *et al.*, 1995), is to represent node types by circles or ellipses for uncertainties, squares for decisions and diamonds for outcomes. Typically, influence diagrams in dam safety analyses do not include decision nodes, but that need not be the case. An example follows. Some authors add a fourth, deterministic node, but this could be viewed as a degenerate case of an uncertain node (Bedford and Cooke, 2001).

A directed arc is used to denote dependence or conditionality in influence diagrams. A directed arc joining node A to node B (Figure 8-16) may denote that: (1) event A

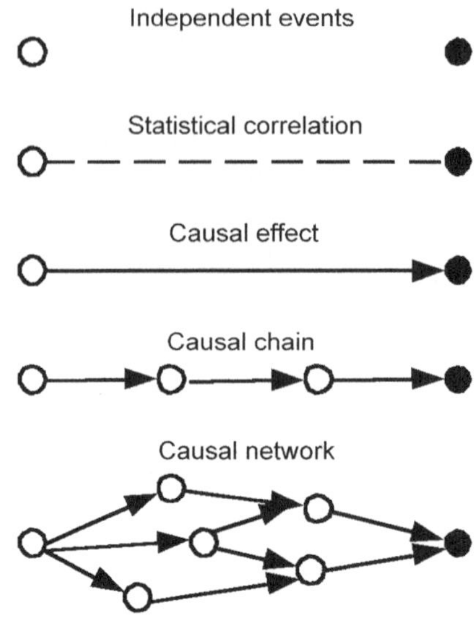

Figure 8-15. Types of relational structures found in event trees, and suggestions from the literature for representing them in event trees. These graphical standards are not used in the current volume

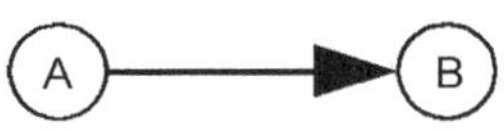

Figure 8-16. Directed arcs in influence diagrams

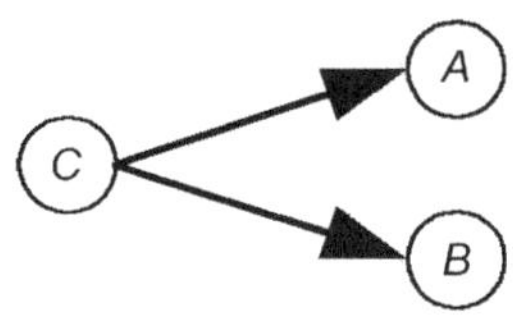

Figure 8-17. Two uncertain variables, A and B, each dependent on a third variable, C

causes event B or influences the way B happens; or (2) knowing the value of A changes the probability or probability distribution of B (i.e. A and B are correlated and A, or the knowledge of A, occurs first in time).

Two events A and B can be probabilistically or statistically correlated and yet not be joined by a directed arc, for example, if they share common dependence upon some third variable, C (Figure 8-17). Knowing the value of C, the uncertainties A and B are probabilistically independent; but if C is unknown, A and B are probabilistically dependent, because each depends on the realisation of C.

Developing an influence diagram

The typical procedure for creating an influence diagram is:

1. Identify the events and uncertain quantities whose outcomes could be important to the safety of the dam.
2. For each uncertain quantity, assign a name and a unit of measure; if decision variables are involved, define specific, quantitative alternatives.
3. Identify influences or dependences among the uncertain quantities, and between each decision and each uncertain quantity; then represent each uncertain quantity and each decision by a node in a graph, aligning them in order of occurrence, if time is relevant.

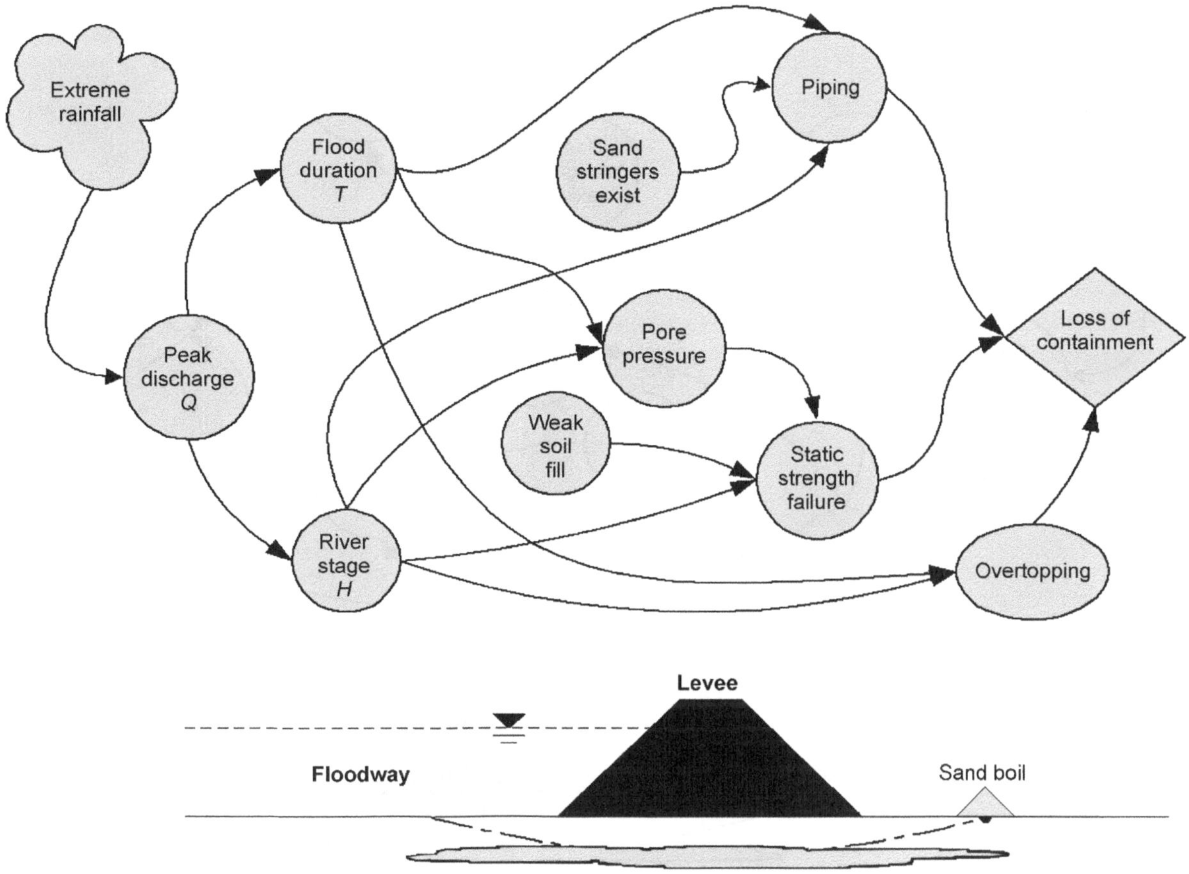

Figure 8-18. Developing an influence diagram for levee failure

4. Assign directed arcs among uncertainty nodes and between decision and uncertainty nodes, with the direction of the arcs indicating the direction of presumed influence.
5. In constructing the influence diagram, there should be no closed loops or cycles among the nodes. That is, there should be no connected path along the directed arcs that leads back to an earlier node.

Section 8.2 introduced the system states and uncertainties associated with a simple floodway levee. These can be combined in an influence diagram to begin to show the interrelationship among the various elements and considerations in beginning a risk analysis of the levee. A first attempt at such an influence diagram is shown in Figure 8-18. This diagram consists both of system states and uncertainties affecting those states.

While it is not common for an influence diagram in dam safety studies to include decision nodes, it is also not unknown. Figure 8-19 shows the same influence diagram, now with the addition of an active decision node representing the option of releasing water upstream of the levee reach into a flood by-pass. The directed arc between flood duration and overtopping is neglected in this diagram as a simplification. If the operators of the floodway recognised that a potentially dangerous flood was heading into the levee reach, they now might make an active decision to intervene. As with the other nodes in the diagram, the behaviour of the decision node may itself be affected by many variables and uncertainties, for example, ones associated with the availability of timely information and with human factors.

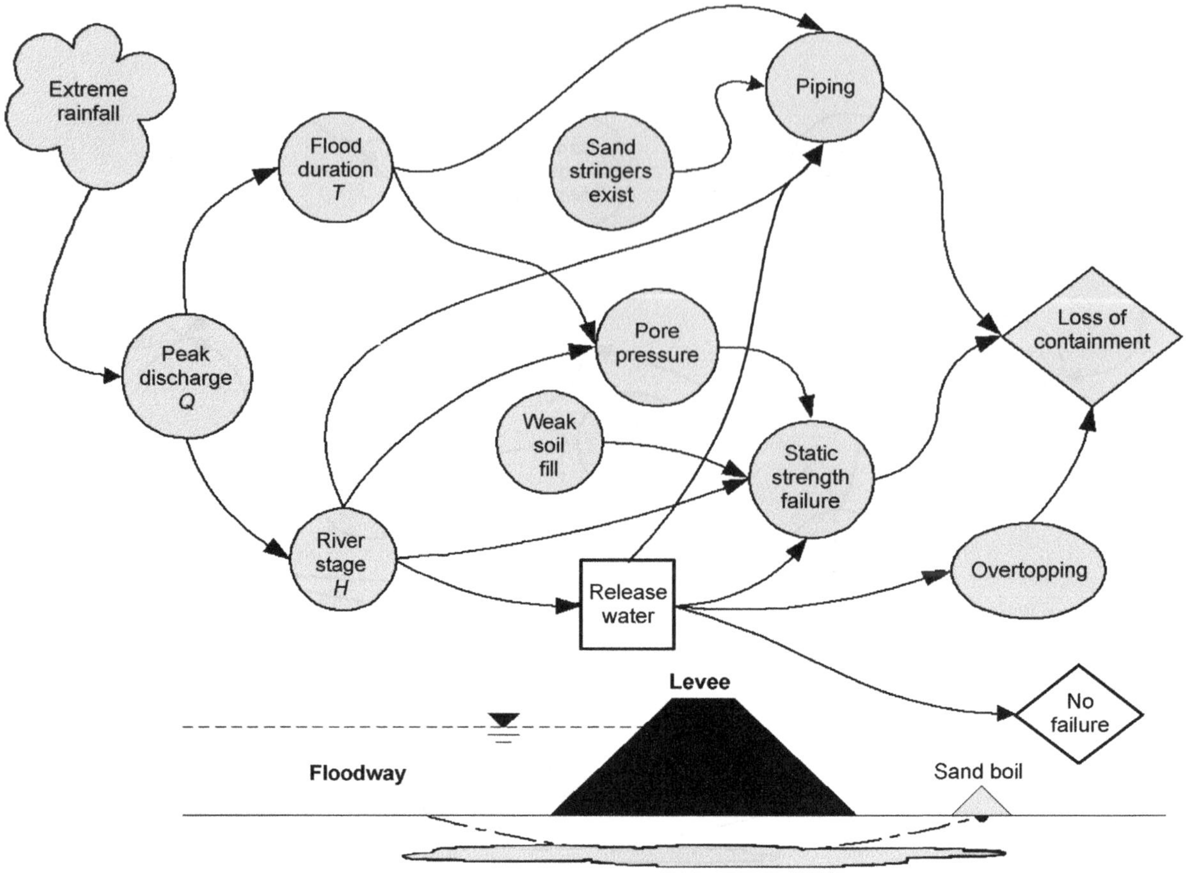

Figure 8-19. Influence diagram for levee failure, including active decision node

Belief networks

Belief networks – also referred to as *Bayesian networks* in the computer science literature and as *Markovian networks* in the probability literature – are graphical tools for probabilistic inference (sometimes called, probabilistic learning) that have developed mostly in the artificial intelligence community (Pearl, 1988). The principal purpose of belief networks is to draw probabilistic inferences about one set of uncertain quantities by invoking Bayes' theorem on observations of some other set of quantities. Clearly, this can be done by direct enumeration within the context of traditional Bayesian statistics, but for the fact that many real diagnostic problems involve a great many variables and thus become too large and complex.

Similar to an influence diagram, a belief network is a graph with nodes representing uncertain quantities and arrows suggesting probabilistic dependence.[4] Unlike an influence diagram, belief networks typically include neither decision nodes nor outcome nodes.

The simple belief network of Figure 8-20 represents the joint probability distribution of a set of uncertain quantities. The specification of probabilities in the network starts

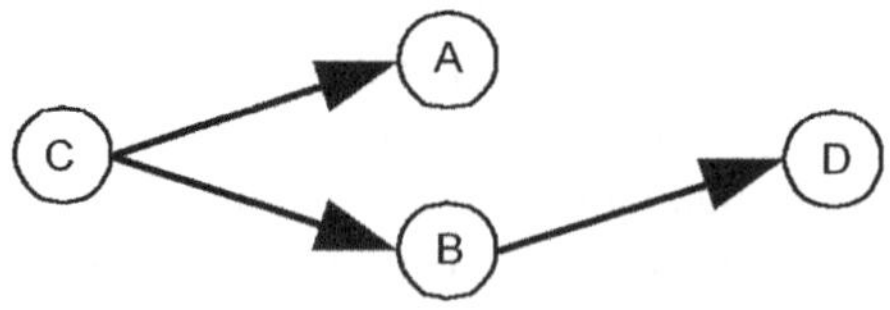

Figure 8-20. Simple belief network

with the marginal probability of C at the LHS, and follows to the conditional probabilities of A and B, progressing to the right. The joint distribution of the two uncertain quantities C and B is:

$$Pr(B, C) = Pr(B|C)Pr(C); \quad (8\text{-}1)$$

and of the three uncertain quantities C, B, and D,

$$Pr(D, B, C) = Pr(D|B, C)Pr(B|C)Pr(C) \quad (8\text{-}2)$$

As a belief network becomes large, the number of terms in equation 8-2 rises correspondingly. Thus, it is convenient if the probability distribution of a variable can be assumed to depend only on the realisation of its immediate predecessor(s) in the network, and not on the realisation of all the uncertain variables preceding it; that is, for Figure 8-20,

$$Pr(D, B, C) = Pr(D|B, C)Pr(B|C)Pr(C) = Pr(D|B)Pr(B) \quad (8\text{-}3)$$

This presumes the uncertainties to exhibit Markovian behaviour, in that they depend only on their immediate predecessors (Papoulis, 1991). This may or may not be a good assumption in practical problems, but it is convenient for rapid calculations in a computer program.

Markovian behaviour in influence diagrams

The Markovian property, so prized for its practical usefulness in belief networks, sometimes runs counter to good practice in influence diagrams involving decision nodes (belief networks typically do not include decision nodes). Consider a site characterisation situation involving the potential for two steps of data gathering (akin to a "second-opinion" decision problem in the clinical literature).

Presume that the performance of the levee with respect to under-seepage depends on the possible but uncertain existence of sand lenses in the substrata. Geological information may lead to a prior probability of lenses existing (Figure 8-21), and based on this information, a decision can be made. The decision might be to choose a remedial option, or to drill more borings to try to gain more information about whether or not sand lenses exist. Having gathered this additional information, a new (posterior) probability of a lens existing is calculated through Bayes' theorem, and a decision on remedial actions revisited. Once a decision on remedial actions is taken, and a lens turns

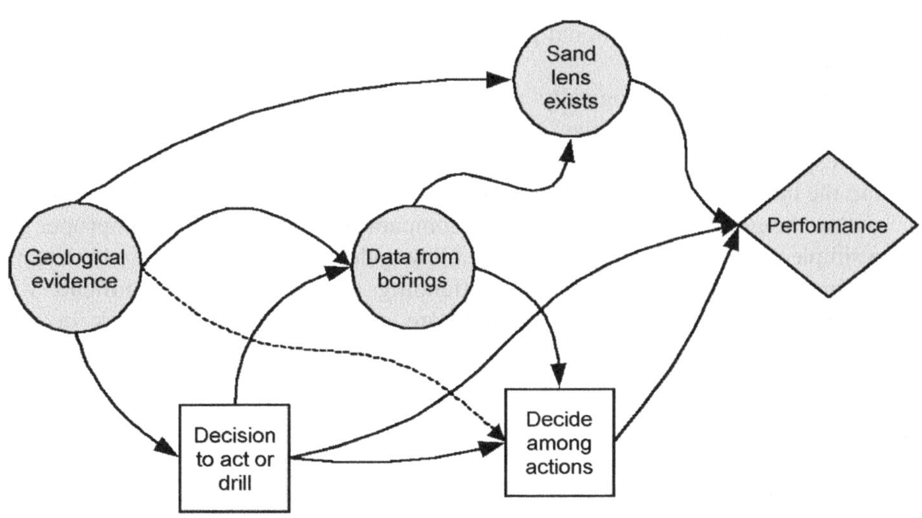

Figure 8-21. Influence diagram for the two-staged exploration decision

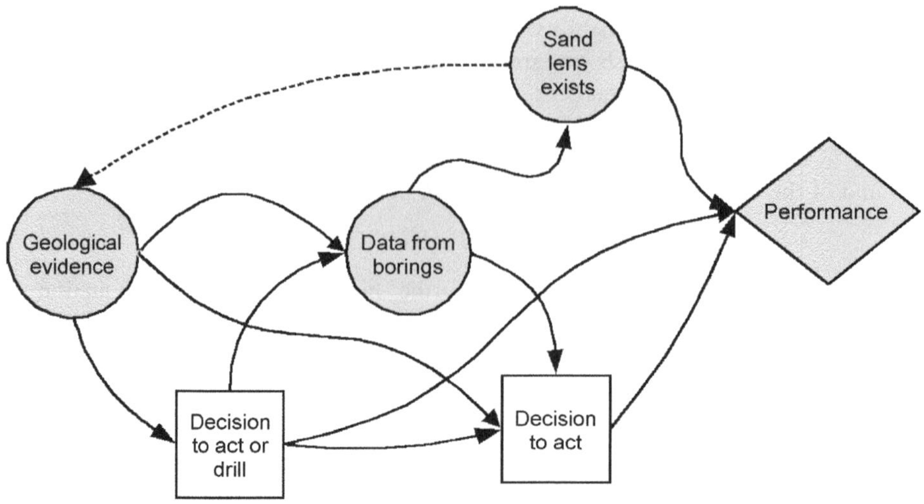

Figure 8-22. Arc reversal in the ID for two-staged exploration

out either to exist or not to exist, some performance result obtains. Before the fact, of course, this performance result can only be described probabilistically.

The directed arc between the geological information and the second decision, denoted by a dotted line in the figure, must be there. The influence diagram cannot describe a meaningful decision problem without this arc. Absent the arc, the geological information known at the time of the first decision would be forgotten by the time of the second decision, and this is not realistic, since the geological information is pertinent to the second decision. A requirement for a proper influence diagram is that information known at an earlier decision is remembered at a later decision. The arc from the "geological evidence" node to the second decision node is sometimes called a no-forgetting arc (Marshall and Oliver, 1995).

The nature of the geological evidence at the LHS initial node leads to a conditional probability of a sand lens existing via Bayes' theorem:

$$Pr(lens \mid geology\ evidence) \propto Pr(lens) L(geology\ evidence \mid lens) \tag{8-4}$$

in which, the LHS is the conditional probability of a lens existing given the geological evidence, and the RHS is the product of the marginal probability of a lens existing, by the likelihood (i.e. conditional probability) of the geological evidence given that a lens exists.

One could just as logically reason that the existence or non-existence of a lens influences the probabilities of different sorts of geological evidence, and the corresponding conditional probabilities are found by reversing equation 8-4. Restructuring the influence diagram in this way (Figure 8-22), however, creates a disallowed cycle within the graph, illustrating that it is comparatively easy to create improper graphs. In a simple influence diagram with only a few nodes, it is relatively easy to detect this sort of inconsistency and reverse the offending arc(s). It is more difficult with large diagrams, and more formal methods are available to detect cycles in such cases (Marshall and Oliver, 1995).

8.4 Event trees

As with influence diagrams, event trees follow standard graphical conventions. A description of the steps in constructing event trees is given in Part II, and is not repeated here. This section attempts to address conceptual issues.

Structure

An event tree is built from a set of events, each represented by a node, usually indicated by either a circle (Figure 8-3) or a point (Figure 8-2). The branches emanating from an event node represent the possible outcomes or realisations of the corresponding event. These branches represent outcomes that are mutually exclusive and collectively exhaustive. Thus, the outcome will follow exactly one of the branches. There may be as many branches as needed to satisfy this condition, including an infinite number if the event has continuous rather than discrete outcomes.

Figure 8-23 shows the initiating event that a flood discharge greater than some value, here 200 cms, is generated by a large storm. The subsequent event, "river stage", is summarised in the upper tree with a dichotomous outcome. In this case, the corresponding river stage is either greater than some elevation, here 48 m, or less than that elevation (by convention, the upper branch at each event usually corresponds to the associated event being "true"). In the lower tree, the event is described by a continuous outcome between lower and upper bounds, here $(0, \infty)$.

A probability distribution is associated with each set of mutually exclusive and collectively exhaustive outcomes at an event node. When the outcomes are defined as discrete, the probability distribution reduces to a set of discrete probabilities (i.e. a probability mass function, PMF). When the outcomes are continuous, the probabilities are represented in a probability density function (PDF). The probabilities in the first case would be described by a simple point value, say 0.1, and its complement, 0.9. The probability is in the second case and would be described by a PDF, presumably some form of extreme value distribution. Calculations in the first case would be made by multiplying branch probabilities by their corresponding consequences, and in the latter case by integrating over the PDF. Issues related to the accuracy of replacing continuous outcomes by discrete outcomes are treated below.

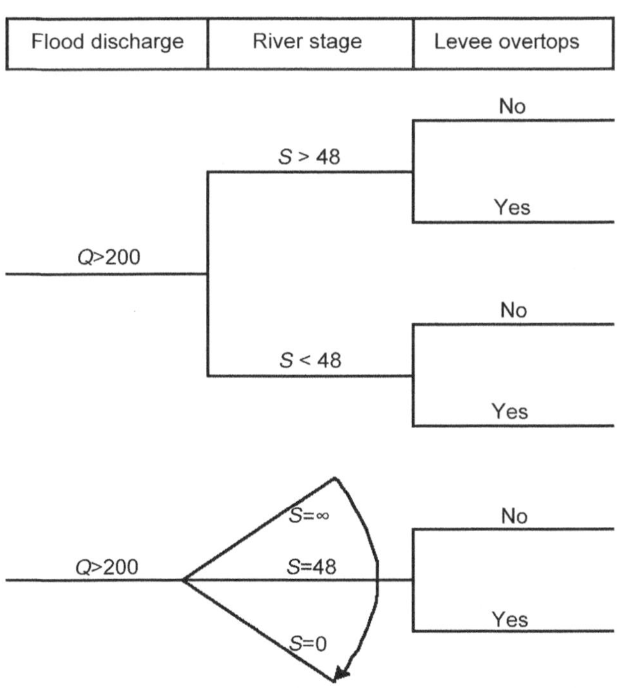

Figure 8-23. Alternative simple event trees for levee overtopping: top showing dichotomous representation of outcome; bottom showing continuous representation of outcome

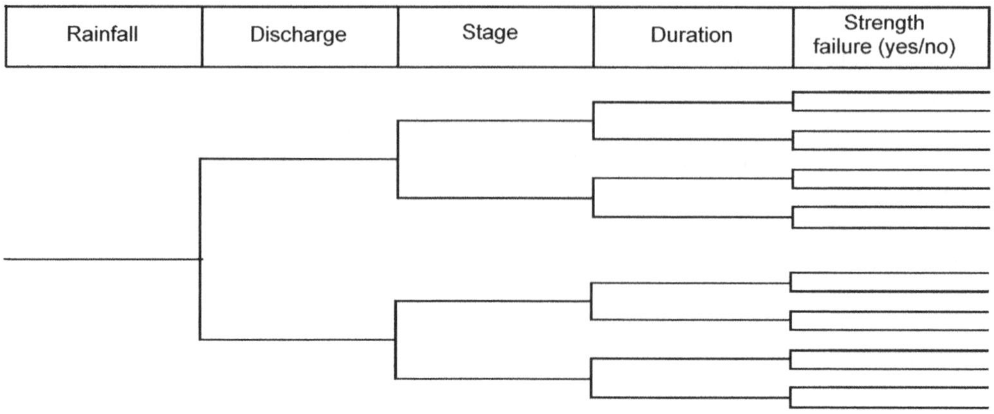

Figure 8-24. Simple event tree for structural/geotechnical strength instability of levee

An event tree can quickly become large. Expanding a simple chain of events leading from an initiating event of high rainfall to structural overloading of the levee could pass through at least the following five events, each of which might have many possible outcomes:

1. An extreme storm generates rainfall of a given amount.
2. Flood flow (discharge) in the river exceeds some value.
3. Stage rises above some critical elevation.
4. High water lasts for a certain critical length of time.
5. The levee fails due to high loads and rising internal pore pressures.

Simplifying – perhaps greatly so – to a dichotomous event yields the tree of Figure 8-24, which is already large.

A more realistic model might include consideration of whether soft zones (flaws) exist within the embankment, how high the pore pressures rise, whether a levee experiencing structural instability might yet retain the flood flow and so on. One problem with any of the simple logic tools used to model risks associated with system performance, such as influence diagrams or event trees, is that the tools are static analyses that decompose events into individual realisations. This simplifies the analysis, but may lose important features, because since the tree grows combinatorially with the number of events, care must be taken to keep event chains as short as can be reasonably accommodated.

Consider the issue of the height and duration of overtopping, each is important for failure and they act jointly. Shallow overtopping, that is not much water volume passing over the levee, may cause failure, but only if the condition lasts for a long time; if the overtopping is brief, the levee may withstand potential erosion. On the other hand, deep overtopping, even if brief, may cause failure. Figure 8-25 shows two cases of overtopping. On the left, the overtopping is shallow but lasts for 20 hours; on the right, the overtopping is deep but lasts a much shorter time, perhaps three or four hours. Both are cases of overtopping, but the levee may respond differently to each. A simple influence diagram or event tree may not capture this combined dependence. Events may need to be defined by depth-duration pairs in order to model the behaviour of the levee.

An event tree is generated from an influence diagram by ordering the nodes according to the way the directed paths of the influence diagram connect the events, and then generating a tree with the event nodes as vertices. Thus, predecessor nodes (parents) precede later nodes (children). At each node, outcome branches are placed for each potential realisation of the associated uncertain quantity. Obviously, these branches do

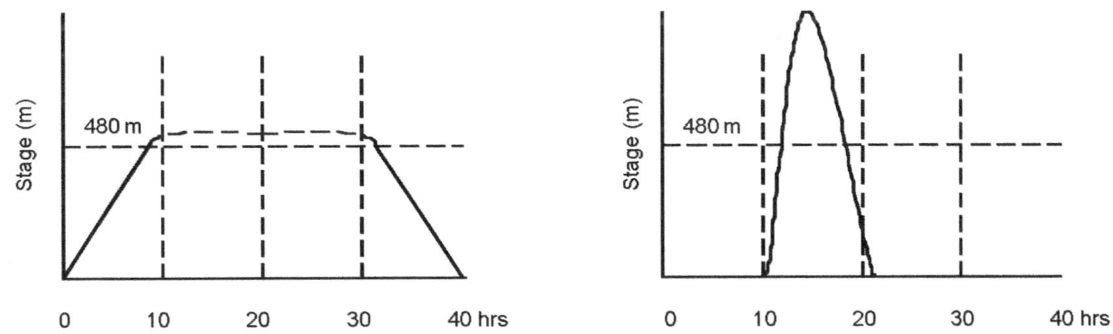

Figure 8-25. Height and duration of overtopping interaction to affect levee performance

not exist in the influence diagram, which shows only a directed path from parent to child.

Chained events, connected by directed paths in the influence diagram, become sequential nodes in the event tree. Parallel events, without connecting paths, can be placed in arbitrary order with respect to each other. Commercially available software applications for risk and decision analysis make this conversion automatically.

For operational convenience, large event trees are usually separated according to the initiating event and according to system failure mode, as shown schematically in Figure 8-26, although this is not a necessity. Thus, in a dam safety study it is common to see three, four or more event trees, each associated with one initiating event: extreme storms, earthquakes, flaws in design or construction (leading to internal erosion), and

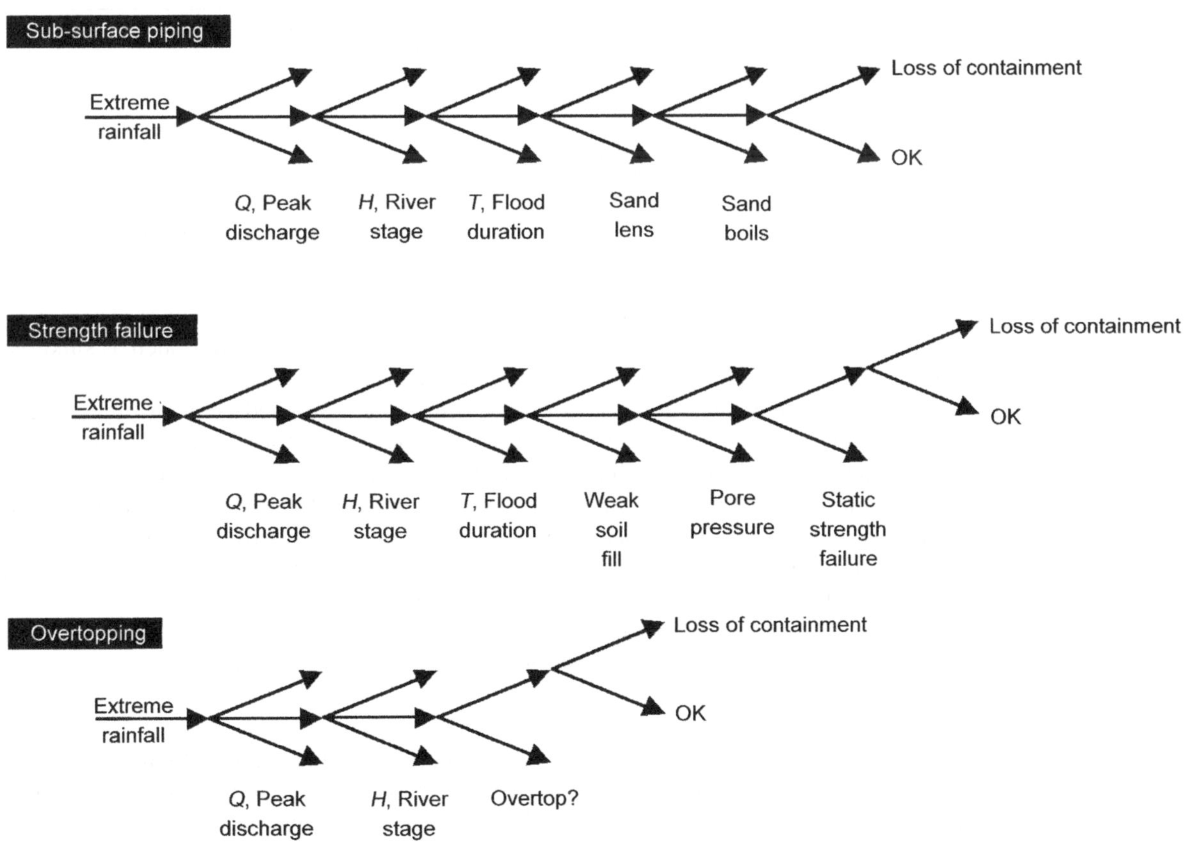

Figure 8-26. Event tree(s) for levee failure due to flood

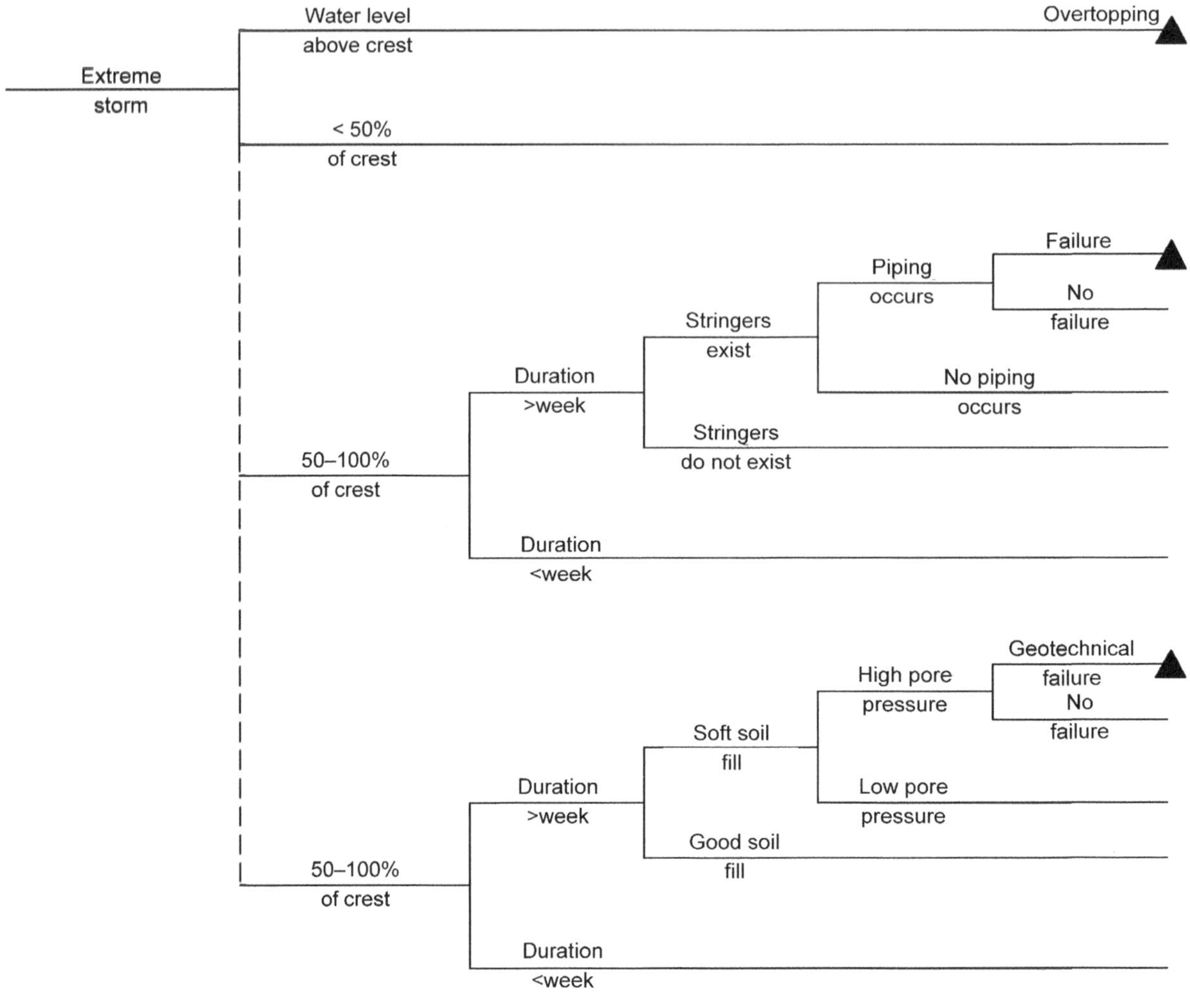

Figure 8-27. Hydrologic, piping and strength failure parts of the levee failure event trees

possibly mechanical or electrical dysfunction of an appurtenant structure. There may also be event trees or fault trees associated with other initiating events, such as sinkhole formation (Vick and Bromwell, 1989), but these tend to be special cases.

These event trees are shown in a more common and detailed form in Figure 8-27. Note that the format of this tree combines states of nature ("soft soil fill exists") with system states of the levee ("high pore pressure in levee"), that is, it does not separate the uncertainties into a logic tree and an event tree, but combines them in one tree. The dashed vertical line indicates separate event trees leading to distinct failure modes, but arising from the same initiating event. The triangle at a terminating end of a path through the tree indicates a "failure" consequence.

Despite the common logical content of influence diagrams and event or decision trees, the practical uses of the two differ, as summarised in Table 8-3.

Initiating events

The identification of initiating events is an important step in the development of event trees. The initiating event could be an extreme load due to a natural hazard, such as a flood or an earthquake, technological failure, such as the dysfunction of the spillway gate, or human error. These may have been identified by some other risk analysis technique, for example preliminary hazard analysis, or failure modes and affect

Table 8-3. A comparison of influence diagrams and decision trees (Marshall and Oliver, 1995)

Model features	Influence diagrams	Event or decision trees
	Modelling and representation	
Timing	Shows timing of all decisions and uncertain events	Shows timing of all decisions and uncertain events
Conditional independence	Shows dependence among uncertain events and decisions	Dependence among uncertain events and decisions not shown
Size	Number of nodes grows linearly with the number of variables	Number of terminal nodes and paths grows exponentially with the number of variables
Data	Identifies dependencies of variables without need for data	Decision, probability and result data shown explicitly
Variable type	Both continuous and discrete decisions and probabilities	Both continuous and discrete decisions and probabilities
Asymmetry	Scenarios with different event sequences not distinguished	Shows asymmetric structure of problem
Modelling usefulness	Most useful in initial stages of modelling; captures interaction between decision-maker and analyst	Useful in depicting detailed uncertain event outcomes and decisions, and model solution; difficult to display large problems
	Solution process	
Bayes' theorem	Indicated by arc reversal, but calculation not shown	Indicated by node reversal (a separate event tree may be used as an aid for calculation)
Method	Reduction by a set of reduction operations possible using advanced methods	Uses simple rollback algorithm

analysis. To serve as the beginning point for an event tree, the initiating event should give rise to a number of subsequent events sequences.

Important initiating events are often anticipated during the design phase of a dam or other structure. In this case, design features or other safety functions have usually been introduced to deal with such initiating events. Nonetheless, different analysts may define initiating events in slightly different ways. The way in which an initiating event is defined and characterised can have an important influence on the subsequent development of an event tree and, by extension, on risk estimates made from the tree.

Table 8-4. Typical initiating events in dam safety event trees

Category	Initiating event
External events	Floods Earthquakes Human agency
Internal events	Flaws (leading to internal erosion) Deterioration Mechanical or electrical failures

Event trees for dam safety typically focus on relatively few initiating events (Table 8-4).

As introduced in Section 1.0, initiating events in most event tree analyses for dam safety are of two types: external events usually associated with natural hazards, and internal events usually associated with flaws or deterioration. The latter are sometimes also called, "failures under normal operating conditions" (Von Thun, 1996). Event tree analysis is especially well suited to representing the potential consequences of natural hazards; it is less well suited for dealing with internal initiating events.

Separability

Whenever possible, it is convenient to separate chains of events into separate trees, thus reducing complexity. Since the size of event trees increases exponentially with the number of events, dividing events into separate trees carries distinct benefits. A set of events can be separated into a distinct event tree, if all the parent events within the corresponding influence diagram having directed paths leading to one or more of the events within the set, are also included in the set. In other words, any event that has an influence on any event in the tree, must itself be in the tree.

Figure 8-28 shows three separable sets of events within the influence diagram introduced earlier for flood levee failure. Stated in a less formal way, the criterion for separability implies that, while directed paths may exit a separable set, no directed path may enter.

The influence diagram in Figure 8-28 is associated with the one initiating event, extreme storm. The same principle of separability applies, however, with more than one initiating event (Figure 8-29). Here, a second initiating event, earthquake, is shown, indicating its influence on events already in the diagram. This new initiating event may also influence events not already in the diagram, but these may be ignored for the moment.

A separable set, as in the earlier case, is one into which no directed path may enter. In Figure 8-29, however, the cross-hatched events emanating from an extreme storm initiating event are also parents to events within the separable set. These would have to be included within the set unless, as in this case, their probability of occurring at the same time as the earthquake initiating event was judged to be negligible; hence, they are excluded.

Event chains may not be separable if dependence exists in the consequences to which they lead, even if the events themselves are independent (or conditionally independent given parent events). For example, consider the case of a reservoir with a saddle dyke at some distance removed from the main embankment dam. Given a large flood, and presuming other things equal, the probability of failure of the main embankment and the probability of failure of the saddle dyke should be independent. Thus, it would seem reasonable to separate the two event trees, the one leading to failure of the dam and the other leading the failure of the dyke. But, the consequences of these two failures are not

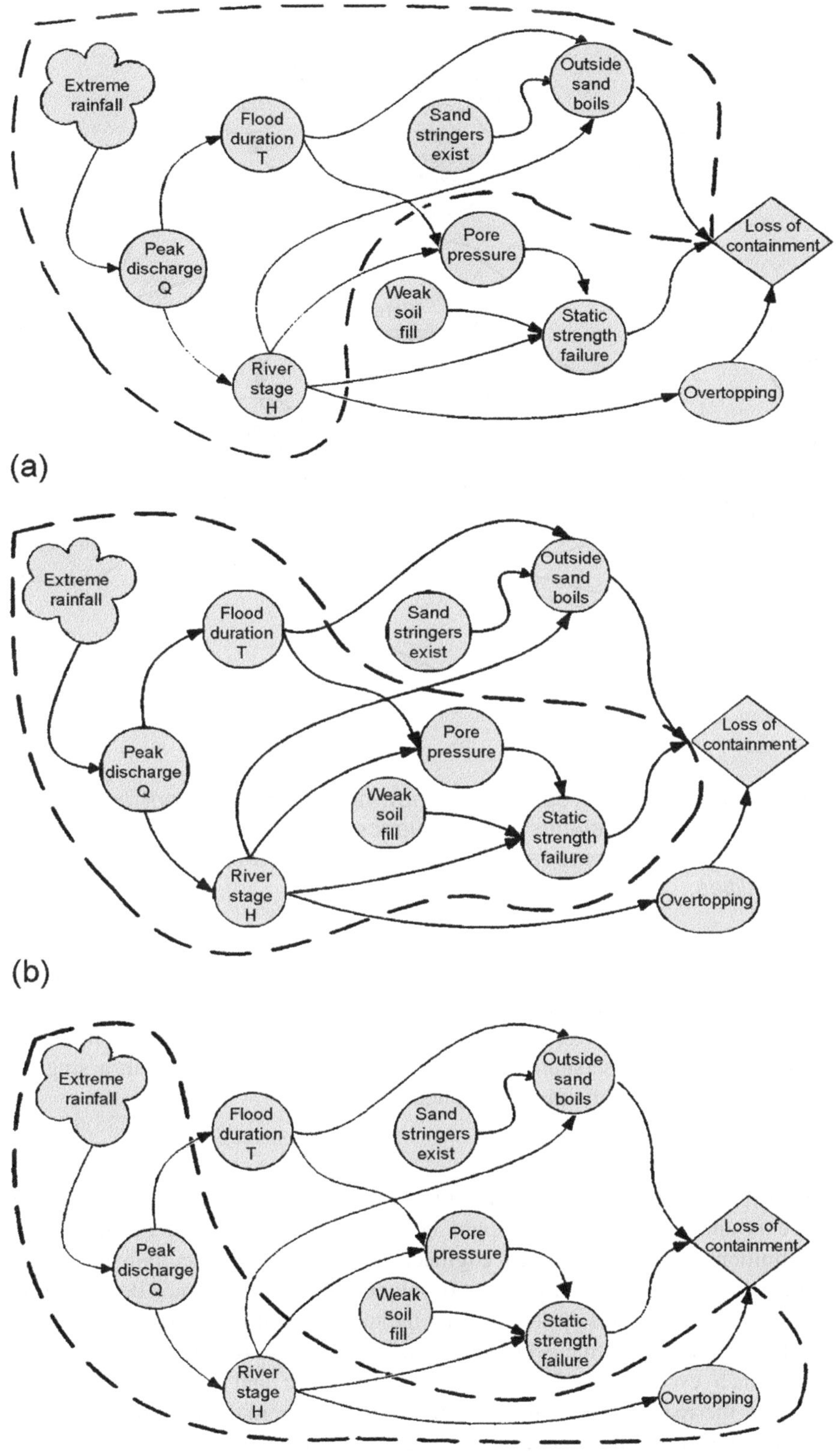

Figure 8-28. Influence diagram for levee failure, showing three separable event sets

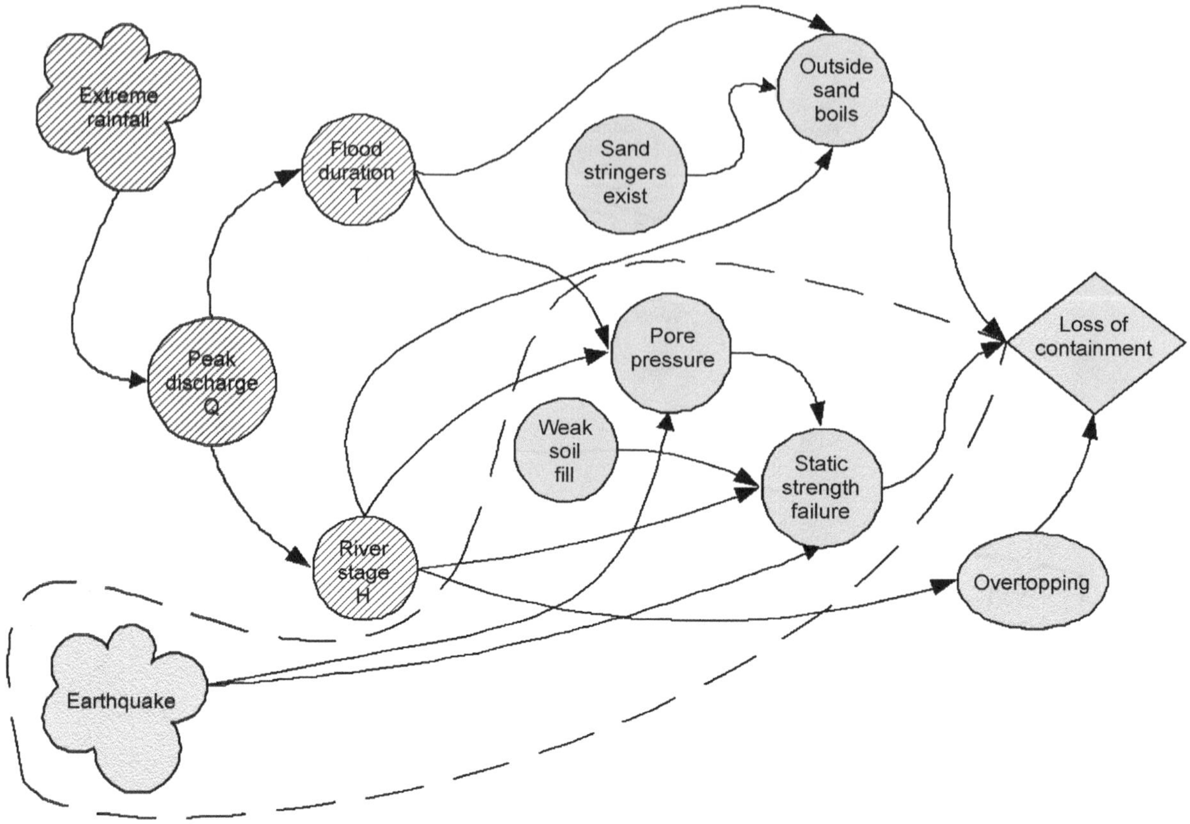

Figure 8-29. Influence diagram for levee failure, showing separable event set for earthquake initiating event. The cross-hatched events also influence the probabilities of events within the main event tree and must be accommodated, possibly within the reliability models generating individual event probabilities

independent. One can imagine a situation in which the saddle dyke fails first, substantially reducing damages downstream when the main embankment fails later. There is no easy way to capture this dependency in the consequences if the event chains are represented as separate trees. They need to be represented in a single tree, even though that tree will be more complicated.

Event chains

Branch events need to be clearly defined in self-evident language. To the extent possible, events and their outcomes should be described in language of sufficient clarity of definition that an uninvolved engineer or scientist can ascertain the meaning, with precision, by the written description alone. The danger in not describing events and outcomes to this level of detail is that the meanings of these things in practice becomes defined in the eye of the beholder and will differ from one person to the next.

Branch events must be collectively exhaustive and mutually exclusive. In other words, the realised pathway through an event node must traverse exactly one path. If common cause failure modes exist (e.g. simultaneous structural failure of the spillway due to earthquake ground shaking and instability of a reservoir slope), they should be represented by different event nodes in the tree. It is generally desirable to place causally dependent events to the right of the "independent" events that lead to them, although from a simple probabilistic view this is not required.

Event tree heuristics and biases

There is a large and well-known literature concerning the heuristics people use and the biases they exhibit in quantifying subjective probabilities. Presumably, similar or perhaps even the same heuristics and biases may operate when people develop event trees. Among the better studied of these heuristics are: (1) availability; (2) representativeness; and (3) anchoring and adjustment.

The way these heuristics might affect event tree construction is, of course, unclear without the insight gained from research that has yet to be done. Yet, it may be instructive to speculate on their impact, given what we know about the impact of such heuristics on subjective probability estimation, such heuristics and their resulting biases operate in traditional deterministic settings as well as in probability estimation and event tree analysis.

Availability refers to the problem-solving technique in which people gravitate to scenarios or descriptions of systems performance that are most easily conceived or imagined. Thus, in the aftermath of a well-publicised failure caused by, say, overtopping, related failure mechanisms might find themselves richly, perhaps too richly, depicted in the event tree for another structure. Psychologists already know that, in the wake of such an event, the subjective probabilities associated with the types of mechanisms observed and recent failures are often over-estimated. In a similar way, scenarios or descriptions of system's performance that are not easily conceived are likely to be under-represented in the event trees. It may be worth speculating that the relative poverty of detail found in the event trees associated with internal erosion could be caused by this heuristic.

Representativeness refers to the problem-solving technique in which people recognise patterns in specific cases which might be associated with pattern types found within some larger population of cases. They then attribute behavioural properties associated with the larger pattern type to the specific case, whether or not these behavioural properties are appropriate to the specific case. In other words, the analyst might perceive that a particular dam fits a pattern typical of some class of dams, and then presume that the many facets of that class of dams also apply to the particular structure in question. Given that almost every structure is unique, this presumption might be far from accurate.

Anchoring and adjustment refers to the problem-solving technique – or perhaps more accurately stated, the estimating technique – of appraising the value of some unknown quantity by first choosing a representative or best value and then adjust up or down from that anchor point. In assessing subjective probabilities, this approach of anchoring and adjustment tends to yield probability distributions which are too tight about the best value, that is over-confident. Such a heuristic might affect event tree construction by narrowing the variety of events and event paths that are considered.

Disaggregating events

The significance of an extreme storm for, say, overtopping depends on things other than just the volume of inflow. One of these, for example, is pool level at the time of the storm. Thus, new events may have to be added to the tree. Similarly, for the initiating event, "earthquake occurs", the significance for a dam will depend not only on ground shaking but also on whether surface faulting occurs (Von Thun, 1996). Thus, it is usually the case that the detail of event chains grows as an event tree analysis progresses.

Salmon and Hartford (1995) provide a more specific example of the gradual increase in event chain complexity as an analysis proceeds. The analysis starts with the initiating event, "peak six-hour inflow" to the reservoir. From this, the event tree analysis intends to lead to the probability of hydrological failure, starting with overtopping. Ultimately, the event tree expands also to include five subsequent failure events: overtopping,

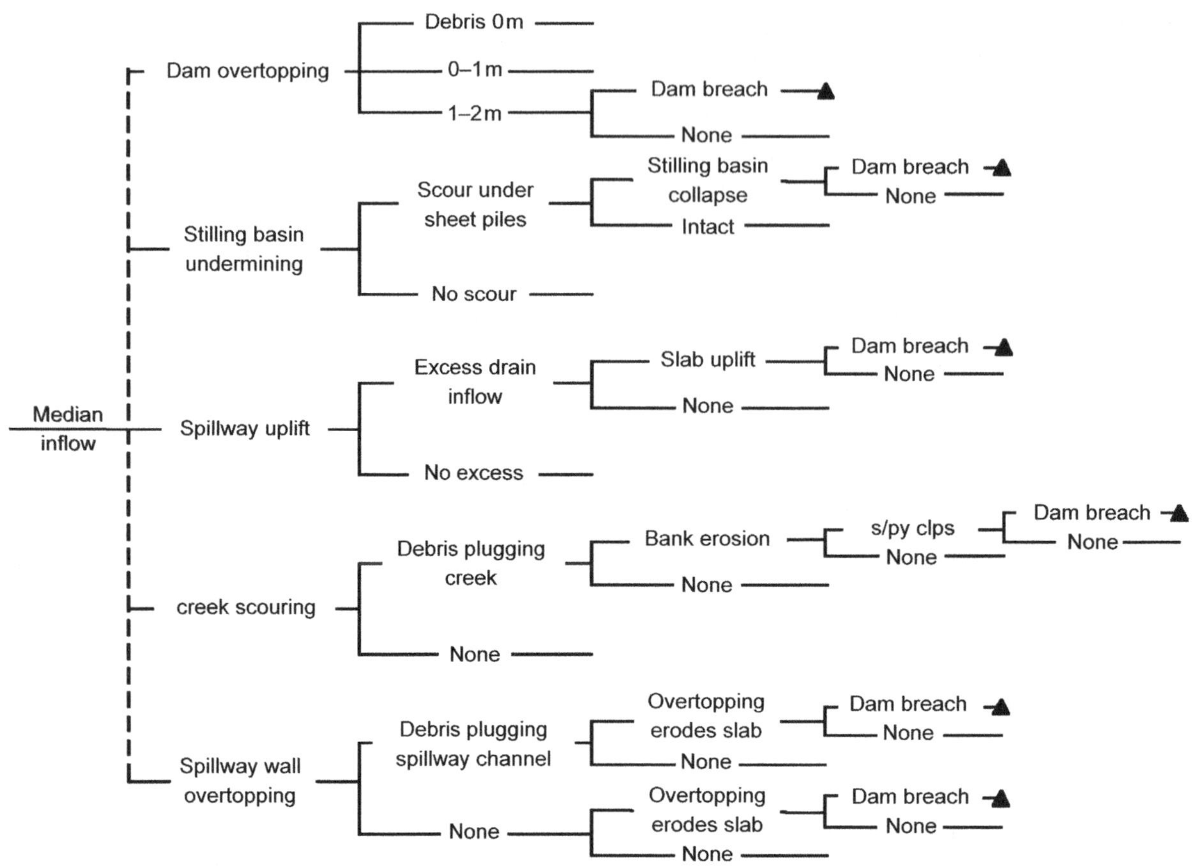

Figure 8-30. Partial event tree for hydrologic failure at Alouette Dam (Salmon and Hartford, 1995)

Note: s/py clps = slipway collapse

spillway uplift, creek erosion, spillway overtopping and stilling basin undermining (Figure 8-30).

A problem noted by Salmon and Hartford (Salmon and Hartford, 1995) is that as a chain of events is decomposed into an ever-increasing number of sub-events, the number of individual probabilities within the chain becomes ever larger, and since these probabilities are each less than one, their product has a tendency – in practice at least – to become ever smaller.

A good deal of work has been done in this literature on the effectiveness of desegregation in estimating subjective probabilities of poorly known quantities. It is somewhat a matter of faith in engineering and reliability circles that disaggregating a problem into its smaller component parts both increases understanding and improves estimation accuracy. Yet, results in the literature are mixed (Morgan *et al.*, 1990). Several studies in the psychology literature (Armstrong *et al.*, 1975; Gettys *et al.*, 1973, Henrion and Fischhoff, 1986) appear to support the contention that desegregation leads to better calibrated probability estimates, But only if subjects know at least as much about the smaller components to be estimated as they do about the aggregated quantity itself.

Poorly understood mechanisms

The increasing use of probabilistic risk analysis in dam engineering has led to a

inadequate. That is, there are event trees of the type illustrated by Figure 8-26 or Figure 8-30 wherein the events may be clearly described, but the probabilities associated with those events cannot be estimated using existing tools. This has led to the realisation that the profession needs to develop models that are more helpful in analysing actual mechanisms of failure rather than focusing on design standards. Traditional civil engineering models are mostly prescriptive, not performance-based, although the state of geotechnical engineering practice is perhaps more performance-based than, say, structural practice.

Piping is one of the mechanisms of failure that is poorly understood. Failures due to piping are not easily modelled, and even developing well-structured event trees is difficult. Nonetheless, all risk analyses for earth- and rock-fill dams need to address piping, and possibly other poorly understood physical processes as well. The propagation of cracking in concrete dams is another poorly understood mechanism of failure.

A simple influence diagram for piping failure within an embankment is shown in Figure 8-31. For piping failure to occur there has to be a sufficient head of water behind the embankment to generate high pore pressures within the embankment, and then there has to be something that allows a sufficiently steep pressure gradient for soil particles to be eroded from the core, implying an inadequate filter zone protecting the core, or there has to be some sort of flaw in the core or filter. Finally, the soil itself has to be erodible.

Figure 8-32 attempts to restructure the influence diagram as a fault tree. Here, for a piping failure to occur, there first has to be a hydraulic pressure gradient within the dam or its foundation, and the fill or foundation soil has to be susceptible to erosion. Clearly, a gradient always exists in the "impervious" dam core, because that is where the barrier to seepage is intended to be but, for a gradient to exist elsewhere, there has to be some sort of flaw, and the core has to erode. Since the core is protected from erosion by the filter, in order for erosion to occur, the core has to be either defective or poorly

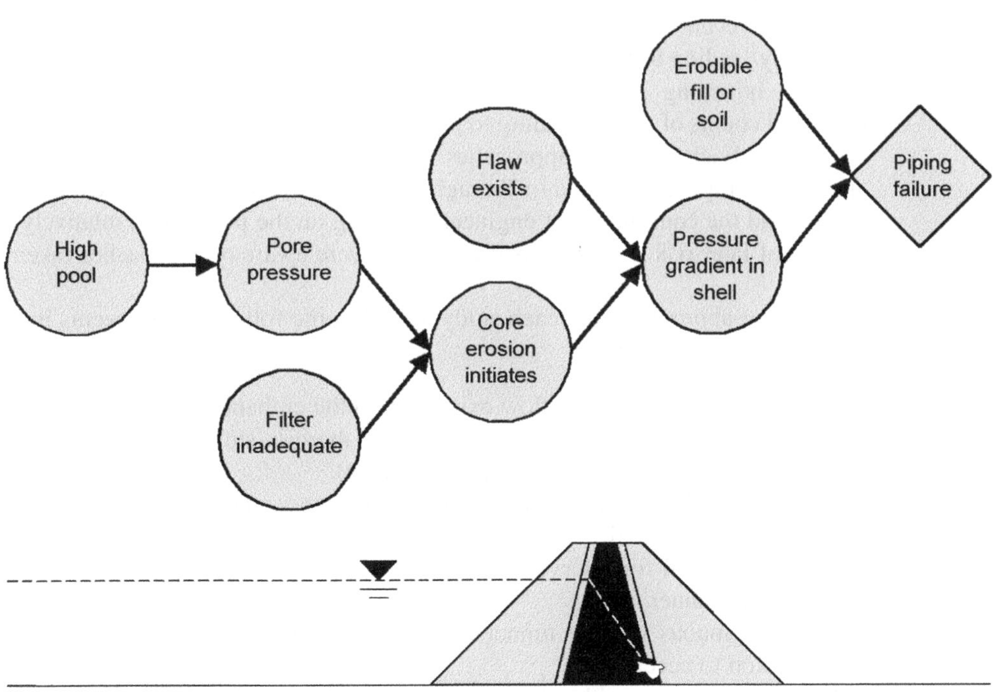

Figure 8-31. Simple influence diagram for piping failure of dam embankment

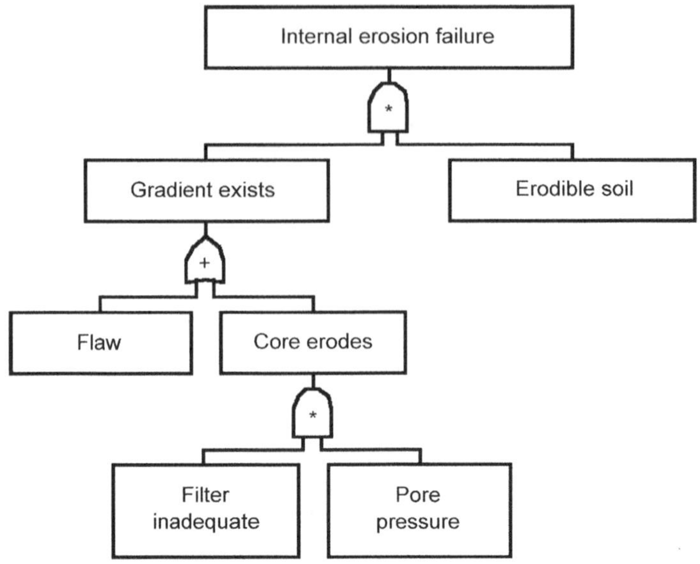

Figure 8-32. Generic fault tree for piping, building on the influence diagram of Figure 8-31

designed. Translating this simple fault tree into an event tree yields something like Figure 8-33.

What lessons can be learned from how analysts have tried to grapple with this issue? Table 8-5 summarises the approaches that a selection of published risk analyses use for modelling chains of events leading to piping failure. Conclusions based on this table need to be made with care: any dam is a complex structure with unique behavioural modes, and unique geological and construction conditions. Also, in each study, the analysts presumably had site specific insight into how initial erosion might begin, how that initial event could continue to erosion, and how that erosion might continue, ultimately leading to failure.

There is strong similarity across the several risk analyses in how the analysts identified chains of events leading to piping failure, and how these chains of events were structured. Each of these approaches is quite different from the prescriptive design approach to piping described in Terzaghi *et al.* (1996). Of course, these are all published studies and the community of engineers working on the problem is relatively small; the authors of later risk assessments presumably were aware of approaches taken in earlier studies.

As a first approximation, each study identifies the following six events in the critical pathway leading to piping:

1. A flaw or concentrated flow exists within the embankment or foundation.
2. Piping starts (sometimes interpreted as: there is sufficient hydraulic gradient or seepage velocity to enable piping to begin).
3. The toe of the embankment erodes, or some other exit point for the eroding seepage is created.
4. The potential for intervention exists, but is not exercised, or is not exercised in a timely manner.
5. A continuous pipe or tunnel forms in the embankment or foundation, allowing erosion to accelerate.
6. A breach occurs either by the formation of a continuous channel or by slumping of the crest and subsequent overtopping.

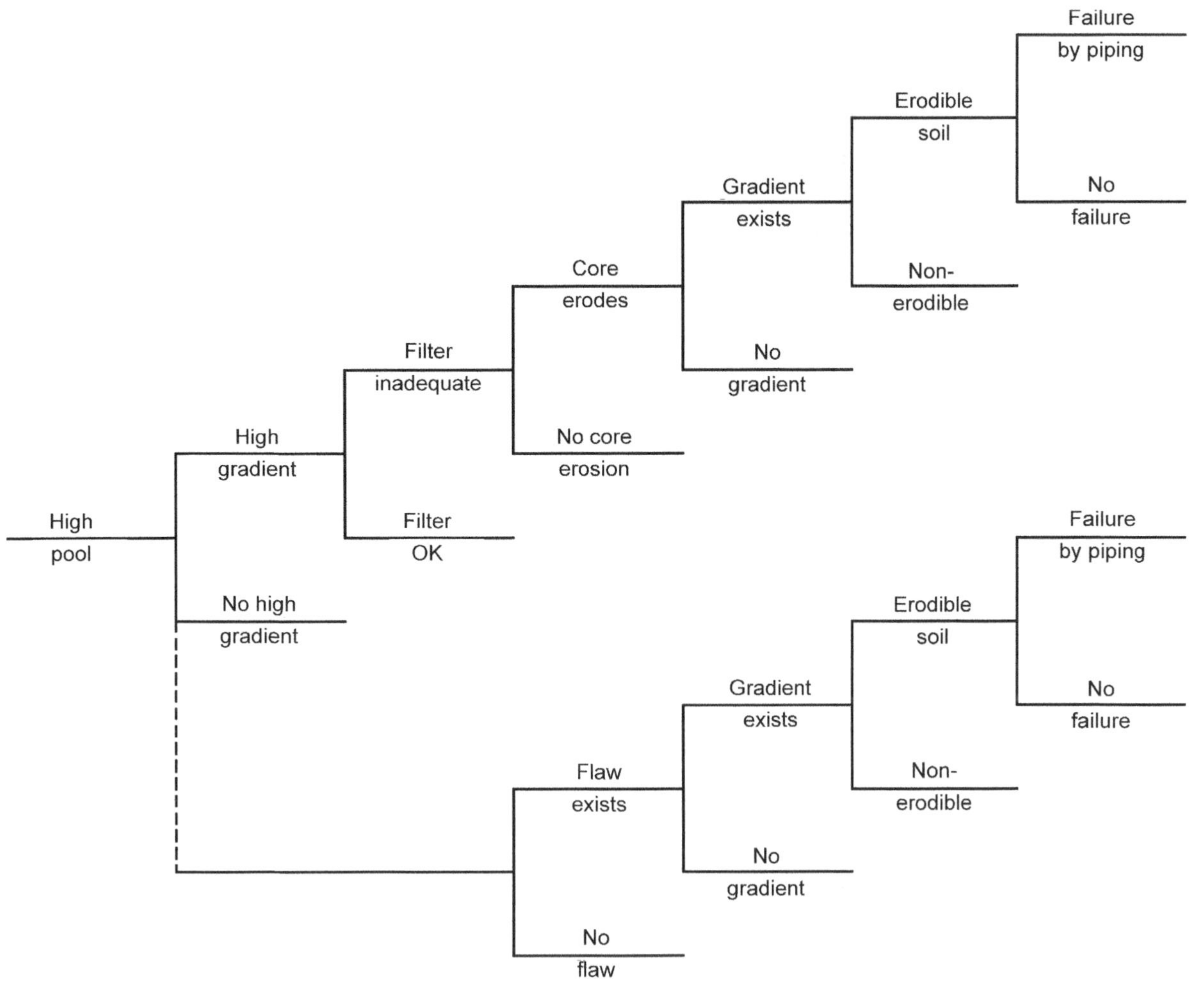

Figure 8-33. Event tree for piping based on the generic fault tree of Figure 8-32

The rudimentary event tree for this typical chain of events is shown in Figure 8-34. This is not dramatically different from the event tree of Figure 8-33. In some cases a variant, not shown, is analysed that involves the potential for human intervention somewhere in the process.

Dam engineering has historically adopted conservative practice as its design approach. As a result, many – possibly most – of the models used in dam engineering focus on upper and lower bounds to behaviour, rather than on best estimates of performance. For example, the strength stability of an embankment is usually calculated using limiting equilibrium models, rather than load-deformation models. In event tree analysis, one often needs more specific performance forecasts than such bounding calculations easily provide.

Consequences

The consequences that result from a dam incident or failure depend on a large number of factors beyond the behaviour of the dam itself. For example, consequences depend on safety functions such as barriers, safety systems, operating procedures, operator actions, and so on, and in how they respond to an initiating event or an indication of dysfunction of the dam proper. Safety functions may include systems that respond automatically to an initiating event or to a failure, alarms that alert operators or other cognizant personnel that an event or failure has occurred, predefined operating procedures that follow an

Table 8-5. Approaches to modelling the chain of events leading to piping failure, summarised from published studies

Coursier dam		**Viddals-, Dravladals-Svartevatn**		**Navajo Dam**		**Como Dam**
BC Hydro		**NGI**		**USBR**		**USBR**
Vick and Stewart (1996)		**Johansen, Vick and Rikartsen (1997)**		**Von Thun (1999)**		**Luehring, Russell, and Mares (2000)**
Internal erosion of embankment		Leak occurs	Initiation of fines migrating	Concentrated flow (flaws or joints)	Concentrated flow path	Concentrated flow path
Loss of core integrity			Development of piping	Initiation of piping	Adequate velocity/gradient	Adequate velocity/gradient
Pipe enlargement		Toe unravelling	Toe unravelling	Piping progresses	Openwork materials	Unprotected exit
Crest subsides	Seepage on downstream face	Self-healing	Self-healing	Potential for intervention	Large voids develop	Potential for intervention
	Slumping of face	Leakage detection	Leakage detection	Breach	Embankment collapse	Piping initiated
	Progressive instability/breach	Drawdown pool	Drawdown pool		Potential for intervention	Erodible materials
		Breach	Breach		Deformation greater than freeboard	Concentrated flow path

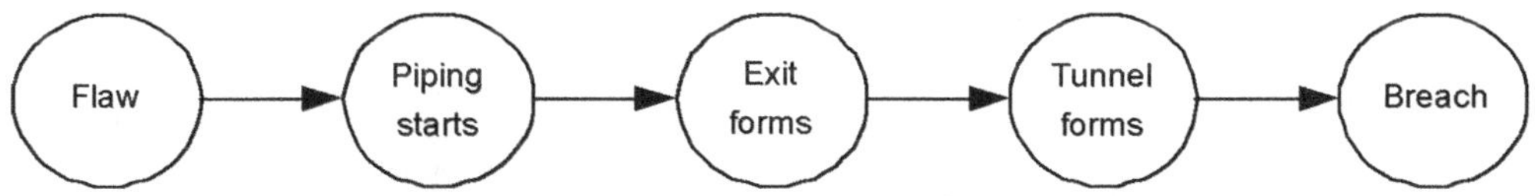

Figure 8-34. Generic event tree for piping as interpreted from six published risk analyses

alarm, or barriers or other containment facilities intended to limit the effects of an initiating event or failure. Other considerations that influence the consequences resulting from an initiating event or failure include time of day, meteorological conditions, downstream warning systems and emergency response procedures to protect life and property.

Typically, these many considerations that are not of the dam system itself, or are not of the uncertainties related to limited knowledge about natural processes or the functioning of the dam, and thus are separated into a consequence model or, more typically, into a separate consequence tree. This tree functions much like an event tree and, indeed, it is a form of event tree, subject to all the considerations discussed above.

Consequence trees

As in probabilistic risk analysis (PRA) of nuclear power plants (McCormick, 1981), a separate event tree is often used to model downstream consequences of a dam failure; that is, to provide an estimate of consequences conditional on loss of pool or some other damage state of the dam (Figure 8-35). This sub-tree incorporates exposure cases and other downstream activities or events, such as the effectiveness of evacuations or other risk mitigation procedures. Usually this downstream sub-tree can be treated independently of the event tree for the dam itself.

Event trees are evaluated by generating a set of probability-consequence pairs for each mutually exclusive end node in the tree. In principle, consequences could be expressed in utilities (Keeney and Raiffa, 1993) rather than direct physical units (e.g. dollars or lives lost), but usually they are not. Multiplying probability by consequence to obtain an expected consequence implicitly treats high-probability low-consequence outcomes as equivalent to low-probability high-consequence outcomes, as long as the product is the same. For government projects this is usually taken to be reasonable, given governments' asset positions and the multiplicity of projects undertaken (Morgan *et al.*, 1990). For a private owner, this risk neutrality may not be reasonable.

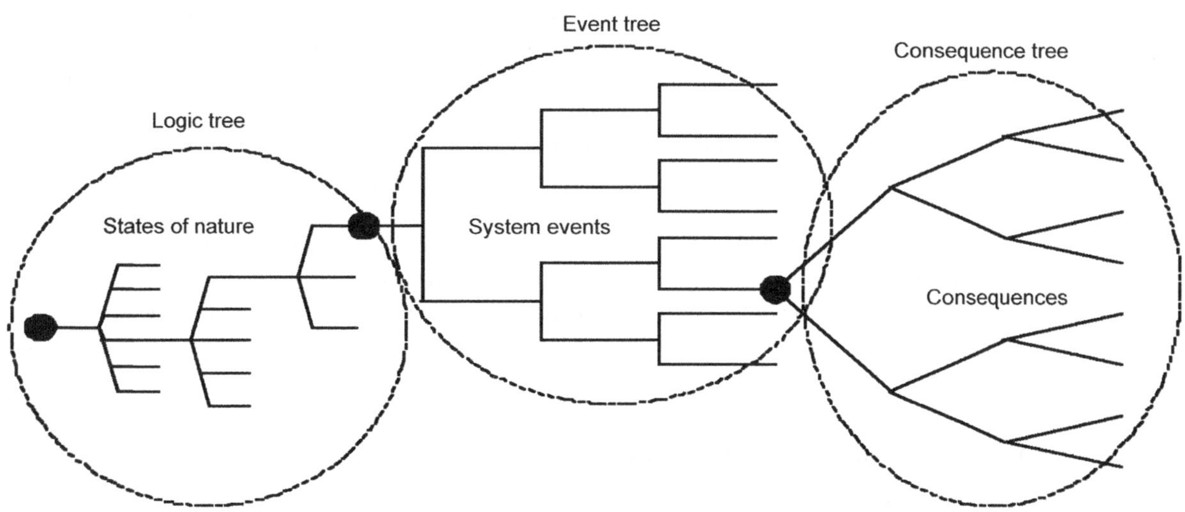

Figure 8-35. Consequence tree used to structure outcomes of a chain of events

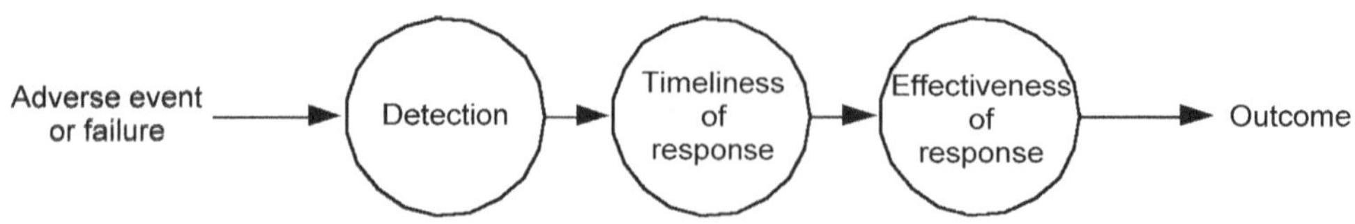

Figure 8-36. Typical model of human intervention uncertainties

Protective (human) intervention

Human intervention to reduce the consequence of some adverse chain of events, such as emergency measures to lower the pool or to place a field expedient filter on an eroding downstream toe, can be included in consequence trees or in the main event trees themselves. On the other hand, following the convention of only including changes of system states in the event tree proper, it would appear to make more sense to include human interventions within a consequence tree rather than an event tree. This provides a clearer separation between events happening within the dam, on the one hand, and cognizant measures taken to mitigate consequences, on the other.

As with all uncertainties in risk analysis, the uncertainties surrounding human interventions to mitigate the adverse effects of events or systems failures are due to limited knowledge. Nonetheless, such uncertainties are often treated as if they were random variables – that is, aleatory uncertainties – since they are outside the control of the emergency planner, let alone the analyst, as noted in Part II. These uncertainties are included in the event tree or in the consequence tree as chance events, and probabilities are assigned to the outcomes leading from those chance events to describe whether interventions are made in a timely and effective manner.

A significant difficulty in attempting to model human intervention is that the nature of those interventions may be impossible for the analyst to foresee. History is complete with stories of heroic efforts made in the face of the insurmountable obstacles, in which people display breathtaking creativity in response to adversity. The risk analyst has no way of foreseeing this extent of human creativity. Thus, greatly simplifying assumptions must be made in trying to include such factors quantitatively.

The literature of risk analysis pertinent to the uncertainties surrounding human response to emergencies is large. The typical event or consequence tree model of these uncertainties is shown schematically in Figure 8-36. To the LHS, some significant event or failure takes place. This occurrence is either detected in a timely manner by the operator, or not. The operator then responds to this information in a timely manner, or not. Depending on what the operator does in response, the responsive action may be effective, or not. This results in some outcome on the RHS.

For each of these three sets of events, a large number of human factors models have been developed and are available to dam safety studies. Whether one incorporates the human intervention events within the main event tree or within a separate consequence tree is ultimately left to the convenience of the analyst. If the human interventions are modelled as random events (i.e. natural variabilities or aleatory uncertainties) as they typically are, they have the same functional form as all the other events within the main event tree.

8.5 Branch probabilities

The preceding section considered the geometry of event trees: how events relate to one another, how they are sequenced, and the consequences to which they lead. The focus

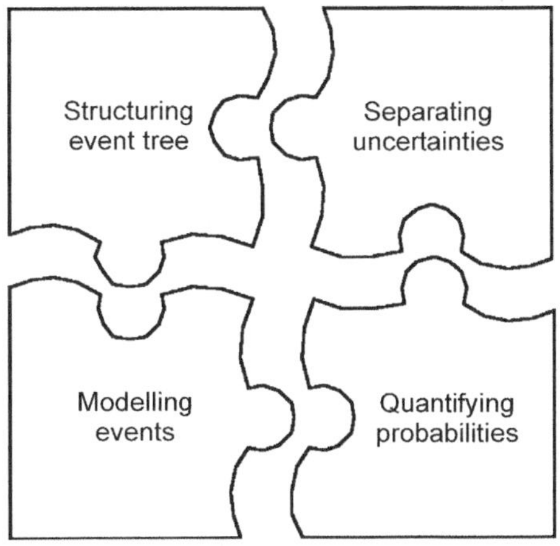

Figure 8-37. Event definition and probabilities are inextricable: the definitions of failure, sequences of events, identification of uncertainties and probability estimates must fit together like the pieces of a puzzle

was qualitative. This section considers how numerical probabilities are associated with individual branches and subsequently with paths. The focus is quantitative.

Defining failure modes, structuring event trees, separating natural variability from limited knowledge, and creating models to calculate probabilities all introduce assumptions. It is necessary for everything to fit together: the way failures are defined, events sequenced, uncertainties identified and probabilities estimated (Figure 8-37). Changing the structure of the event tree, changing the way uncertainties are separated, or changing the way uncertainties are modelled affects the numerical estimates of branch probabilities appropriate to the risk analysis. Change one of these, and the others change in response.

Conditionality of branch probabilities

Each event node in an event tree generates some number of outcome branches representing the mutually exclusive and collectively exhaustive ways in which that event might obtain. To calculate the risks represented in the tree, some quantitative value of probability must be associated with each branch. These probabilities are conditional on all preceding events in the tree leading to the node from which they emanate, and their sum over all sibling branches (Figure 8-38) equals one (by virtue of being exclusive and exhaustive).

In the event tree of Figure 8-39, the probabilities associated with the branches for "embankment cracking" differ depending on the branch leading to that node. The branches leading into the nodes for embankment cracking represent the possible ways "foundation liquefaction" can be realised. Successful performance of the foundation system is defined as "no liquefaction behaviour" (or possibly, "no liquefiable deposits present"); unsuccessful performance is defined as the complement (Von Thun, 1999). The events associated with the embankment structure itself are defined as no significant cracking in the embankment, and unsuccessful performance as the complement. Thus, the conditional probability of successful performance of the embankment structure depends on whether the first system, the foundation, performs successfully or not, that is, on whether there is liquefaction (or liquefiable soil) in the foundation. The numerical values of probability shown in the figure suggest possible differences.

Note that at any node, the sum of the probabilities of the possible events (i.e. sibling branches) is 1.0. They must include all possible ways the events could obtain. The total probability for a particular chain of events or path through the tree is found by

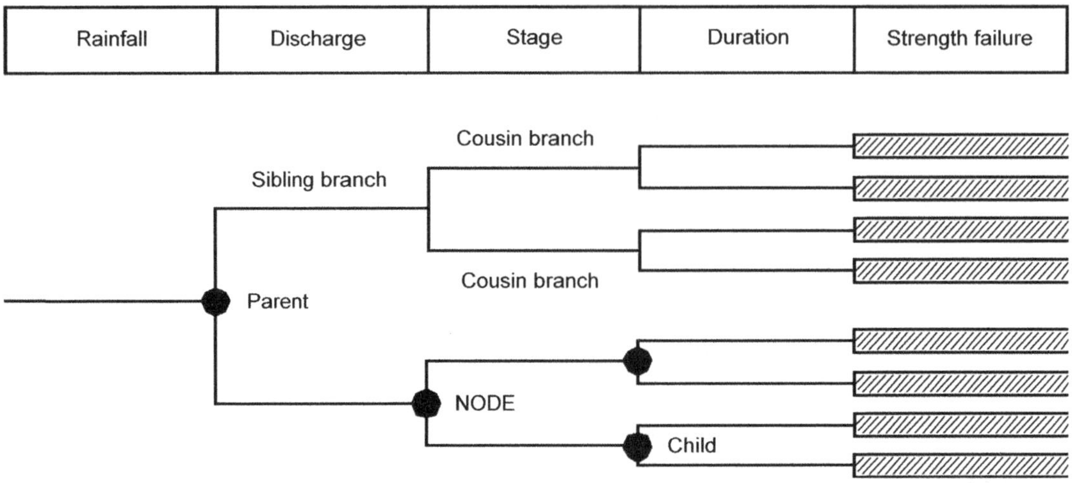

Figure 8-38. Relationships among nodes in a simple event tree, relative to the node shown hatched

multiplying the sequences of conditional probabilities. Thus, the probability of the uppermost consequence on the RHS of Figure 8-39 is:

$$\begin{aligned}&Pr[\textit{no cracking AND no liquefied soil AND earthquake}]\\&=Pr[\textit{no cracking} \mid \textit{no liquefied soil AND earthquake}]\\&\quad\times Pr[\textit{no liquefied soil} \mid \textit{earthquake}]\\&\quad\times Pr[\textit{earthquake}]\\&=0.95\times 0.9\times 0.01\\&=8.55E-03\end{aligned} \qquad (8\text{-}5)$$

Note, the sum of the joint probability of all chains of events emanating from a particular initiating event must equal the marginal probability of the initiating event. Thus, the sum of the probabilities of the consequence nodes on the RHS of Figure 8-39 is 0.01, as this is the marginal probability of the initiating earthquake event.

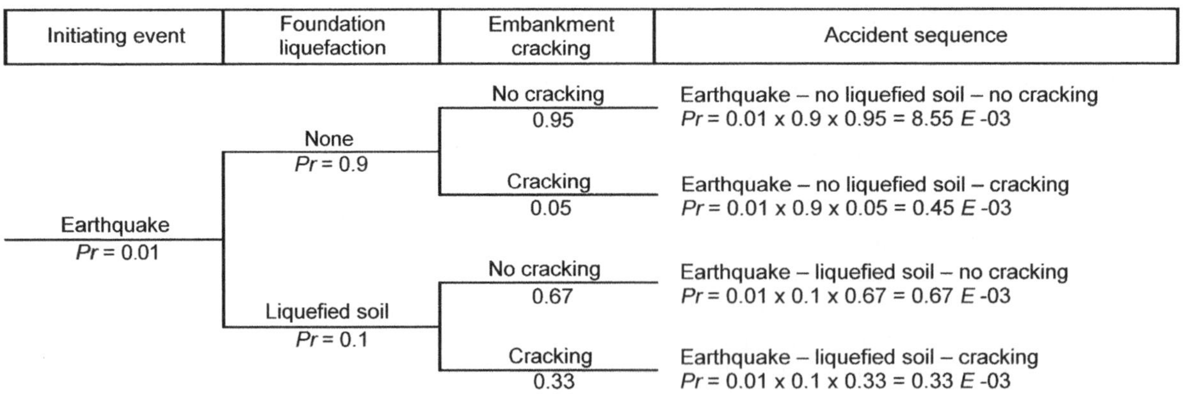

Figure 8-39. Illustration of a simple event tree for an embankment dam, showing dependence of probabilities on the outcomes of earlier events in the tree

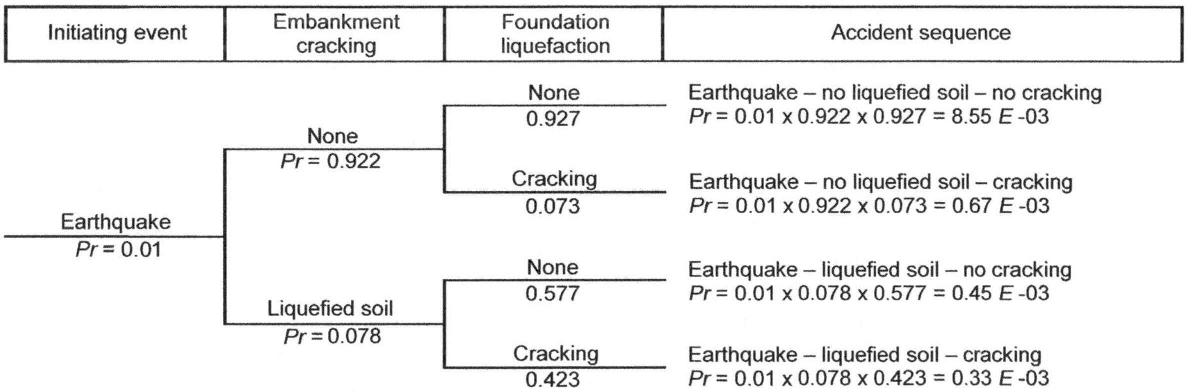

Figure 8-40. Illustration of a simple event tree for an embankment dam, showing dependence of probabilities on the outcomes of earlier events in the tree

By the total probability theorem (Benjamin and Cornell, 1970), the sequencing of the conditional probabilities can be inverted (Figure 8-40), such that the joint probability could also be written as:

$$\begin{aligned} & Pr[\textit{no liquefied soil AND no cracking AND earthquake}] \\ & \quad = Pr[\textit{no liquefied soil} \mid \textit{no cracking AND earthquake}] \\ & \quad \quad \times Pr[\textit{no cracking} \mid \textit{earthquake}] \\ & \quad \quad \times Pr[\textit{earthquake}] \\ & \quad = 0.927 \times 0.922 \times 0.01 \\ & \quad = 8.55E-03 \end{aligned} \quad (8\text{-}6)$$

in which, 0.927, the conditional probability of no liquefied soil given that there is no cracking as a result of the earthquake is determined from Bayes' theorem. Thus, the sequencing of events in the tree itself can be rearranged as suggested by Figure 8-40, as long as the corresponding conditional probabilities assigned to the branches are similarly rearranged.

Estimating branch probabilities

An individual branch probability is typically estimated by one of several procedures: (1) statistical (i.e. empirical) estimates; (2) reliability models; (3) fault tree analysis; or (4) expert opinion. Statistical estimates and reliability models are discussed in this section. Fault trees were discussed earlier in the present chapter; expert opinion is discussed in Chapter 7.

Statistical estimates

Among the more common places that statistical estimates are used is in estimating probabilities of initiating events such as extreme floods or earthquakes. The occurrence of initiating events is typically modelled as a stationary Poisson process, with a single parameter, λ, describing the number of occurrences per unit time, usually a year. This model is:

$$f_n(n \mid \lambda) = \frac{\lambda^n e^{-\lambda}}{n!} \quad (8\text{-}7)$$

in which n is the number of occurrences, and λ is the rate per unit time. Stationary means that λ is constant over time. The number of occurrences in time interval t is:

$$f_n(n \mid \lambda, t) = \frac{(\lambda t)^n e^{-\lambda t}}{n!} \tag{8-8}$$

for which the mean of n is $E[n] = \lambda t$ and the variance, $Var[n] = (\lambda t)$. Other point process models could also be used to represent the occurrence of initiating events, for example if one wanted to introduce probabilistic dependence among the event occurrences (Cox and Lewis, 1966).

The most important assumption of the Poisson distribution is that events occur independently of one another. The occurrence (or lack of occurrence) of any one event in no way influences the possible occurrence of another event. Thus, if a large storm occurs this year, the probability of a similar storm occurring next year is unaffected: in principle, one could have two $Pr=0.001$ events (1000-year storms) back-to-back. The exception to independence in the Poisson model is that, for most natural hazards, the length of historical record is short. New information, such as the occurrence of an extreme storm or a large earthquake, affects the statistical estimate of the parameter λ. Thus, the occurrence of an extreme event changes the parameter of the model of the process itself, and consequently influences the estimated probabilities of future events. For example, in constructing the Oroville Dam in California, two presumably 250-year ($p=0.004$) storms and one 1000-year ($p=0.001$) storm occurred during the few years of construction (Gilbert, 2002). With the occurrence of each of these storms the rate parameter λ for the site changed, as did subsequent estimates of the annual exceedance curve.

Consider some infrequent event that has occurred four times in the past 100 years. An initial estimate of the rate of occurrence of this event is $\hat{\lambda}=4/100=0.04$ per year. With time (Figure 8-41), having observed no further events of this size or larger, the estimate $\hat{\lambda}$ begins to decline. After an additional 100 years without such an event, the rate will have declined to $\hat{\lambda}=4/200=0.02$ per year. On the other hand, should additional large storms occur during the subsequent 100 years of observation (shown for illustration at

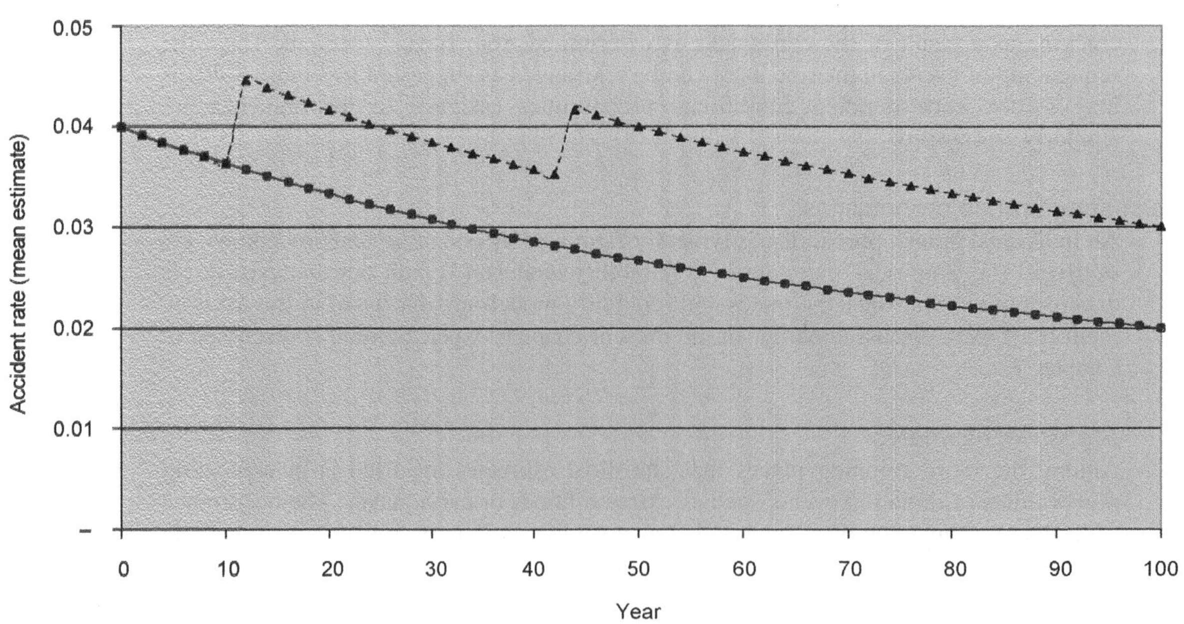

Figure 8-41. Estimate of the rate of occurrence of some infrequent event over time, accounting for the occasional occurrence of subsequent events of a similar character

12 years and 44 years in Figure 8-41), the estimated rate $\hat{\lambda}$ will show a sharp rise immediately following the observation(s), followed again by a period of slow decay while no further large storms are observed.

Among the most important limitations to the statistical approach to estimating probabilities for initiating events is short length of record. For example, historical stream flows are usually the basis for estimating exceedance probabilities of annual flood discharge, but in most regions of North America it is unusual for there to be more than perhaps 50 years of record. Estimates of the exceedance probability of rare events, such as the $p=0.01$ (100-year) event are prone to error. Estimates of yet lower probability events are subject to even more error.

Common practice – presupposing a degree-of-belief view of probability – is to represent statistical uncertainty in a model parameter such as λ by a probability distribution, $f_\lambda(\lambda)$. Then a *predictive probability distribution* on the number of events n in a unit length of time is found by integrating the model forecast over the uncertainty in the parameter (Aitchison and Dunsmore, 1975),

$$f_n(n)=\int_\lambda f_n(n,\lambda)d\lambda=\int_\lambda f_n(n\,|\,\lambda)f_\lambda(\lambda)d\lambda \tag{8-9}$$

This predictive distribution is simply the marginal distribution on n having integrated out uncertainty on λ. Similar results obtain for the case when more than one uncertain parameter is involved, that is for vector λ.

Note that the predictive distribution combines uncertainty due to natural variability, that is, the number of occurrences n given the parameter λ, with uncertainty due to limited knowledge, that is, the uncertainty in the model parameter itself. If more than one aspect of natural variability depends on the same uncertain parameter (e.g. number of occurrences in two separate periods, n_1 and n_2) then forming the predictive distribution of each in isolation from the other will mask the implied correlation caused by their shared dependence on the same uncertain realisation of λ. This is one of the arguments for separating out natural variability from limited knowledge using logic trees.

The usual approach to quantifying the probability distribution for an uncertain model parameter, such as λ, based on statistical data is to start from some uniform or a non-informative probability density function (PDF) on the parameter, presumed to exist prior to having observed the data, and then to update that probable distribution using Bayes' theorem. For the case of Poisson processes, standard practice is to presume that the prior distribution on λ follows the gamma form,

$$f_\lambda(\lambda)\propto\lambda^{\alpha-1}e^{-\lambda\beta} \tag{8-10}$$

with $\alpha=0$ and $\beta=0$, implying a prior pdf proportional to λ^{-1}, sometimes called the non-informative distribution (Jeffreys, 1948, p. 138). If k extreme events are now observed in r years, then by Bayes' theorem the statistical uncertainty in λ can be estimated as:

$$\begin{aligned} f_\lambda(\lambda\,|\,k,r) &\propto f_\lambda(\lambda)f(k,r\,|\,\lambda) \\ &\propto[\lambda^{\alpha-1}e^{-\lambda\beta}]\left[\frac{(\lambda r)^{-k}e^{-\lambda r}}{k!}\right] \\ &\propto[\lambda^{\alpha-(k+1)}e^{-\lambda(\beta+r)}] \end{aligned} \tag{8-11}$$

which conveniently continues to have the form of a gamma distribution. Any pdf on the parameter values which has this property of closure under updating by Bayes' theorem is said to be a *natural conjugate distribution* (Raiffa and Schlaifer, 2000).

Modelling the occurrence of initiating events as a Poisson process typically presumes the uncertainty of those events to be due to natural variability in time or space. Given the average rate of occurrence specified by the parameter λ, the actual occurrence of initiating events is a purely stochastic process. The only place that knowledge uncertainty enters the formulation is in the statistical estimate of the parameter λ itself (which may be far from negligible).

The assumption that initiating events are stochastic in time or space is convenient but not necessary. In the case of extreme storms, one could model storm events based on first principles of meteorology. Given this approach, the occurrence and magnitude of storm events are results of the modelling activity, and are not necessarily treated as random in time or space. Uncertainty enters this approach to modelling extreme events in the statistical estimates of the parameters of the meteorological model and in model error. This approach trades off the uncertainty due to natural variability for uncertainty due to limited knowledge.

This example of changing the way uncertainty is modelled between natural variation and limited knowledge is conceptually similar to an example from seismic hazard analysis that was cited in a recent National Research Council study (National Research Council, 2000),

Epistemic uncertainty would be much greater if, in the assessment of seismic hazard at an eastern US site, instead of representing random seismicity through homogenous Poisson sources one used a model with an uncertain number of faults, each with an uncertain location, orientation, extent, state of stress, distribution of asperities, and so forth. As little is known about such faults, the total uncertainty of future seismicity and the calculated mean hazard curves would be about the same, irrespective of which model is used. However, the amount of the epistemic uncertainty would be markedly different; it would be much greater for the more detailed, fault-based model. Consequently, the fractile hazard curves that represent epistemic uncertainty would also differ greatly.

The chapter on Essential Elements of Probability Theory argues that all uncertainty is fundamentally due to limited knowledge, and that we use natural variability only as a modelling convenience. The problem that arises from this arbitrary separation of natural variability and knowledge uncertainty is that as the model changes so, too, does the relative contributions of these two types of uncertainty. Where fractile curves are used to represent "probability distributions of frequencies", these too are affected by the way natural variability and limited knowledge are separated (Veneziano, 1994).

Usually, an event tree analysis presumes stationarity for stochastic initiating events, but for events such as flood volumes or river stages the assumption may not be good (NRC, 1995). Changing upstream conditions in land use, channelisation, and other factors may cause flood frequencies or rating curves to gradually change with time. This means that the probability distributions used to represent random variables, such as peak flood discharge, may change.

Reliability model estimates

Probabilistic engineering models – sometimes called reliability models – reason from first principles of mechanics or natural science to calculate uncertainties in the performance of specified variables. For example, such models might start from uncertainties in soil engineering properties in an embankment and calculate probabilities of excessive settlement of the crest of the embankment, or probabilities of strength failure. Conceptually, they are similar to the predictive distributions discussed in the previous section, in which statistical uncertainties in parameter values are integrated out to give an aggregate uncertainty in some random variable. With probabilistic engineering models, uncertainties in input parameters such as, say, undrained soil strength, are integrated over to yield implied uncertainties in model predictions that might otherwise have been deterministic.

Presume an engineering model, perhaps deterministic, that relates some dam performance variable x to a parameter or set of parameters θ through the equation(s):

$$x = g(\theta) \tag{8-12}$$

Presume also that uncertainty in θ can be expressed in a pdf $f_\theta(\theta)$. Then uncertainty in the performance variable x can be expressed as

$$f_x(x) = \int_\theta g(\theta) f_\theta(\theta) d\theta \tag{8-13}$$

This is akin to the predictive PDF of equation 8-9, in that uncertainty in the parameter is integrated out. If the engineering model $g(\theta)$ is deterministic, then $f_x(x)$ is a simple probability density function reflecting the effect of uncertainty in the model parameter on uncertainty in the performance variable. If the engineering model is itself a stochastic relationship, then $f_x(x)$ will have a more complicated form, and some of the model parameters of $g(\theta)$ may themselves be an expression of spatial or temporal variability.

Whether the prediction $f_x(x)$ reflects natural variation or limited information, or some combination of both, depends on the nature of the uncertainty in the model parameters, and on whether the model itself is deterministic or stochastic. If the model is deterministic and the uncertainty in θ is statistical, then $f_x(x)$ reflects limited knowledge. If the model is deterministic and the uncertainty in θ is spatial or temporal, then $f_x(x)$ reflects to natural variation. If the model is stochastic, then $f_x(x)$ will typically reflect both natural variation and limited knowledge.

The other source of uncertainty in a prediction of the performance variable, in addition to that due to parameter uncertainty, is that due to uncertainty in the model formulation itself. This model uncertainty manifests in many ways. For example, a model may itself be statistical, as in the case of models based on regression analysis. Many current soil liquefaction models are of this variety. For such statistical models there may be well founded estimates of the uncertainty in model predictions made when using reasonably well identified input parameters. These uncertainties have to do with the set of calibrating data used to develop the original model, and may reflect both natural variability and limited knowledge but usually in unknown proportions.

Another way model uncertainty manifests is in models that are approximations to complicated physical conditions. For example, Mohr–Coulomb strength theory is a linearisation to what is more commonly a nonlinear strength relationship between normal and shear stresses. Model uncertainties based on such approximations are due to limited knowledge and may be systematic, that is, they may tend consistently to bias predictions in one direction or another.

How to handle model uncertainty in the probabilities assigned to event tree branches? No model is "true" in the sense of being perfectly in accord with fact or reality, and wholly free from error. Any model is an imitative representation of the functioning of one system or process by means of the functioning of another. Today, this "other process" is usually mathematical, although it could be physical, as in the case of scale models of natural hydraulic systems. Thus, model uncertainty is not a question of the truth of a model, for all models are false in the sense that they are inaccurate.

Model uncertainty is treated in risk analysis in one of three ways, none of which is wholly satisfactory. The first approach is to apply an uncertain bias term to a model

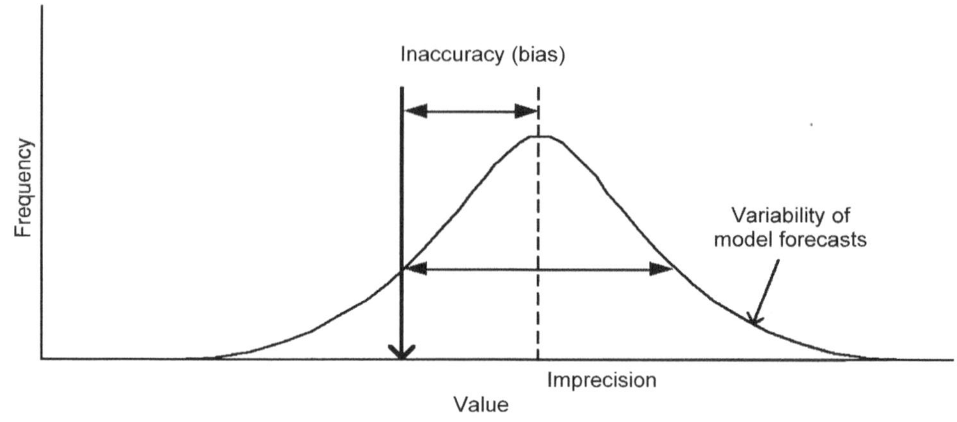

Figure 8-42. Inaccuracy (bias) and imprecision in model predictions

prediction (Figure 8-42). Thus, the model prediction of some performance variable x becomes,

$$x = bg(\theta) + e \tag{8-14}$$

in which b is a systematic bias with mean, $E(b)$, and variance, $Var(b)$; and e is a zero-mean random error with variance, $Var(e)$. The values of $E(b)$, $Var(b)$, and $Var(e)$ are inferred from historical experience with the model, or are subjectively estimated by a knowledgeable practitioner. This essentially treats models as inaccurate and imprecise transducers. The inaccuracy reflects bias; the imprecision reflects lack of replicability. This is much the same as the question in applied statistics of noisy observations.

The second approach is to apply weights to model predictions. This implicitly supposes that there is more than one model, and that one of the models is correct or true. The weights, which are forced to sum to 1.0, function as probabilities: the probability of a particular model is "right". But none of the models is "right" in a philosophical sense, so this approach has been subject to criticism, even though it is widely used (Bedford and Cooke, 2001; Morgan *et al.*, 1990). Since none of the models is actually right, the weights assigned to them cannot be thought of as probabilities, and might not add to 1.0. If, on the other hand, one interprets "right" to mean approximately correct, then more than one of the models might satisfy this criterion, and thus the weights might reasonably add to more than 1.0. So, this approach has conceptual limitations.

The third, and arguably most successful, approach is to build upon the notion of models as transducers, but instead of applying simple bias and random error terms, to treat model predictions as information in a decision theoretic or Bayesian sense (Morris, 1974). This approach works for one or multiple models, and makes no presumption that any model is "right". It asks only, what is the likelihood that a model would give the result it does if one knew the correct value? Given a model forecast $\hat{x} = g(\hat{\theta})$, Bayes' theorem is used to update a probability distribution on the actual value of the performance parameter in question, x, in the familiar form,

$$f_x(x|\hat{x}) \propto f_x(x)L(\hat{x}|x) \tag{8-15}$$

In the case of multiple models, the joint likelihood function – which is a conditional probability density of the multiple model predictions as calculated, given the (unknown) correct value of the performance parameter – is a joint probability distribution and automatically accounts for correlations among the model predictions. This approach has been applied to both model predictions and aggregating expert opinion (Morris, 1977). While the approach is theoretically rigorous, a difficulty lies in knowing the joint likelihood, which presumably is either inferred statistically from historical data or estimated subjectively.

Discretising continuous events

When dealing with mechanical and electronic equipment, there is often a distinct demarcation between "failure" and "non-failure" conditions. There may be a body of data for such equipment, from which probabilities can be inferred using statistical tools either of a frequency nature or a Bayesian nature (Martz and Walker, 1982). But for many questions about the geotechnical, hydraulic or structural behaviour of dams, such historical performance information is not available, and strong databased statistical conclusions cannot be drawn. The subtle definition of potential outcomes, as well as the assignment of probabilities to those outcomes, is strongly subjective. Again, event trees have proven to be tools for eliciting knowledge about the structure of these latter types of failures, and for identifying the chains of events and considerations that go into their analysis.

Unlike fault tree representations, events within an event tree need not, and generally are not, represented as having dichotomous (Boolean) outcomes. "Failure" may not be a simple matter of yes or no, but a matter of degree. Indeed, event tree analysis has proven especially suitable for analysing situations wherein adverse consequences are subtle and possibly depend on sub-events in the tree, rather than a simple failure vs non-failure. Vertical slumping of a dam crest due to liquefaction or other strength instabilities within an embankment might have a range of consequences depending not only on how great the ultimate slumping but also on the geometric and temporal pattern of the resulting deformations. The outcomes of individual events within an event tree may have multiple possibilities, and it may be necessary to further dissect an apparently straightforward event into a number of sub-events. The question here is, how to provide adequate accuracy when discretising continuous processes?

Once probabilities start to be assigned to event outcomes, the need for further clarity about what the events mean becomes apparent. What exactly is an "extreme storm"? Total rainfall and hence runoff are continuous variables, but for computational reasons they would most likely be discretised in an event tree. In principle, an event outcome can be described over a continuous range and assigned a probability density function rather than point probabilities (i.e. rather than a probability mass function), but this requires large numbers of calculations of the subsequent events and, at some point, a (usually) numerical integration to combine all those results.

Peak annual flood discharge into a reservoir is a continuous variable with realisation ranging from zero to some large value, nominally the "probable maximum flood". Figure 8-43 shows the flood frequency curve estimated from historical data for the Sacramento River. Instead of using a flood frequency curve to describe the distribution of probability over flood event outcomes, presume the outcomes might be discretised as shown in Figure 8-44. In this tree, continuous outcomes of flood discharge are replaced by discrete values, the mean and the mean ± 1, 2 and 3 standard deviations, respectively.

Presume, hypothetically, that the consequences of a flood are proportional to discharge squared. That is, a flood of twice the size has four times the consequences.

Then "risk", calculated as probability times consequence and integrated over the range of possible flood discharges $(0, \infty)$, is:

$$E[C] = \int_0^\infty x^2 f_x(x \mid \boldsymbol{\theta})dx \qquad (8\text{-}16)$$

in which C=consequences, x=flood discharge, $f_x(x \mid \boldsymbol{\theta})$=the pdf of discharge, and $\boldsymbol{\theta}$ is a vector of parameters of the probability distribution. Presuming for computational

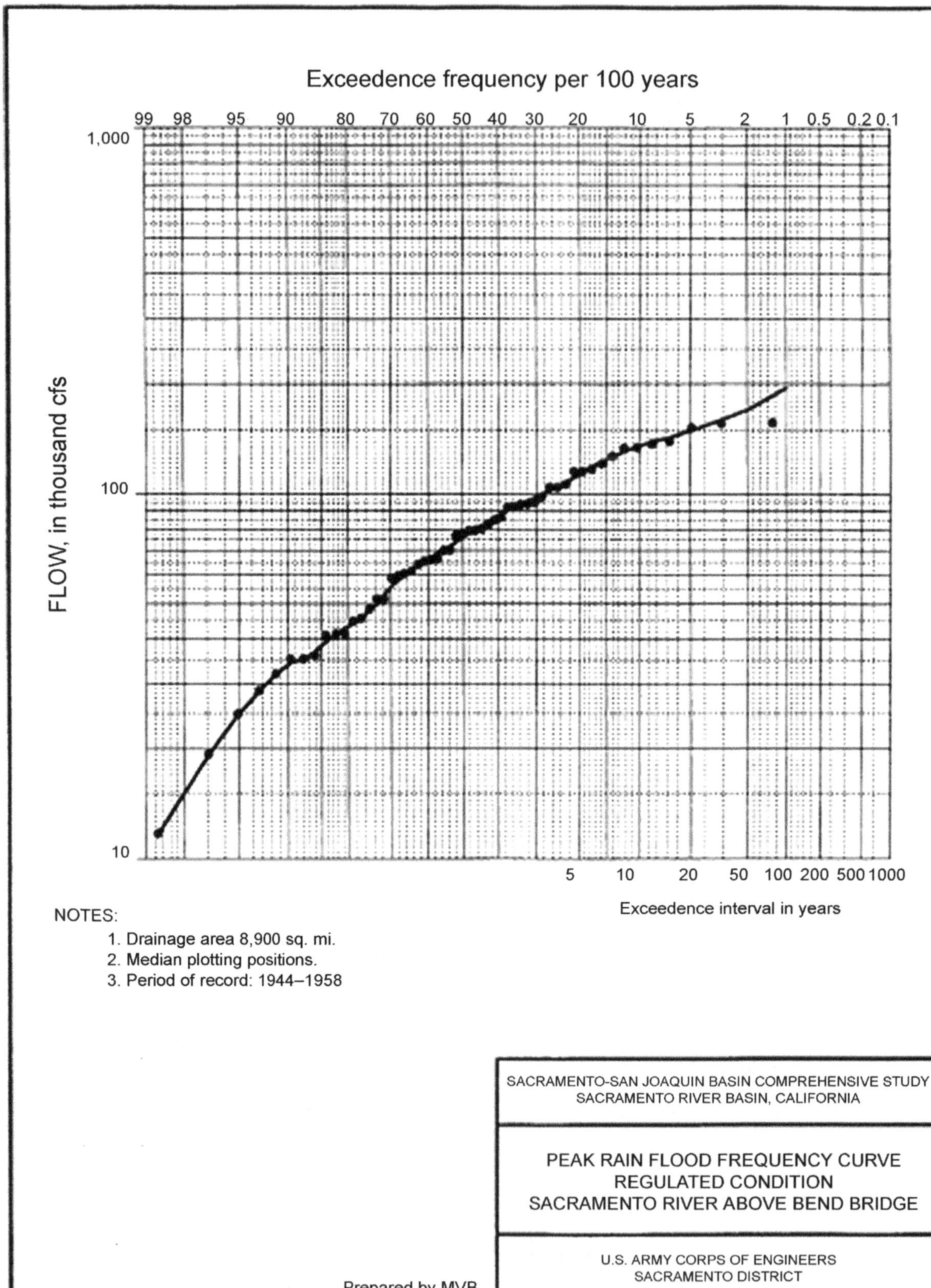

Figure 8-43. Flood frequency for the Sacramento River above Bend Bridge (USACE, 1995)

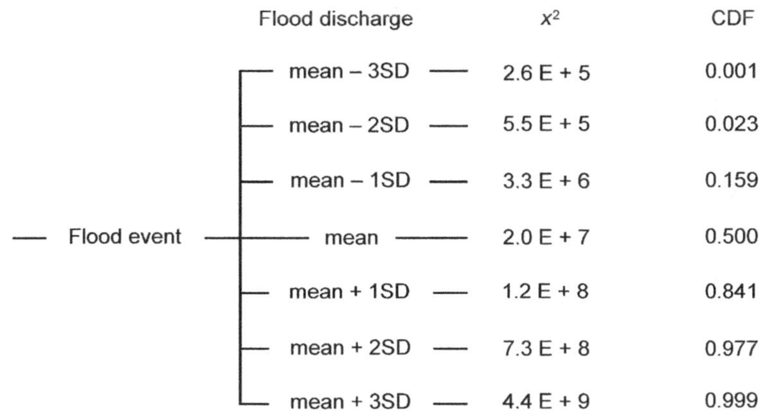

Figure 8-44. Discretised representation of "flood event"

convenience that the pdf of flood discharge is logNormal with parameters $\boldsymbol{\theta}=\mu, \sigma$, in which μ=mean of the log-discharge and σ=standard deviation of log-discharge, then the solution to equation 8-16 is (Aitchison and Dunsmore, 1975):

$$E[C]=\exp(n\mu+\tfrac{1}{2}n^2\sigma^2) \tag{8-17}$$

For μ=4.18 and σ=0.39, as shown in Figure 8-45, the risk is proportional to 5.93×10^3.

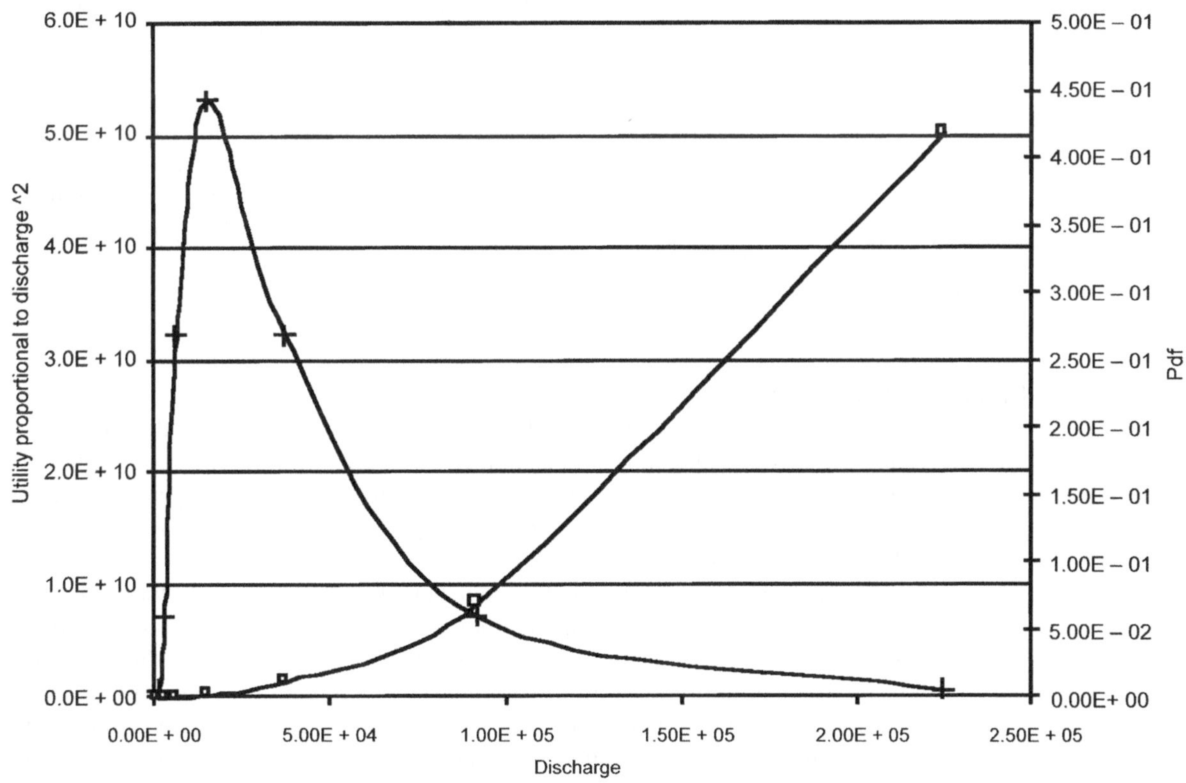

Figure 8-45. PDF of flood discharge and corresponding consequences proportional to the square of discharge

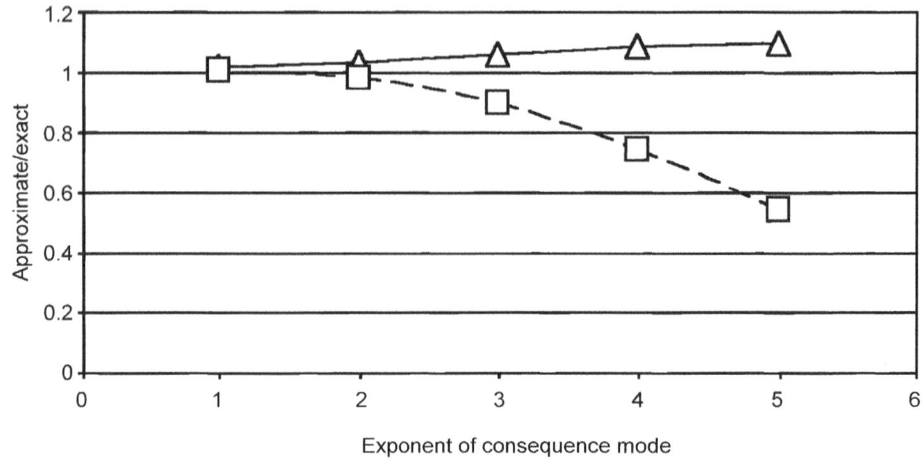

Figure 8-46. Error ratio (approximate to exact) as a function of the exponent of the consequence model, Consequence=x^n, for five-point approximation (mean, ±1SD, ±2SD, ±3SD) and two-point approximation (±1SD)

In contrast, for each of the discretised values x_0, take as the corresponding probability the area under the PDF between x_0 and plus or minus one-half the interval to the next highest and lowest values of discharge in the tree (i.e. the differences in the CDFs), and take as the consequence the value at x_0, squared. The risk is then proportional, approximately, to 5.79×10^3, which is only about 2% different, and given the many approximations of a risk calculation, is entirely acceptable.

On the other hand, discretising affects calculations, even when a relatively large number of discrete outcomes is used, as the non-linearity of the consequence or subsequent calculation increases (Figure 8-46). Approximating an event by a simple dichotomous outcome, say plus and minus one standard deviation away from the mean, requires more care and may introduce more error. For a linear or low-power outcome, the continuous distribution and its discretised approximation give nearly similar results.

Probabilistic dependence

Correlations among the branch probabilities associated with separate nodes in an event tree can be critical to obtaining proper numerical results in an event tree analysis. To complicate matters, these correlations can also be both subtle to identify and difficult to estimate. It is imperative that correlations among probabilities be dealt with properly, as the following examples show.

Presume that one is concerned both about potential liquefaction in the foundation soils of an embankment during a major earthquake, and also about potential overtopping after a major storm. The probability of liquefaction happening sometime during the life of the dam, accounting for the various earthquakes that might occur but marginal of all other uncertainties, is estimated as 0.09. Similarly, the probability of overtopping happening sometime during the life of the dam, accounting for the various storms that might occur but marginal of all other uncertainties, is estimated as 0.076. Then, were the liquefaction and overtopping independent of one another the probability of one or the other occurring during the life of the dam would be:

Table 8-6. Conditional probabilities of liquefaction and overtopping given the existence or non existence of low-density, soft lenses in an embankment

	Pr	Liquefaction	No liquefaction	Overtopping	No overtopping
Low-density lenses exist	0.1	0.9	0.01	0.67	0.33
No low-density lenses exist	0.9	0	1.0	0.01	0.99

$$\begin{aligned} Pr[liquefaction\ OR\ overtopping] &= Pr[liquefaction] \\ &\quad + Pr[overtopping] \\ &\quad - Pr[liquefaction]Pr[overtopping] \\ &= (0.09) + (0.076) - (0.09)(0.076) \\ &= 0.16 \end{aligned} \tag{8-18}$$

Consider, however, that both liquefaction of the foundation and overtopping – due in part to settlement of the embankment – are influenced by the possible existence of low-density lenses of soil under the embankment. The respective conditional probabilities are given in Table 8-6, from which the marginal probabilities in equation 8-18 can be verified.[5]

Given that the two events depend in common on the existence of low-density zones in the foundation, the actual probability of one or the other happening during the life of the dam is the weighted sum of the calculation of equation 8-18, taken once conditioned on the zones occurring and a second time conditioned on the zones not occurring. This yields an actual probability of 0.11, about a third less than the calculation above.

$$Pr[liquefaction\ OR\ overtopping] =$$

$$Pr[zones]\left\{\begin{array}{l} Pr[liquefaction \mid zones] \\ + Pr[overtopping \mid zones] \\ - Pr[liquefaction \mid zones[Pr[overtopping \mid zones]] \end{array}\right\} \tag{8-19}$$

$$+ Pr[no\ zones]\left\{\begin{array}{l} Pr[liquefaction \mid no\ zones] \\ + Pr[overtopping \mid no\ zones] \\ - Pr[liquefaction \mid no\ zones]Pr[overtopping \mid no\ zones] \end{array}\right\}$$

The reason is that the existence of the soft zones makes both events simultaneously more likely, while the non-existence makes both simultaneously less likely. Since the two events are no longer independent, their marginal probabilities can no longer be simply summed to give the probability that either one or the other occurs. The correlation due to a common dependence reduces the risk to the structure in this case, but the outcome could be the reverse in other circumstances.

Branch probabilities associated with separate event nodes can be correlated through any of the following:

- *Causal dependence*, meaning that one event physically causes another; for example, liquefaction-induced settlement may directly lead to overtopping of an embankment, thus the liquefaction event and overtopping event would not be independent of one another. If the liquefaction settlement occurs, the probability of overtopping might be greatly enhanced.

- *Probabilistic correlation*, meaning that two uncertainties may share a common dependence on a third uncertainty, as in the case of the low-density soil lenses in the example above. The outcome of whether the low-density soil lenses exist or not simultaneously changes the probability of liquefaction cracking and of overtopping.
- *Spatial or temporal autocorrelation*, meaning that two uncertainties depend on the spatial or temporal realisation of some third uncertainty which itself exhibits stochastic dependence in space or time. The performances of two sections of a long levee may depend on soil engineering properties in the naturally occurring valley bottom, which, when modelled as a stochastic (aleatory) process, exhibit a long-wave length of correlation in space; thus adjacent sections will exhibit similar settlements or factors of safety against slope instability.
- *Statistical correlation*, meaning that two uncertainties are simultaneously estimated from a fixed set of data and therefore influenced by a common sampling variability error. In soil mechanics, a common – and almost always overlooked – statistical correlation is that between soil cohesion, c, and soil friction angle, ϕ, which being regression parameters of the Mohr–Coulomb model, are negatively correlated, given a finite number of test data.

Levee safety example revisited

Figure 8-47 shows the event tree for levee failure during an extreme flood that was earlier developed in Section 8.4, but now with branch probabilities. Table 8-7 shows the justification for each probability estimate, the estimation approach, and the sources of uncertainty due to natural variation and limited knowledge.

8.6 Calculations on event trees

The computational issues of event trees are reasonably straightforward. Except in unusual circumstances, a simple multiplication of branch probabilities along any path yields the probability associated with that terminal node. For computational reasons, sometimes calculations are made only along those paths leading to "failure" outcomes. This section considers certain details of the computation of event trees and mathematical procedures that have been suggested for those computations.

Direct calculation

The common calculation procedure for event trees, as outlined in Part II, can be summarised in the following steps:

- Perform event tree calculations to obtain terminal node probabilities.
- Prune extremely low probability paths.
- Verify the reasonableness of results against historical rates of dam incidents and failures, and for internal consistency.
- Perform sensitivity studies.
- Post process results for summary and presentation.

Calculation

Calculation of the event tree results, even in reasonably complex cases, is usually carried out by direct enumeration. That is, branch probabilities are assigned to all branches in the tree, or at least all those branches leading to consequences of concern, and then a complete multiplication of the (conditional) branch probabilities along each path is made. This leads to a complete set of path probabilities at each consequence node. The path probabilities leading to "failure" consequences are summed to give an overall probability of failure for the dam. Similarly, paths from probabilities leading to other

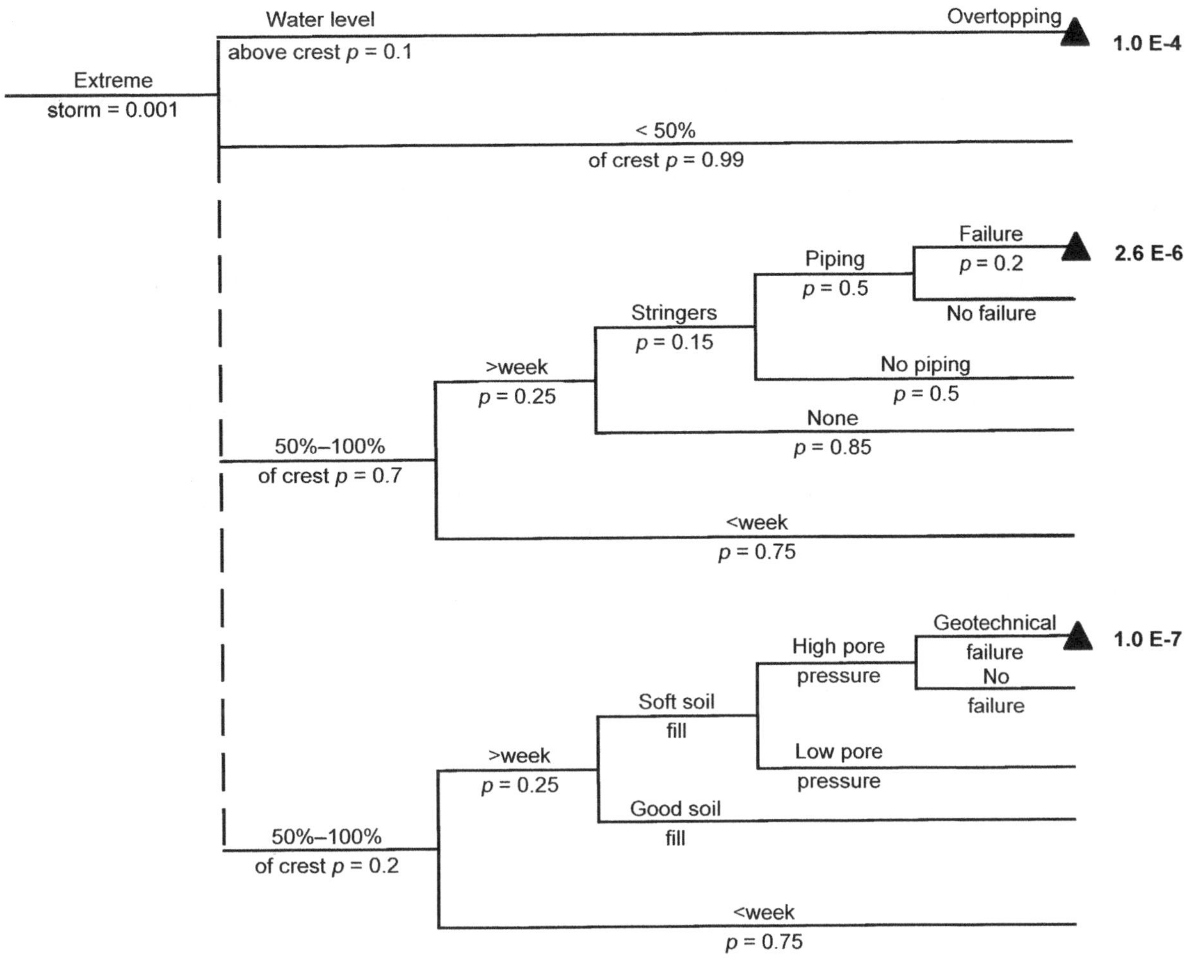

Figure 8-47. Event tree for levee failure during extreme storm, with estimated branch probabilities as detailed in Section 8.5. The pentagons beneath event branches show where a fault tree or reliability model is used to calculate the corresponding branch probability, and may introduce additional parameters

identifiable types of consequences are summed to give the corresponding probabilities of those consequences. This is a simple and exact procedure, which is practical in most cases.

Pruning

The issue with pruning an event tree is how much the tree can be simplified without affecting calculated probabilities of failure. It is relatively easy to eliminate branches that lead to outcomes of little importance. Another relatively easy procedure is to reduce the number of branches at an individual node. If the variability among the branches is small, this procedure is straightforward. If the variability is not small, sensitivity studies may need to be conducted to ensure that significant error is not introduced.

As a rule of thumb, it is generally reasonable to use the total number of terminal nodes in the tree as a factor for pruning event chains. If the total number of terminal nodes is of order N, then no event chain leading to an undesirable outcome should be pruned unless its terminal risk (probability times consequence) is of order $1/N$ smaller than that of the dominating event chain. Using this rule of thumb, the largest error that could result would be of order twice the risk value of the dominating event chain, but

Table 8-7. Branch probabilities and the basis for their calculation

Event node	Conditioning event	Event branch	Pr*	Method	Natural variability	Limited knowledge
Extreme storm		Yes	0.001	Flood frequency curve	Stochastic process	Distribution type, mean, variance and skew
Water height	Extreme storm	Water level above crest	0.1	Rating curve	Bottom conditions, debris/ice, etc.	Statistical regression parameters
	Extreme storm	50%–99% of crest	0.7	Rating curve	Bottom conditions, debris/ice, etc.	Statistical regression parameters
	Extreme storm	less than 50% of crest	0.2	Rating curve	Bottom conditions, debris/ice, etc.	Statistical regression parameters
Duration	Extreme storm	More than a week	0.25	Flood frequency curve	ditto	ditto
	Extreme storm	Less than a week	0.75	Flood frequency curve	ditto	ditto
Stringers	None	Yes	0.15	Poisson process	Age number per reach and size	Chance exploration detected
Sand boils	Stringers exist	Yes	0.5	Seepage model	Spatial variability of permeability	Soil parameters and spatial model parameters
Failure	Sand boils	Yes	0.2	Slope failure model	Spatial variability of soil parameters	Statistics of soil properties, model uncertainty
Soft soil	None	Yes	0.1	Poisson process	Age number per reach and size	Chance exploration detected
High pore pressure	Duration	Yes	0.5	Seepage model	Spatial variability of permeability	Soil parameters and spatial model parameters
GT failure	Soft soil, pore pressure	Yes	0.2	Slope failure model	Spatial variability of soil parameters	Statistics of soil properties, model uncertainty

Note: * Conditional probability given earlier events in tree.

would normally be much smaller than this limiting value. To be conservative, the rule may be modified to order $1/(1+N)$.

Partitioning the event tree

One way to reduce computations and increase confidence in calculated results is to reduce the size of the event tree by collapsing sub-parts of the tree or by separating pieces of the tree into independent elements. For example, the performance of a dam in the case of high pool level and possibly other circumstances may depend on the actions of a number of gate assemblies. Rather than fully enumerate the combinatorially large number of ways those gates structures might perform, the actions of the gate structures might be collapsed into a separated event tree, or possibly into a fault tree or structural reliability analysis, and then treated as a "black box" in the main event tree. Such substitutions are specially helpful when the same sub-event tree appears more than once in the main tree.

Most event tree analyses separately consider the sets of events associated with principal initiating events. For example, the probability of an extreme flood and an extreme earthquake happening at the same time may be negligibly small. In this case the two initiating events and the event trees resulting from them can be treated separately from one another. This is the common practice. Indeed, it is unusual that event trees for different initiating events are combined with one another.

Computational step size

In general, the range of loading due to initiating event is broad, and the characterisation of the probability distribution of that loading in a continuous analytical function (PDF) can make the subsequent event tree intractable. Event trees use discrete mathematics and are not easily amenable to continuous variables except in simple situations. Thus, initiating events are usually divided into computational steps, sometimes also called partitions (although the latter term may be confusing).

The question of dividing initiating events into computational steps is the same as that about discretisation event nodes further along in the tree. Fundamentally, the issue is how nonlinear the response of the system is to levels of the initiating event. If response is reasonably linear, calculated results are insensitive to step size and to the way initiating loads are computationally divided. If the system response is nonlinear, calculations may be sensitive to the way the initiating loans are divided.

A reasonable way to divide initiating events into computational steps is to ascertain threshold levels of the initiating event at which systems performance exhibits an inflection. A simple example is the level of floodwater inflow at which reservoir pool exceeds crest elevation. Inflows lower than this value might be lumped together with some associated point probability, while inflows larger than this value might be more finely divided, since their consequences are of greater import.

Verify reasonableness of results

Event trees can become large and unwieldy, so computational and other checks are useful to verify the reasonableness of results. One such check is to verify internal consistency, another is to compare results to historical data or trends.

Computational checks

An event tree is the graphical enumeration of a probability sample space. As such, it should be congruent with basic properties of sample spaces, for example, that the sum of probabilities over all outcomes of an experiment equal 1.0 (Figure 8-48). If such simple properties do not manifest in the tree, an attempt should be made to determine

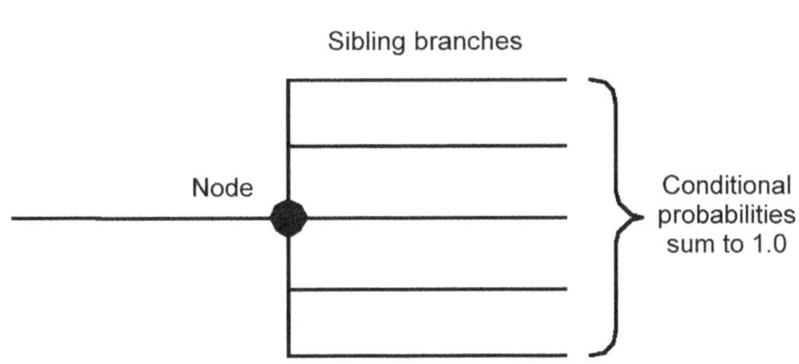

Figure 8-48. Computational check that the sum of the probabilities of sibling branches is 1.0

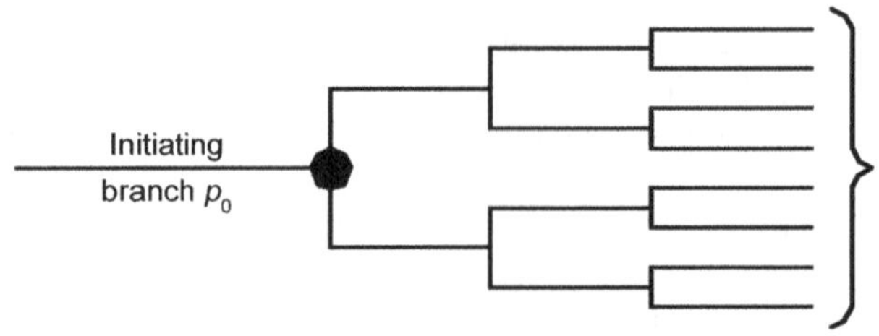

Figure 8-49. Computational check that the sum of the path (total) probabilities from a given initiating branch or event equals the probability of the initiating event or branch

why and to correct the deficiency. A few more obvious properties to check for are as follows:

The branches emanating from an individual node represent mutually exclusive and collectively exhaustive events. The probabilities of these sibling branches – conditional on the path of events leading to that node – should sum to 1.0 at every node in the tree.

Given the property above, for any single event branch or initiating event in the tree, the total path probabilities of chains of events starting at that branch and ending in terminal nodes should sum to the marginal probability associated with the initiating branch (Figure 8-49).

Table 8-8. Overall failure statistics for large embankment dams up to 1986, excluding dams constructed in Japan pre-1930 and in China (Foster et al., 2000)

	No. of cases		**% failures (where known)**		**Average frequency of failure (E-03)**	
	All failures	Failures in operation	All failures	Failures in operation	All failures	Failures in operation
Overtopping and appurtenant						
Overtopping	46	40	35.9	34.2	4.1	3.6
Spillway gate	16	15	12.5	12.8	1.4	1.3
Subtotal	62	55	48.4	47	5.5	4.9
Piping						
Through embankment	39	38	30.5	32.5	3.5	3.4
Through foundation	19	18	14.8	15.4	1.7	1.6
From embankment into foundation	2	2	1.6	1.7	0.18	0.18
Subtotal	60	58	46.9	49.6	5.38	5.18
Slides						
Downstream	6	4	4.7	3.4	0.54	0.36
Upstream	1	1	0.8	0.9	0.09	0.09
Subtotal	7	5	5.5	4.3	0.63	0.45
Earthquake liquefaction	2	2	1.6	1.7	0.18	0.18
Unknown mode	8	7				
Total no. of failures	139	127			12.2	11.1
Total no. of failures where mode of failure known	131	120				
No. of embankment dams	11,192	11,192				

Comparison against historical performance

As in any probabilistic or risk estimation enterprise, the reasonableness of a calculation of dam safety risk needs to be viewed within the context of historical experience. A number of authors have attempted to summarise the historical record of dam incidents and failures. Such efforts go back several decades (Baecher *et al.*, 1979; Foster *et al.*, 2000; Gruetter, 1982; Gruner, 1967; Jansen, 1980; Serafim, 1984). Recent summary results are shown in Table 8-8. Such comparisons, however, are difficult, perhaps providing at most an order-of-magnitude comparison.

Endnotes

1. There is debate in the risk analysis literature over the inductive vs deductive nature of event trees, but it is not clear what important lessons are to be taken from this debate. Some authors suggest that event tree construction is deductive (Bier, 1997; Pate-Cornell, 1984), while others suggest it to be inductive (Gruetter and Schneiller, 1982; Leveson, 1995; McCormick, 1981; Turney and Pitbaldo, 1996). Clearly, the building of event trees is a creative act that shares little with logical deduction. On the other hand, once an event tree is constructed, calculations of its results are purely deductive.
2. The present usage is more restrictive than that common in the decision analysis literature (Raiffa, 1968), where the meaning of *states of nature* also includes what are here called events.
3. It is interesting to speculate whether these two analysts would end up with the same final estimate of risk. On the one hand, the total uncertainty should be unchanged by the assumption that one part thereof is epistemic and the other aleatory. On the other hand, we use the separation between aleatory and epistemic uncertainties to exploit differences in the power that statistical and probabilistic models bring to the analysis. In a specific case there may well be an optimum way of dividing the two types of uncertainty, although it would be difficult to prove.
4. The computer science literature refers to these as *directed, acyclic graphs*, but for simplicity we avoid this use.
5. The calculations are:

$$\begin{aligned} Pr[liquefaction] &= Pr[liquefaction \mid lenses]Pr[lenses] \\ &\quad + Pr[liquefaction \mid no\ lenses]Pr[no\ lenses] \\ &= (0.9)(0.1) + (0)(0.9) \\ &= 0.09 \end{aligned}$$

$$\begin{aligned} Pr[overtopping] &= Pr[overtopping \mid lenses]Pr[lenses] \\ &\quad + Pr[overtopping \mid no\ lenses]Pr[no\ lenses] \\ &= (0.67)(0.1) + (0.01)(0.9) \\ &= 0.076 \end{aligned}$$

9 Fault tree analysis

In Part II, the discussion of FTA focused on the graphic nature and modelling features of fault tree construction. There is, of course, an analytic side that provides the tools for fault tree evaluation and quantification. In this module an overview of the mathematical foundation of FTA is presented. The discussion is fundamental, highlighting basic principles.

9.1 Mathematics of fault tree analysis

As a graphic display of a system, a fault tree is visual and provides an easy way to follow the logic flow as it relates to the occurrence of the top event (e.g. system failure). At the same time, however, the picture of a fault tree cannot, in general, be readily assimilated to the point that the analyst can identify the modes of system failure (i.e. combinations of events that lead to the top event occurring). For large or complex systems, it is next to impossible to visually identify the unique failure modes for a system. Furthermore, a fault tree is also not amenable to estimating the probability of the top event.

To transform a fault tree graphic into an analytical form that can be manipulated and ultimately quantified to estimate the probability of the top event, we rely on set theory, Boolean algebra and probability. These analytic tools provide the foundation to convert a fault tree into equation form, making it possible to algebraically manipulate and simplify. Once expressed in equation form, a fault tree can be evaluated to determine a system's failure modes and to quantify the probability of occurrence of the top event.

Set theory is a field of mathematics for organising items of like features. As such, it provides a foundation for organising the events in a fault tree and, ultimately, the occurrence of the top event. Boolean algebra is the algebra of variables that take on values of 0 or 1; or in fault tree terms events that occur or do not occur. A fault tree can be viewed as a collection of events that either occur and contribute to the occurrence of the top event or do not occur and therefore are not part of a failure sequence. Probability theory provides the analytic tools to quantitatively evaluate events which are the fundamental component of fault trees. The following subsections discuss the basics of set theory, Boolean algebra and probability.

Set theory

One of the valuable features of FTA is its organisation of events and their interrelationship. Set theory is a general approach to the organisation of events of like features. The capability to systematically organise events in a fault tree is key to determining the probability of the top event.

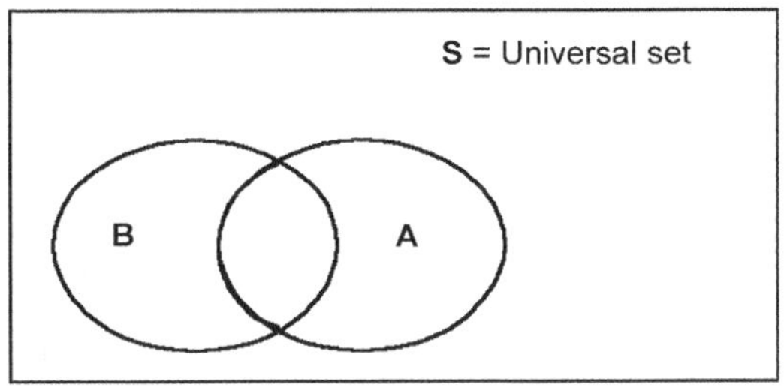

Figure 9-1. Venn diagram

For purposes of this module, basic set theory concepts and definitions are presented. These establish the foundation for Boolean algebra, the simplification and manipulation of fault trees, and ultimately the estimation of event probabilities.

The universal set is defined as the set of all possible events. It can be illustrated using a graphical technique called a Venn diagram. Figure 9-1 shows a Venn diagram with the universal set noted. All events or sets which are subsets of the universal set are shown inside the rectangle that defines the universal set.

In the figure, the following are shown, **S**=universal set, **A**=a set (or event), which is a subset of **S**, and **B**=a set (or event), which is a subset of **S**. If **A** is a set, it is denoted:

$$\mathbf{A}=\{a_1, a_2, a_3, a_4\} \tag{9-1}$$

where the braces denote the set and the quantities inside are the elements or members of the set.

Set union

A union of sets is the set that contains the elements of all sets that comprise the union. For two sets **A** and **B**, where:

$$\mathbf{A}=\{a_1, a_2, a_3, a_4\} \tag{9-2}$$

$$\mathbf{B}=\{b_1, a_1, a_4, b_2\} \tag{9-3}$$

the union of **A** and **B** is:

$$\mathbf{A} \cup \mathbf{B}=\{b_1, a_1, a_2, a_3, a_4, b_2\} \tag{9-4}$$

The symbol "∪" denotes the union operation. The union can also be illustrated in a Venn diagram as shown in Figure 9-2.

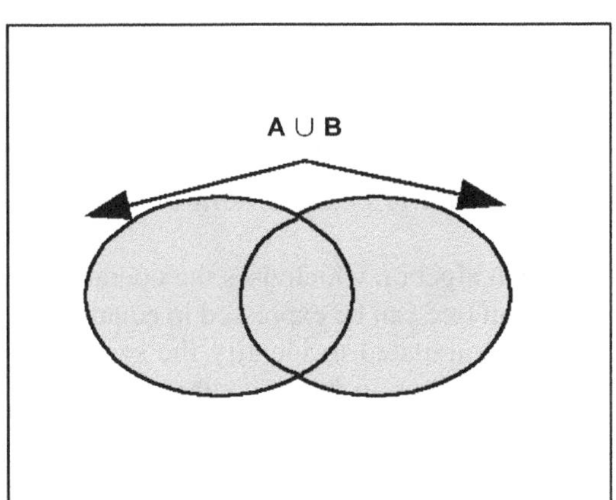

*Figure 9-2. Illustration of the union of sets **A** and **B***

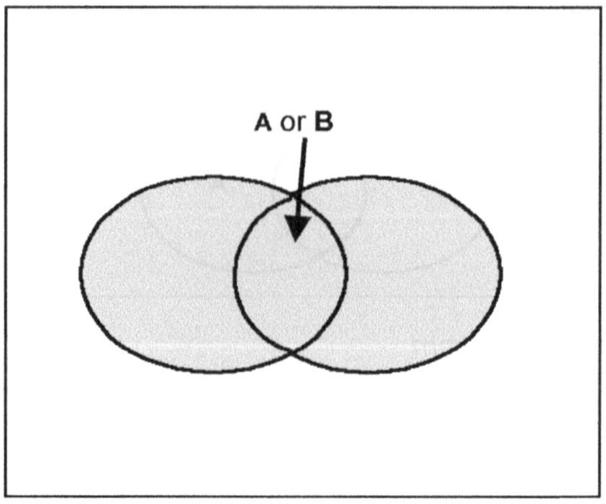

*Figure 9-3. Illustration of the intersection of sets **A** and **B***

Set intersection

Is the set which contains all elements that are common between sets. For example, the intersection of sets **A** and **B** (see equations 1-2 and 1-3) is:

$$\mathbf{A} \cap \mathbf{B} = \{a_1, a_4\} \tag{9-5}$$

where "$\cap$" denotes intersection. The intersection of these sets is shown schematically in the Venn diagram in Figure 9-3.

Set complementation

For a set **A**, the set complement contains all elements or members not in **A**. The complement of a set is denoted $\bar{\mathbf{A}}$. The complement of **A** is shown in the Venn diagram in Figure 9-4. Concepts of sets and set operations are a fundamental first step to fault tree evaluation and quantification. As will be shown shortly, the failure of a system will be represented as a "union of intersections", which define the system failure modes. Thus, one of the products of a FTA is a determination of the set of event combinations (e.g. intersections) that lead to failure.

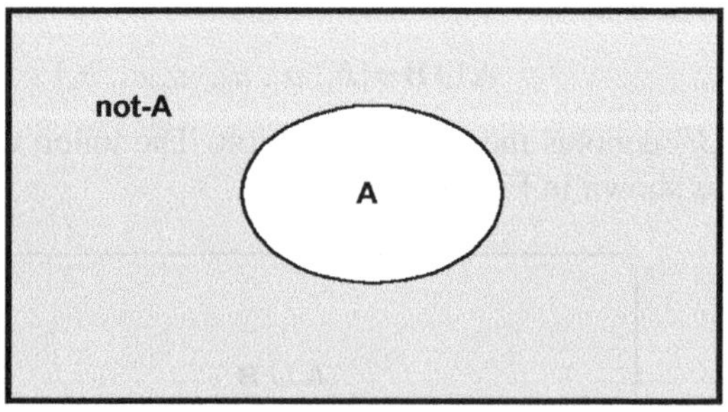

Figure 9-4. Illustration of the complement of set 4A

By means of Boolean algebra, which uses the operations of unions, intersections and complementation, a fault tree can be expressed in equation form. Once in equation form, the fault tree can be manipulated to identify the system failure modes. The basics of Boolean algebra are touched on in the next subsection.

Boolean algebra

Boolean algebra is the algebra of events and their relationship. As noted above, the graphical display of a fault tree cannot be manipulated, nor can the characteristics of a system in terms of the collection of events that lead to the occurrence of the top event

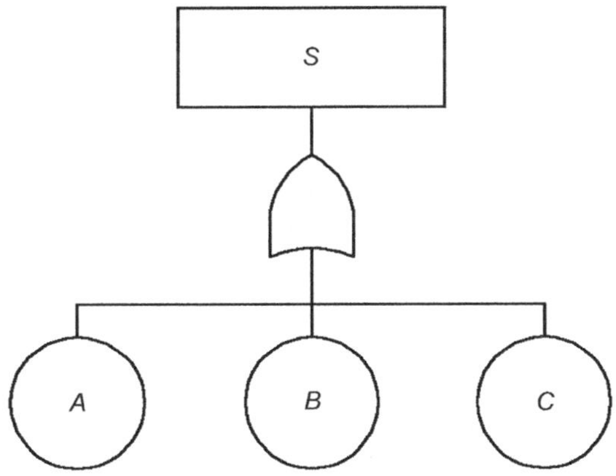

Figure 9-5. Fault tree for a simple series system

be easily determined. Boolean algebra is particularly suited to these tasks. Furthermore, Boolean algebra is particularly suited to binary events, such as is usually the case in a fault tree where an event has occurred or not, a switch has failed or not, a diesel generator fails to start or not.

A Boolean variable is a two-valued variable, where when 1s and 0s are used to indicate the occurrence or non-occurrence, respectively, of an event. The presentation in this section is basic and to the point, at the expense of mathematical rigor. The reader more mathematically inclined can consult USNRC (1991) or texts on Boolean algebra.

There are definitive rules of Boolean algebra that define the relationship between events. For our purposes, we relate the basic elements of Boolean algebra as they apply to fault trees and their analysis. To begin, consider a simple system (S) whose failure occurs if any one of three events (A, B or C) occur. The fault tree for the top event, failure of S, is shown in Figure 9-5. An OR-gate denotes the top event occurs if any one of the three events, A, B or C occurs. In this case, the top event is represented by the union of the three events that are input to the OR-gate.

In Boolean algebra, the union of events is denoted by the "+" symbol. With this symbology the fault tree in Figure 9-5 can be represented in equation form by:

$$S=A+B+C \tag{9-6}$$

If the OR-gate in Figure 9-5 is replaced by an AND-gate, the system would fail if events A, B and C occurred.

The Boolean symbol for the AND-gate is "*". The AND-gate denotes the intersection or simultaneous occurrence of events. The Boolean expression for the system in this case is:

$$S=A * B * C \tag{9-7}$$

For any fault tree, the top event can be expressed in Boolean form. As a fault tree has more events and logic gates, the complexity and size of the Boolean expression likewise becomes larger and more complex. However, once in equation form, a Boolean expression for the top event can be simplified (using the rules of Boolean algebra). In particular, one of the advantages of the Boolean representation of a fault tree is the fact that "failure modes" for a system can be determined.

The failure modes for a system are referred to as the minimal cutset for the top event. Formally defined, a minimal cutset is the smallest combination of events (e.g. component failures) which, if they all occur, will cause the top event to occur. A fault tree will consist of a finite set of minimal cutsets which are unique for the top event and any one of which leads to the occurrence of the top event.

In Boolean form, the expression for a top event can be written in terms of the minimal cutsets as:

$$TE=C_1+C_2+\ldots+C_k \qquad (9\text{-}8)$$

where TE is the top event and the C_i are the minimal cutsets. By definition, each minimal cutset is a specific combination of events (e.g. component failures). A minimal cutset can be expressed as:

$$C_i=X_1 * X_2 * \ldots * X_n \qquad (9\text{-}9)$$

where the X_i are basic events.

From equations 9-8 and 9-9, we see that when a top event is expressed in terms of its minimal cutsets, it is a union (the top event occurs if any one of the minimal cutsets occurs) which is unique and finite.

To determine the minimal cutsets for a fault tree, the tree is first converted to Boolean form (usually by computer). Next, the fault tree (as defined in a Boolean equation), is expanded using the rules of Boolean algebra. The expansion is carried out by performing the operations required by the logic gates and substituting input events (e.g. inputs to a logic gate) for the corresponding outputs. This expansion is continued until the top event is defined in terms of primary events.

This process is easily illustrated in an example. Consider the fault tree in Figure 9-6. Proceeding in a top-down manner, the Boolean expression for the top event and its expansion is shown in the following equations. To begin, the top event is expressed in Boolean form as,

$$TE=S_1 * S_2 \qquad (9\text{-}10)$$

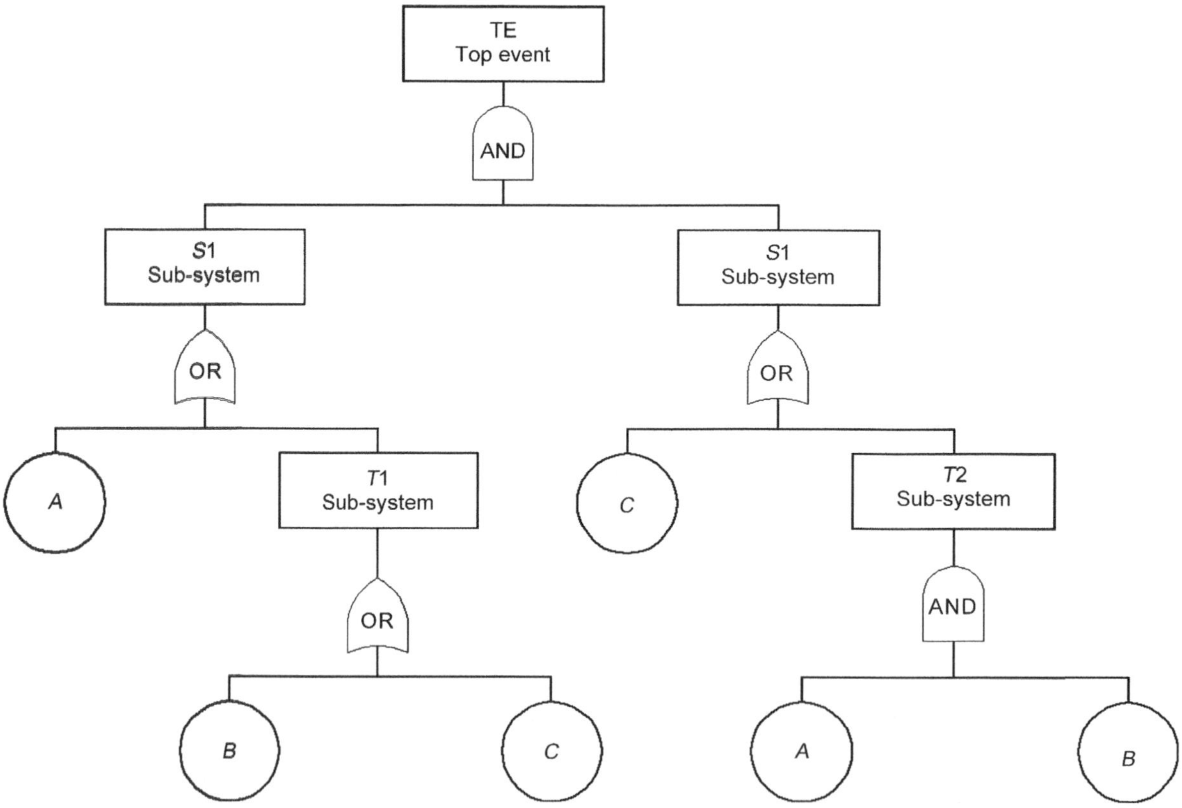

Figure 9-6. Fault tree for a simple system for which Boolean equation is determined in the text

S_1 and S_2 are intermediate events (see Table 9-1) and might be redundant sub-systems in a larger system. Proceeding, the intermediate events, which are outputs of logic gates, can be written in terms of the gate inputs as:

$$TE=(A+T_1) * (C+T_2) \quad (9\text{-}11)$$

Continuing down through the fault tree, and substituting for intermediate events T_1 and T_2, the expression for the top event becomes:

$$TE=(A+B+C) * (C+(A * B)) \quad (9\text{-}12)$$

Algebraic expansion of equation 9-13 and use of basic rules of Boolean algebra lead to:

$$TE=C+A * B \quad (9\text{-}13)$$

Equation (9-14) is the product of the expansion and subsequent reduction of the Boolean expression for the top event in Figure 9-6. The result produces the minimal cutsets, C and $A * B$. The top event occurs if C occurs or if A and B both occur. These are the system failure modes.

While the derivation of the minimal cutsets is essentially an algebraic process of expansion and reduction, it is a vital step in the qualitative and eventual quantitative assessment of fault trees. Qualitatively, the minimal cutsets identify the different ways there are for a system to fail and which events (e.g. component failures or faults) that must occur. For complex systems, this qualitative insight is vital for the analyst looking to develop an understanding of system performance or trying to improve system

Table 9-1. List of importance measures

No.	Importance measure	Type[1]	Description
1	Fussell–Vesely	BE	Measures the relative contribution of a basic event to the system unavailability
2	Birnbaum	BE	Measures the sensitivity of a system's unavailability with respect to changes in the probability of a basic event (e.g. from certainty to cannot occur)
3	Criticality	BE	Probability that an event is critical to a system, given that the system has failed
4	Risk worth reduction	BE	Shows the relative benefit to the system failure probability if the probability of a basic event (e.g. component failure probability) is set to zero (i.e. perfectly reliable)
5	Risk achievement worth	BE	Shows the relative impact on the system failure probability, if the probability of a basic event (e.g. component failure probability) is set to one (i.e. failure is certain)
6	Fussell–Vesely	CS	Measures the relative contribution of cutsets that contain a basic event to the system unavailability

[1] Type indicates whether the importance measure applies to basic events (BE) or cutsets (CS).

reliability. The determination of the minimal cutsets is also a useful, although not necessary, step to evaluate the probability of the top event.

For small fault trees, the derivation of the minimal cutsets can be determined by hand. (In some cases, the experienced analyst is able to identify the minimal cutsets by inspection.) However, in short order as the number of logic gates and events increases, the algebraic operations to determine the minimal cutset becomes extensive and must be performed by computer. One can imagine the operations involved to determine the minimal cutsets for a fault tree with hundreds of logic gates and a comparable number of events.

9.2 Fault tree evaluation

Qualitative evaluation

The qualitative evaluation of a fault tree is a straightforward process. Using rules of Boolean algebra and substitution, a top event is expressed in terms of basic events and their combinations. Once all Boolean reductions have been carried out, the minimal cutsets for the top event are determined. Recall, a minimum cutset is a combination of events (one or more) which, if all events in the cutset occur simultaneously, the top event occurs. Alternatively, if one or more events in a cutset does not occur, the cutset is no longer minimal and the top event does not occur.

The qualitative evaluation of a fault tree is carried out with FTA software. While the minimal cutsets for simple fault trees can be determined by inspection, today's FTA software serves a number of functions (e.g. documentation, data management) that support a risk analysis. As a result, it is recommended that all fault trees be constructed and evaluated using FTA software.

Once a fault tree is converted to Boolean form, the top event is expressed as a union of minimum cutsets (e.g. a union of intersections). The top event occurs if any one (or more) of the minimum cutsets were to occur. Cutsets are often referred to in terms of the number of basic events they contain. For instance, a cutset with just one event (i.e. the occurrence of just the one event leads to system failure) is known as a "single". Cutsets with two events are referred to as "doubles", three event cutsets are called "triples" and so on.

In Boolean form, the top event is denoted:

$$TE = C_i + C_2 + \ldots + C_n \tag{9-14}$$

where C_i is the ith minimum cutset. Each cutset is an intersection of events, such as:

$$\begin{aligned} C_1 &= A \\ C_2 &= B * C \\ C_n &= B * C * H * R * S \end{aligned} \tag{9-15}$$

Combining, the top event becomes:

$$TE = A + B * C + \ldots + B * C * H * R * S \tag{9-16}$$

Figure 9-7 shows an example of the minimum cutsets for a fault tree. The cutset listing is typical of the output that would be provided by FTA software. The table is annotated to illustrate the results.

At this point in the FTA the analyst can examine the minimal cutsets to develop insights into the system failure modes, the importance of different events or cutsets, etc. From a review of the cutsets, the analyst will be able to determine if a component or a cutset plays a significant role in determining the satisfactory performance of a system. For instance, if a review of the minimum cutsets for a system reveals that one component shows up in the majority of the cutsets, it is reasonable to expect that its

```
Cutset probability

Topevent :SYS_A
Topprob.:1.94E-01

Top Event      Symbolic
Cutsets with single events
1.5200E-04     DCC216/116N
1.5200E-04     DCC119/219N
1.3700E-04     DCC123/223N
1.3700E-04     DCC124/224N
1.0000E-04     DHFDADS
8.5900E-04     DCC125/225N
Cutsets with multiple events
2.7500E-04     DCC121/2/221/2N
5.0715E-04     DMM1FTO123/4/03     DMM2FTO223/4/03
2.8900E-04     ECC220V_ACCU1/2     SEISLOSP
3.3330E-05     DMM1FTO123/4/03     DMM2FTOBPSOLFTO
3.3330E-05     DMM1FTOBPSOLF       DMM2FTO223/4/03
1.0000E-05     SEISE220V_ACCU1     SEISE220V_ACCU2                SEISLOSP
2.9100E-03     EMM220V_ACCU1       SEISE220V_ACCU2                SEISLOSP
2.9100E-03     EMM220V_ACCU2       SEISE220V_ACCU1                SEISLOSP
1.8700E-05     EMMSR1              SEISE220V_ACCU1                SEISE220V_ACCU2
2.9800E-04     EC DR697OR1BGR37    SEISDACOMOUNT                  SEISE220V_ACCU1    SEISLOSP
2.9800E-04     ECDR697OR1AGR16     SEISDACOMOUNT                  SEISE220V_ACCU2    SEISLOSP
2.9800E-04     ECDR697OR1 BGR37    SEISDGBTASEISE220V_ACCU1       SEISLOSP
2.9800E-04     ECDR697OR1AGR16     SEISDGBTASEISE220V_ACCU2       SEISLOSP
1.5234E-05     ECDR 697OR1BGR37    EDGF697DACO                    SEISE220V_ACCU1    SEISLOSP
1.5234E-05     ECDR697OR1AGR16     EDGF697DACO                    SEISE220V_ACCU2    SEISLOSP
```

Figure 9-7. Example of the minimum cutsets for a system. Each cutset is a mode of failure for SYS_A

reliability will play an important role in the overall reliability of the system. Similarly, if a basic event is a "single" (e.g. it makes up a minimum cutset by itself), it will clearly be important (see Table 9-1).

The cutset listing will also provide the analyst with a qualitative measure of the importance of individual components based on the frequency of their occurrence in multiple cutsets. For example, in Figure 9-7 the event SEISLOSP shows up in a number of cutsets, suggesting its failure could lead to the occurrence of one or more cutsets. A review of the cutset listing can also be a step to determine the susceptibility of a system to dependent failures.

Number of cutsets

Depending on the size and complexity of a fault tree, there may be hundreds or thousands of cutsets. Typically, FTA software will list cutsets by size (e.g. singles are listed first, followed by doubles, triples and so on) or by cutset probability. In many applications it is neither necessary or informative to determine all of the minimal cutsets for a fault tree. This occurs, for example, when the probability of the top event can accurately be estimated when only a fraction of the total number of cutsets are determined and quantified. As a result, various truncation methods are available to limit the number of cutsets that are retained.

Cutset truncation

For systems with a large number of cutsets, it is not practical to review and quantify all of the minimum cutsets. Computer programs to evaluate fault trees typically offer two methods to truncate the generation of minimum cutsets,

1. Probability cutoff – cutsets whose probability is greater than a cutoff (e.g. truncation limit) specified by the analyst are determined. Cutsets with smaller probability are not calculated.
2. Order cutoff – the order of a cutset refers to the number of events in the cutset. This truncation method limits the size of the cutsets that are generated. If an order cutoff of four is specified, cutsets with five or more events are not determined.

To apply a probability cutoff, the analyst must have some idea of the order of magnitude of the final result. If this is not known, a series of calculations using different cutoff levels can be used to determine the sensitivity of the result. As the cutoff probability is lowered and/or the order cutoff size is increased, the number of cutsets that are generated will increase as well.

9.3 Quantitative evaluation

There are two parts to the quantitative evaluation of fault trees. The first is the quantification of the probability of the top event. The second is the quantitative measure of importance of events to the top event probability. These topics are discussed in the following subsections.

Quantifying the top event

In equations 9-14 to 9-16, the top event is defined in Boolean form as a union of intersections. To quantify the probability of the top event, most FTA software use a min-cut upper bound procedure. This is an approximate procedure that is exact if each cutset consists of a single event, which is generally not the case. Otherwise it produces a conservative estimate of the top event probability. Using this approach, the probability of the top event is calculated according to:

$$P(TopEvent) = 1 - \Pi(1 - P(C_i)) \qquad (9\text{-}17)$$

where n is the number of cutsets considered and $P(C_i)$ is the probability of the ith cutset.

The probability of the ith cutset is simply the product of the probabilities of the basic events in the cutset:

$$P(C_i) = \Pi P(BE_j) \qquad (9\text{-}18)$$

where $P(BE_j)$ is the probability of the jth basic event. For applications where the event probabilities are small, $\sim 10^{-3}$ or less, this is a pragmatic and numerically acceptable alternative.

For certain applications, however, the min-cutset upper bound approximation can introduce errors into the assessment of the top event probability. This occurs in cases where the basic event probabilities are not small, such as in a seismic or flood risk analysis, where basic event probabilities are conditional on the level of the seismic or flood load and as a result will vary from relatively small conditional probabilities of failure (i.e. near 0) for small loads to 1.0 as the loads increase beyond the capacity of the component. In this case an approach is required that conducts an appropriate mapping from the Boolean expression for the top event (as given by the minimum cutsets) to the probabilistic equation for the top event. A more exact method accounts for the cutset intersections that are not considered on the min-cutset upper bound approximation.

Event/cutset importance

One of the primary advantages of FTA is the fact that it provides the capability to quantitatively estimate the reliability/unreliability of a system. As described above, certain insights into the vulnerability of a system can be gained from a qualitative review of the minimum cutsets. However, a quantitative estimate of the importance of individual basic events and cutsets can be obtained through the use of a set of numerical tools called importance measures.

The importance of a component to system performance depends on its position or role in the system and its reliability. In most cases, however, it is useful if not necessary to have a systematic, numeric means to understand the role/importance that events and cutsets have in a system. These questions become particularly important when the reliability of a system must be improved or design changes are required. Importance measures can be used to rank order components taking into account their role in the system (e.g. its functional role) and reliability.

A number of importance measures have been developed and are in standard use in FTA. Each method provides an alternative view of a system and the role that an event or cutset has. This section describes some of the importance measures that are in use. Simply stated, the purpose of an importance measure is to provide a quantitative measure of the relative importance (a quantitative ranking) of the basic events (e.g. component failures) or cutsets in terms of their contribution to the occurrence of the top event. Table 9-1 provides a list of a number of the importance measures that have been developed. The table lists each measure by name and provides a brief description of what it measures and whether it applies to Basic Events (*BE*) or to Cutsets (*CS*).

The Fussell–Vesely (*FV*) measure of importance is a frequently quoted measure of event importance. It provides a relative measure of the contribution of a component failure to the top event probability. The Fussell–Vesely measure of importance for an event is defined as:

$$FV_A = \frac{P_{TopEvent} - P_{TopEvent}(A=0)}{P_{TopEvent}} \tag{9-19}$$

where $P_{TopEvent}$ is the probability of the top event (e.g. system failure) and $P_{TopEvent}(A=0)$ is the probability of the top event given the probability of basic event A is zero (never occurs).

The Birnbaum (*BE*) importance measure shows the change (increase) in the top event probability attributed to variations in the probability of a basic event. This importance measure is defined by:

$$Birnbaum_A = P_{TopEvent}(A=1) - P_{TopEvent}(A=0) \tag{9-20}$$

where $P_{TopEvent}(A=1)$ is the probability of the top event given the probability of the basic event is 1 (e.g. it is certain to occur).

The criticality (*BE*) measure of basic event importance computes the probability that an event has occurred and is critical to the system, given that the system has failed. It is defined as:

$$Criticality_A = \frac{(P_{TopEvent}(A=1) - P_{TopEvent}(A=0))P_A}{P_{TopEvent}} \tag{9-21}$$

where P_A is the probability of event A.

The risk reduction worth (*RRW*) measures the relative change in the top event probability assuming a basic event cannot occur. This is defined by the ratio:

$$RRW_A = \frac{P_{TopEvent}}{P_{TopEvent}(A=0)} \tag{9-22}$$

Similarly, the risk achievement worth (*RAW*) shows the relative increase in the probability of the top event, assuming a basic event is certain to occur. This is defined by the ratio:

$$RAW_A = \frac{P_{TopEvent}(A=1)}{P_{TopEvent}} \tag{9-23}$$

Fussell–Vesely (*FV*) provides a relative measure of the contribution of a cutset that contains a basic event to the top event probability. It is defined as:

$$FV_{CS} = \Sigma_i \frac{P(CS_i)}{P_{TopEvent}} \tag{9-24}$$

where the sum is carried out for all cutsets (CS_i) that contain a specified basic event.

Importance measures can be particularly valuable for complex systems and/or systems with a large number of basic events. The information and ranking provided by importance measures can be used as input for system design or upgrade diagnosis (e.g. failure detection investigation), and planning of inspection and maintenance activities.

9.4 Component considerations

Recall that early in the discussion of fault tree construction, the top event is defined in specific terms (e.g. environmental conditions, system status). For the same reasons that it was important to clearly define a top event, it is important that the assessment of basic event probabilities be carried out for these same conditions. These conditions include the environment, status of the system, etc.

The assessment of event probabilities is also made in the context of how an event is defined in a fault tree. For example, the probability of an event which depends on (is conditional) the occurrence of another event is likely different than the unconditional probability of the event.

Environmental factors

In the discussion of fault tree construction, the importance of environmental conditions was identified. An even greater level of importance is assigned to environmental conditions when it comes to the estimation of event probabilities. This is true for a number of reasons:

1. Environmental factors will impact the number and type of failure modes that are modelled (this was considered in the construction of the fault tree).
2. As the environmental conditions get more complex (e.g. seismic, extreme flood, extreme wind) the level of ongoing interaction that is required between the system analysts (e.g. the fault tree builders) and those responsible for estimating basic event probabilities increases.
3. Environmental loads may vary substantially in magnitude, leading to a corresponding variation in basic event probabilities.
4. The assessment of rational and defensible basic event probabilities may be expensive (e.g. time consuming, require detailed assessment). As a result it is prudent to use a staged approach to conduct the analysis. In a staged approach, initial probability estimates are used to conduct the initial risk quantifications. Once it is determined which basic events are important to the results of the risk analysis, more effort is focused on them.
5. Environmental factors often produce dependencies or common cause effects that should be considered in the construction of the fault tree and in the assessment of basic event probabilities.

Table 9-2. Methods for estimating event probabilities

Event type	Method
Mechanical, electrical equipment – normal conditions	Reliability analysis based on experience or test data Generic reliability data Component specific reliability/experience or test data
Mechanical, electrical equipment – external	Engineering analysis and probabilistic or engineering reliability methods
Environmental	Environment and component specific reliability data
Civil structures – all conditions	Engineering analysis and probabilistic or engineering reliability methods
Natural structures (e.g. reservoir slopes)	Engineering analysis and probabilistic or engineering reliability methods
Operator actions – all conditions	Human error analysis methods Simulation studies Experience data

In summary, the local or global environment that a component is in has a primary impact on the analysis of the probability of events in the fault tree.

Sources of component probability data

This section discusses, in general terms, sources of event probabilities and identifies some of the basic resources available to obtain/estimate event probabilities for mechanical and electrical components. Basic events in a fault tree will include:

1. Failure of mechanical and electrical equipment items.
2. Operator actions.
3. Structural failures of civil structures and mechanical and electrical items due to applied loads (e.g. loads exceeding capacity).
4. Random events (e.g. time of day).

The methods to estimate the probability of these events will vary from case to case. For example, an assessment of the probability of failure of a motor to operate during normal conditions can be estimated from manufacturer or industry reliability data. However, an assessment of the failure probability of the same motor during a seismic event will be very different in terms of the basic data that is used, the failure modes of the component that are examined, and the method of calculation. Table 9-2 provides a summary of the methods/resources that are used to estimate event probabilities.

Table 9-3 provides a list of available resources of reliability data for mechanical and electrical components in normal operating conditions.

In many cases the assessment of basic event probabilities for different environmental conditions will require a component specific assessment. This is discussed in the following sub-section.

Component specific probability assessments

The uniqueness of dam structures, their support systems and operations, etc. and the environmental conditions they must survive in, will require component-specific assessments in many cases. For example, the performance of a hoist house and all the equipment it supports during high wind conditions will require a structure specific

Table 9-3. Sources of reliability data for mechanical and electrical components

Resource	Component	Source
Project or utility specific experience	Mechanical, Electrical, Electronic	Ontario Hydro CEA reports
MIL-HDBK-217	Electronic	Reliability Analysis Center PO Box 4700 Rome, NY 13440
IEEE Guide to Reliability Data for Nuclear Generating Stations, ANSI/ IEEE Std 500-1984	Mechanical, Electrical, Electronic	Institute of Electrical and Electronic Engineers, Inc. 345 East 47th Street New York, NY 10017
Nonelectronic Parts Reliability Data (NPRD-3)	Electrical	Reliability Analysis Center PO Box 4700 Rome, NY 13440
Vendor proprietary databases	Mechanical, Electrical, Electronic	Reliability Analysis Center PO Box 4700 Rome, NY 13440
Handbook of Human Reliability	Operator Action	See Swain and Guttman (1983)

assessment (i.e. there are no data resources that provide the probability of failure of a hoist house). As a result, component specific analyses of basic event probabilities will be required in many cases. Environments or cases where this will be true, include:

1. All external events that impose loads on the dam system (i.e. seismic, inflow floods, wind, waves, etc.).
2. Reservoir loads (static and hydraulic) on structures.
3. All structures for all environmental conditions.
4. Operator actions during external events.

Other circumstances not covered by the above list may also exist.

9.5 Operator actions

As part of a fault tree model for a system, there may be certain basic events that correspond to actions of support staff and/or operators. The subject of assessing the reliability of operator actions (often referred to as human reliability analysis (HRA)) has been studied extensively. Various methods of analysis and estimates of error rates for different circumstances and conditions have been developed. One of the widely used methods, the technique for human error rate predictions (THERP), is documented in handbooks that can be used to derive error probabilities and uncertainty estimates (see USNRC, 1991).

Operator actions can fall into two general categories. First, there are actions that occur prior to an accident or initiating event (e.g. an earthquake or flood). An example might be an error made during maintenance activities that make a component unavailable. The second type of action is one committed during or after an accident occurs. In both cases, these errors may be ones of commission or omission. The importance of operator actions will, of course, vary depending on the system design (e.g. degree of redundancy), the effect of an operator error, etc.

Factors that are considered in a HRA to evaluate pre-accident errors include:

1. Procedures followed during routine maintenance, testing and surveillance activities.
2. Dependencies that exist as a result of one person performing a routine task (e.g. maintenance, testing and surveillance activities).
3. Verification activities (e.g. post-maintenance testing).
4. Automatic systems or back-ups that identify if a component is unavailable.

In the case of operator actions that occur during or following an accident, factors that are considered include:

1. Procedures followed during accident, conditions or emergencies.
2. Availability of annunciators or indicators.
3. Recovery options.
4. Time availability.
5. Accessibility of the site, or other location, where recovery actions are carried out.
6. Complexity of procedures (e.g. are they written?).
7. Complexity/difficulty of actions (e.g. require physical activity such as lifting heavy equipment or difficulty in diagnosing a problem or course of action during an event).
8. Level of stress involved.

The process of evaluating operator error probabilities can be difficult and time consuming. As an early first step in the fault tree quantification process, the analyst can consider using "place holder" probabilities (e.g. 0.50 or 1.0) for operator error events. Early quantifications of the top event probability and importance calculations (with the place holder values) will give the analyst a measure of how important operator actions are. If the importance of the operator actions is low with the place holder values, it will not be necessary to perform a detailed HRA. If the preliminary calculations draw the opposite conclusion, the analyst must then undertake an analysis of the important operator actions.

Resources that can be consulted to carry out a HRA are Swain and Guttman (1983) and Erickson *et al.* (1990). These references define comprehensive procedures to access operator error probabilities. In addition, they provide a means to access the uncertainty (epistemic) in these estimates.

9.6 Probability

To this point the discussion has addressed the basics of fault trees, the representation of a fault tree in Boolean equation form, and the derivation of the failure modes of a system (e.g. the minimal cutsets). Probability theory completes the foundation of FTA and provides the tools to quantify the probability of a top event. This section builds on set theory and Boolean algebra concepts discussed above and provides a basic discussion of probability concepts. The interested reader can consult USNRC (1991) or other probability texts for a more detailed, theoretic discussion of probability and set theory.

Events are the primary elements of a fault tree. Whereas Boolean algebra is used to express a top event in equation form to qualitatively evaluate a fault tree, probability theory provides the analytic foundation for the algebra of probabilities and their quantitative evaluation. The following presents a summary of probability as used in fault tree analysis.

The three basic axioms of probability apply (Chapter 7). Using the foundation of set theory discussed above and the rules of probability, basic concepts of probability that have direct application to the quantitative evaluation of fault trees (e.g. determination of the probability of the top event) are addressed.

Union of events (events in series)

In the discussion of set theory, Venn diagrams were used to illustrate basic relationships between events (e.g. union, intersection). Taking Venn diagrams one step further, they can also be used to visualise basic probability concepts. The Venn diagram in Figure 9-2 showed the union of two events, *A* and *B*. In the figure the union corresponds to the area of the two events. As these areas are added, the area where they intersect is added twice. Assume now that the area of each event is proportional to the probability of an event. The probability of the union of *A* and *B* (i.e. $A+B$ in Boolean form) is:

$$P(A+B)=P(A)+P(B)-P(A * B) \tag{9-25}$$

where $P(\,)$ denotes the probability of an event and $A * B$ is the Boolean expression for the intersection of *A* and *B*. Viewing the probability terms in equation 9-1 as areas, the intersection of *A* and *B* is subtracted from the sum of the areas to avoid double counting. If *A* and *B* are independent events, equation 9-1 becomes:

$$P(A+B)=P(A)+P(B)-P(A) * P(B) \tag{9-26}$$

If *A* and *B* are mutually exclusive, their intersection is the null set (see Rule 3 above). This is shown in Figure 9-8. In this case the union of *A* and *B* is:

$$P(A+B)=P(A)+P(B) \tag{9-27}$$

since the probability of *A* intersection *B*, $P(A * B)$, is 0. The Venn diagram is a useful tool to graphically display the interaction of events and certain basics of event probabilities.

We can relate these concepts to the fault tree in Figure 9-5. The top event occurs if event *A* or *B* or *C* occurs. The probability of this union is:

$$P(TE)=P(A)+P(B)+P(C)-P(A)P(B)-P(A)P(C)-P(B)P(C)+P(ABC) \tag{9-28}$$

where $P(A * B * C)$ is the probability of the intersection of events *A*, *B* and *C*. The expression in equation 9-28 can be determined by drawing a Venn diagram for the case of three events. As one might expect, even for three events the union and intersection of events can be difficult to track.

Event intersection (events in parallel)

When the top event is defined by an AND-gate(see Figure 9-6), the probability of the top event is simply the product of the event probabilities:

$$P(S)=P(A)\times P(B)\times P(C) \tag{9-29}$$

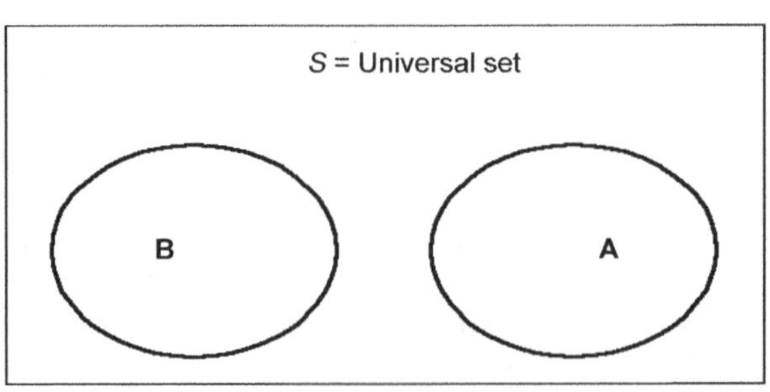

*Figure 9-8. Venn diagram for the events **A** and **B** which are mutually exclusive*

Consider how the expressions in equations 9-25, 9-28, and 9-29 relate to the calculation of the top event probabilities. Assume that the probability of the events is equation 9-26 is on the order of 10^{-3}. Evaluating equation 9-2, we have:

$$P(S) = 10^{-3} + 10^{-3} - 10^{-3} * 10^{-3} \qquad (9\text{-}30)$$

$$P(S) = 2 \times 10^{-3} - 10^{-6} \qquad (9\text{-}31)$$

This result suggests that the intersection of independent events is rare given the basic event probabilities are small (less than about 10^{-3}). Furthermore, it suggest the intersections need not be considered when calculating the probability of the top event. As a result, the probability of the union of events can be calculated as the simple sum of the event probabilities. This is known as the rare event approximation.

Event complements

The complement of an event is the non-occurrence of the event. The complement of event A is its non-occurrence. This is denoted $\bar{A}$. The union of an event and its complement define the universal set. That is:

$$S = A + \bar{A} \qquad (9\text{-}32)$$

In probability terms, this union is:

$$P(S) = P(A) + P(\bar{A}) \qquad (9\text{-}33)$$

since A and A are mutually exclusive. Because the probability of the universal set is 1.0, we have:

$$1.0 = P(A) + P(\bar{A}) \qquad (9\text{-}34)$$

$$P(\bar{A}) = 1.0 - P(A)$$

When a series system consists of many components (i.e. more than three), the calculation of the probability of the union can be lengthy (even for the computer for large systems). Alternatively, the probability of the top event can also be determined as the probability that none of the elements of the series systems occurs. For the three component system, this is expressed by:

$$P(S) = 1.0 - P(\bar{S})^{1} \qquad (9\text{-}35)$$

$P(S) = 1.0 -$ probability that A does not occur, and B does not occur, and C does not occur which can be expressed as:

$$P(\bar{S}) = 1 - \{P(A) \times P(B) \times P(C)\} \qquad (9\text{-}36)$$

where the over-bar denotes the complement of the event.

Conditional probability

To this point independent events and their combination (e.g. unions and intersections) have been considered. There are, of course, events that are not independent. To treat events of this nature, the concept of conditional probability is introduced. Conditional probability is an essential and fundamental part of probability that is used extensively to model dependent events.

Consider two events: A and B. These events are independent if the probability of event A, given we know event B has occurred, is unchanged. This is expressed as:

$$P(A \mid B) = P(A) \qquad (9\text{-}37)$$

where $A \mid B$ is read, event A given the occurrence of event B (or simply A given B). If equation 9-36 is not true, events A and B are dependent. Referring to the Venn diagram in Figure 9-1, the conditional probability of A can be related to the areas of the two events. Given the occurrence of event B, the probability of A will simply be the proportion of event A that intersects with (overlaps) B. This can be expressed as the following ratio,

$$P(A \mid B) = \frac{P(A \cap B)}{P(B)} \tag{9-38}$$

This ratio simply expresses the probability that A will occur, given that we know B has already occurred. From this expression, we can derive the probability of the intersection of A and B in terms of the conditional probability of $A \mid B$. This is given by:

$$P(A \cap B) = P(A \mid B)P(B) \tag{9-39}$$

This expression applies if A and B are dependent. If A and B are independent, then $P(A \mid B)$ will equal $P(A)$ and equation 9-14 becomes the same as equation 9-5.

The above discussion covers only the basics of set theory, Boolean algebra and the probability of events. They are straightforward and set the foundation for fault tree quantification.

10 Consequence assessment

The purpose of this chapter is to provide background of a more theoretical nature than appropriate to Part II on the methods and procedures of consequence analysis. The materials presented in the present section discuss *why* consequence analyses are done in a certain way, rather than *how* they are done. Published work on consequence assessment for civil works and natural hazards is vast: the present summary is short. An intent of this summary is to provide an overview of conceptual or theoretical issues involved, and to provide an entree to the literature where more detail is available.

The literature of quantitative benefit – cost evaluation for public investments in water resource development projects commences in the post World War II era with the Harvard Water Project of the 1950s and 1960s, but harkens back to the great civil works development projects of the 1930s. This is largely a literature of economics, as it should be. The assessment of dam failure consequences in their economic aspects builds on this large literature. The current summary, which is largely an engineering view, does not attempt to comprehensively summarise the subtleties of that literature, especially regarding indirect costs, but only to provide points of entrance into it.

Today, the assessment of the consequences of dam failure is incomplete without consideration of potential loss of life and of potential environmental changes brought on by a failure. Prediction of potential loss of life due to dam failure – or due to failures of other civil infrastructure – is a topic that does not enjoy the breadth of literature associated with the assessment of economic benefits and costs. On the other hand, much of the development of assessment approaches to loss of life has, in fact, appeared within the dam safety and natural hazards literatures and it is, therefore, reviewed here in more detail than might otherwise be the case.

The assessment of environmental consequences of dam failures – or of floods, or natural hazards, or civil infrastructure more generally – is an emerging "science". It has become evident in recent years that the environmental impacts of large floods can have beneficial as well as adverse aspects, and the assessment of these benefits and costs presents challenges more complex than in the relatively simple concerns of economics and loss of life. The assessment is also inherently multidisciplinary, adding to the difficulty. Environmental expertise is a necessary component of a dam safety risk analysis that hopes to capture quantitative environmental consequences within its formulation. The summary in this section attempts to present an overview of current approaches to addressing quantitative environmental consequences, how those consequences are structured and measured and the difficulties facing current approaches.

10.1 Importance of consequence analysis to risk

The notion of risk in general use is broad and inconsistent, given the many endeavours of human enterprise to which the term is applied.

> **Risk** n. 1. The possibility of suffering harm or loss; danger. 2. Factor, thing, element, or course involving uncertain danger; a hazard: 3 (a) The danger or probability of loss to an insurer; (b) The amount that an insurance company stands to lose. 4 (a) The variability of returns from an investment; (b) The chance of non-payment of a debt. 5. One considered with respect to the possibility of loss: a poor risk. (Merriam-Webster, 2000)

However, when we speak of risk in an engineering context we mean: (1) a quantification of the probabilities of a set of undesired events; and (2) a quantification of the consequences should those events occur. In this sense, risk is a formal combination of two sets of numbers: probabilities and consequences. This meaning of risk differs somewhat from uses of the term in other aspects of life or in other industries such as finance and insurance; for example, the insurance industry speaks of risk simply as the consequence of an adverse event, but is common across engineering.

The economic and social losses of natural disasters and major accidents are not consistently calculated. Different agencies and organisations calculate damage estimates, but the techniques used to make them differ widely, cover a variety of inconsistent costs and change through time (Table 10.1). The way risk calculations are used in categorising the severity of consequences, and thus categorising specific structures as low, intermediate or high risk, also differs from organisation to organisation. The hub of the dilemma is that there appears to be no broadly accepted framework for estimating losses of natural disasters and related accidents such as dam failures (NRC, 1999b). This is a recurring theme. In addition, public safety impacts, specifically potential loss of life, carry ethical questions for which there is a lack of theoretical framework.

In the course of daily life, situations are encountered that involve some event that might occur and which, if it did, would bring important consequence. If we assign a numerical probability to the event's occurrence, and a numerical value to the cost of the adversity associated with the event's occurrence, then this combination of uncertain event probability and adverse consequence cost becomes the determinant of risk.

Kaplan and Garrick (1981) say that formal risk analysis requires three things: (1) a set of scenarios or events, E_i, $i=1, \ldots, n$; (2) a probability associated with each scenario or event, p_i, $i=1, \ldots, n$; and (3) consequences associated with each scenario or event, c_i, $i=1, \ldots, n$. Described another way, these three things pertain, respectively, to what can happen, how likely it is and what the consequences are (Bedford and Cooke, 2001).

In engineering, risk is usually defined as the product of probability and consequence, in other words, as the expectation of adverse outcome:

$$Risk \equiv \{probability \times consequence\} = pc \qquad (10\text{-}1)$$

This is the way most dam safety risk analyses use the term *risk*. When more than one event may lead to an adverse outcome, the definition is extended to be the expected value of consequences over the set of events:

$$Risk \equiv expected\ consequences = \sum_{i=1,n} (p_i c_i) \qquad (10\text{-}2)$$

in which p_i is the probability of consequence c_i. Note, this is not the same as the expected value of the consequences given a dam failure, which is a conditional probability that assumes dam failure has already occurred. The risk or expected

Table 10-1. Examples of downstream hazard classifications (after Almeida and Viseu, 1996)

	USBR	**FERC**	**Portugal**	**Spain**
High	Dam failure will cause excessive economic loss (urban area, industry, agriculture or outstanding natural resource), and more than six lives in jeopardy	Dams whose failure may cause serious damage to homes, agricultural, industrial and commercial facilities, important public utilities, main highways or railroads, and there would be danger to human life	Dam failure provokes a large number of life losses and severe property damages	Dam failure, or malfunction, can severely affect urban areas or essential services, and provoke very important environmental and property damages
Significant	Dam failure will cause appreciable economic losses (rural area with notable agriculture, industry or work sites, or outstanding natural resources) and one to six lives in jeopardy	Dams whose failure may damage isolated homes, secondary highways or minor railroads, cause interruption of use or service of relatively important public utilities, or cause some incremental flooding of structures with possible danger to human life	Dam failure provokes a few life losses and moderate property damages	Dam failure, or malfunction, provokes important environmental and property damages or affects a reduced number of dwellings
Low	Dam failure will cause minimal economic losses (undeveloped agriculture, occasional uninhabited structures, or minimal outstanding natural resources with no lives in jeopardy)	Dams whose failure may damage some buildings, limited agricultural land, or township and country roads. Low hazard potential dams have a small storage capacity, the release of which would be confined to the river channel in the event of a failure and therefore would represent no danger to human life	Dam failure provokes no life losses and small property damages	Dam failure, or malfunction, provokes moderate environmental and property damages as well as small danger to human life

consequence is the expected value of the consequences given a dam failure, multiplied by the probability of the dam failure. The probabilities in equation 10-2 combine both the probability of dam failures of different types or with different characters and the probabilities of downstream consequences when a failure of a certain type does occur.

The point here is that the effort invested in quantifying the consequences of failure in a quantitative risk assessment is just as important in the final estimate of risk as is the effort invested in quantifying probabilities. The two quantities multiply by one another, and thus their errors compound. To a first-order approximation, the variance in the estimate of risk is the weighted sum of the variances contributed by errors in consequences and by errors in probabilities, according to a relationship of the form:

$$Var[Risk] = \sum_{i=1,n} (p_i^2 Var[c_i] + c_i^2 Var[p_i]) \tag{10-3}$$

in which p_i and c_i are the mean values of probability and consequence, respectively, *Var* is the variance in the respective terms.

It is worthwhile to note that dam failures are low-probability high-consequence events. The annual probability of catastrophic failure of a modern, well-constructed dam is likely to be lower than, say, 10^{-4}, and possibly much lower. At the same time, the corresponding consequences, should failure occur, can be immense. Thus, the quantitative calculation of risk involves multiplying a diminishingly small number by a correspondingly large number. Mathematically, this is an unstable operation that can lead to erratic results. Of course, dam safety studies share this property with all risk analyses of catastrophic events.

In some cases, a separate event tree is used to model downstream consequences of a dam failure; that is, to provide an estimate of consequences conditional on loss of pool or some other damage state of the dam occurring in some specific way (Figure 10-1). This sub-tree can incorporate exposure cases and other downstream activities or events, such as the effectiveness of evacuations or other risk mitigation procedures. Usually this downstream sub-tree is treated independently of the event tree for the dam failure itself, although both the probabilities and consequences in the consequence tree depend on the event tree leaf describing the failure event. For example, a leaf on the event tree from which the consequence tree begins might be, "spillway erodes and washes out". This describes the failure type. The consequences that result from this failure type depend on

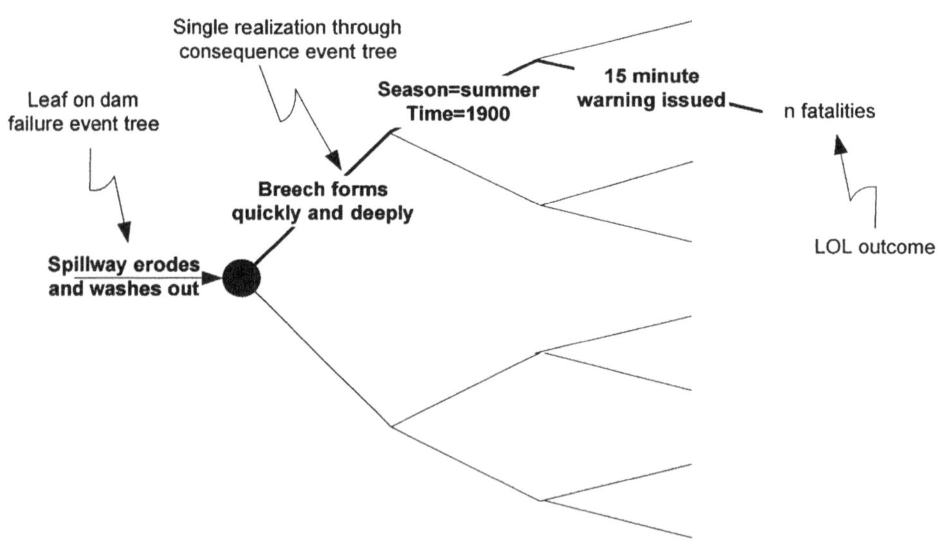

Figure 10-1. Example of a partial event tree of consequences starting from one leaf on the dam failure event tree

the events leading up to the failure, as well as the events within the consequence tree that follow failure. This and related issues are discussed in more detail in Chapter 8.

Organisation of the section

The section begins by discussing concepts behind the methodology of consequence assessment, including analysing and predicting consequences, describing empirical–statistical approaches, mathematical modelling from first principles, and systems approaches such as event tree analysis. It proceeds to describe how consequences are organised around a hierarchy of consequence types and attribute measures through which consequences are quantified. Following this general discussion, the section goes on to address the three major categories of consequences in dam safety studies: public safety, economics and environmental. The section limits discussion to direct local effects of dam failure, ignoring regional and national impacts, for example costs to national economic development attributable to unique resources lost within the impacted area, which are important but outside the scope of the present guide.

Methodology of consequence assessment

This section addresses the general methodology of consequence assessment. Discussion of the specific considerations of public safety (potential loss of life), economics and the environment are deferred to Sections 10.4 to 10.6. As in the case of probabilistic analysis of failure potential, the forecasting of consequences is, at its most basic, a modelling activity and is therefore subject to all the limitations, both theoretical and practical, that modelling entails. Consequence assessment uses much the same approach as does failure probability modelling. The principal difference, perhaps the only important difference, is that consequence modelling involves both technological aspects of engineering and sociological or human factors aspects of human behaviour. In addition, consequence modelling involves a great deal of data collection and database manipulation. As in the other chapters of Part III, the focus here is on concepts rather than details and the discussion refers back to Chapters 7 and 8 rather than repeat earlier materials.

Identifying consequences

The first step in consequence assessment is identifying the consequences and sub-consequences that make up the hierarchy of concern. Figure 10-2 suggests such a hierarchy of consequences, sub-consequences, and attributes (measures) that might compose the target of attention in a risk analysis for dam safety. This hierarchy will change from situation to situation and reflects project-specific considerations.

The breadth of possible consequences that could be considered in a risk analysis is large. Some of these have to do with public safety, such as deaths, injuries and illnesses as a result of the flooding. Others have to do with economic costs incurred by the loss of the dam itself and its productive value, property losses downstream caused by the destructive forces of the flood and by the loss of productive value of the resources (infrastructure, land, industrial facilities) damaged by the flood. Still others have to do with environmental degradations, for example, the loss of habitat and wildlife. Specialists – for example, economists, fish and wildlife biologists, emergency response planners and others – are needed on many of these specific topics for the consequence assessment to be done well. This section cannot address each of the many consequences that could appear in a consequence tree, but rather focuses on a few, to suggest principles and theoretical considerations that are common to any consequence assessment. Specifically, the section calls attention to loss of life within the public safety category, loss of capital property within the economic category, and species and habitat loss within the environmental category.

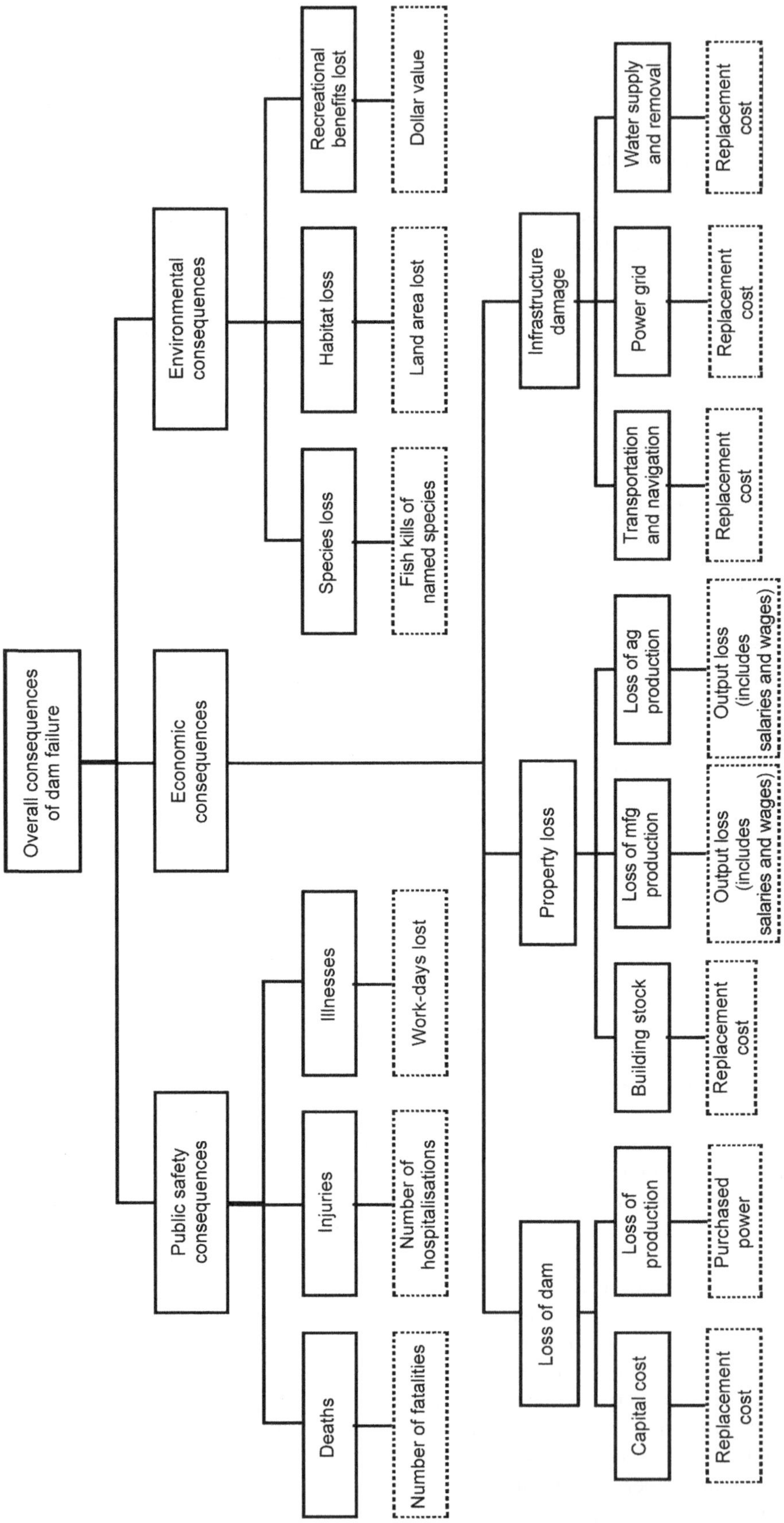

Figure 10-2. Example hierarchy of direct consequences and sub-consequences (consequence and sub-consequence boxes shown in full line; attribute boxes shown in dashed line)

Consequence identification is a process of exploration and thus there is no deductive procedure by which one can assure that all important consequences are identified and nothing of importance has been left out. A variety of techniques of consequence identification are used in routine practice. In a conceptual way, such consequence identification techniques build upon the spirit of impact assessment outlined by Leopold *et al.* (1971) and the Multi-agency Task Force under USBR (1972). These efforts organise consequences by sector (e.g. economic, environmental, public safety, legal, political, etc.) in a reductionist way, dividing the sectors first and then looking for consequences. Other approaches are to divide out consequences by the stakeholder group affected, or by the economic interest affected. These are also reductionist in their approach, but organise categories in a different way at the start.

One way to categorise identification techniques is to distinguish between scanning and tracing approaches. Scanning techniques use a single-step identification of consequences, relying on intuition and insight, which may be beyond the experience of analysts who have not experienced dam failures. Tracing techniques emphasise the structure within the consequence hierarchy as a road map for identifying consequences. This structure could be a formal model, such as the event tree of Figure 10-1, or a causal chain of events leading to consequences, or a chronology, or a combination of all these. In practice, there is no fine dividing line between scanning and tracing approaches, and perhaps not a difference in theory between them. Usually, there is a temptation to develop an exhaustive list of consequences across sectors, but this may be tempered by the availability of resources. Generally, a small number of significant and well-chosen consequences suffices to capture the important concerns of dam safety analyses.

It is worthwhile to note that risk analysis is an aid for making decisions and not an end in itself. The important concerns of identifying consequences should be bounded by the type of decision to be made. For example, a risk analysis conducted to optimise a technical decision may focus only on the direct costs related to the dam. Environmental consequences, potential loss of life, and indirect consequences may only be taken into account in decision-making by legislatures or in other public decision environments. Narrowing the extent of consequences considered in the analysis may help focus attention on the critical, analysable and relevant aspects of a decision.

Scanning for consequences

The most common approach to consequence identification is a checklist, which is a simple form of scanning. Checklists attempt to guarantee, and often assure, some degree of exhaustiveness in seeking a complete list of consequences. Leopold *et al.* (1971) provide a checklist of some 100 items to look for in considering the environmental consequences of possible actions or policies (Table 10-2). This is a purely observational, and somewhat rudimentary process, but nevertheless commonly used.

Matrix approaches are similar to checklists but add a second dimension to the taxonomy. This provides more flexibility in characterising interdependencies among consequences and the causal relationship to the initiating failure events. Table 10-3 shows an assessment of potential environmental consequences of a phosphate mining operation on public land (Rossini and Porter, 1983). Other matrix representations of consequences could include a stakeholder list along the vertical axis combined with environmental or economic consequences along the horizontal axis. This could be used to shed light on the distribution of costs and benefits across affected groups downstream of a project. Matrices of higher than two dimensions are possible. Matrices can also be used to identify more complex interactions among sub-consequences, impacted stakeholder groups, and project alternatives which can then be a starting point for economic or other modelling.

Table 10-2. Potential environmental consequences of government actions or policies (Leopold et al.*, 1971)*

Media	Consequence	Media	Consequence
Physical environment		2. Fauna	a. Birds
1. Earth	a. Mineral resources		b. Land animals
	b. Construction material		c. Fish and shellfish
	c. Soils		d. Benthic organisms
	d. Landform		e. Insects
	e. Force fields and background radiation		f. Micro fauna
	f. Unique physical features		g. Endangered species
2. Water	a. Surface		h. Barriers
	b. Ocean		i. Corridors
	c. Underground	*Cultural*	
	d. Quality	1. Land use	a. Wilderness
	e. Temperature		b. Wetlands
	f. Recharge		c. Forestry
	g. Snow, ice and permafrost		d. Grazing
3. Atmosphere	a. Quality (gases, particulates)		e. Agricultural
	b. Climate (micro, macro)		f. Residential
4. Processes	a. Floods		g. Commercial
	b. Erosion		h. Industrial
	c. Deposition (sedimentation, precipitation)		i. Mining and quarrying
	d. Solution	2. Recreation	a. Hunting
	e. Sorption (ion exchange, complexing)		b. Fishing and boating
	f. Compaction and settling		d. Swimming
	g. Stability (slides, slumps)		e. Camping and hiking
	h. Stress – strain (earthquake)		f. Picnicking
	i. Air movements		g. Resorts
Ecological		3. Aesthetics	a. Scenic views and vistas
	a. Salinisation of water resources		b. Wilderness qualities
	b. Eutrophication		c. Open-space qualities
	c. Disease – insect vectors		d. Landscape design
	d. Food chains		e. Unique physical features
	e. Salinisation of surficial material		f. Parks and reserves
	f. Brush encroachment		g. Monuments
	g. Other		h. Rare and unique species or ecosystems
Biological			i. Historical or archaeological
1. Flora	a. Trees		j. Presence of misfits
	b. Shrubs	4. Cultural status	a. Cultural patterns (lifestyle)
	c. Grass		b. Health and safety
	d. Crops		c. Employment
	e. Microflora		d. Population density
	f. Aquatic plants	5. Man-made	a. Structures
	g. Endangered species		b. Transportation network
	h. Barriers		c. Utility networks
			d. Waste disposal
			e. Harriers
			f. Corridors

Systematic tracing techniques are commonly based on an enumeration of the consequence-and-sub-consequence hierarchy of Figure 10-2. That is, they systematically start with high-level, broad consequences, and then try to work toward ever more detail to graphically represent the interrelationships among the different types of consequences of dam failure (Eccleston, 2001; Warner and Preston, 1974). Structuring consequences in a tree-like hierarchy facilitates determining whether the set possesses the desirable properties one seeks in a set of sub-consequences and attributes, as discussed in the following section: completeness, operationalness, decomposability,

Table 10-3. Potential environmental consequence matrix for phosphate mining operations, number in cell represents severity of consequence / consequence importance (Rossini and Porter, 1983)

	Possible actions			
Possible environmental consequences	**Highways and bridges**	**Surface excavations**	**Mineral processing**	**Emplacement of tailings**
Water quality		2/2	1/1	2/2
Erosion	2/2	1/1		2/2
Aquatic plants		2/2		2/3
Rare species	2/5	2/4	5/10	

non-redundancy and minimalness. A consequence hierarchy tree is not an event tree. One tries to make the sub-consequences at any node exhaustive in the sense of completeness, but the sub-consequences are not mutually exclusive. Similarly, the same sub-consequence may have more than one cause.

Tracing (modelling) consequences

Other tracing methods, although less common in dam safety risk analysis, include input–output analyses of the type developed by Leontief (Haimes, 1990). Input–output analysis involves structuring a set of linear equations relating interconnected variables, and then using matrix algebra to perform calculations on those interrelationships. These calculations can, in principle, lead to improved understanding of the implicit relations among failure mode characteristics, downstream conditions and consequences, in that the calculations simultaneously account for large numbers of interconnected variables.

Scope of the analysis of consequences

Just like decision trees or event trees, the list of consequences and sub-consequences identified as potentially resulting from dam failure can become large quickly. The goal of consequence identification is to ensure that the consequences of major importance are all represented in the analysis. Time is never sufficient to identify all possible consequences of a dam failure and, even if it were, time would not be available for the even more consuming task of quantitatively analysing all possible consequences.

Bounding the extent of consequences to be retained in an analysis is similar to the problem of pruning an event tree, and the argument is sometimes made that such bounding is difficult to do without first extending beyond what the final analysis will consider. That is, one must consider a large sweep of potential consequences before narrowing the scope of analysis to a workable level. Several factors need to be considered. First, are the consequences in question critical to the owner and to affected parties? Are they the consequences the parties involved care most about? Is the product of probability and importance or severity above some threshold level of perceptibility? Second, are the consequences analysable? Can the line of causality or logical association from dam failure to detailed consequence be traced? Are the consequences measurable? Is there sufficient information with which to statistically estimate parameters or, if not, is there relevant expert opinion on the issue? Third, are the consequences relevant to the scope of the risk analysis? Usually an engineering risk analysis deals with direct, local and immediate costs and benefits, not with indirect, national, or far in the future consequences.

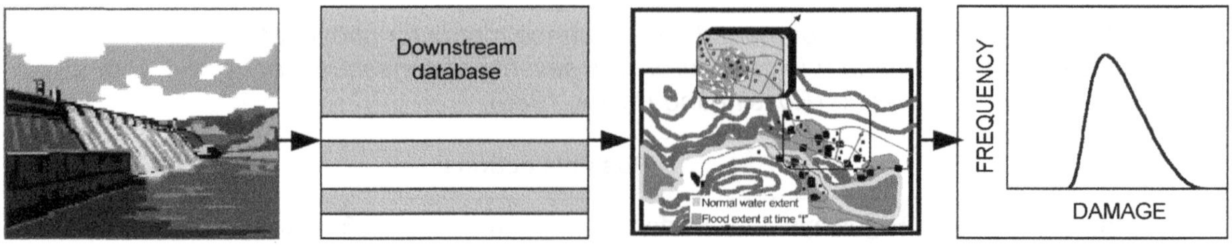

Figure 10-3. Four phases of consequence modelling: physical modelling, land-use characterisation, forecasting response and valuation

10.2 Modelling consequences

By its nature, consequence modelling is a broader activity than the engineering performance modelling discussed earlier. Typically, consequence modelling involves four phases:

1. Characterising the breach of the dam.
2. Characterising the downstream flood routing of the flood wave.
3. Assessing the impact of the flood wave downstream.
4. Assigning numerical values for lives lost, dollar economic loss and environmental damage.

Thus, consequence modelling has four parts (Figure 10-3): (1) physical process modelling of the dam breach and flood routing; (2) characterisation of land use, economic activities and population within the affected area; (3) forecast of physical responses and of individual and social response to the inundation; and (4) valuation either in monetised units or other measures.

Validation and verification of consequence models

Models are used in each part of consequence assessment. These models attempt to mimic natural processes and human behaviour and, thus, are never closed in the sense of deductive logic. They are almost never based on simple first principles or theorems, in the way that theoretical physics is, or from limited data, and other approximations to the natural and human systems they attempt to represent. The literature often describes the adequacy of such models by two criteria of assurance: verification and validation.

Verification is a check of the veracity or truth of a model. It deals with the consistency between model and reality. Because engineering and economic models are always simplifications and approximations, they can never be "true" in this sense, for example, in the way that a geometric proof can be true. In engineering or economic practice the term verification is sometimes weakened from this philosophical meaning to take on the meaning of calibration against reality; that is, a model is verified if it accurately and precisely forecasts the natural processes it attempts to mimic. This notion of calibration, widely used in modelling, is distinct from the notion of calibration in laboratory testing, which has to do with conformance with a standard.

Validation, in contrast, is a check of the validity of the internal logic of a model. It deals with the consistency of the arguments made and relationships invoked within the model. Even though engineering and economic models are simplifications and approximations, they can still be internally consistent in their representations, for example the outputs are uniquely related to inputs through a chain of transparent reasoning. In engineering or economic practice the term validation is sometimes weakened from this strict cause-and-effect meaning to take on a statistical meaning; that is, a model is validated if it correctly reflects empirical data as summarised in statistical relationships such as regression equations.

While verification and validation are widely used terms in modelling they are often used loosely, inconsistently and interchangeably. More useful and specific terms are *correspondence* and *coherence*. Correspondence means that predictions are, in fact, close to what is observed in reality. This is the same concept as calibration as used above, but semantically more precise. Coherence means that the model is internally consistent in a logical way. This is similar to the concept of validation as used above, but more precise.

Dam breach modelling

When downstream populations are far from the dam site, the specific details of a dam failure have little influence on public safety, evacuation planning and other consequences of the failure. In these cases, flood routing, travel times and attenuation dominate the calculation of consequences. On the other hand, closer to the dam site the specific details – breach shape, width and time for the breach to develop – do affect both planning and the magnitude of the consequences. The important outputs of dam breach modelling for consequence prediction are the arrival time of the flood wave (which affects warning time), depth of the flood and velocity of the flow.

A hypothetical outflow hydrograph for a breach developed over various hours is shown in Figure 10-4. The important factor for downstream consequences is the form of this outflow hydrograph: how quickly the breach develops, how quickly the discharge builds up, how large the peak discharge becomes, and how quickly the pool drains. Dam breach modelling attempts to predict all these conditions. Dams can fail slowly or suddenly depending on the physical mechanisms of failure and the remedial actions taken by dam operators to inhibit failure. Downstream consequences depend in large measure on the rapidity with which the failure occurs, and on corresponding warning times and the force of flooding. Typically, the outflow hydrograph peaks quickly within one or a few hours, and then more slowly drains the reservoir over several hours.

In practice, dam break models normally require *a priori* estimates of the time for a breach to form, and of the shape and width of the breach (Figure 10-5). These factors typically are not outputs of dam break models but rather inputs. Uncertainties in these

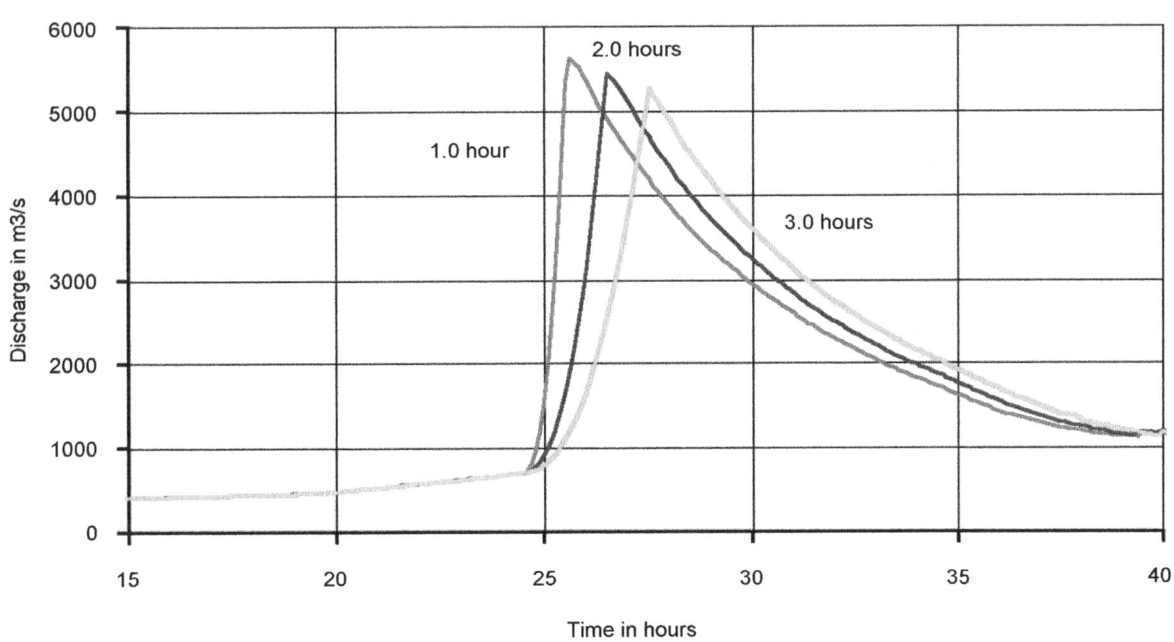

Figure 10-4. Simulated dam breach outflow hydrographs, with various breach formation times (Zielinski, 2003)

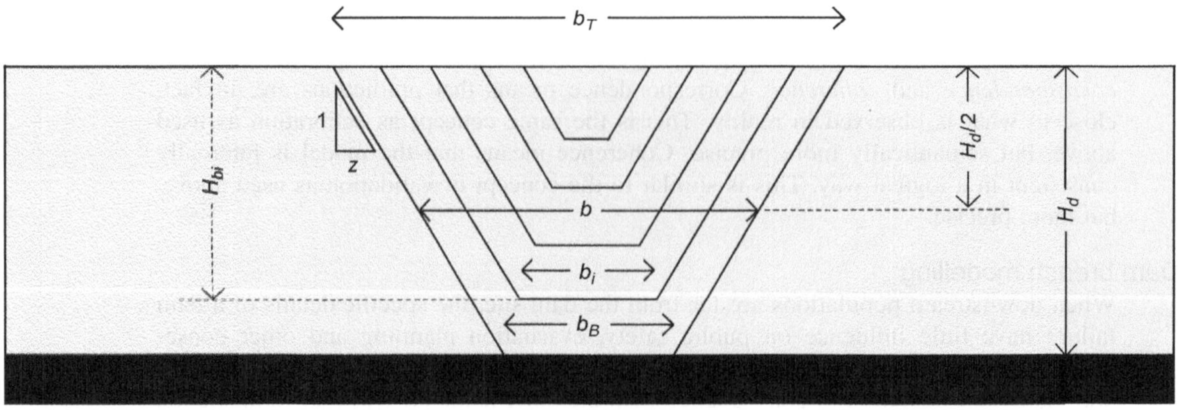

Figure 10-5. Parametric representation of dam breach formation (after Fread and Harbaugh, 1973)

parameters lead to uncertainties in the outflow hydrograph of Figure 10-4 and, consequently, to uncertainties in the downstream factors that are critical to analysing consequences, specifically, flood wave travel time, population at risk, warning time and flood severity. The *a priori* estimation of these inputs is typically based on experience and judgement, and is subject to considerable uncertainty. Empirical data on average width and time of breach are summarised by Singh (1996). Wahl (1997) provides data comparing observed and predicted breach widths and times to breach formation for several prediction formulae.

In cases of reservoirs having two or more structures (e.g. two different dams blocking the two arms of a river and having the same population at risk downstream, or a major reservoir having many large dams and each having different population at risk (PAR) and monetary consequences), the choice of which structure to "fail" is also an important *ex ante* consideration.

Breach formation is usually modelled either: (1) empirically, using statistical data from field observations and test fills; or (2) analytically, using engineering models based on hydraulics and the erosion properties of soils. Wurbs (1987) and Wahl (1997) have each surveyed the literature of modelling breach formation. Empirical models structure the key parameters of an expanding breach into a predictive equation. Typically, these equations involve, at least, the average breach width, *b*, and the breach formation time, *t*. Sometimes, these parameters are treated parametrically to develop *pro forma* calculations and tables or charts of results.

The hydraulic outflow hydrograph is computed using a numerical time-stepping procedure, or is sometimes simplified to an equation relating peak breach discharge, average width, dam height, and surface area (Fread, 1998). The most common models – for example, DAMBRK and its updated version FLDWAV – use this approach. Other workers have attempted to use regression analysis to compute peak discharge using reservoir volume and dam height as parameters, usually in a logarithmic relationship (Costa, 1988; Evans, 1986; Froehlich, 1995; Hagen, 1982; Walder and O'Connor, 1997).

Relations such as these incorporate uncertainty due to the scatter in the empirical data record, especially regarding the duration of breach formation and the side slope angles of the breach (*z* in Figure 10-5). The historical record of dam failures includes few large dams or large reservoirs, so the degree of applicability of the resulting statistical models to large structures is unclear. Singh and Snorrason (1984) and Petrascheck and Sydler (1984) note that the variation of breach parameters can strongly affect peak discharge and inundation heights near the dam site. For locations far downstream, the timing of the flood wave alters with changes in breach formation time. Variation in other parameters (storage behind the dam affecting the speed of head drop, channel and valley properties affecting attenuation) may substantially impact the peak outflow and the effects of discharge on far downstream locations.

Physics-based models use soil mechanics, hydraulics and sediment erosion equations to develop time-stepping numerical solutions of the physical breaching process and the hydraulics of the outflow hydrograph (Bechteler and Broich, 1993; Fread, 1984, 1987; Macchione and Sirangelo, 1988; Ponce and Tsivoglou, 1981). A variety of physically-based breach models have been proposed (Table 10-4). As noted by Wahl (1997), several of these models are fit against limited data. The physics-based models are mostly based on tractive–stress erosion models, and ignore the commonly observed geotechnical failure modes of breach development, namely, head cut erosion and slope failure.

In forecasting the consequence of rapid failure, the assumption is sometimes made of instantaneous removal of the dam, and the sudden release of floodwater to make its way downstream. This assumption is considered to represent the most severe condition. Thus, it is used to estimate an upper bound on consequences (Singh, 1996). It also simplifies the modelling of breach behaviour since complications surrounding the expanding size of the breach due to erosion and other engineering considerations can be neglected. While instantaneous failure may approximate the failures of certain types of dams failing in particular modes (e.g. catastrophic failure of an abutment, as at Malpasset, or failure of a reservoir slope and subsequent overtopping, as at Vaiont),

Table 10-4. Embankment breach models (Singh and Scarlatos, 1988; Wahl, 1997; Wurbs, 1987)

Model	Year	Sediment Transport	Breach Morphology	Parameters	Citation
Christofano	1965	Empirical	Constant breach width	Angle of repose, others	Christofano (Christofano, 1965)
Harris and Wagner, BRDAM	1967 1977	Schoklitsch formula	Parabolic breach shape	Breach dimensions, sediment properties	Harris and Wagner (Harris and Wagner, 1967)
Lou, Ponce and Tsivoglou	1981	Meyer–Peter Mueller formula	Regime type relation	Critical shear stress, sediment, tailwater	Lou (Lou, 1981) Ponce and Tsivoglou (Ponce and Tsivoglou, 1981)
BREACH	1985	Meyer–Peter Mueller formula modified by Smart	Rectangular, triangular, trapezoidal	Critical shear, sediment, tailwater, dry slope stability	Fread (Fread *et al.*, 1985)
BEED	1985	Einstein–Brown formula	Rectangular or trapezoidal	Sediment, tailwater, saturated slope stability	Singh and Scarlatos (Singh and Scarlatos, 1988)
FLOW SIM 1 and 2		Linear erosion of Schoklitsch formula	Rectangular, triangular, trapezoidal	Breach dimensions, sediment properties	Bodine (Bodine, undated)

most embankment dams fail gradually. An embankment is overtopped or internal erosion creates a piping condition, and the embankment begins to erode, finally releasing its pool over time (Figure 10-4). If the duration of time required for this erosion is short, the breach can be modelled as if instantaneous; but, if not, an engineering hydraulics model is needed to predict the characteristics of the flood wave as it leaves the dam site.

Flood routing

Flood routing models the downstream hydraulics of the flood wave as it moves downstream of the dam (Figure 10-6). The objective of these models is an accurate prediction of flow discharge, velocity and depth as a function of time and distance downstream after the formation of a breach. Flood routing is important because these unsteady flows from dam breach experience significant peak attenuation as the flood wave moves downstream.

Three approaches are used to route flood waves downstream: Numerical codes for the complete one-dimensional St Venant equations of unsteady flow (Fread, 1977, 1988, 1993); peak (breach) discharge attenuation curves coupled with Manning's equation to calculate flow depth (Wetmore and Fread, 1984); and Muskingum–Cunge routing with Manning's equation for flow depth (Singh and Scarlatos, 1988). Numerical solution to the St Venant equations provides the most accurate predictions and, thus, the second and third approaches using Manning's equations to approximate flow depth introduce additional error. Fread *et al.* (1991) and Fread and Hsu (1993) discuss the extent of these errors. For simple downstream routing conditions, the error with respect to the St Venant equations solution is thought to be less than 10%; however, for low river gradient conditions ($<3/1000$) the errors can be large. In addition to uncertainties in the dam breach modelling, uncertainties also apply to flood routing. Little work has been done on probabilistic modelling of dam breach and downstream flood routing, although it is conceptually straightforward to add parametric and model uncertainty to existing

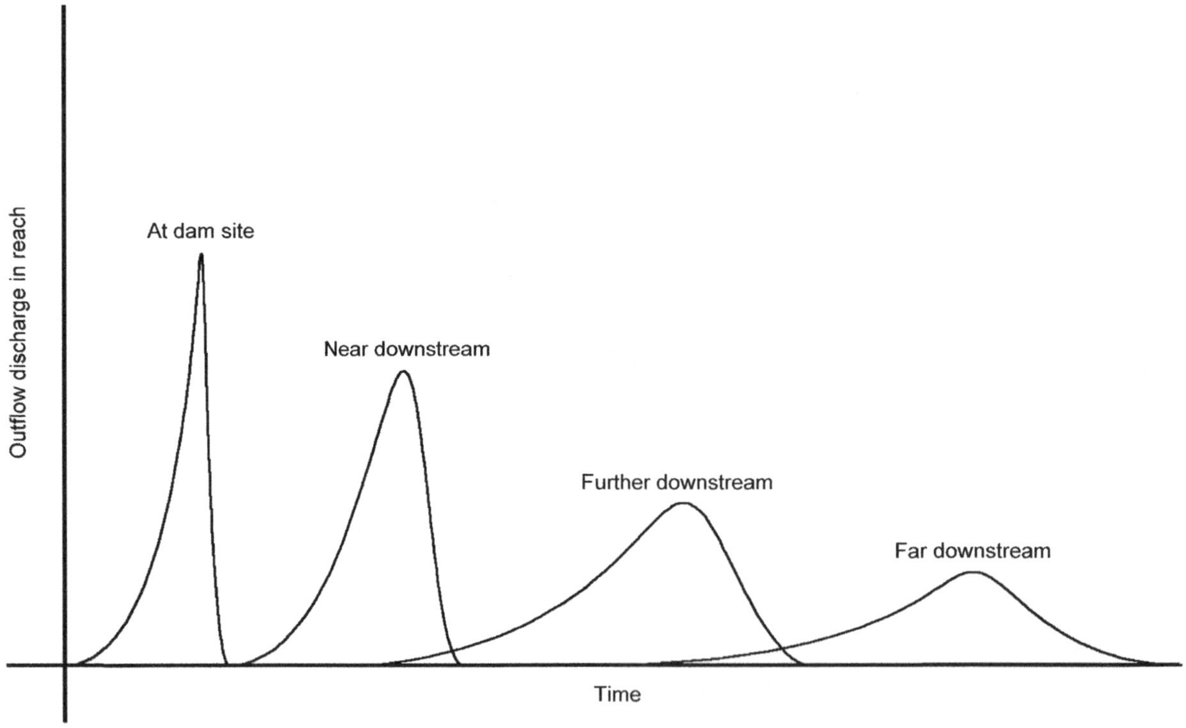

Figure 10-6. Flood hydrographs following dam failure in various downstream reaches

modelling approaches. The biggest challenge is the spatial variability in channel and floodplain geometry and their physical properties. Since all available models are discrete and have size restrictions, a conceptually simple and straightforward solution is difficult to implement. In addition, significant uncertainties in flood routing are generated by debris load, channel erosion, sediment transport and a variety of considerations usually not included in the hydraulic routing model.

Downstream data mapping

As much as system models and event trees for the dam itself, consequence forecasting depends on downstream characteristics, land uses and exposures. These are typically represented in downstream demographic and land-use databases, often incorporated in geographic information systems (GISs). These data can be of many types, but should include: (1) topography; (2) land-use categories; (3) population; (4) critical facilities and infrastructure; (5) economically important facilities; and (6) environmental or culturally important sites.

Consequences event tree

The theory and concepts of event tree models of sequences of events leading to dam failure were described in Chapter 8. Event trees graphically organise plausible chains of events that could lead to loss of containment and hence "dam failure". Starting from a terminal node on an event tree indicating failure, a further chain of events and outcomes can be used to trace the consequences of the failure downstream; that is, to provide an estimate of consequences conditional on loss of pool or some other damage state of the dam (Figure 10-7). This sub-tree incorporates exposure cases and other downstream activities or events, such as the effectiveness of evacuations or other risk mitigation procedures. Sometimes this downstream tree can be treated independently of the event tree for the dam itself and sometimes not.

Consequences event tree

Chains of subsequent events and outcomes when taken as a whole constitute what might be called a consequences event tree. Such a consequences event tree is the same in principle as the original event tree that models aspects of dam failure, only now the events that compose the consequences event tree have to do with possible mitigation

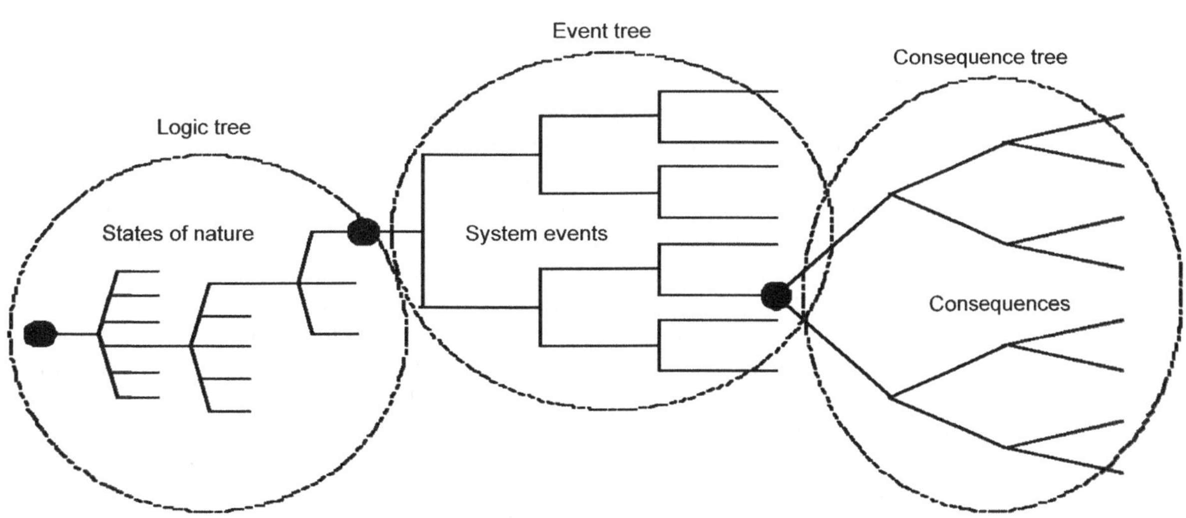

Figure 10-7. Complete event tree representation of dam failure and resulting consequences: logic tree of knowledge uncertainties (model parameters, model error), event tree of failure mechanisms and consequence event tree of outcomes

activities and downstream occurrences once failure is initiated. Example events that might populate a consequence event tree are: (1) operator intervention to draw down the pool once failure is initiated; (2) timing of the failure (season of the year, time of day); (3) warning time provided downstream riparians; or (4) heroic actions taken to evacuate riparians.

The terminal node at the end of any chain of events through an event tree model of dam failure is referred to as a leaf. Each leaf in the event tree has an associated set of potential consequences, each with some probability. The consequences are the costs or benefits accruing should the particular chain of events leading to that leaf obtain. In most cases, these consequences themselves are complex and analysed by constructing a consequence tree. A consequence tree structures the set of considerations involved in estimating consequences of failures in the same way that an event tree structures chains of events leading to possible systems failure.

The consequences that result from a dam incident or failure depend on a large number of factors beyond the behaviour of the dam itself. For example, consequences depend on barriers, safety systems, operating procedures, operator actions, and so on, and how they respond to an initiating event or an indication of dysfunction of the dam proper. Safety functions may include systems that respond automatically to an initiating event or to a failure, alarms that alert operators or other cognizant personnel that an event or failure has occurred, predefined operating procedures that follow an alarm, or barriers or other containment facilities intended to limit the effects of failure. Other considerations that influence the consequences resulting in failure include time of day, meteorological conditions, downstream warning systems and emergency response procedures to protect life and property. Reliability models of individual threatened structures or other components of infrastructure in the dam failure floodplain may accompany a consequence event tree, as illustrated by the use of fault tree analysis to assess probabilistic predictions of bridge failures due to scour by the flood flow (Johnson, 1999). Event trees are evaluated by generating a set of probability–consequence pairs for each mutually exclusive end node in the tree.

Human intervention

Human intervention to reduce the consequence of some adverse chain of events, such as emergency measures to lower the pool or to place a field-expedient filter on an eroding downstream toe, can be included in consequence trees or in the main event trees themselves. On the other hand, if following a convention of only including changes of system states in the event tree proper, it may make more sense to include human interventions within a consequence tree rather than an event tree. This provides a clear separation between events happening within the dam, on the one hand, and measures taken to mitigate consequences, on the other.

As with all uncertainties in risk analysis, the uncertainties surrounding human reactions to mitigate the adverse effects of events or systems failures are due to limited knowledge. Nonetheless, such uncertainties may be treated as if they were random variables – that is, as aleatory uncertainties – since they are outside the control of the emergency planner or analyst. These uncertainties are included in the event tree or in the consequence tree as chance events, and probabilities are assigned to the outcomes leading from those chance events to describe whether interventions are made in a timely and effective manner. This can also be treated as a "base case", which presumes no intervention or response.

A difficulty in modelling human intervention is that the nature of interventions may be impossible for the analyst to foresee. History tells many stories of heroic efforts made in the face of insurmountable obstacles. The risk analyst has no way of foreseeing this extent of human creativity. Thus, simplifying assumptions must be made in trying to include human factors quantitatively.

The typical event or consequence tree model of these uncertainties includes three levels of action: first, some significant event or failure takes place, and this occurrence is either detected in a timely manner by the operator or it is not. Second, the operator responds to this information in a timely manner, or he or she does not. Third, depending on what the operator does in response, the responsive action may be effective, or not be. This results in some final outcomes and associated consequences. For each of these three sets of events, a large number of human factors' models have been developed and are available to dam safety studies. Whether one incorporates the human intervention events within the main event tree or within a separate consequence tree is ultimately left to the convenience of the analyst. If the human interventions are modelled as random (i.e. aleatory) events as they typically are, they have the same functional form as all the other events within the main event tree.

10.3 Measuring consequences

This section considers how the broad consequences of dam failure are structured into a framework for analysis, and how measurement scales are chosen against which to quantitatively describe consequence predictions. The section considers three broad categories of consequences: public safety related, economic and environmental.

Attribute hierarchies

The consequences of dam failure come in many varieties, typically categorised within groups that include public safety, economic loss, environmental degradation, socio-economic impact, reputation and political impacts, and possibly other things. Public safety includes human injury and loss of life. Economic loss includes direct infrastructure and property damage. Environmental degradation includes ecosystems disruption and loss of habitat, among other things. Consequence assessments for an engineering risk analysis typically include only near-term and direct consequences of a failure leading to downstream inundation, plus a limited number of indirect consequences having to do with the loss of productive capacity of assets, are considered. These are limited typically to three categories: (1) public safety (injury or loss of life); (2) direct economic cost (infrastructure and property damage); and (3) environmental degradation (natural populations and ecosystem destruction).

Each of these high-level categories may comprise sub-categories of finer detail. For example, injuries might be divided by severity level or to whom they occur, economic costs might be divided by loss of the dam vs downstream property damage, and environmental degradation might be divided by wildlife loss or habitat destruction. These sub-categories of consequence might, in turn, be further subdivided if the analysis warrants.

Attributes

The high level or broad categorisation of consequences is useful in thinking about dam failures, but it is not useful in making quantitative predictions. Broad categories of consequences are descriptive and usually not measurable. For modelling and analysis, a set of summarising measures or statistics is needed, and this necessitates specific scales against which to measure impacts. Specific scales, in turn, require specific sub-categories of sufficient precision to allow numbers to be assigned to them. Thus, the first step in quantifying consequences is establishing a set of subcategories of consequences that are specific enough to allow specific measurement scales to be developed. Choosing these subcategories is not a trivial task, because whatever final measurement scales are chosen they become the quantified consequences used in the risk analysis.

Once specific subcategories of consequences are selected, the next step is to associate a scale of measure with each. Such a scale of measure is typically called, an

Table 10-5. Example of consequence categories and associated attributes

Consequence Category	Subcategory	Attribute
Public safety	Injuries	Working days lost
	Loss of life	Number of deaths
Economic cost	Loss of dam	Dollars
	property damage	Dollars
	Infrastructure lost	Dollars
Environmental degradation	Habitat lost	Hectares
	Wildlife lost	Number of individuals

attribute (Keeney and Raiffa, 1993). This serves as the quantitative measure of the qualitatively described consequence. For example, the attribute associated with loss of life is most likely to be, "number of deaths". The attribute associated with economic costs is usually some monetary unit.

For categories of consequence other than loss of life or economic cost, where lives and dollar values are obvious metrics, the choice of attribute may often be made in differing ways, since the best attribute may not be obvious. The choice of one attribute rather than another may affect the quantification of impacts because two attributes may not be one-to-one related, or because the relationship between two attributes may be non-linear.

Consider the attribute that might be associated with human injury. One attribute for injury might be "number of working days lost by victims", but others could be, "financial cost of hospitalisation and recovery", or "number of people suffering injuries regardless of severity" (Table 10-5). In specific instances, these somewhat different measures lead to somewhat different orderings of the consequences of failure among different scenarios. For "environmental degradation", one of the more specific subcategories might be "habitat loss". An attribute such as "area lost in hectares" could be specified to make predictions of habitat loss, but a number of other specific attributes could be chosen as well, and these other choices might yield differences in the way different outcomes compare with one another.

To be practically useful in risk analysis, attributes such as measurement scales for consequences should have a number of well-behaving mathematical properties, of which the more important are: comprehensiveness, measurability and comparability (Fishburn, 1964; Keeney and Raiffa, 1993). An attribute is comprehensive if it captures the full extent to which the corresponding consequence outcome is valued, that is the full extent to which the consequence occurs so that it might be evaluated against other categories of cost or benefit. An attribute is measurable if a prediction can be made over the attribute for each scenario of dam failure under consideration in the risk analysis. An attribute is comparable if increments of value can be assigned to levels of the attribute such that those values can be compared across different types of consequences to draw conclusions on net cost or benefit, that is, preferences can be assigned to levels of the attribute.

Comprehensiveness refers to the conceptual appropriateness of an attribute: does it capture all the information; measurability refers to the practical appropriateness, can we actually analytically determine numbers over it; and comparability refers to the operational appropriateness, can we use these attribute values to make comparisons with other consequences? However, we do not usually trade off one attribute against another in a dam safety risk analysis. At the most, we aggregate to these three categories of

Table 10-6. Desirable properties of a set of consequence attributes

Property	Meaning
Individual attributes	
Comprehensive	Attribute captures the full extent to which the corresponding consequence outcome is valued
Measurable	A prediction can be made over the attribute for each scenario of dam failure under consideration in the risk analysis
Comparable	Increments of value can be assigned to levels of the attribute such that those values can be compared across different types of consequences to draw conclusions on net cost or benefit
Set of attributes	
Complete	Measurements along individual attributes when, taken as a whole, fully capture preferences for high-level consequences
Operational	Meaningful to the decisions that are intended to result from the risk analysis
Decomposable	Set can be separated into subsets for comparing trade-offs among preferences for outcomes
Non-redundant	Avoids double counting of the same impacts across sub-consequences
Minimal	Smallest dimension possible while still satisfying other criteria

consequence. Desirable properties of attributes and sets of attributes are summarised in Table 10-6.

The categories of Table 10-6 can be used to illustrate these properties of attributes. If a risk analysis is concerned with consequences to public safety, then the attribute, "number of lives lost", may not be comprehensive, depending on the pertinent legal structure or value system, because it does not capture the number or extent of non-fatal injuries which may be important to a decision about dam safety. In a similar way, an attribute of non-fatal injury dealing with the suffering of the victims may not be measurable because there is no objective way to measure the pain of the injured.

Sets of attributes

In the last section we introduced the concepts of comprehensiveness, measurability and comparability as desirable properties for individual attributes. Related desirable properties also apply to the set of attributes in respect to the consequences-and-attributes hierarchy to which they belong. Important among these are completeness, operationalness, decomposability, non-redundancy and minimalness (Table 10-6) (Keeney and Raiffa, 1993).

A set of attributes is complete if measurements along the individual attributes of the set, when taken as a whole, fully capture the extent to which a high-level consequence is deemed favourable or unfavourable (adverse). For this to be true, the set of sub-consequences must span all the major concerns associated with the high-level consequence they derive from, and the individual attributes associated with sub-consequences must themselves be comprehensive in the sense used above.

A set of attributes is operational if it is meaningful to the decisions that are intended to result from the risk analysis, that is if the set of attributes is useful in evaluation alternatives. For this to be true, the set of sub-consequences and their attributes must be intuitively understandable to the analysts and communicable to other affected parties and the public.

An attribute is decomposable if it can be broken down into subsets (e.g. pairs) for the purpose of making comparisons and trade-offs among preferences for outcomes, that is, for the purpose of being able to get at measurable entities.

A set of attributes is non-redundant if it avoids double counting of the same impacts across sub-consequences, and thus avoids inflating the importance of particular types of consequences relative to the other. The example set of attributes for environmental degradation used above, which included wildlife loss and habitat loss, may suffer some redundancy in that one immediate effect of habitat loss could be the loss of wildlife resident in the affected areas; thus, there would be some double counting as the attributes were summed across sub-consequences.

A set of attributes is minimal if it is of the smallest dimension possible while still satisfying the other, above, criteria. Minimalness is desirable because each time a consequence is further sub-divided there is some potential of losing important qualities or facets of the consequence in question. Furthermore, since the risk analysis intends to lead to a probability distribution over the set of attributes, the smaller that set is the easier the task of modelling and calculating the required joint distributions.

Qualitative attributes

Most attributes dealt with in dam safety risk analysis involve objective measures. The cost of damaged downstream properties is assigned a dollar value attribute and can be objectively, if not precisely, measured. But some attributes are not of this objective type, for example, the impact to the reputation of the dam owner in the event of a failure. For such attributes no objective scale exists, so either a proxy scale or a subjective scale must be used instead. A proxy attribute is one that is correlated with the degree to which a consequence is valued but does not actually measure the consequence itself. A subjective attribute is typically composed on an ordinal rather than cardinal measure. For example, the verbal descriptions of likelihood used in a Kent chart (Cooke, 1991) form a subjective attribute; similarly, the reputation of the dam owner might be measured against an attribute consisting of verbal descriptions of relative but not absolute relation to one another: "despised" on the low end to "exalted" on the high end. The potential consequences of specific dam accident or failure scenarios would be projected against this set of scenario descriptions and relative preference values assigned to the relative levels of the subjective scale.

Proxy attributes

Proxy attributes are correlates to the consequence impacts of importance in a risk analysis yet which are not actually direct measures of the consequence, but which can be used as "stand ins" for a more direct attribute that is not possible to identify. Proxy attributes seldom arise in association with economic consequences, for which dollar measures are obviously comprehensive, measurable and comparable, but they are common for environmental and sometimes public safety consequences.

Consider, for example, the high-level consequence of environmental degradation. We may be concerned with how a dam failure destroys environmental conditions downstream to the extent that permanent changes are effected to ecosystem health. But how can this be made more specific, and how can the potential for permanent change be measured? Generally, it cannot. So, we might adopt attributes that we know or suspect to be closely related to ecosystem health, for example, the land area of a riparian zone destroyed by the uncontrolled release of water, measured in hectares. This is not exactly the same as ecosystem health, but it is probably correlated with whatever ideal attribute it might be that directly measured ecosystem health. If we can place preference values over the area of ecosystem destroyed as a proxy attribute, that might be accurate enough to make a risk-informed decision in lieu of a more precise measure of ecosystem health.

Consequence–attribute hierarchy

The consequences of dam failure, for the purposes of risk analysis, reside in a hierarchy with high-level but vague and qualitative consequences at the top, and low-level but explicit and measurable attributes at the bottom. This is an analytical structure that risk analysis imposes on the otherwise unstructured considerations about consequences should accidents or failure occur, and is a common approach of decision analysis and utility theory (Gál *et al.*, 1999).

How detailed should the consequences-and-attributes hierarchy be? This is a general question of modelling and analysis, not just risk analysis, and is a decision generally left to the analysts, depending on reasonableness; there is no technocratic way to make the decision. A balance needs to be struck between the advantages of more detailed decomposition of the hierarchy, and the corresponding simplifications of identification and measurement that result, compared to the disadvantages of the ever increasing width of the tree as the levels increase, and the corresponding loss of significance of the lowest level sub-consequences or attributes. An ultimate test of whether or not to eliminate a level of the hierarchy is whether, eliminating the layer carries the potential of altering the best course of action indicated by the risk analysis. This criterion is discussed by Keeney and Raiffa (1993).

The more detailed the consequences–attribute hierarchy, the easier to identify objective attributes to associate with sub-consequences. Using "environmental deterioration" as an example again, at the highest level this consequence can only be quantified with a subjective attribute. Even if the highest-level objective is subdivided into, say, destruction of riparian ecosystem and reduction in productivity of flooded lands, there are no obvious and simply measurable attributes to associate with these sub-consequences. Only if we further subdivide the hierarchy into even more detailed consequences such as wildlife loss or specific types of habitat loss, or specific types or contamination do we get to a level where clearly objective and measurable attributes can be associated with the sub-consequences.

The discussion should make clear that any consequences-and-attribute hierarchy is necessarily non-unique. Two groups of analysts will likely arrive at different hierarchies for the same problem, and both hierarchies may be "right" to the extent that they capture the essence of the risk analysis problem, and to the extent that the attributes are comprehensive, measurable, and comparable for the high-level consequences identified by the analytical teams. For example, to measure the extent of injuries to downstream residents resulting from a dam breach we might use the attribute, as above, "total numbers of days lost from work due to injuries" which is denominated in "days", or we might use the attribute, "insurance claims paid to downstream residents in compensation for injuries" which is denominated in "dollars." Each attribute attempts to measure the same sub-consequence, and there may be strong similarity in the ordering of outcomes measures along these two dimensions, but there is likely not to be a one-to-one correspondence and likely not to be even a linear relation between them (and if linearity is weak, even the correlation between the two may be weak).

Direct consequences of dam failure

Most dam safety studies focus on three broad categories of consequence: public safety, economic consequences and environmental consequences.

Public safety

Arguably the most important consequence of dam failure is the effect on public safety, specifically loss of life and non-fatal injuries to downstream victims. Table 10-7 shows fatality numbers for individual dam failures for North American and selected international cases.

Table 10-7. Selected dam failures in the US 1894–1995, year of failure and lives lost (data from McCann, 1995)

Named dams								
Name	**Year**	**Lives**	**Name**	**Year**	**Lives**	**Name**	**Year**	**Lives**
Mill River, MA	1874	143	East Lee, MA	1968	2	DMAD, UT	1983	1
South Fork, PA	1889	2209	Lee Lake, MA	1968	2	Bass Haven, TX	1984	1
Walnut Grove, AZ	1890	150	Virdert Creek, IA	1968	1	Little Falls, DC	1984	5
Mountjoy Hill, ME	1893	4	Anzalduas, TX	1972	4	Kendall Lake, SC	1990	4
Angles, CA	1895	1	Black Hills, SD	1972	245	Shadyside, OH	1990	24
Melzingah I & 2, NY	1897	7	Buffalo Creek, WV	1972	25	Timber Lake, VA	1995	2
Austin, TX	1900	8	Canyon Lakes, SD	1972	33	**Unnamed dams**		
Bayless (Austin), PA	1911	80	Lake of the Hills, AK	1972	1	*Location*	Year	Lives
Lyman, AZ	1915	8	Lakeside, SC	1975	1	Winston, NC	1904	9
Lower Otay, CA	1916	30	Asheville, NC	976	4	Kansas River, KS	1951	11
Mammoth, UT	1917	1	Bear Wallow, NC	1976	4	Denver, CO	1965	1
St Francis, CA	1928	450	Teton, ID	1976	11	Black Hills, SD	1972	245
Wagner, WA	1938	1	Evans & Lockwood, NC	1977	2	Big Thompson R, CO	1976	144
Schoellkopf Sta, NY	1956	1	Kelly Barnes, GA	1977	39	Newfound, NC	1976	4
Baldwin Hills, CA	1963	5	Laurel Run, PA	1977	40	Kansas City, MO	1977	20
Little Deer Creek, UT	1963	1	Sandy Run, PA	977	5	Texas Hill Co, TX	1978	25
Moohegan Park, CT	1963	6	Lake Keowee, SC	1978	7	Austin, TX	1981	3
Swift, MT	1964	19	Swimming Pool, NY	1979	4	Northern, NJ	1984	2
Two Medicine, MI	1964	9	Eastover Mining, KY	1981	1	Allegheny Co, PA	1986	9
Skagway, CO	1965	2	Lawn Lake, CO	1982	3	Americus, GA	1994	3

Table 10-8. Factors influencing the number of fatalities resulting from dam failure (Graham, 1999)

Factors influencing fatality numbers
Cause and type of dam failure
Number of people at risk
Timeliness of dam failure warning
Flood depths and velocities in the downstream floodplain prior to dam failure
Flood depths and velocities resulting from dam failure
Availability of sensory clues (sight of floodwater or sounds created by rushing floodwater) to the people at risk
Time of day, day of week, and time of year of failure
Weather, including air and water temperatures
Activity in which people are engaged
General health of people threatened by floodwater
Type of structure in which people are located
Ease of evacuation

Characterising the number of potential fatalities due to a dam failure is complicated by the large number of factors that influence the death rate within a population potentially exposed to inundation. Graham (1999) cites the factors shown in Table 10-8. Graham concludes that, of these, the most important are: (1) the number of people occupying the dam failure flood plain; (2) the amount of warning provided to the people exposed; and (3) the severity of flooding. As a result, uncertainty enters any forecast of fatalities due to dam failure because the time of day and season of the year is *a priori* unknown, the extent of effective or usable warning is unknown, and the exact conditions in the flood plain are unknown, as are the human reactions.

In some cases, public safety consequences are limited to lives lost, and do not include injury and disability, or pain-and-suffering. From a policy point of view, public safety consequences short of death are sometimes treated as economic variables in that their impact can enter consequence accounts through litigation and financial settlements. This does not obviate the need for quantitative estimates of the extent of these impacts, but merely shifts them to another account.

Economic consequences

Estimates of economic consequences are in most cases easier than corresponding estimates of public safety impacts. Buildings and most other infrastructures in the floodplain (e.g. highways, bridges, utility networks) are stationary, and good data are often available with which to forecast damage and impairment. A long history of benefit–cost analysis in public investment theory provides economic and analytical guidance on how to quantify and compare economic costs (Mishan, 1988), even in the absence of markets. Nonetheless, the economic consequences of failure are presumably more complex to estimate than the economic costs and benefits projected during project planning.

The principal categories of economic consequence of dam failure are direct consequences and indirect consequences. Direct consequences are the immediate impacts of the failure or being in contact with floodwaters. Direct economic consequences include, for example, loss of the capital investment in the dam, appurtenant structures and downstream improvements; destruction of downstream property: buildings, equipment, land improvements; and loss of seasonal crops. Indirect consequences are the subsequent impacts that cascade from the direct impacts of inundation, principally including the loss of use of resources. Indirect economic consequences – which economically may be even more important than direct impacts – include, for example: loss of productive use of land for agriculture, industry, and recreation; loss of power production; and loss of use of residential property and substitution of other housing for that loss. In some ways, modelling and forecasting economic consequences of failure differ little from modelling and forecasting of costs and benefits for routine projects. However, in projecting economic consequences given a failure, issues of depreciation, book value and replacement costs arise, that do not enter consideration with new projects.

Environmental consequences

Until recently, environmental costs and benefits of water resource projects have not in the past factored quantitatively in project investment decisions for most dam projects. This being the case, the calculus for evaluating environmental consequences of dam failures is inadequately developed. Increasingly, public opinion and government policy seem to indicate that environmental consequences should be accounted for. Yet, including environmental consequences within risk analyses for dam safety raises a number of challenges: how comprehensively to account for environmental consequences many of which are subtle; how methodologically to estimate environmental impacts when understanding of ecosystem response is nascent; how to economically value those impacts in the absence of markets? The uncertainties, both of predicting quantitative environmental consequences of dam failure and of valuing those consequences, are great, probably greater than corresponding uncertainties in public safety or economic consequences.

One approach is that environmental and ecological costs and benefits not be quantitatively incorporated in a consequence assessment unless a unique downstream situation demands inclusion: examples might be: specifically protected habitat or wildlife under government regulation, national heritage sites, or hazardous installations whose flooding would lead to the spread of contaminants.

Watersheds are hydrologically bounded ecosystems, and therefore a logical unit for environmental and ecosystem analysis. Yet, incorporating science consistently in the environmental models and predictions of watershed response to catastrophic assaults, such as dam failure floods, is challenging and we have little experience by which to validate predictions. Trade offs among environmental, political, economic and social factors based on subjective value judgements are common in environmental analysis. It is often difficult to reconcile the desire to make scientifically supportable predictions with the complexity of how local watershed hydrology, hydraulics and ecology work. As a result, scientific understanding may be under-utilised in such analysis in favour of simple accounting schemes.

The USEPA (1998) has concluded that:

> Assessment is one of the most critically important parts of watershed management because it attempts to transform scientific data into policy-relevant information that can support decision-making and action. Many other definitions and methods of environmental assessment are in use, but none has been widely adopted for incorporating science into watershed management. Ecological risk assessment may be particularly useful in watersheds as a scientific method that includes steps for integration with planning, priority-setting, and decision-making.

Such risk assessment techniques, now being developed for other applications, could come to play a major role in dam safety risk assessments in coming years, although they are embryonic today (Sherbot, 1997).

Socio-economic and other consequences of dam failure

While public safety, economics and environmental impacts are the consequences of principal concern in dam failures, other and secondary consequences may inform decisions. Important among these are the effect of dam failure on the reputation of the owner and operator of the dam, political or regulatory repercussions of a failure, and the distribution of costs across affected parties or victims.

Critical infrastructure may exist downstream of a dam which, if destroyed or damaged sufficiently to impair service, could seriously affect public safety or economic well-being outside the zone of flooding. Such infrastructure includes transportation facilities and structures (highways, railroads, navigation works, bridges), utility networks (power transmission and distribution, pipelines), telecommunications networks and structure (relay towers), water supply and waste-water removal systems, and other lifelines upon which society depends.

Damage to these infrastructure components can have regional or national economic impacts, for example, causing a redistribution of services through other regions, that would not normally be included in a dam safety risk analysis. However, they may also have identifiable direct impacts, typically of an economic nature, and beyond the simple cost of reconstruction, that would be included in a risk analysis. For example, the disruption of electrical power to an identifiable factory or other installation might have to be made up for from alternative, higher priced, sources.

10.4 Public safety modelling and assessment

Work on estimating loss of life due to dam failures have been undertaken by several workers and agencies. The techniques resulting from this work build on one of two approaches: (1) statistical estimates based on observed fatalities in historical failures; or (2) physical process models of inundation and evacuation, including the use of event-driven consequence trees. The latter two may be augmented by hydraulic models attempting to capture the relationship between water velocities, water depths, and fatality rates for people caught in the flow, or of the relationships among human factors, sociological considerations, and evacuation conditions. The development of event-driven consequence trees has been discussed above, and event trees more generally have been discussed in Chapter 8. This is not repeated here.

Loss of life in floods is determined in large measure by factors of two sorts: (1) demographic and land-use characteristics of the inundated region, including population, land uses, civil infrastructure, warning systems and emergency response systems; and (2) hydraulic characteristics of the flood event, including peak discharge, water velocity, depth of inundation and related factors (Figure 10-8). The first set of factors, to a large extent, influences the number of people caught in a flood. The second set of factors, to a large extent, influences the fatality rate of people caught. The population at risk and the fatality rate combine to give an estimate of the loss of life.

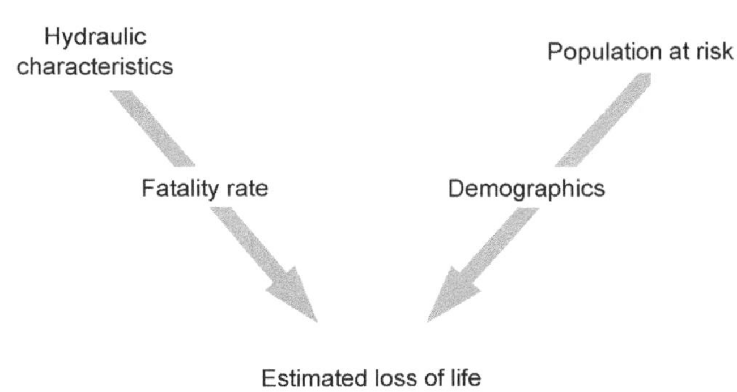

Figure 10-8. Determining factors for loss of life estimation

Statistical estimates from historical failures

Statistical estimates use observed fatalities or other consequences of historical dam failures as the basis for forecasting fatalities or other consequences of potential future failures. Such statistical approaches enjoy the benefit of a large literature on actuarial risk assessment (see, e.g. Bühlmann, 1970) and a history of methodological acceptance in government and insurance. McCann (1997) reports a list of selected dam failures in the US between 1894 and 1995 with associated loss of life (Table 10-7). The absolute number of dam failures per year is high, but most failures result in no loss of life. According to the National Performance of Dams Program, seven US dams failed in 2001, mostly small, privately-owned dams. In the 12-year period between 1990 and 2002, 502 dams failed, with little loss of life.

Statistical approaches to estimating loss of life are correlations and, thus, are limited by the degree to which calibrating data are appropriate to the population of dams for which predictions are sought. That is, the normal limitations of sampling theory pertain to the calibrating data. One should think of the set of dams that have failed, leading to loss of life, as a sample from some population. Conceptually, from this sample we are trying to estimate the characteristics of the sampled population using statistical theory. However, the sample has not been taken according to some *a priori* plan, but comes about according to which dams happened to have failed. Obviously, that is always the case with historical data, but it means that the set of failures is not a random sample of dams, but disproportionately includes small and poorly engineered or poorly maintained structures, often not maintained by a competent authority (Table 10-9).

The objective of the statistical analysis is to estimate the characteristics of some target population of dams that are the subject of the risk analysis. In contrast to the sampled

Table 10-9. Dam failure and flash flood cases (Brown and Graham, 1988)

Location	Year	Population at risk	Loss of life	Hours warning
Baldwin Hills, CA	1963	16,500	5	1.5
Bearwallow, NC	1976	4	4	0
Big Thompson, CO	1976	2,500	139	<1.0
Black Hills, SD	1972	17,000	245	<1.0
Buffalo Creek, WV	1972	4,000	125	<1.0
Bushy Hill Pond, CT	1982	400	0	2.3
Denver, CO	1965	3,000	1	3
DMAD, UT	1983	500	1	1.12
Kansas City, KS	1977	1,000	25	<1.0
Kansas River, KS	1951	58,000	11	> 1.5
Kelly Barnes, GA	1977	250	39	<0.5
Laurel Run, PA	1977	150	40	0
Lawn Lake, CO	1982	5,000	3	<1.5
Lee Lake, MA	1968	80	2	<1.0
Little Deer Creek, UT	1963	50	1	<1.0
Malpasset, France	1959	6,000	421	0
Mohegan Park, CT	1963	500	6	0
Swift and Two Medicine Dams, MT	1964	250	27	<1.5
Northern NJ	1984	25,000	2	>2
Prospect Dam, CO	1980	100	0	>5
Teton, ID (dam through Wilford)	1976	2,000	7	<1.5
Teton, ID (Rexburg to American Fans)	1976	23,000	4	> 1.5
Texas Hill Country	1978	1500	25	<1.5
Vega DeTera, Spain	1959	500	150	0

population, these are likely to be modern, well-engineered dams, operated by a competent authority. In principle, we can estimate the characteristics of the sampled population from characteristics of the observations, and the reliability of these estimates quantified with statistics. On the other hand, the characteristics of the target population can only be estimated from the characteristics of the sampled population by judgement, and the reliability of these estimates can only be determined subjectively. This represents a conceptual limit to which judgement must be brought to bear in interpreting the results of the statistical analysis.

The reliability of statistical estimates depends on, among other things, the number and variability of the data from which the estimates are made. For example, the variation of the mean of a sample across repetitive sampling has standard deviation approximately equal to the standard deviation of the sampled population divided by the square root of the number of observations ($\sigma_{\bar{x}} \approx \sigma_x / \sqrt{n}$, in which $\bar{x}$ is the sample mean, σ_x is the standard deviation of the sampled population of x, and n is the sample size). Thus, if the variability of the sampled population is large and the number of observations is small, the variability of estimates made from sample statistics is high. Correspondingly, the confidence bounds (in a frequentist sense) or the probability distribution (in a belief sense) are broad. This is certainly the case with estimates of loss of life from dam failure, as noted by everyone who has studied the data.

Brown and Graham (1988)

Brown and Graham (1988) use the data of Table 10-9, a subset of 24 dam failures in the US and Europe between 1950 and 1988 resulting in one or more immediate fatalities and meeting certain other criteria, as the basis for identifying a statistical model of life loss. The dams within this data set are for the most part low (<15 m), small drainage area, earthfill embankments. None failed due to earthquake ground shaking. There simply is not a large number of failures of high, modern dams to include in the database.

Brown and Graham developed a simple summary of the empirical data, noting the critical importance of warning time to loss of life (Figure 10-9). This led to a set of three equations relating loss of life to population at risk (PAR), where PAR is the total number of people occupying the flooded area. Note, this is not the number of people "who get wet", and as a result may be much larger than the number of people truly placed at risk.

The three equations apply to different warning times: less than 15 minutes, between 15 and 90 minutes, and more than 90 minutes. These equations were subsequently incorporated in the USBR's, 1989, *Policy and Procedures for Dam Safety Modification Decision Making*.

DeKay and McClelland (1993)

DeKay and McClelland (1993) expanded on Brown and Graham, using the same set of event data, but expanding the number of parameters for each data record to include hours of warning treated as a continuous variable (whereas Brown and Graham binned warning time into categories), and flooding forcefulness treated as a dichotomous variable (high force wherein greater than 20% of the residences are destroyed or heavily damaged, vs low force wherein fewer that 20% are destroyed or damaged). The original purpose of the study was to set hazard classes for dams rather than to estimate numerical consequences for risk analysis, thus subsequent uses of the approach for risk analysis are outside the scope originally envisioned for the study.

In contrast to the earlier study that used direct curve fitting to obtain results, McKay and McClelland fit the data by logistic regression. This approach regressed the log-odds ratio $ln(LOL/PAR)$ (the so-called logit) against warning time, WT, for each category of

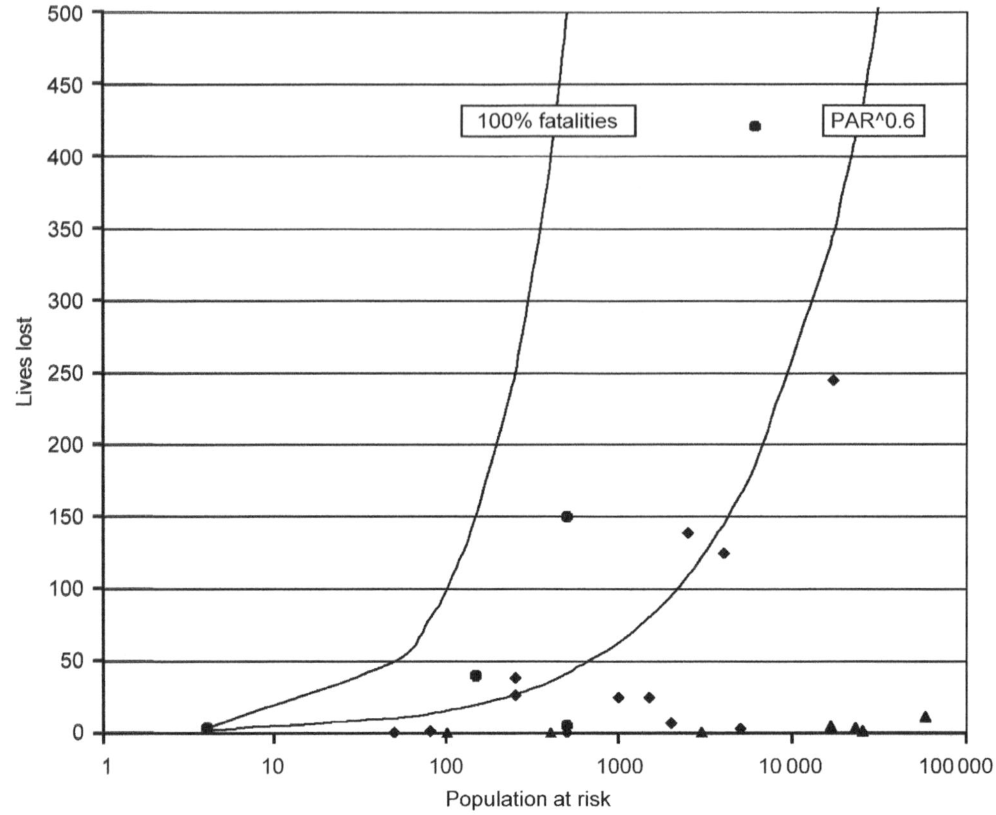

Figure 10-9. Effects of warning on fatalities after dam break showing only data of Table 10-9 (after format of Von Thun, 1984). Circles = less than $\frac{1}{4}$ hour warning time; diamonds = $\frac{1}{4}$ to $1\frac{1}{2}$ hours warming time; triangles = greater than $1\frac{1}{2}$ hours warning time

flood force. This is a mathematical approach similar to that of Lee *et al.* (1986) in predicting loss of life from floods, and of Liao *et al.* (1988) in analysing liquefaction data. This resulted in two statistically estimated predictive formulae, for high-force and low-force situations, respectively, showing the characteristic form of logistic regression expressions.

The authors caution that the predictive equations do not apply to situations wherein dams fail without warning above high populations at risk (e.g. the Vaiont Dam failure, the most severe in the database, was excluded from the analysis). As with Brown and Graham, the authors note that warning time is a critical variable in reducing fatalities. It should be noted, however, that the study uses average warning time across whole basins, and not warning time for specific settlements at risk. Also, the data are averaged across season of the year, time of day, and other conditioning variables that are often separated in event trees. Thus, the numerical results do not apply to these conditional events.

The conclusions of the study, beyond the statistical predictive equations, were that: (1) loss of life is a non-linear function of the number of people at risk, and (2) loss of life is greater in cases where the floodwaters have high force (i.e. deep and swift conditions). Other conclusions of empirical studies of fatalities in dam failure incidents and flash floods, as summarised in BC Hydro's Keenleyside Dam risk analysis (BC Hydro, 1997), are that over half of all fatalities are associated with automobiles getting caught in the floodwaters, and that a disproportionate number of victims are children and the elderly.

Graham (1999)

Graham (1999) provides a reinterpretation and expanded analysis of the data considered in the two earlier studies. The report notes that the data set is deficient in that it does not include important types of events and warning scenarios, most of the dams in the data set are small, severe flooding events are under-represented, and none failed due to earthquake and so with little warning. Thus, the conclusions may not be applicable across dam sizes, types, failure causes, flood severity or warning scenarios. The report concludes that the earlier predictive equations are deficient for high lethality events following "truly catastrophic flooding" similar to St Francis or Vaiont.

The report also observes that the declining fraction of fatalities with increasing PAR observed in the database may be due to the general trend for larger PARs to be spatially more spread out than small PARs, and thus less vulnerably exposed to severe flooding or the sort that might occur within the confines of a narrow canyon. The report, therefore, questions the low LOL/PAR ratios predicted by the earlier studies for high PAR.

The recommendation of the study is that the first five steps of the statistical procedure continue to be used, but that the final step of empirically relating LOL to PAR be modified. The proposed method determines the ratio LOL/PAR based on flood severity, amount of warning time, and a "measure of whether people understand the severity of the flooding". The data set is expanded beyond the 24 cases of Table 10-9 to include approximately 40 records, most, but not all, from dam failures. Other, non-dam-failure, flood events were added to the data set to more fully cover ranges of flood severity, warning and flood severity understanding (Table 10-8). Verbal descriptions are used to categorise the three independent variables in the forecast. Fatality rates (LOL/PAR) are estimated by inspection for each resulting category, and presented in a summary table in the original report.

Physically-based modelling of public safety

In contrast, the statistical approach, potential loss of life can also be modelled analytically using physically-based models. A relatively small amount of work has been done on this approach of directly modelling the dynamics of inundation, evacuation and human factors in predicting fatalities. These physically-based models complement the statistical approach, which is hampered by the limited numbers of large dam failures on which to base empirical estimates.

Loss of life in physically-based models is again determined, as suggested in Figure 10-8, by two considerations: demographic and land-use characteristics of the inundated area, and hydraulic characteristics of the flood event. The first set of factors influence the number of people caught in a flood; the second influence the fatality rate of people caught. The population at risk and the fatality rate combine to estimate the loss of life.

As with statistical methods, physically-based models are not without difficulties and limitations. To begin, one can never know whether all important factors are included in a model, or even the most important chains of events. Also, physically-based modelling can be expensive. Nonetheless, such modelling promises to go beyond just the raw statistics of historical failures to attempt to understand how fatalities occur and to provide an explicit and transparent accounting.

Demographic and land-use information

The demographic model is primarily a geographic database of the valley downstream of the dam (Figure 10-10), showing the distribution of land uses, location of structures and critical infrastructure, and populations of residents or other people who could be at risk in the event of a flood. The geographic database may also contain downstream

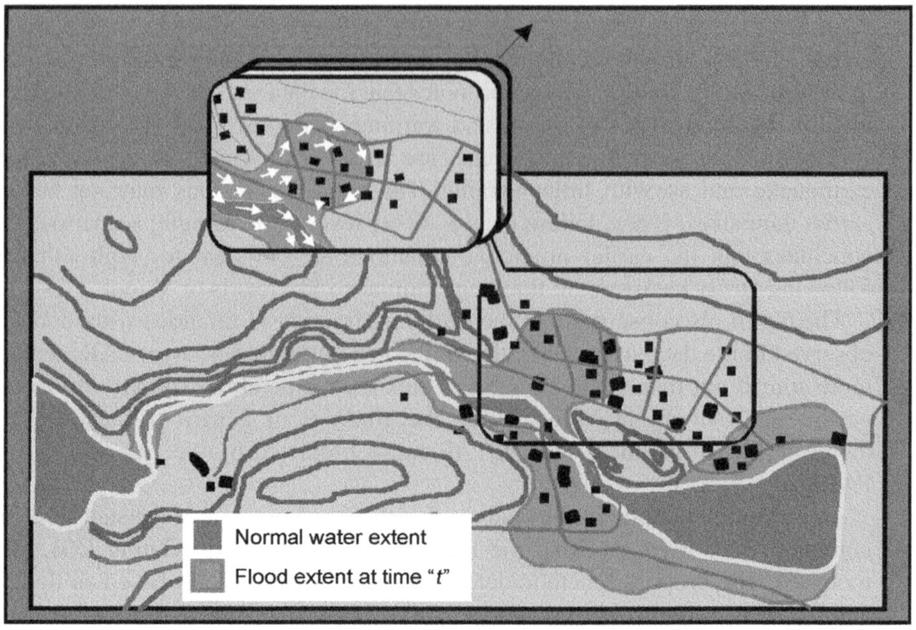

Figure 10-10. World model of a dam breach flood (Hartford, 2002)

topographic information to serve as input information from which hydraulic characteristics such as inundation areas, heights and velocities of flow are calculated.

Because the downstream region is a dynamic entity, the number of people potentially exposed to a flood, even ignoring warning systems and evacuation processes, depends on a number of dynamic factors, such as the time of day and season of the year. These numbers all have uncertainties associated with those in Figure 10-11. These factors can be combined in an event tree representation along with the many other factors that control exposure, such as warning times and evacuation policies, to generate a probabilistic description of the potential number of people at risk in the flooded areas

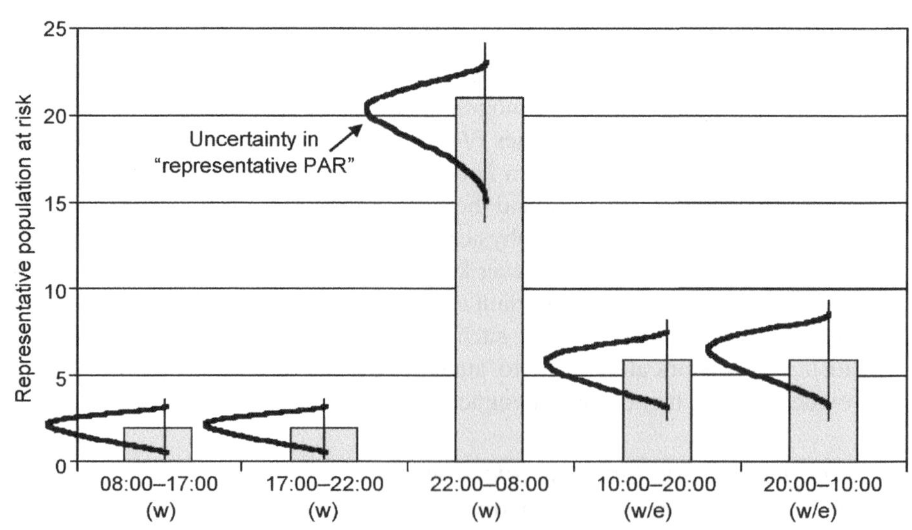

Figure 10-11. Uncertainty in population at risk (Hartford, 2002)

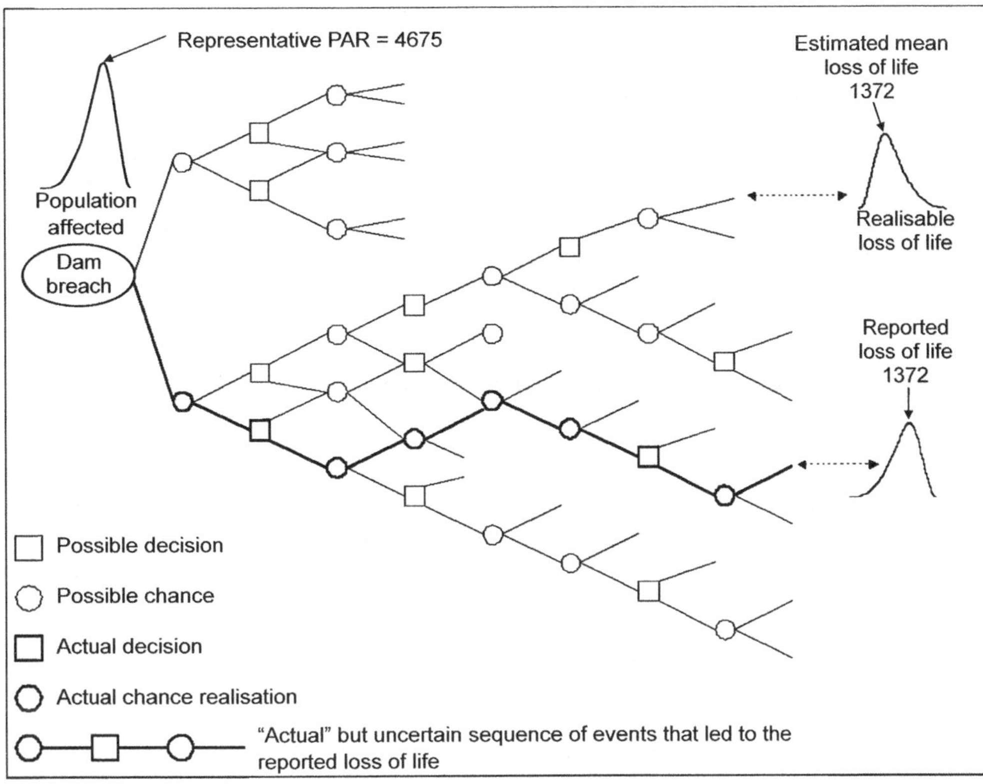

Figure 10-12. Event tree representation of the multiplicity of loss of life outcomes using a physically-based model (Hartford and Stewart, 2002)

(Figure 10-12). Depending on the extent of existing information, such probabilistic characterisations of the downstream population at risk may also include the spatial distribution of that population.

Fatality rate model

The hydraulic characteristics of the flood event (peak discharge, velocity, depth and the spatial distribution of those factors) influence mortality, the chance that a person caught in the flood becomes a casualty. The principal factors in most such models of mortality are warning time (permitting evacuation), water depth and water velocity.

Jonkman *et al.* (2002) report two statistical relationships between water depth and mortality from Waarts for coastal flooding. These are derived from data for the 1953 flood that caused 1800 flood deaths in the Netherlands, and thus the generality of the relations may be questioned. The current method for predicting loss of life in coastal floods in the Netherlands is an outgrowth of these relationships. An extension of the model considers mortality from three causes: collapse of buildings near a breached dyke, collapse of buildings due to wave action and other drowning.

A second approach to mortality estimation is based on predictions of the water height and flow velocity needed to sweep people away, thereby causing them to become casualties. Apt *et al.* (1992) conducted flume tests from which to predict critical water heights and velocities. Based on these, Jonkman *et al.* have proposed a probabilistic relationship based on flow velocity, describing the probability of mortality in a cumulative–Normal probability distribution with a mean value correlated to water velocity. Apt *et al.* report a predictive relationship based on depth and velocity, of the exponential form, correlated to water depth and velocity, the weight of the person, and the height of the person. Lind and Hartford (2000) have expanded on this relationship by the use of mechanical and physical models representing the human form in flowing water.

Except for dams having large population at risk, the main question is often, will there be (any) loss of life or not? The modelling is needed to help answer this question and not necessarily to try to determine a specific loss-of-life value. The decision following a risk analysis may be the same if the expected loss of life is 12 instead of 10. However, after answering this question, the modelling is also useful as a sensitivity analysis to determine the main parameters affecting loss-of-life values. A second benefit of modelling is, therefore, to help implement emergency preparedness plans.

10.5 Economic modelling and assessment

Estimating economic losses due to flooding is a simpler problem than estimating fatalities, and one for which experience exists in both benefit cost analysis and flood damage estimation. On the other hand, many economic losses caused by dam failures are intangible or difficult to quantify, including the disruption of daily life, lost time from work, and the loss of personal belongings having sentimental value. In sum, these intangibles may be of comparative value to the more easily quantified loss of physical assets, such as buildings, equipment and infrastructure.

One usually distinguishes between the *direct loss* of assets and the *consequent costs* caused by those losses. Physical loss of assets means the destruction of physical property and goods, or the rendering of those properties and goods economically valueless. Physical loss is the loss of asset value; consequent cost is the loss of income or production of those assets. A second distinction is between reimbursed losses and unreimbursed losses. Reimbursed losses, or costs, are those losses paid for by the dam owner, government or insurance. Unreimbursed losses are the uncompensated impacts on victims. These may include loss of assets that are not insured or not compensated for by the dam owner, deductible amounts on insurance payouts, and may include undocumented losses.

Categories of economic costs

The principal categories of economic loss due to dam failure are shown in Table 10-10. Direct losses are principally physical damage to property by floodwaters. They are measured by the cost of restoration or replacement. They also include agricultural losses measured by the change in net revenue to farmers. The principal categories of direct loss include: residential loss, commercial loss, public property loss and agricultural loss (Eckstein, 1958). Indirect losses are the net economic losses of goods and services to society due to interruptions of industry, commerce, traffic, communications and other activities, both within and outside the area subject to inundation. These also include the

Table 10-10. Categories of economic loss due to dam failure (NRC, 2000)

Category of direct loss
• Physical property: including structures, contents, infrastructure and utilities, vehicles • Agricultural products: crops and livestock • Economic compensation for injury or death: injuries and death, life insurance payouts, medical expenses • Response and cleanup: emergency services, remediation, temporary housing • Loss of the services of the dam itself, including: power generation, flood control, navigation, recreation, water supply and irrigation

cost of emergency measures, relief, care and rehabilitation of victims. Among other things, indirect losses include: loss of goods and services in the area caused by cessation of production, loss of wages and other incomes (sometimes used as a surrogate for the value of lost production), loss of stock due to spoilage, increased cost of business operations, including increased transport costs, and costs of evacuation, reoccupation, temporary living, emergency work and relief of victims.

The owners of these assets include individuals, businesses and government at various levels. In an actual disaster, the accumulation of loss data may take months to account for. Victims may not know the extent of their losses for weeks or months. Initial estimates may be off by 100% or more compared with later estimates. In natural hazard assessments, such as those following hurricanes or earthquakes, initial estimates can be a fraction of later and more accurate accounts. This suggests that the uncertainties and potential for error in before-the-fact damage forecasts can be large. The NRC (1999b) report proposes a standard accounting matrix for aggregating estimates of direct economic damage from natural hazards, and this matrix is a good starting point in making a forecast of potential economic costs of dam failures (Table 10-11).

Lessons from benefit-cost analysis

Benefit-cost analysis (BCA) is well established in both theory and practice. Applications to water resource planning, especially in the post World War II period, were formative to the development of BCA and influential in the evolution of major water resources and development agencies around the world, including Canadian and US agencies, the UN and World Bank (Dasgupta *et al.*, 1972; Eckstein, 1958; Maass, 1962). Benefit-cost analysis has become the dominant methodology in water and energy resources planning programmes.

Benefit-cost analysis accounts for consequences in monetary terms. Determining the monetary value of a consequence begins with the premise that people are the best judges of the value of a consequence – or a good or service – to themselves. Individuals take their personal resources and decide how best to spend them and, in doing so, they make trade-offs. The more something is valued, the more a person is willing to give up for it. From this idea comes the concept of willingness-to-pay (WTP) as a measure of value. This is a measure, in money, of what a person would be willing to give up in exchange for some good or service, or to avoid some negative outcome. WTP is the most that a person would be willing to pay, whether or not he or she actually has to pay it.

Several characteristics of WTP as a measure of value are worth emphasising: (1) it is based entirely on people's preferences; (2) it is not limited to marketed goods; and (3) while it is denominated in money, the concept itself has little to do with money. To the extent people care about things outside of markets (e.g. environmental or social impacts), they will be willing to give up other goods to attain them. Thus, within BCA, valuing environmental goods is consistent with valuing goods traded in regular markets. The key idea is that value can be measured by how much of something else a person is willing to do without.

When assessing the consequences of a dam failure, values must be aggregated across individuals. The use of a single metric, such as money, makes the summation easy. By summing across people to generate a single consequence "cost", all dollars are treated equally regardless of the source or who is impacted. This equal weighting may not be consistent with social objectives. For example, low income or minority groups might bear most of the costs, more expensive property is valued higher than less expensive, and the costs may be geographically concentrated.

The cost of failure consequences usually contain more than financial costs. The full opportunity cost is the measure used in most BCA (Mishan, 1995). "The opportunity cost of a project represents what society overall gives up to obtain something, including both the direct financial costs and the indirect costs, generally described as the benefits

Table 10.11. Sample data sheet on direct impacts from a major natural disaster event (adapted from NRC, 1999b)

Type of loss			Who bears the cost?				
			Insurers	**Govern-ment**	**Busi-ness**	**Indi-viduals**	**NGO**
Property	Government	Structures					
		Contents					
	Business	Structures					
		Contents					
	Residential	Structures					
		Contents					
		Landscapes					
Vehicles	Autos						
	Boats						
	Planes						
Infrastructure	Utilities						
	Transportation						
	Other						
Agricultural products	Crops						
	Livestock						
	Other						
Human losses	Deaths						
	Injuries						
	Illnesses						
Cleanup and response costs							
Adjustment costs, temporary living, aid							

Note.
NGO – non-government organisation

foregone, associated with the failure." For example, the full opportunity cost of a failure might include reconstruction costs, administrative costs, the value of lost environmental benefits that the dam provided, and the economic value of discomfort or inconvenience suffered by anyone as a consequence.

Discharge-damage functions

Economic flood damage estimates today tend to be done by full enumeration of the structures at risk downstream. This has been greatly facilitated by the advent in recent years of GISs and spatial databases. Economic assessments can be relatively complex, involving the integration of information at several spatial scales. For example, in many projects three main spatial scales are the project scale in which all economic analysis is summarised, the damage-reach scale used for most hydraulic analyses and the structure scale where assessments of damage to individual structures are made. At the project scale the calculation is typically divided into many damage reaches, each containing many structures. Within each damage reach there may be many hydraulic cross sections for determining water surface profile.

Al-Futaisi and Stedinger (1999) have proposed an approximate analytical damage function, reasoning on the basis of Manning's equation. Approximating the flood channel as a regular shape, presuming that damages start at a threshold water height, and presuming that above the threshold the damages accrue quadratically with water height, then the stage-damage function can be approximated analytically.

Perhaps the most detailed methodology for developing stage-damage functions for riverine floods is that developed by the US Army Corps of Engineers for flood damage reduction studies (USACE, 1996, 1998). This methodology is built around a (flood) stage-damage function which is a summary statement of the direct economic loss due to flood inundation within reach of a river as a function of a rising stage. The methodology uses a full enumeration survey of structures within the potentially inundated area (Figure 10-13). The standard procedure for conducting these surveys is outlined in Table 10-12.

The principal pieces of information in translating survey results into damage estimates are: (1) first-floor elevation of the individual structures; (2) structure value; and (3) content value or content-to-structure value ratio. Estimation of first-floor elevation, which vertically locates a structure with respect to flood height, may be based on existing topographical maps, aerial or satellite imagery, or on-site field inspection. Each is associated with a different level of probable error, and can be modelled using standard least-square error analysis techniques from land surveying practice.

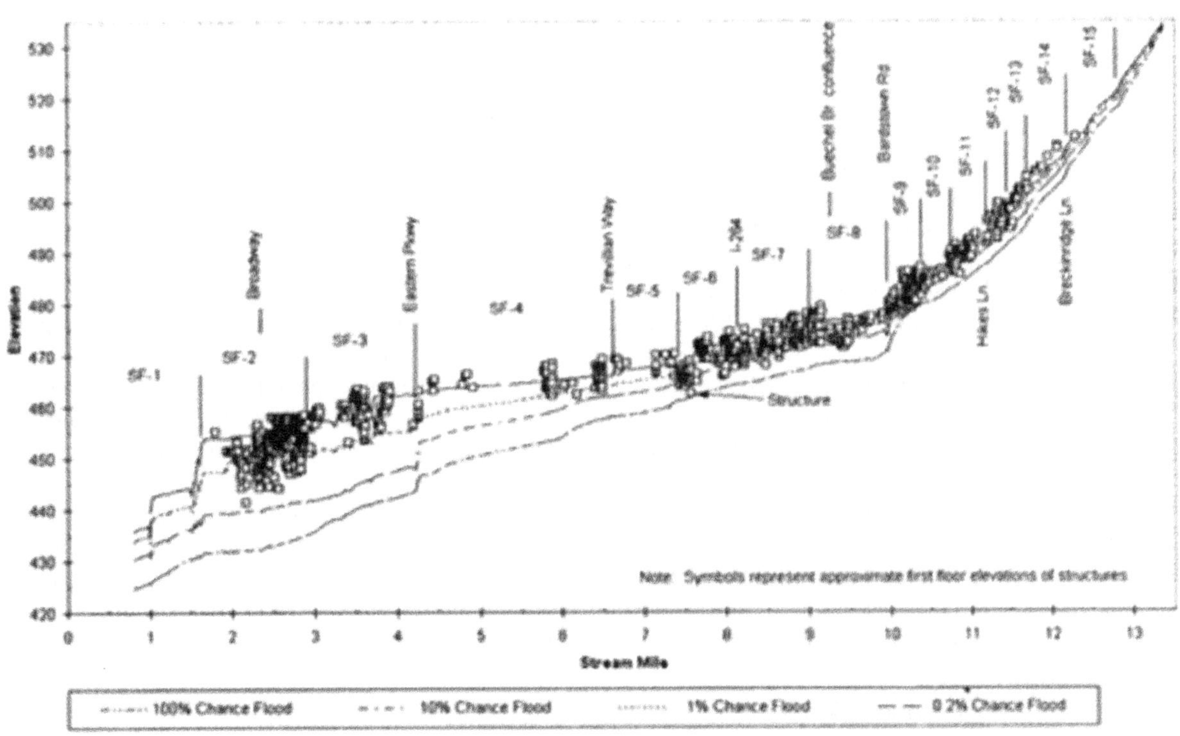

Figure 10-13. Locations of structures on floodwater surface profiles along the damage reaches of the South Fork of Beargrass Creek (reproduced from USACE, 1997)

Table 10-12. Traditional USACE procedure for development of stage–damage function (USACE, 1996)

Step	Task
1.	Identify and categorise each structure in the study area based upon its use and construction
2.	Establish the first-floor elevation of each structure using topographic maps, aerial photographs, surveys and/or hand levels
3.	Estimate the value of each structure using real estate appraisals, recent sales prices, property tax assessments, replacement cost estimates or surveys
4.	Estimate the value of the contents of each structure using an estimate of the ratio of contents value to structure value for each unique structure category
5.	Estimate damage to each structure due to flooding to various water depths at the structure's site using a depth-percent damage function for the structure's category along with the value from Step 3
6.	Estimate damage to the contents of each structure due to flooding to various water depths using a depth-percent damage function for contents for the structure category along with the value from Step 4
7.	Transform each structure's depth-damage function to a stage-damage function at an index location for the floodplain using computed water-surface profiles for reference floods
8.	Aggregate the estimated damages for all structures by category for common stages

The sources of uncertainty in the resulting stage–damage function arising out of survey and measurement issues, and for the purposes of flood risk analysis, are outlined in Table 10-13. In addition to these survey issues, a number of hydraulic, construction and warning time factors also affect the uncertainty in damage estimates (Table 10-14). These are highly pertinent to estimates of damage associated with dam failures, as the resulting floodwaters may rush down the flood plain with velocities and rates of water rise much faster than for normal floods.

Given these large data sets and the very large number of individual structures, each with uncertain elevation, uncertain value of contents, uncertain value of the structure itself, among other certainties, Monte Carlo simulation is often used to determine results. The results of Monte Carlo simulations may be aggregated by damage category, for example by building or use type, to form a damage-water stage function with uncertainty for each category of damage. The category damage functions are aggregated to form a damage-stage function for each reach. As can be seen, this process generates a large number of simulated realisations. Once all the realisations are obtained they are statistically summarised in probability curves.

This modelling process introduces a number of interesting theoretical issues that could be discussed here. Among these are the sampling statistics of gathering data on the flood plain, correlations among modelling uncertainties at different levels, spatial and temporal correlations introduced by the nested scales of the analysis, and how the uncertainties in the predicted economic losses are quantified.

Hazus flood risk methodology

The US Federal Emergency Management Agency (FEMA) has developed a multi-hazard loss estimation methodology for natural hazards under the name, HAZUS. One part of this methodology addresses flood losses. Early versions of the flood

Table 10-13. Components and sources of uncertainty in stage-damage function (USACE, 1996)

Parameter/model	Source of uncertainty
Number of structures in each category	Errors in identifying structures; errors in classifying structures
First-floor elevation of structure	Survey errors; inaccuracies in topographic maps; errors in interpolation of contour lines
Depreciated replacement value of structure	Errors in real estate appraisal; errors in estimation of replacement cost estimation–effective age; errors in estimation of depreciation; errors in estimation of market value
Structure depth-damage function	Errors in post-flood damage survey; failure to account for other critical factors: floodwater velocity; duration of flood; sediment load; building material; internal construction; condition; flood warning
Depreciated replacement value of contents	Errors in content-inventory survey; errors in estimates of ratio of content to structure value
Content depth-damage function	Errors in post-flood damage survey; failure to account for other critical factors: floodwater velocity; duration of flood; sediment load; content location; flood warning

Table 10-14. Factors other than depth that influence damage (USACE, 1996).

Factor	Effect
Velocity	Major factor aggravating structure and content damage. Limits time for emergency flood proofing and evacuation. Additional force creates greater danger of foundation collapse and forceful destruction of contents
Duration	May be the most significant factor in the destruction of building fabric. Continued saturation will cause wood to warp and rot, tile to buckle, and metal objects and mechanical equipment to rust
Sediment	Can be particularly damaging to the workings of mechanical equipment and can create cleanup problems
Frequency	Repeated saturation can have a cumulative effect on the deterioration of building fabric and the working of mechanical equipment
Building material	Steel frame and brick buildings tend to be more durable in withstanding inundation and less susceptible to collapse than other material
Inside construction	Styrofoam and similar types of insulation are less susceptible to damage than fibreglass and wool fibre insulation. Most drywall and any plaster will crumble under prolonged inundation. Waterproof drywall will hold up for long periods of inundation. Panelling may be salvageable when other wall coverings are not
Condition	Even the best building materials can collapse under stress if the construction is poor or is in deteriorated condition
Age	May serve as an indicator of condition and building material
Content location	Important factor, as small variations in interior location of contents can result in wide variation in damage
Flood warning	Major reduction in both content and structural loss can be made through flood fighting and evacuation activities when there is adequate warning

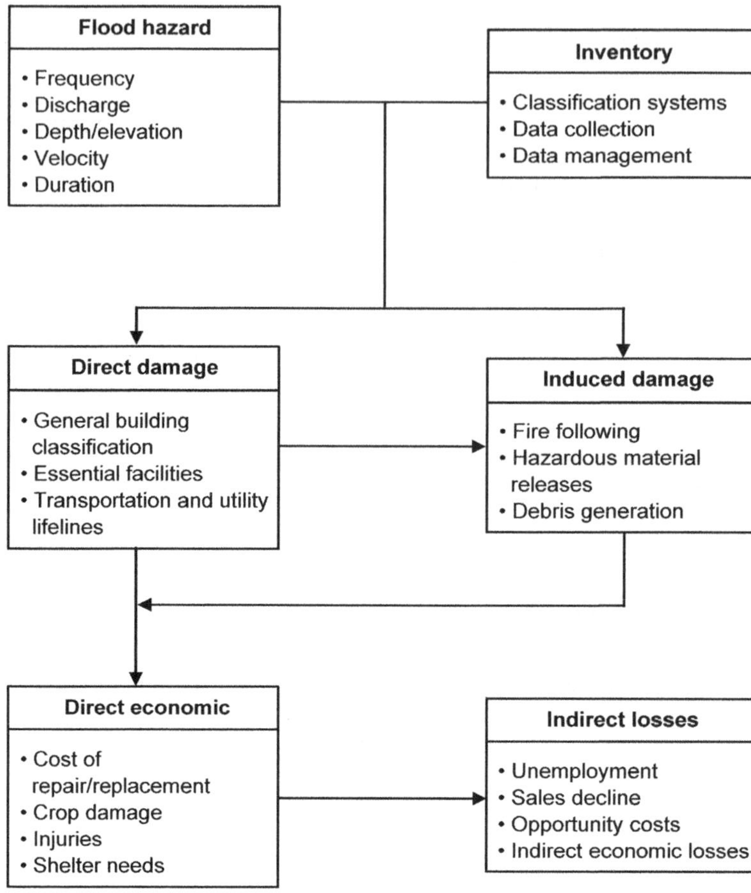

Figure 10-14. Elements flood loss estimation model (http://www.nibs.org/hazusweb/methodology/flood.php)

methodology are available as of early 2002, with a final delivery date of early 2003. The model addresses both river and coastal flooding (Figure 10-14).

Based on national and local data populating a GIS database, the methodology intends to estimate "potential damages to all classes of buildings, essential facilities, transportation and utility lifelines, and agricultural areas. The model estimates debris, shelter and casualties. Direct losses are estimated based on physical damage to structure, contents, and building interiors." The methodology will also incorporate the effects of flood warning and velocity (NIBS, 2003). The methodology is also planned to enable users to estimate the effects of flooding on populations.

10.6 Environmental modelling and assessment

Many significant non-market consequences accrue from dam failures, among the more important of these is degradation of the downstream environment. Some environmental consequences of natural flooding – as opposed to dam failure flooding – can actually be benefits rather than costs, in that riparian areas have evolved with floods and may be rejuvenated by flooding. This is less so the case with dam failure floods, since they are more extreme than normal flooding, and rare in the evolutionary record (if not wholly unknown in geological history). In the long term, downstream environments equilibrate to post-flood conditions, establishing a new base level. For this reason, the focus of attention in assessing environmental consequences of dam failures is on immediate or short-term consequences.

While a great number of literature references can be found on the environmental consequences of dam construction, operation, decommissioning and removal, few if any published accounts can be found of ecological risk assessment for evaluating the damages due to dam failure. As a result, this section draws heavily from the literature on assessing the consequences of natural hazards such as floods and hurricanes, on the literature of ecological risk assessment pertinent to regulatory action (especially the US Environmental Protection Agency's *Framework for Ecological Risk Assessment*, 1998), and on the limited number of reports that have been produced on dam failures, in particular the environmental risk assessment for the BC Hydro Keenleyside Dam. Canadian guidelines on environmental risk assessment found in *ERA of Priority Substances Under the Canadian Environmental Protection Act Draft 2.0* (Environment Canada, 1996) are restricted to chemical hazards.

An NRC report on losses from natural disasters (NRC, 1999b) discusses principles that apply to the assessment of costs and benefits of extreme events:

- Costs and benefits of consequences to the natural environmental and ecosystems are considerably less tangible than direct economic effects, and difficult or even impossible to quantify with precision. Even when physical consequences can be measured, commensurate monetary values are difficult to assign. In considering and forecasting the consequence of dam failure floods on the environment, it is useful to distinguish between the natural environment and the man-made landscape.
- Ecological systems have evolved over time in response to extreme natural events like floods, droughts and fires. This process has occurred over thousands of years. Such extreme geophysical events may provide long-term benefits to natural ecosystems (Sparks, 1996). The critical factors are frequency, intensity and spatial extent. While dam failure floods are more extreme than natural floods, their impact on the environment is complex.
- Extreme events, like dam failure floods, can produce different and mixed types of consequences in different parts of the floodplain.
- The costs and benefits to the natural environment may take years or decades to play out. For example, agricultural chemicals washed off the land during planning season may flow to lakes or reservoirs far downstream, with corresponding contamination effects on the ecosystems of those lakes or reservoirs in the future.
- Long-term environmental outcomes can be subtle and nearly impossible to foretell. The 1993 Upper Mississippi floods washed the zebra mussel, an environmental pest, from the Upper Illinois River downstream into the Mississippi, allowing it to establish colonies in upstream tributaries backed up by the flooding.

Modelling environmental consequences

The downstream environment of any dam is highly textured. In modelling environmental consequences only a limited number of attributes of this system can be analysed, and only a limited number of indicator attributes can be formed into the accounting scheme. The question is, which of these attributes are critical for evaluating consequences? The choice of what to measure is critical for environmental consequence analysis, and the challenge is different than in the case of public safety or economic consequences. Because the downstream environmental system is inherently complex, much of the effort in modelling environmental consequences of dam failures lies in structuring the problem and deciding upon how to simplify and quantify potential impacts so that they can be dealt with operationally.

The approach to modelling environmental consequences of dam failure can be built upon the USEPA framework, which can be reformulated in the following way, building upon Sherbot's (1997) interpretation:

- Identify potential liabilities for the environmental risk assessment.
- Summarise dam failure scenarios.
- Define assessment metrics ("endpoints", below).
- Develop a conceptual model.
- Identify metrics of principal concern and assess associated consequences.
- Assign monetary values to damages and liabilities.
- Incorporate these values and liabilities in overall dam safety risk analysis.

Assessment endpoints

In presenting guidelines for ecological risk assessments, the US Environmental Protection Agency (USEPA, 1998) suggests the following considerations:

- What are the environmental values of concern?
- What are the critical ecological (or assessment) endpoints?
- How likely is recovery and how long will it take?
- What is the state of knowledge of the issues surrounding these endpoints?
- What data and data analysis are reasonably available?

In analysing the public safety and economic consequences of a dam failure, the objectives that drive the selection of measurable attributes of impact are reasonably straightforward: to minimise injuries and the loss of life, and to minimise direct financial cost. In analysing environmental consequences, on the other hand, the objectives that drive the selection of measurable attributes are less well-defined. These objectives are statements about the desired condition of the environmental values of concern, for example, "ensure a sustainable riverine community of flora and fauna", or "minimise the loss of riparian wetland", or "minimise hazardous or toxic contamination downstream of the dam". The choice of these objectives dictates the choice of measurable attributes.

Environmental objectives may come from law, regulatory interpretations of or guidance about the law, the desires of locally affected parties, organisational principles, or other sources. More so than with public safety and economic consequence evaluations, the identification of environmental values and objectives may rely on public and local community participation. As a practical matter, the number of environmental objectives for consequence evaluation should be kept to a minimum. One reason is that at least one attribute must be chosen to measure the consequences of dam failure against each objective. As the number of objectives increases, so do the number of attributes whose values must be modelled and the number of cross comparisons of preferences among attributes. As a result, the consequence-attribute hierarchy becomes ever more "bushy".

Specific environmental objectives and their associated measurable attributes are characterised as assessment endpoints (sometimes called ecological endpoints in the ecological risk literature). Assessment endpoints bring together two things: (1) a specific environmental or ecological sub-category that is considered valuable and which may be adversely affected by dam failure; and (2) an attribute (i.e. measurable characteristic) of that entity that is at risk and potentially important with respect to the entity. For example, the specific valued entity might be a species such as salmon, a group of species such as anadromous fish, a community such as benthic invertebrates, an ecosystem such as a downstream lake, a valued habitat such as riparian wetlands, a protected place such as a listed endangered area, or something else. The measurable characteristic is a quality of potential impact that is important to the entity at risk. For example, what is important for the impact on downstream salmon populations may be population size, and thus the characteristic could become number of fish killed. What is important for impact on a downstream lake may be contamination by toxic sediments, and thus the characteristic could become tons of sediment deposited in the lake. Taken together, the entity and measurable characteristic are an environmental attribute upon which the consequence evaluation is based.

Assessment endpoints are distinguished from environmental objectives in that they are specific and value-neutral. They pertain to identifiable species or conditions, and do not contain words such as, minimise, protect or prohibit. As with the consequence-attribute hierarchies discussed above – of which environmental objectives and attributes are a part – individual higher-level environmental objectives may be too broad to be operationally useful, and may need to be broken down into smaller sub-objectives. In this case, multiple endpoints may ultimately be needed to represent the consequences against one higher-level objective, and in the ecological risk analysis literature these are sometimes called *measurement endpoints*. Nonetheless, determining what measures to use to categorise environmental consequences is both challenging and important to the success of the endeavour.

As with all attributes, but perhaps more importantly for environmental consequences, it is desirable for practical reasons that ecological endpoints have well-established test methods, measurement protocols and predictive models. Similarly, for some endpoints, direct measures of impact are available (number of fish killed, or wetland area destroyed); while for others, direct measures may not be available, so proxy or subjective measures may be needed (recovery potential, or ecological integrity).

Conceptual models

A conceptual model is often a combination of written description and graphical representation to summarise the impact of dam failure on downstream entities and measures. The written description summarises a set of risk hypotheses relating direct impacts of downstream flooding due to dam failure to endpoint responses. An accompanying graphical representation diagrams these relationships. Conceptual models may represent many relationships that qualitatively link dam failure flooding to ecological processes and then to attributes. They may represent both primary and secondary processes, and multiple conceptual models may be developed to represent different types of downstream impact. The process of developing conceptual models of environmental impact is similar to the process of developing event trees for dam performance. The complexity of the conceptual model depends on the complexity of the issue being analysed, and conceptual models can become very complicated. Yet, the conceptual model also serves as an effective communication tool for describing potential consequences. Early in the process of modelling environmental consequences, the conceptual models may be broad. As more information is collected, the reasonableness of the hypotheses becomes clearer, and the analysis narrows.

The graphical representation of a conceptual model is in essence an influence diagram, as discussed in detail in Chapter 8. These influence diagrams describe the flow of influence or relationship from one outcome to another (e.g. Figure 10-15). Although there is no standard format for influence diagrams, especially in environmental applications, their usefulness is enhanced by the extent to which the influence relationships within the diagrams are explicated by detailed written descriptions for the relationships portrayed. Factors to be considered when building conceptual models are the complexity (number) of linkages included, the completeness of the influencing factors in determining consequences, the uncertainty in a linkage, and the prospect for obtaining field data from which to make quantitative estimates.

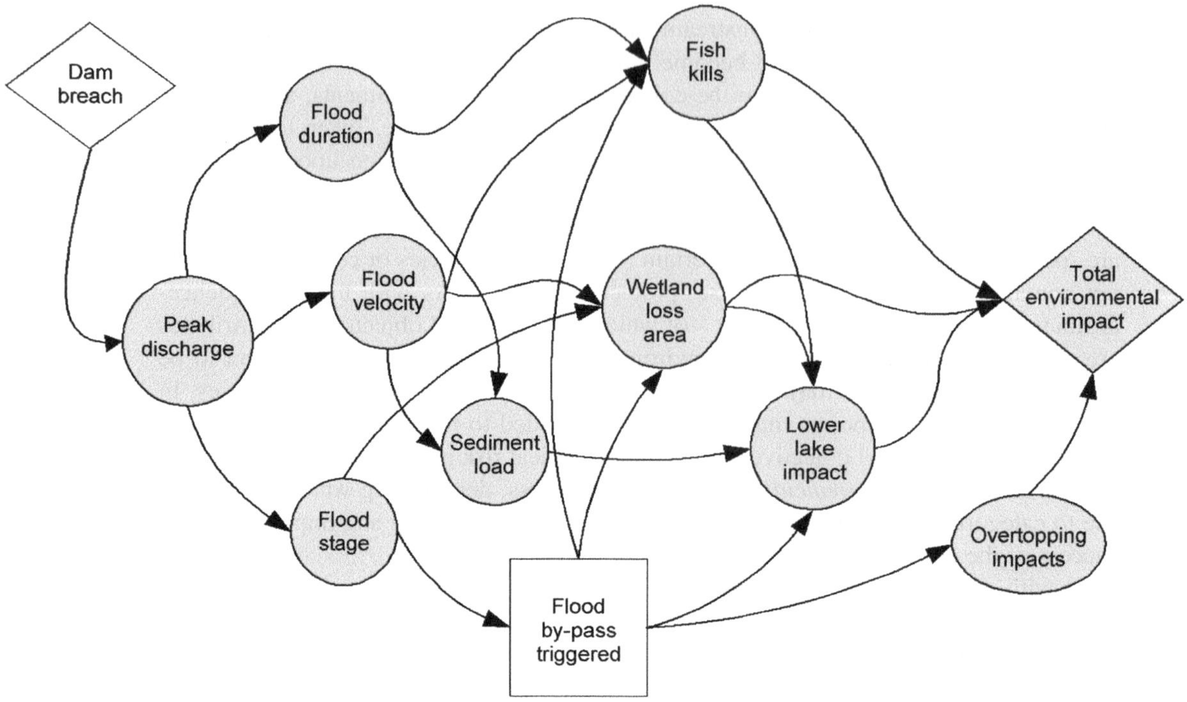

Figure 10-15. Example of an influence diagram for the conceptual model of environmental impact

Uncertainty enters the conceptual model from a lack of knowledge about how ecosystems function and the potential exclusion of secondary effects. In contrast to public safety and economic consequence modelling, multiple influencing factors and multiple pathways to environmental outcomes are the norm rather than the exception. In some cases, little may be known about how identifiable triggering events influence ecological response. Because uncertainties cannot be avoided, USEPA guidance (USEPA, 1998) for ecosystem risk analyses mandates that: (1) ecological endpoints be explicitly defined by both an entity and measurable attribute; (2) the boundaries of the analysis be specifically defined; (3) strengths and limitations of relationships included in the conceptual model be explicitly described; (4) the rationale be established for key assumptions regarding model simplification, approximation, and extrapolation; and (5) data limitations be documented.

Assessing environmental consequences

The analysis of environmental consequences, as public safety and economics, examines two primary things: exposure and effect. For environmental consequences the relationship may be more complex than for the other principal categories, and the relationship also depends on other ecosystem characteristics. A principal challenge in environmental consequence evaluation is to identify specific assessment endpoints of interest and to narrow the breadth of analysis to something that can be accomplished within the planned scope, while yet doing justice to the nature of the problem.

The assessment endpoints and conceptual models developed in the initial phase of the risk analysis provide the structure for the analysis. The product of the analysis phase is a set of predictions made against the scales of the assessment endpoints, quantifying the degree of impact potentially caused by a dam failure. Upper bound estimates of environmental consequences within some reach of downstream riparian zone are a standard approach. For example, given a dam breech model and the flood wave it predicts, assessment endpoints within some reach of river may be assumed to be entirely lost. For example, fish and other fauna populations within the floodway from the dam site to whatever natural stilling basin (if any) exists downstream would be assumed to be entirely lost.

Table 10-15, adapted from Sherbot (1997), shows the set of assessment endpoints for a preliminary dam safety environmental risk analysis for the Hugh Keenleyside Dam. In this preliminary study, three categories of assessment endpoints were identified: a set dealing with fish kills (stranding and flushing), a set dealing with habitat loss (loss of riverine habitat, vegetation loss and loss of reservoir habitat), and one dealing with hazardous and toxic contamination. A conceptual model corresponding to these endpoints, and interrelating the modelling activities necessary to develop quantitative impact assessments, is shown schematically in Table 10-15 and Table 10-16, adapted from Sherbot.

Dam failure cannot always be modelled with confidence and, therefore, one may resort to probabilistic models of environmental impact, or to quantitatively assessing expert opinion. These approaches are similar to those discussed elsewhere in this guide, and are not separately discussed here. Table 10-17, taken from USEPA (1998), summarises an approach to uncertainty analysis used in ecosystem risk analysis practice.

The failure of a dam may release hazardous or toxic sediments downstream. Hazardous and toxic materials such as polychlorinated biophenyls (PCBs), heavy metals, agricultural chemicals, dioxins, and others, may settle or precipitate out of the

Table 10-15. Assessment endpoints for a dam safety environmental risk assessment (after Sherbot, 1997)

Category	**Assessment endpoint**	**Measures**	**Parameters**
Fish kills	Stranding	Population exposed Stranding Population change	Population, species, season, Diel timing, habitat type State rate, habitat susceptibility Population exposed, stranding, density dependent effects
	Flushing	Population exposed Fish passage Population change	Population, species, season, Diel timing, habitat type Flow, duration Population exposed, fish passage
Habitat loss	Riverine habitat loss	Bed-load transport Riparian bank erosion Habitat loss Population change	Depth, slope Flow, duration, inundation area Bed-load transport, riparian erosion Spawning, rearing, feeding, habitat loss
	Riparian vegetation loss	Vegetation loss Riparian bank erosion Riparian loss	Inundation area, depth, slope Flow duration, inundation area Vegetative loss, riparian bank loss
	Reservoir habitat loss	Habitat loss Population change	Littoral zone loss, pelagic volume reduction, tributary barriers Spawning-rearing-feeding habitat loss
Contamination	Hazardous and toxic contamination	Population change	Dose response data, concentration, exposure

Table 10-16. Assessment endpoints and modelling of consequences for a dam safety environmental risk assessment (adapted from Sherbot, 1997)

<table>
<tr><th>Models</th><th colspan="2">Fish kills</th><th colspan="2">Riparian environment</th></tr>
<tr><td rowspan="2">Inundation flood wave (BOSS DAMBRK)</td><td>Flood sweep</td><td rowspan="2">Fish kill</td><td colspan="2" rowspan="2"></td></tr>
<tr><td>Stranding</td></tr>
<tr><td rowspan="2">Scour sedimentation (HEC-6)</td><td>Short-term habitat loss</td><td rowspan="4">Natural recovery potential</td><td>Short-term bank erosion</td><td rowspan="3">Natural recovery potential</td></tr>
<tr><td>Long-term habitat loss</td><td>Long-term bank erosion</td></tr>
<tr><td>Inundation mapping (GIS)</td><td rowspan="2">Hazardous chemical contamination</td><td>Vegetative destruction</td></tr>
<tr><td>Chemical fate and form (ECOFATE)</td><td colspan="2"></td></tr>
<tr><td></td><td colspan="2">Assessment endpoint: sport fish population</td><td colspan="2">Assessment endpoint: riparian habitat</td></tr>
</table>

water column in the still waters of a reservoir, and accumulate in bottom sediments. In the event of a rapid dam failure, these sediments may be quickly eroded and carried downstream, where they expose flora and fauna to toxicity, endanger fresh-water supplies, and possibly subject human populations to health risks. Even the cold waters stored in the reservoir, depleted of dissolved oxygen, may cause environmental damages beyond those forecast on the basis of physical flooding alone. There have been four European tailings dam failures in the past five years, two in Spain, and one each in Sweden and Romania.

Monetising environmental consequences

Common practice in benefit–cost analysis (BCA) is to equate environmental impacts to financial costs. Dollar values based either on market values, for goods and services which have markets, or on contingent valuation surveys to subjectively assess willingness-to-pay for non-market goods and services, or are used to assess unit dollar values. For example, fish kills downstream of a dam failure might be valued by assigning a dollar value to units of fish, and the environmental impact found by multiplying by the forecast numbers of fish expected to be lost. Approaching environmental impacts in this way commensurates the environmental impacts as economic ones and, in effect, moves them into the economic account. Implicitly, this treats environmental impacts as economic costs.

Under current environmental law, for example the *Natural Resource Damage Assessment*[1] (NRDA) rules in the US, the monetary costs for environmental damages consist of two things: (1) the direct cost of restoration and remediation of the environment; and (2) damage costs for the loss of the environmental resources themselves during the period of restoration and remediation. Thus, in addition to restoring the downstream riverine environment and restocking important fish species, the environmental consequence must also include a cost associated with the loss of those

Table 10-17. Uncertainty evaluation in the analysis phase of ecological risk assessments (USEPA, 1998)

Sources of uncertainty	Example analysis phase strategies	Specific example
Unclear communication	Contact principal investigator or other study participants if objectives or methods of literature studies are unclear	Clarify whether the study was designed to characterise local populations or regional populations
	Document decisions made during the course of the assessment	Discuss rationale for selecting the critical toxicity study
Descriptive errors	Verify that data sources followed appropriate QA/QC procedures	Double-check calculations and data entry
Variability	Describe heterogeneity using point estimates (e.g. central tendency and high end) or by constructing probability or frequency distributions	Display differences in species sensitivity using a cumulative distribution function
	Differentiate from uncertainty due to lack of knowledge	
Data gaps	Collect needed data	Discuss rationale for using a facto or ten to extrapolate between a lowest-observed-adverse-effect level (LOAEL) and a NO AEL
	Describe approaches used for bridging gaps and their rationales	
	Differentiate science-based judgements from policy-based judgements	
Uncertainty about a quantity's true value	Use standard statistical methods to construct probability distributions or point estimates (e.g. confidence limits)	Present the upper confidence limit on the arithmetic mean soil concentration, in addition to the best estimate of the arithmetic mean
	Evaluate power of designed experiments to detect differences	Ground-truth remote sensing data
	Collect additional data	
	Verify location of samples or other spatial features	
Model uncertainty (process models)	Discuss key aggregations and model simplifications	Discuss combining different species into a group based on similar feeding habits
	Compare model predictions with data collected in the system of interest	
Model uncertainty (empirical models)	Evaluate whether alternative models should be combined formally or treated separately	Present results obtained using alternative models
	Compare model predictions with data collected in the system of interest	Compare results of a plant uptake model with data collected in the field

fish species and other environmental benefits until they are restored. These costs, today, are interpreted as including both environmental goods and services for which markets exist and goods and services for which markets do not exist (e.g. species existence, cultural use, genetic preservation). Previously, damages to non-market goods and services were not monetised, and as a result demanded little compensation in the aftermath of a dam failure. Today, compensation for non-market goods and services can be a significant liability.

In current practice, BCA uses two principal methods for assigning monetary costs to the loss of environmental resources: market prices, and contingent valuations. Market prices apply to those goods and services for which markets exist. A typical example is commercial fish. Fish have a commercial price in the market and, thus, the loss of some estimated number of fish can be valued in dollars by resorting to their market price. There is a large literature in the public economics literature on procedures for making such estimates (e.g. Mishan, 1988).

Contingent valuations apply to those goods and services for which markets do not exist. A typical example is a non-commercial and non-sport fish. Such a species does not have a commercial price in the market, and thus the loss of some estimated number of fish cannot be valued by resorting to market prices. Some indirect approach is needed. Contingent valuation methods (CVMs) which survey public opinion to obtain estimates of willingness-to-pay for environmental goods and services are increasingly seen as a reasonable way to obtain monetary equivalents, and a large literature has also developed with respect to them (US Department of the Interior, 1987). However, CVM is not without its critics. CVM valuations from surveys can be much larger than corresponding market values when the latter exist – by an order of magnitude or even more. CVM valuations are also subject to the vagaries of survey preparation and can be volatile.

An implicit result of monetising environmental impacts is that the preference structure for money is imputed to the environmental impacts, and this may or may not reflect the preferences of affected parties. For example, BCA presumes a linear preference structure over dollars; thus, ten times as many fish killed, when converted to dollars, is valued as ten times as great an impact. This may incorrectly reflect the actual consequences, if so, dramatically increasing the number of fish killed threatens the recovery of a species in the river. That is, there may be ranges of impact within which preferences are more or less proportional to unit losses, but these may be separated by points of inflection, or even trigger points, across which preferences change fundamentally.

Similarly, BCA presumes preference independence among different categories of impact, in that financial accounting schemes simply add marginal monetised values. For example, if in addition to loss of fish in a reach of river, there is also a loss of riparian zone area adjacent to the river. BCA monetises the fish kill, monetises the area lost, and calculates the combined impact by adding the dollar values of the two losses, fish kills and riparian area. In fact, the two elements of the downstream environment interact. Riparian areas may serve as nurseries for small fish and may enhance the survival of larger fish. Post-flood recovery may depend on an interaction between these factors, and thus the simultaneous loss of fish population and riparian area is worse than the sum of the individual losses. The preference structure may be multiplicative rather than additive.

Endnote

1. Section 301(c) of the Comprehensive Environmental Response Compensation and Liability Act of 1980 (CERCLA) (42 USC 9601 *et seq.*), modified by the Superfund Amendments and Reauthorization Act (SARA) of 1986 and Section 1006 of the Oil Pollution Act of 1990 (OPA) directs the development of regulations for assessing natural resource damages from oil spills and hazardous substances.

11 Methods for probability assignment

At some point in any risk analysis, numerical probabilities must be assigned to event tree branches, fault tree nodes or other models used to calculate risk. Earlier chapters have discussed the philosophical meanings of probability, the logical structure of risk models and the importance of judgement. This chapter addresses less philosophical and more practical engineering approaches to developing the numbers associated with probabilities in a risk analysis.

This section considers three questions: (1) What do assigned probabilities in a risk analysis mean regarding dam safety? (2) How do we assign these numerical probabilities using statistics, engineering models and judgement? (3) What can we do with such probabilities to understand dam safety? Some engineers and managers are reluctant to use risk analysis, not because they question the logical foundations of the methodology, but because they doubt the way numerical probabilities are assigned to event trees and other risk models. This section addresses the concepts behind practical ways of assigning probabilities, and shows that such methods are closely related to the many other engineering models and procedures used in dam safety studies. They are sound or unsound not in their logic but in their application, just as any other engineering procedure.

The verb "to assign" is used throughout this section in referring to numerical probabilities – rather than, to estimate or to assess – emphasising that probabilities reside in models and not in the world (Matheron, 1989). Probabilities reflect a level of understanding of geological, hydrological, human intervention and other processes that impact the safety of a dam. They are measures of knowledge, not measures innate to the physics of the world. They change as a function of how much the engineer or risk analyst knows about a specific situation, and as a function of how that understanding is abstracted in an engineering model. Changing a model means changing – or at least revisiting – the probabilities assigned within it. This view presumes that the world is not a random place but that it follows natural laws, has deterministic properties, and behaves in a predictable way: the world of Laplace (Chapter 7). Uncertainty arises due to limited knowledge of those natural laws, incomplete information about material and other properties and inadequate models with which to combine laws and properties and thus to make predictions.

In order to make this chapter self-contained, it reviews some material covered earlier, particularly in Chapter 7, "Essential elements of probability", Chapter 8, "Event trees", and Chapter 9, "Fault trees". The reader is directed at various places in the present text to those other chapters for more complete treatment.

11.1 Quantifying probabilities

A core task in any risk analysis for dam safety is assigning quantitative probabilities – sometimes referred to as "*p*-values" – to the branches of an event tree. Regardless of the effort invested in understanding the structure of events and the vagaries of dam performance, if the numerical values of the probabilities that enter the analysis are poorly assigned, the calculated results will be a poor representation of the risks faced. This chapter describes the types of uncertainty that appear in dam safety risk studies, discusses how probabilities are used to describe those uncertainties, and surveys common approaches to assigning numbers to those probabilities.

In risk analysis for dam safety, probability assignments are required primarily for three things:

1. *Loading conditions*, typically floods, earthquakes and operating loads, although more exotic loading conditions (e.g. terrorist attack) may be of concern.
2. *Systems response* of the dam and its associated structures, typically strength, deformation, seepage and erosion, and including the response of mechanical equipment such as spillway gates.
3. *Consequences* should failure occur, typically loss of life, economic costs and environmental degradation.

In the mathematics of risk analysis, each of these three is of equal importance. This chapter makes no distinction among these three areas of probability assignment. The engineering profession, arguably, has most experience with the first of these, and least with the last, but the methodologies of probability assignment are similar for each.

In practice, dam safety studies usually consider scenarios for loading conditions, systems response and consequences. This is because the conditions under which the dam operates are more complex than a simple load and a simple response. As noted by McDonald (2002):

> A typical [loading] scenario is a particular flood discharge state associated with a prior reservoir storage state, with a concurrent wind velocity state, a wind direction state and a spillway gate operating state. The effect of the flood load on the dam may depend on all of these conditions. Consequently, it is necessary to assign the probabilities of all of the associated condition states as well as that of the primary load state.

The methods of probability assignment for all of these things are the same.

How are probabilities assigned?

The common approach to dam safety risk analysis is by means of event trees (Chapter 8). Branch probabilities in an event tree are based on data, models and experience. They are assigned in one of four ways (or in some combination of these ways):

1. Statistical estimates based on empirical data.
2. Engineering models based on physical processes.
3. Fault tree analysis based on logical constructions.
4. Judgement.

The second and third of these are normally grouped together under the name, engineering reliability models.

Statistical estimates are characterisations or summaries of past observations. For example, flood–frequency relationships are usually just a summary of historical data, plotted according to agreed-upon conventions. If in the past, only 10% of peak annual floods exceed some discharge, then the conclusion might be drawn that this discharge has a 10% probability of being exceeded in a particular future year, and the probability of exceeding this discharge is assigned a value, $p=0.1$. Statistical estimates reflect the frequency of occurrence of some condition in time or space. As such, they deal with properties of the world, that is, they deal with the rate at which certain conditions obtain in nature, as in the case of flood exceedances. Often, probability (i.e. frequency) distribution models of regular mathematical form, such as Normal, logNormal, or Poisson functions, are used to approximate historical data frequencies. This is mathematically convenient and usually allows stronger inferences to be drawn than does non-parametric (i.e. distribution-free analysis) inference.

Engineering models are constructs based on reasoning from first principles of physics. For example, embankment stability might be analysed using limiting-equilibrium models of soil mechanics having to do with stresses and strains. Strength parameters, such as cohesion and friction angle, are estimated for the soils of which the embankment is constructed, and then the stability of the embankment is calculated using the model. Uncertainties in parameter values and possibly in the model itself are propagated through the calculation to establish probabilities that the embankment can resist certain loading conditions. Engineering model estimates reflect a combination of frequencies of occurrence in the world (e.g. the spatial variation of soil properties) and degrees of belief (e.g. in the adequacy of the model and in model parameters).

Fault trees are also reliability models but of a particular type. They differ from engineering models in that they model the logic of a system (e.g. how component malfunctions combine to cause failure of an overall system) rather than model the physics of the system. Fault trees, like engineering models, reflect a combination of frequencies in the world and degrees of belief in models and parameters. Fault tree analysis is common in applications to mechanical systems, and is often associated with assigning p-values to event tree branches having to do with spillway gates, turbines or other mechanical equipment. Fault trees are considered in more detail in Chapter 8.

Judgement is based on intuition and qualitative reasoning. For example, the probability that hydraulically connected discontinuities are present in a rock abutment might be assessed by asking a geologist familiar with local conditions. Based on experience and an understanding of the geological history of the dam site, the geologist might be asked to compare his or her degree of belief about the existence of discontinuities to his or her degree of belief about other uncertainties (e.g. frequencies of other geological features, or common random events). By making such comparisons, an approximate numerical value for degree of belief can be obtained. There are at least two schools of thought on what constitutes judgement: one holds that judgement is intuition (i.e. the recognition of patterns in a situation, possibly subconsciously); the other holds that judgement reflects a base of knowledge and an analytical process of reasoning that manifests in quantitative estimates based on logic. These are considered in more detail below and in Chapter 7.

Each of these approaches to assigning probabilities – statistics, reliability models, and judgement – provides information for a risk analysis. Each provides a set of valid procedures in that they follow internally consistent logic. They provide equally important information – but usually different information – and usually apply to different circumstances or different parts of a risk analysis. This is shown in Figure 11-1, in which the influence of data completeness and the influence of model completeness are shown against a backdrop of the different approaches to assigning probabilities.

Probability assignments are based on both data and models. With extensive data but no model, that is, with only rudimental theoretical understanding of a phenomena, statistical prediction is usually the best (or only) approach to probability assignment. In contrast, with good theoretical understanding and a good model but few data, a purely theoretical prediction is the common approach. In the ideal case, one has both good data and a good theoretical understanding, and a robust, model-based approach can be used in which the parameters of the model are identified by extensive observation. In the less-than-ideal situation, one has neither good data nor a good theoretical understanding, and

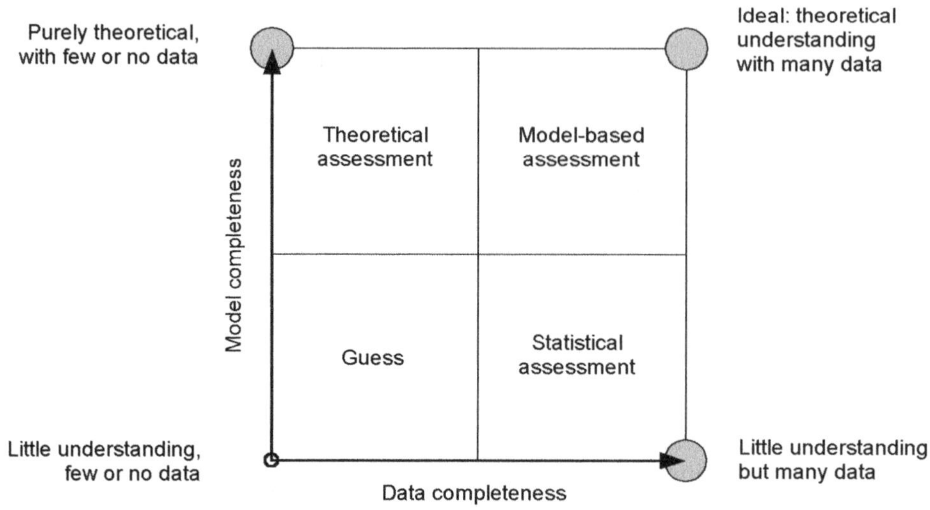

Figure 11-1. Considerations of predictability

at most a guess can be made at assigning probabilities. Judgement is an overlay across all of these, and the quality of judgement increases as both data and model completeness increase.

Uncertainty in assigned probabilities

Whether probabilities are assigned by statistical analysis, engineering modelling, judgement, or some combination of these approaches, they are almost never specified precisely. There is always a degree of uncertainty about *p*-values, and how to treat this uncertainty has sparked philosophical debate. On the one hand, de Finetti (1937, 1972) has argued that uncertainty about probability is meaningless because, in subjective theory, individuals are always able to develop unique and precise judgements of probability based on their willingness to act upon those judgements (Ramsey and Braithwaite, 1931). On the other hand, many engineers and risk analysts find the notion that probabilities are uncertain to be both appealing and useful, as do many statisticians (Edwards, 1975; Good, 1962, 1983; Howard, 1988; Savage, 1954). This question of "uncertainty about probability" is sometimes referred to as higher-order uncertainty (Lehner *et al.*, 1996) or as Type II probabilities (Good, 1962).

The practical importance of higher-order uncertainty in risk analysis comes to the fore only when performance measures are non-linear in the *p*-values. When they are linear, the value of a risk – the integral of probability times consequence – is the same whether an expectation is taken of the probability before the integration is performed, or of the risk afterwards. In much of engineering risk analysis for dam safety, linearity in probability prevails, but sometimes non-linearity can be subtle. Consider, for example, the situation in which a performance measure depends on the product of two probabilities. If the probabilities are conditionally independent given some third variable, the condition is linear. However, if the conditioning variable is itself uncertain, then the two probabilities forming the product are dependent and the condition is no longer linear. This could occur, for example, in an analysis of embankment deformations in which both loads and pore pressures are affected by an uncertain pool level.

Mosleh and Bier (1996) have suggested dividing higher order probability into two sorts: that based on uncertainty about underlying events or conditions upon which probabilities are conditioned, and that based on cognitive imprecision. They argue that the former is consistent with the axioms of degree-of-belief probability, whereas the latter is not. One example of a probability that is uncertain because of uncertainty in underlying events or conditions could be seismic ground shaking. The level of ground shaking at a dam site may depend on the proximity of a fault, the location of which is known only in probability. Were the fault closer to the site, the probability distribution (exceedance probabilities) of ground shaking would favour higher values; whereas, were the fault more distant, the probability distribution would favour smaller values. Since the location is imprecisely known, the corresponding probability distribution is uncertain and describable itself only by probabilities. If the performance function of the dam is non-linear in the exceedance probabilities, then the higher-order uncertainty could not be finessed by taking an expectation *ex ante*.

Degree-of-belief probability theorists, like Mosleh and Bier, discount the existence of frequency-based probabilities, but most engineering risk analysis for dam safety divides uncertainty into aleatory and epistemic parts, and presumes some level of uncertainty due to spatial or temporal variations, modelling these as aleatory probabilities. Within the aleatory models, probability is defined as frequency of occurrence, which is presumed to be a state of nature not a degree of belief. In this case, it is perfectly consistent with subjectivist theory for there to be a probability (belief) distribution defined over a poorly known frequency probability. This is conceptually the same as saying that the frequency is a parameter of the model, and then specifying a probability distribution over that parameter. For example, the average recurrence rate of earthquakes on a particular fault, which yields probabilities of occurrence based on frequency, could be described by a probability distribution based on limited information and reflecting degrees of belief in a particular value for the recurrence rate. Within classical frequentist theory, on the other hand, such uncertainty could only be described by confidence limits or similar measures, since classical frequentist theory does not admit degrees of belief (see Chapter 7).

The problematic case of higher-order probability from a theoretical view is that caused by cognitive imprecision. Cognitive imprecision means that, even though – in theory – a person ought to be able to specify a precise number to reflect his or her degree of belief, in practice this may not be the case. Many or most probability assignments are subject to imprecision. People are unable to distinguish among fine differentiations in degrees of belief, and even experts rebel at being asked for the next decimal place in the assignment of a subjective probability. As noted above, many highly-regarded statisticians and decision theorists discount the theoretical limitations and treat imprecisions about degrees of belief also as higher-order probabilities. This seems a reasonable approach for dam safety studies.

Verification and validation of models

Models and logical constructions – chains of reasoning – are used in each part of the probability assignment. As described in more detail in Chapter 10, these models attempt to mimic natural processes and human behaviour and, thus, are a reflection of pure deductive logic. The adequacy of models is reflected in two considerations, verification and validation. Verification is a check of the veracity of a model. Validation is a check of the internal logic of a model.

The way a model or a chain of reasoning is verified is to compare its predications with observations in the field or laboratory. These comparisons yield two measures: an accuracy and a precision. Accuracy is the degree to which, on average, the predictions of the model or chain of reasoning agree with observations. Precision is the degree of variability of the predictions around their own average (Figure 11-2). This is sometimes illustrated by a target metaphor, as suggested in Figure 11-3.

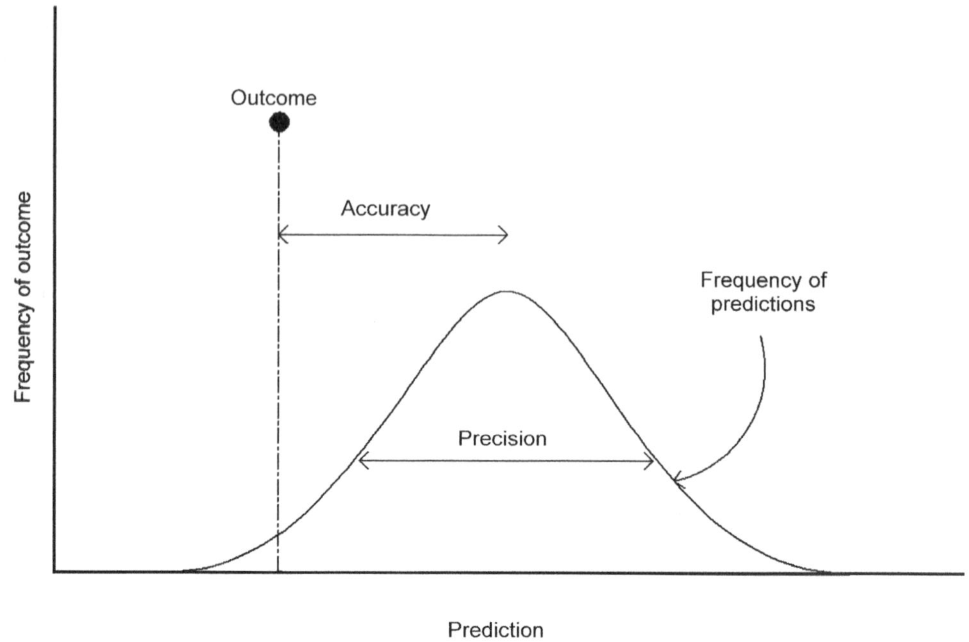

Figure 11-2. Notions of accuracy and precision in model verification

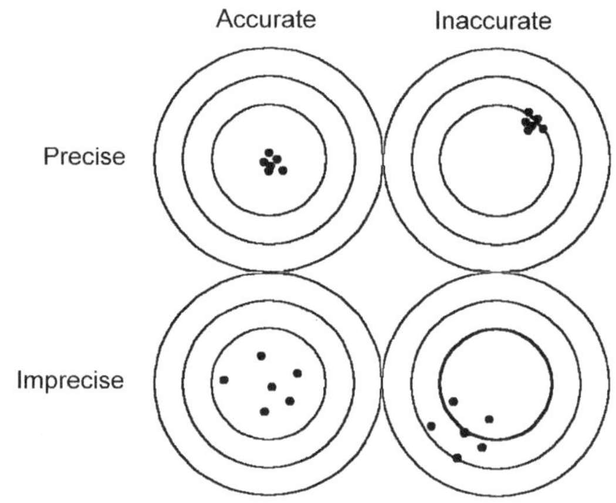

Figure 11-3. Target metaphor for accuracy and precision

11.2 What do assigned probabilities mean?

In assigning probabilities to the branches of an event tree, we again encounter the distinction between natural variation and limited knowledge, discussed at length in Chapter 7. As we found in Chapter 7, what these assigned probabilities mean – both philosophically and operationally – depends on the type of uncertainty they attempt to quantify. The philosophical meaning dictates how the probabilities are assigned specific values; the operational meaning dictates what manipulations can be performed on the resulting numbers.

The notion of natural variation has to do with processes that are inherently unpredictable: the role of a dice, patterns of the weather, occurrences of an earthquake and similar phenomena. Flood frequencies, for example, are often treated as a natural variation, as inherently random. These are also described as aleatory uncertainties. The notion of limited knowledge, in contrast has to do with uncertainty about models, parameters and other states of nature: the order of cards in a deck, the truth of a proposition, whether or not climate will change and similar things. The likelihoods of geological anomalies (faults, soft lenses), for example, are often treated as reflecting limited information. These are sometimes described as epistemic uncertainties.

While natural variation and limited knowledge are clearly different considerations in a risk analysis model, the way that the whole of uncertainty is divided into these two components is simply a modelling decision. However, the implication of dividing uncertainty in this way is to cause those probabilities assigned to natural variations to be statements about frequencies of occurrence in time or space, and those assigned to limited knowledge to be statements about degrees of belief. Both are measured as mathematical probabilities.

Probabilities of dam failure are beliefs

Within an event tree there will be some *p*-values that reflect frequencies and some that reflect beliefs. These can be mixed, because they both adhere to the formal rules of probability theory. The end result, however, the calculated probabilities over how a dam performs, are not "objective" but a result of analysts' assignments. They are conditional on available data, the models used to interpret those data, and the intuitive judgements made in performing the analysis. The importance of this distinction between frequencies and belief for dam safety is that dams are unique structures. A dam site is a unique geographical and geological entity. While the loads a dam is exposed to may exhibit variation in time or space, for example, the variations of hydrological loads, the response of a particular dam to a particular loading condition is unique: it either "fails" (by whatever definition) or it does not. The probabilities associated with these possible responses have to do with beliefs as reflected in probability assignments. There is no "objective probability" of the dam's response, and thus no objective probabilities of dam safety, because the performance of a dam is not part of a long series of responses under identical conditions.

In this regard, Figure 11-4 is instructive. It shows estimated probabilities of structural response to increasing flood levels for a concrete gravity dam. Following the example

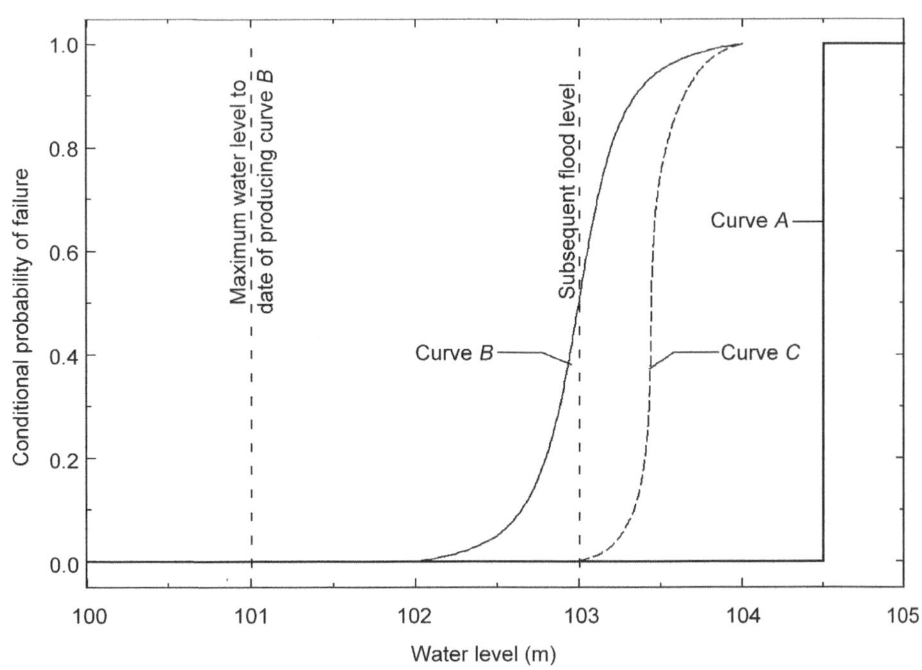

Figure 11-4. System response of a concrete gravity dam to pool level (McDonald, 2002)

discussed by McDonald (2002), if the pool were steadily to rise under specified conditions, a point would presumably be reached where the loads exceed the resistance of the dam and sliding begins, leading to failure. At lower pools failure would not occur and at higher pools it would already have done so. This response is shown by Curve *A*. There are no probabilities: the dam either fails or it does not. This is the system response curve of the dam, but it is *ex ante* unknown.

A risk analysis might be made of the safety of the dam – having noted that a maximum pool of 101 m had been historically sustained – and this analysis might lead to the estimated response shown as Curve *B*. Curve *B*, in essence, expresses the analysts' uncertainty about the location of the step in Curve *A*. Where a subsequent flood with pool level 103 m is also to be sustained, then the risk assessment might be updated, leading to Curve *C*. The height of a future pool in such risk analyses might be taken as a frequency of occurrence, but the response of the dam is unique and, thus, the probabilities resulting from the risk analysis are beliefs not frequencies.

Herein lies the important issue of "first filling". Many dams that fail by piping or because of design and construction errors, for example, do so shortly upon first filling of the reservoir. Historical data suggest that about half of all dam failures occur within the first five years after construction, when the reservoir is filling for the first time (Baecher *et al.*, 1979; McCann *et al.*, 1985). This is, in essence, the proof testing of the dam, just as any product might be proof tested. If the dam does not fail (by whatever definition) upon first filling, then the probabilities assigned to its failing in the future should be reduced because some of the uncertainties contributing to those future probabilities have been eliminated. Thus, the risks associated with an existing dam that has withstood the vagaries of time are different from those of a new dam that is yet to be built. On the other hand, many flood control dams, despite decades of service, may have never experienced the design pool and, thus, have not been "proof tested" in the same way.

Some observations follow from this example (McDonald, 2002):

- First, there is presumably an objective, real performance of the dam, but in practice this performance is not known.
- Second, the probabilities resulting from a risk analysis are beliefs about that objective, real performance of a dam.
- Third, these probabilities are based on available knowledge, both in the sense of observed data and phenomenological models.
- Fourth, probabilities of failure can change over time as a result of newly acquired information or improved knowledge and, thus, are defined at a particular time.

Are probabilities reproducible?

If probabilities of failure are based on beliefs and are related to a point in time, are they "reproducible?" The answer is yes, but perhaps not in the way that one might naïvely assume. Risk analysis is no more than an accounting scheme. Assumptions, data and models are laid out in an orderly way, and conclusions are calculated from them according to an explicit procedure. The numerical results are a function of those assumptions, data and models. The assumptions depend on who makes them, the data depend on the point in time, and the models depend on professional understanding of the physics of dam behaviour. If any of these things change the numerical output changes. Yet, the structure of the analysis does not necessarily change. It is the structure of the risk analysis for a particular dam that is reproducible, not necessarily the numerical results.

Absolute vs annual probabilities

Many empirical studies of the statistics of dam failure (e.g. Baecher *et al.*, 1979), as well as of failures of other engineered systems such as buildings or pavements, report failure rates on an annual basis, that is, a typical conclusion might be, "on average there is one failure for every so many dam-years". This is consistent with the property and casualty insurance business and suggests probabilities that are frequencies rather than beliefs, that is, uncertainties of an aleatory rather than epistemic quality. Most of the naturally occurring loads on dams, such as floods and seismic ground shaking, are indeed modelled as frequencies; but uncertainties about the response of a dam, such as its strength, deformation or seepage behaviour, have to do with limited knowledge. Probability arising out of limited knowledge is usually not annualised, but absolute – either the dam fails under a given load condition or it does not – there is nothing to do with frequency in this. It is important to keep the dimensions of a derived number correct in calculating risk: annualised and absolute probabilities cannot be combined without the consideration of time intervals.

When a failure condition is triggered by a process modelled as random in time, such as pool level, the issue of annualised probabilities is moot. Random loads in time occur with frequencies that can be expressed as a function of exposure duration, and conditioned on a given load condition, either a failure state occurs or it does not. This conditional probability of failure need not be annualised, that is it can be an absolute probability, for the joint probability of load and effect to be a frequency. The dimensionless absolute probability multiplied by a frequency yields a frequency.

When a failure condition occurs under normal operating loads, which are not modelled as random in time, the issue of annualised probabilities is vexing. The obvious example is piping. What, if anything, does it mean for piping to be expressed as an annual rate? The driving force for piping is pool level, because without water pressure against the upstream face of a dam there would be no internal pore pressure gradients, no seepage and consequently no piping. But it is not the fluctuation of pool level through time that causes piping; a constant high pool would just as likely, or even more likely, lead to piping.

The uncertainties inherent to the analysis of piping in an earth dam have somewhat to do with frequencies, and somewhat to do with knowledge uncertainties. On the first count, many models of piping start with the random presence of a "flaw". This flaw can be one of design or one of construction. It is often taken to occur randomly in space within the dam. Thus, the flaw is partly an epistemic uncertainty (does a flaw exist?), and partly an aleatory uncertainty (if so, where?). On the second count, the process of internal erosion, even given a flaw, is poorly understood. There are significant limitations to our understanding of the physics of internal erosion, and thus to the models and material properties that apply. These have little to do with randomness and almost nothing to do with frequencies in time.

The ability to characterise risks involved in piping will not improve until an understanding of the physics of piping improves. This is not a risk analysis issue, it is a mechanics issue. In the meantime, the best that can be done is to try to understand the nature of time-dependency in describing the piping risk (and risks similar to it, in this sense). The loads driving a potential piping failure (i.e. pool levels) may or may not have a frequency aspect to them but if they do it can be ignored. The process of piping depends on physics and material values about which there is limited knowledge and thus considerable uncertainty, but piping does not occur instantaneously. It is a time-dependent physical process.

From a risk point of view, two things are uncertain: first, will a particular dam under particular load conditions fail by piping if left forever to do so, and second, if the answer to the first question is, yes, then over what time period does this failure unfold? The probability that the dam will fail by piping in a given year, t_0, is the product of an absolute probability, the probability that it can fail by piping at all, and of a time-dependent probability of how long the process will take (Figure 11-5). All these uncertainties may be epistemic, and yet they may lead to "annualised" probabilities

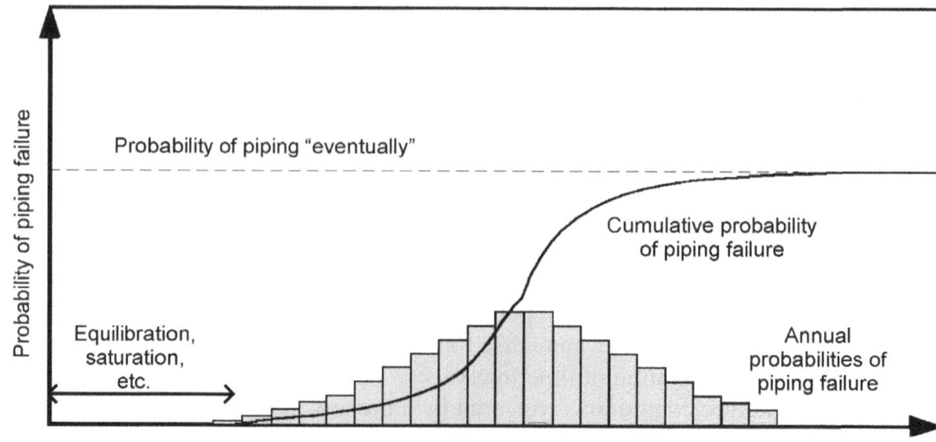

Figure 11-5. Time-dependent "annualised" probabilities of piping caused by epistemic uncertainties in both the absolute chance of piping and in the time the process takes

having to do with uncertainties over how long the process takes to complete. These annualised probabilities (i.e. the derivatives of the cumulative curve) change with time, rising slowly from zero, cresting at the most likely time to failure (presuming a failure occurs at all), then falling off again. The asymptote of the cumulative probability is the absolute probability that piping failure occurs at all, which may be less than 1.0.

11.3 Reasoning with probabilities

Probability theory provides a way of thinking about what is known of the world, for example about the confidence that can rationally be placed in a future course of events. In particular, the logic of probability theory gives a basis for inductive reasoning, that is, for calculating degrees of confirmation based on observations or other evidence.

Inductive reasoning, unlike deductive reasoning, supports its conclusions only to a degree; a set of premises does not necessitate a conclusion, but changes the extent to which the conclusion is confirmed. This use of probability as a logic structure for inductive reasoning has been widespread since Laplace (Laplace and Dale, 1814) and can be traced back at least as far as Bayes (1763).

As discussed in Chapter 7, consider the syllogism: if *A* is true, then *B* is true; *A* is true, therefore, *B* is true. This is deductive logic. In contrast, consider the syllogism: *A* is true, then *B* is true; *B* is true; therefore, *A* is more plausible. This is inductive reasoning. *A* does not become true if *B* is true, only more likely. This can be reformulated using conditional probabilities. From the premise, *A* implies *B*, the conditional probability of *B* given *A* must be 1.0 (Figure 7-3):

$$P(B|A)=1 \tag{11-1}$$

Then, from the total probability theorem:

$$P(A|B)=\frac{P(A)P(B|A)}{P(B)}=\frac{P(A)}{P(B)} \tag{11-2}$$

Since $P(B)\leq 1.0$, it must be that $P(A)\leq P(A|B)$. Thus, if *B* obtains, the credibility of *A* must stay the same or increase; it cannot decrease. Further, if $P(B)<1.0$, then $P(A|B)$

must increase over $P(A)$, and the less probable B is to begin with, the greater the confirmation afforded A, since:

$$\frac{P(A \mid B)}{P(A)} = \frac{1}{P(B)} \tag{11-3}$$

If A implies some almost certain event B, such as spring floods appearing this spring, when that almost certain event occurs, little increase in confirmation is afforded to A. In contrast, if A implies some improbable event B, such as no spring floods coming, then should that improbability occur, it affords strong confirmation to A. This is consistent with Popper's concept of corroboration: "In appraising the degree of corroboration of a theory we take into account its degree of falsifiability. A theory can be the better corroborated the better testable it is" (Popper, 1968). In other words, a theory or premise that implies something unexpected, becomes all the more corroborated or confirmed when that unexpected thing actually happens.

The analogy to dam safety is straightforward. Suppose that a detailed geotechnical analysis of a dam foundation concludes that liquefaction during an earthquake of given severity has probability, p. Then, an earthquake occurs and liquefaction ensues. If the numerical value of p had been large, the credibility we impute to the geotechnical analysis would increase; but if p had been small, the credibility would decrease. The same is true of expert opinion.

Inductive reasoning

Bayes' theorem is the formal vehicle used to draw inductive inferences from possibly incomplete evidence. Bayes' theorem is a relationship among conditional probabilities, derived by rearranging the total probability theorem as:

$$Pr(A \mid B) = \frac{Pr(A)Pr(B \mid A)}{Pr(A)Pr(B \mid A) + Pr(\bar{A})Pr(B \mid \bar{A})} \tag{11-4}$$

in which A is some event or proposition, and $\bar{A}$ is its complement (i.e. "not-A"); B is some event correlated with A. The conditional probability of B given A is called the likelihood of B given A. Note that the denominator of the right-hand-side (RHS) is simply a normalising constant ensuring that the sum of $P(A \mid B)$ and its complement $P(\bar{A} \mid B)$ is one. This is a generalized restatement of equation 11-3.

The interest in Bayes' theorem derives not from its simple statement about the relationship among conditional probabilities but from its use as a vehicle for inductive reasoning. That is, given some value $P(A)$, and some (new) information related to A (e.g. whether some event correlated to A occurs), Bayes' theorem can be used to modify the probability of A from $P(A)$ to $P(A \mid B)$. Thus, for example, given some background rate on the probability of liquefaction in the foundation materials of dams, and given that low Standard Penetration Tests (SPTs) blow count materials are more likely to liquefy than are high blow count materials, then the information that SPT blow counts in a particular foundation are low can be propagated through Bayes' theorem to calculate an increase in the probability that this particular foundation will liquefy. Bayes' theorem is used as a vehicle for drawing inductive conclusions from observations.

In its inductive capacity, Bayes' theorem is sometimes written:

$$P'(A \mid B) \propto P^0(A)\mathrm{L}(B \mid A) \tag{11-5}$$

in which $P'(A \mid B)$ is called the posterior (or *a posteriori*) probability of A, $P^0(A)$ is called the prior (or *a priori*) probability, and $L(B \mid A)$ is called the likelihood (Figure 11-6). The likelihood is the conditional probability of the new information were A true. The terms posterior and prior imply the notions of after and before the data B are known. The proportional symbol is used to avoid the necessity of writing out the sometimes long

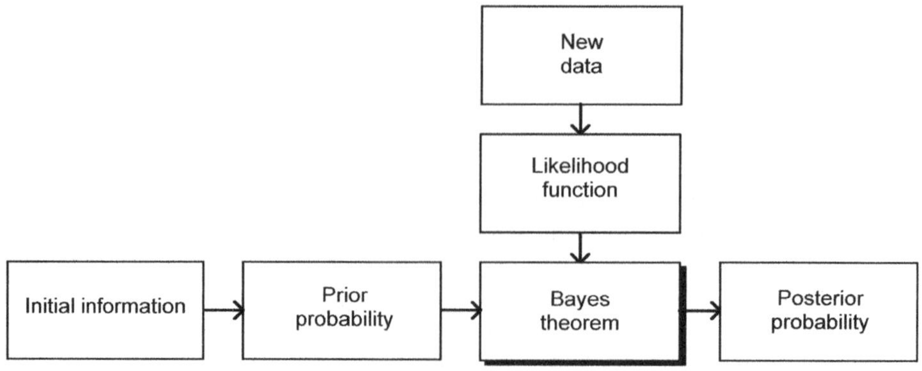

Figure 11-6. Inductive reasoning with Bayes' theorem

integral expression for the normalising constant implied by the denominator of Equation (11-4).

The vehicle for evaluating the magnitude of this change – that is, the strength of evidence in the data – is the likelihood function. Bayesian methods, associated with degree of belief probability, use likelihood as the basis for inference. The Likelihood Principle, fostered by Fisher and Bennett (1971), says the weight of evidence in a set of data in favour of some parameter value θ is proportional to the conditional probability of the data given θ. Parameter values for which the observed data are probable are given more weight than parameter values for which the observed data are improbable.

We have dealt with the case in which A implies B, and then B obtains and, in so doing, increases the confirmation of A. What happens in the case that A implies B, but A does not obtain? Does this change the degree of confirmation of B? The answer is yes. Again rearranging Bayes' theorem and noting that in this case, $P(B|A)=1$,

$$P(B|\bar{A})=\frac{P(B)-P(A)}{P(\bar{A})} \quad (11\text{-}6)$$

from which, $P(B|\bar{A})\leq P(B)$. Thus, the degree of confirmation of B can only diminish if a premise leading to it does not obtain. Furthermore, the greater the degree of confirmation in the premise itself (i.e. $P(A)$), the greater the decrease in confirmation of B when that premise fails to obtain.

Inferring probabilities

Consider a simple site characterisation example in which standard penetration test (SPT) borings are used to search for liquefiable layers in the foundation of a dam site (Baecher and Christian, 2003). The tests are taken every 1.5 m vertically, and the target layers are assumed to be 0.6 m thick. Given the data that result, what is the probability that a liquefiable layer exists?

If a layer exists, then the conditional probability of finding it with a single boring is $p=0.6/1.5=0.4$, but engineers sometimes mistake a low blow count as an outlier. Thus, assume the conditional probability is 0.3. Also, it is possible to record a false positive when no liquefiable material exists, so an assumption is made that, the probability of "finding" a lens that is not really there is still 0.1. Then, from Bayes' theorem:

$$P[E|F]=\frac{P[E]P[F|E]}{P[E]P[F|E]+P[\bar{E}]P[F|\bar{E}]} \quad (11\text{-}7)$$

in which E=layer exists, F=positive result ("find"), $\bar{E}$=layer does not exist. Presuming no prior information on whether a layer exists, the prior probability is presumed to be 0.5. There is no more reason to suspect a lens exists than that it does not.

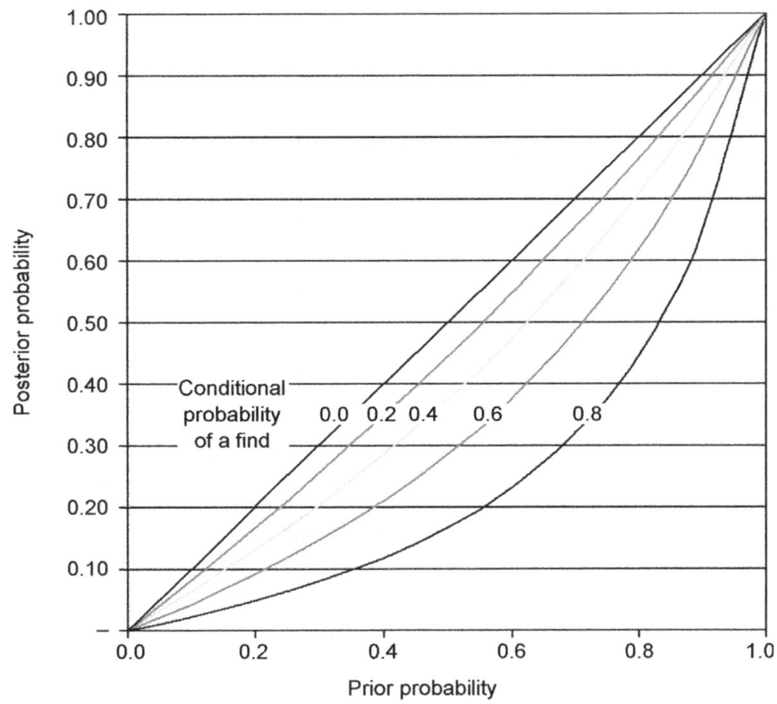

Figure 11-7. Probability of target layer existing, given that exploration has not found it, as a function of the prior probability that it existed. Curves show parametric results for different conditional probabilities of finding an exiting target with the same exploration effort

If one boring is drilled, and shows a hit, then:

$$P[E|F] = \frac{0.5 \times 0.3}{0.5 \times 0.3 + 0.5 \times 0.1} = 0.75 \tag{11-8}$$

If the probabilities of hitting a layer are independent from one boring to another, the results of Figure 11-7 are obtained showing the posterior probability that a target exists, given that it was not found during exploration, as a function of its prior probability of existing.

Log-odds chart

The sequential revision of probabilities as information accumulates can be charted as in Figure 11-8. This plot shows changes in the log odds-ratio of a proposition and its complement as more and more evidence is collected. Bayes' theorem provides that the probability of a proposition after certain evidence is observed is proportional to the probability before, times the likelihood of the proposition in light of the data. Dividing this posterior probability by the probability of the complementary proposition gives the odds-ratio:

$$\frac{P'(A|B)}{P'(\bar{A}|B)} = \frac{P^0(A)}{P^0(\bar{A})} \frac{L(B|A)}{L(B|\bar{A})} \tag{11-9}$$

and taking the logarithm of each side gives:

$$\log \frac{P'(A|B)}{P'(\bar{A}|B)} = \log \frac{P^0(A)}{P^0(\bar{A})} + \log \frac{L(B|A)}{L(B|\bar{A})} \tag{11-10}$$

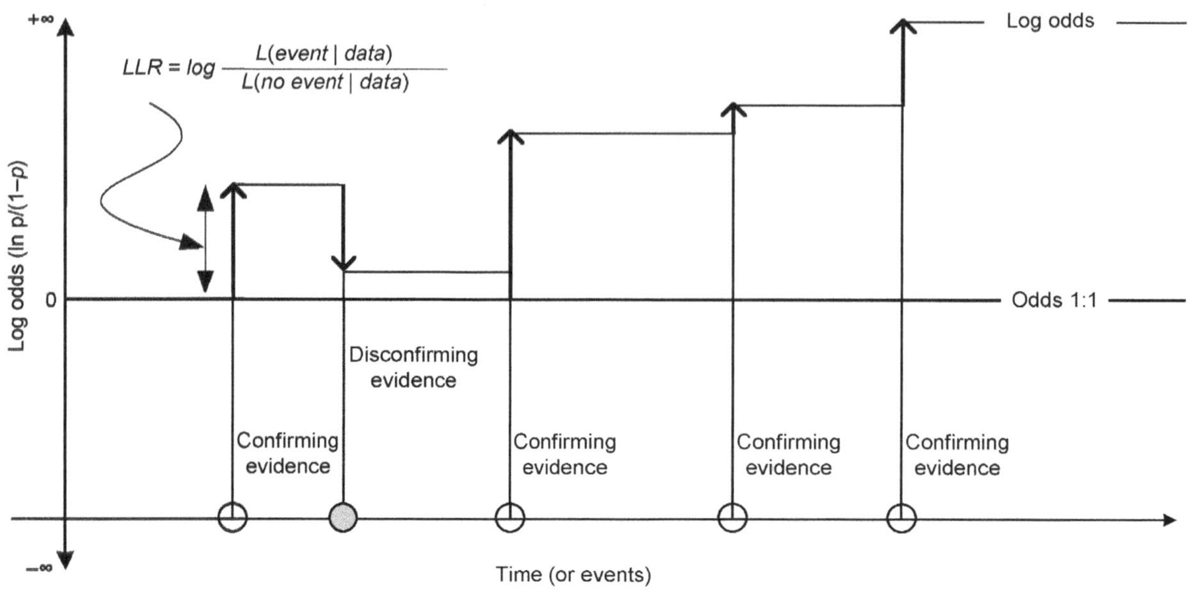

Figure 11-8. Sequential revision of probabilities as observations accrue

Thus, the log odds-ratio after the evidence *B* is observed equals the log odds-ratio before, plus the log likelihood ratio (LLR) of the proposition and its complement in light of the evidence. This provides a convenient way of tracking information and its impact on the probabilities associated with particular events, hypotheses or propositions. The likelihood ratios can be based on statistical models, or judgement or a combination. This approach is particularly effective in dealing with qualitative evidence, in which the likelihood ratios might be based on judgement. It is conceptually the same as the base-rate adjustment approach suggested for evaluating the potential for internal erosion (Foster *et al.*, 2000; McCann, 1997), and can serve as the graphical representation of that method.

Inferring probability distributions

The case of inferring probability distributions over a discrete or continuous random variable is analogous to the preceding case. Starting from the marginal probability distribution (PMF or PDF) over a state of nature, $f_\Theta(\theta)$, the likelihood function of the data is used to update the distribution via Bayes' theorem to a posterior probability distribution conditioned on the observations:

$$f_\Theta(\theta \mid \mathbf{z}) = N f_\Theta(\theta) L(\theta \mid \mathbf{z}) \qquad (11\text{-}11)$$

in which N is a normalising constant ensuring that the posterior distribution is proper, that is, that it integrates to one, $N^{-1} = \int_{-\infty}^{+\infty} f_\Theta(\theta L(\theta) \mid \mathbf{z}) d\theta$. Often, this is simply written as a proportionality.

In the example of finding liquefiable layers in SPT borings, we assumed a value for the detection probability, that is we assumed a probability that an existing layer would be detected in an SPT boring log if the boring in fact penetrates the layer. If empirical data were available from other sites or from other areas of the present site, we might instead have tried to use those data to infer the detection probability statistically. Consider that we have a set of twenty (20) SPT borings that we know *ex ante* to have intersected a liquefiable layer, and that in six (6) of those the layer was detected. What can we say about the value of the detection probability?

Assume that the observations are independent of one another, and that the probability of detection in any one boring is π. The individual outcomes are Bernoulli random

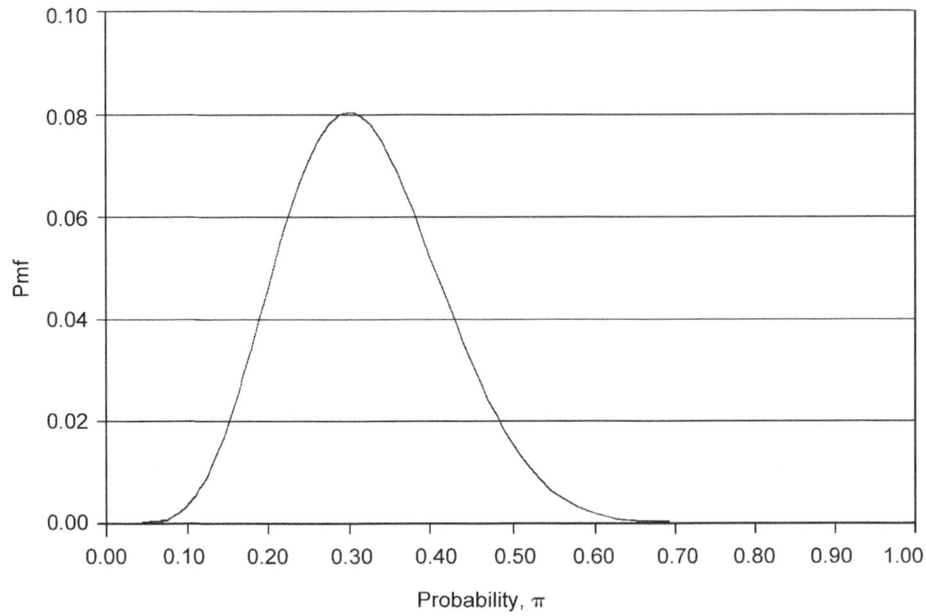

Figure 11-9. Posterior probability distribution over the probability of detection, π

variables (i.e., zeros or ones with some fixed probability), and thus the likelihood of observing six "hits" out of twenty attempts can be modelled by a binomial PMF:

$$f(x|n,\pi)=\binom{x}{n}\pi^{x}(1-\pi)^{n-x}$$
$$=L(\pi|x,n) \qquad (11\text{-}12)$$

in which $x=6$, the number of "hits", and $n=20$, the number of attempts. From Bayes' theorem:

$$f_{\pi}(\pi|x,n)\propto f_{\pi}(\pi)L(\pi|x,n)$$
$$\propto(1.0)\pi^{x}(1-\pi)^{n-x} \qquad (11\text{-}13)$$

in which the prior PMF is taken to be uniform over [0,1], $p_{\pi}(\pi)=1$, and terms not involving π are ignored (Figure 11-9). The maximum (mode) of the posterior PMF is at $\pi=6/20$, corresponding to the naïve estimate, although the distribution is not perfectly symmetrical.

Prior probabilities

A criticism of probabilistic induction is that the probabilities must start from somewhere, so what to do in the case of ignorance? A sticking point in inductive inference is specifying a prior distribution in this case. Bayes (1763) introduced a postulate that says, in the case of ignorance, every possible value of an uncertain quantity should be given equal probability. This suggests, for example, a uniform prior on a parameter such as π in the example above. Jeffreys (1948) called this the principle of indifference.

A difficulty with this approach is that the resulting prior probability distributions expressing total ignorance are not invariant to transformations of variable. For example, if one is ignorant of the relative probabilities of values of π, one would be equally ignorant of the relative probabilities of values of π^2, so why not specify a uniform prior on the latter? Doing so results in a slightly different posterior on π. This problem is discussed at length by Jeffreys (1948) and Zellner (1971), and a variety of pragmatic justifications have been suggested by other authors (Box and Tiao, 1992; Lindley, 1971). Given the great many assumptions in any modelling activity – for example, the assumption that a random variable is Binomial, that the observations are independent, or even that the process is stationary – the invariance introduced by the flat prior seems inconsequential in this mix. In practical situations with little prior information, as the number of observations grows, the posterior PMF or PDF rapidly approaches the likelihood function.

Non-informative priors

The case of assigning prior probabilities to discrete variables when no information is available – such as "presence" vs "non-presence" of a geological anomaly – is conceptually straightforward. There are n categories, each is assigned the same prior probability, $1/n$, and the sum is 1.0. But, what to do about variables with infinite domain?

Consider an event such as extreme flood or severe earthquake. This is usually modelled as a Poisson process with parameter, λ, describing the number of occurrences per unit time:

$$f_n(n\,|\,\lambda, t)=\frac{(\lambda t)^n e^{-\lambda t}}{n!} \tag{11-14}$$

in which n is the number of occurrences, and t is duration. The parameter λ has domain $[0,\infty]$. What prior distribution should be used? A "flat" PDF, such as $f_\lambda^0(\lambda)\propto k$, k=constant, is improper because its area is infinite; and as before, one could just as reasonably make the PDF flat over λ^2, or λ^3, and so on.

Several authors, notably Jeffreys (1948), have considered the best choice of non-informative prior distributions. Jeffreys argues that prior probabilities should be taken as uniform over the parameter for variables with domain $(-\infty, \infty)$, and uniform over the logarithm for variables with domain $[0, \infty)$. The latter becomes proportional to the inverse of the parameter in arithmetic space:

$$f_X^0(x)\propto\begin{cases} k & -\infty\le x\le\infty \\ x^{-1} & 0\le x\le\infty \end{cases} \tag{11-15}$$

Jeffreys presents a long and subtle discussion of the issues surrounding PDFs to represent ignorance. He notes that both Bayes and Laplace used uniform prior distributions in problems relating to sampling, but that this choice is not always satisfactory.

The so-called Jeffreys prior, $f_x(x)\propto 1/x$, is widely used in practice. Jeffreys argues for this prior on the basis of power transformation invariance. Sivia (1996) argues for it on the basis of scale transformation invariance. Zellner (1971) summarises much of the Jeffreys argument in favour of the prior. Suggestions of other non-informative priors are tabulated in Baecher and Christian (2003).

Another line of reasoning in selecting non-informative priors is based on the information content in the PDF. One measure of information common in modern communications theory is Shannon's Entropy (Shannon, 1948):

$$H=\int f_X(x)\log f_X(x)dx \tag{11-16}$$

in which $f_X(x)$ is the PDF of x, and H is said to be the informational entropy. This approach has been championed by Jaynes (Jaynes, 2003; Sivia, 1996) and by Harr

(1987). Baecher and Christian (2003) summarise maximum entropy priors for various constraints and levels of information.

Jeffreys made a broader observation respecting the non-invariance of prior distributions based on Fisher's information matrix. Fisher (1925) proposes a measure of the information contained in a set of observations $\mathbf{y}=\{y_1, \ldots, y_n\}$ with respect to a parameter θ to be:

$$I_\theta=E_{\mathbf{y}|\theta}\left(\frac{\partial^2 \log f(\mathbf{y}|\boldsymbol{\theta})}{\partial\theta_i\partial\theta_j}\right) \tag{11-17}$$

Note, that $f(\mathbf{y}|\boldsymbol{\theta})$ is the likelihood of $\boldsymbol{\theta}$. Jeffreys observed that taking the non-informative prior proportional to the square root of the determinant of the information matrix:

$$f(\boldsymbol{\theta})\propto|i_\theta|^{\frac{1}{2}} \tag{11-18}$$

provides invariance to any one-to-one differentiable transformation $\eta=g(\theta)$. That is, similarly taking the prior $f(\boldsymbol{\eta})\propto|I_{\boldsymbol{\eta}}|^{\frac{1}{2}}$ leads to a consistent posterior probability for both cases (Zellner, 1971).

Consider again sampling from a binomial process with parameter π. The likelihood of an individual observation is:

$$f_x(x|\pi)\propto\pi^x(1-\pi)^{1-x} \tag{11-19}$$

in which x is the observation (either zero or one), and π is the probability of success on the individual trial. Thus, since $E[x|\pi]=\pi$:

$$I_\pi=E_{x|\pi}\left(-\frac{\partial^2 \log f(x|\pi)}{\partial\pi^2}\right)$$

$$=\pi^{-1}(1-\pi)^{-1} \tag{11-20}$$

and the non-informative prior PDF on π should be taken proportional to:

$$f_\pi(\pi)\propto\pi^{-\frac{1}{2}}(1-\pi)^{-\frac{1}{2}} \tag{11-21}$$

Note, this differs from the naïve prior, $f_\pi(\pi)\propto k$, a constant. Box and Tiao (1992) discuss the justification for this rule in more detail.

Informative and conjugate priors

It is sometimes the case that prior information exists about the value of a parameter or other uncertainty so that the prior distribution is informative rather than non-informative. This is the approach, for example, used by McCann *et al.* (1985), McDonald for Hume Dam (2002) and Foster *et al.* (2000) of adopting statistically observed rates of failure as a prior or base-rate probability from which to adjust up or down using site specific information.

While prior information can be modelled by any probability distribution, in practice it is convenient to choose a functional form for the prior distribution that simplifies multiplication by the likelihood function in Bayes' theorem. The most common choice is a distribution closed under multiplication by the likelihood function. This is called the conjugate (or natural conjugate) distribution.

Continuing the example of binomial sampling, the likelihood of an individual observation x_i is as equation 11.14, thus the inference from a set of observations $\mathbf{x}=\{x_1, \ldots, x_n\}$ on the parameter π by Bayes' theorem is:

$$f_\pi(\pi|\mathbf{x}) \propto f_\pi(\pi)L(\pi|\mathbf{x})$$
$$\propto f_\pi(\pi)[\pi^m(1-\pi)^{n-m}] \tag{11-22}$$

in which $m=\Sigma x_i$. Taking a prior PDF in the form of a beta distribution:

$$f_\pi(\pi) \propto \pi^{a-1}(1-\pi)^{b-1} \tag{11-23}$$

with parameters (a, b) yields a posterior PDF of the same beta form:

$$f_\pi(\pi|\mathbf{x}) \propto [\pi^{a-1}(1-\pi)^{b-1}][\pi^m(1-\pi)^{n-m}]$$
$$\propto \pi^{a+m-1}(1-\pi)^{b+(n-m)-1} \tag{11-24}$$

but with updated parameters, $a'=a+m$ and $b'=b+(n-m)$. More detailed discussions of conjugate distributions are presented by Jeffreys (1983), Raiffa and Schlaifer (1971), Zellner (1971) and Box and Tiao (1992).

Base-rate adjustments

For the simpler case of updating from a base rate, the Bayesian calculation is correspondingly simpler. Presume that the statistical base rate of some performance variable, y, for some class of structures, S, is 10%. That is, 10% of that class of structures exhibits the performance; thus the prior probability of y is assigned the value, $p(y|S)=0.1$. Now, say that some site-specific condition, x, is observed. For example, this might be the presence of dispersive soils, or a fractured rock foundation, or some other geotechnical factor. The base-rate probability can be updated by Bayes' theorem in the normal way,

$$p(y|S, x)=\frac{p(y|S)L(y|S, x)}{p(y|S)L(y|S, x)+p(\bar{y}|S)L(\bar{y}|S, x)}$$
$$=\frac{(0.1)(0.2)}{(0.1)(0.2)+(0.9)(0.033)}=0.4 \tag{11-25}$$

in which $\bar{y}$ is the complement of y, and the numerical values are taken from Table 11-1. So, knowing that this particular structure is associated with the factor, x, increases the probability of the performance variable, y, from the base rate 0.1 to a posterior probability 0.4. Note, that *ad hoc* scoring rules that are not based on likelihood ratios are likely to lead to incorrect results.

The only difficulty of importance in adjusting probabilities of failure from historical failure rates resides in the quality of the historical data record (McDonald, 2002). The problem is twofold: first, the number of dam failures in the historical record is small; and second, the failures that have occurred have disproportionately been among older, poorly engineered dams, not operated by a competent authority. Thus, the failures may not be representative of the population of dams for which *ex ante* assignments of probability are desired.

Table 11.1. Example frequency data for updating from a base-rate

	Exhibiting Y	**Not exhibiting Y**
Base rate	10%	90%
Condition exists	20%	80%
Condition does not exist	3.3%	96.7%

Conceptually, this adjustment method is a standard approach, widely used by the insurance industry for establishing premiums on property and casualty and on life insurance policies (Bühlmann, 1970). In insurance industry risk assessment, the adjustment approach has led to a widely used procedure called the credibility formula which combines historical frequencies with case-specific data to arrive at an adjusted premium.

Predictive distributions

The inferences from sample observations discussed above lead to posterior PDFs on the parameters of the underlying processes. Often, however, we are not so concerned with uncertainty in the parameters as we are with the uncertainties in future outcomes of the process itself. For example, from the historical record of flood or earthquake events, we might use an exponential distribution and infer a PDF on the rate of occurrence, λ. The question becomes, what uncertainty does this imply about the number of future events, and what time duration can we expect between events?

Common practice is to integrate out the parametric uncertainty to obtain a predictive probability distribution. For example, the PMF of n marginal of uncertainty in λ is found by integrating the model forecast over the uncertainty in the parameter (Aitchison and Brown, 1969):

$$f_n(n) = \int_\lambda f_n(n, \lambda) d\lambda = \int_\lambda f_n(n \mid \lambda) f_\lambda(\lambda) d\lambda \tag{11-26}$$

Thus, presume that we have observed $\mathbf{n} = \{n_1, \ldots, n_k\}$ numbers of events in k unit time periods. From equation 11-14, starting with a non-informative prior on λ, the posterior PDF on λ is gamma with parameters $(\Sigma n_i, k)$, and thus the predictive PMF on the future number of events per unit of time, n, is:

$$f_N(n) = \int_\lambda f_N(n \mid \lambda) f_\lambda(\lambda \mid n_0) d\lambda$$

$$\propto \int_\lambda \left(\frac{\lambda^n e^{-\lambda}}{n!} \right) (\lambda^{\Sigma n_i - 1} e^{-(k+1)\lambda}) d\lambda \tag{11-27}$$

which, upon integrating, becomes the Negative-Binomial PMF (Aitchison and Dunsmore, 1975):

$$f_N(n \mid \Sigma n_i, k) = \binom{n + \Sigma n_i - 1}{\Sigma n_i} \left(\frac{1}{k+1} \right)^n \left(1 + \frac{1}{k+1} \right)^{\Sigma n_i} \tag{11-28}$$

Note that the predictive distribution may combine uncertainty due to natural variability, for example, the number of flood occurrences in a period of time, with uncertainty due to limited knowledge, that is, the uncertainty in the rate parameter. If more than one aspect of natural variability depends on the same uncertain parameter (e.g. number of occurrences in two separate periods, n_1 and n_2) then forming the predictive distribution of each in isolation from the other may mask an implied correlation caused by their shared dependence on the same uncertain realisation of λ.

A point of caution in dealing with predictive distributions is that, if the conditional distribution $f_X(x \mid \theta)$ changes rapidly within the interval of interest of θ, then reliance is being placed on the precise shape of the PDF of θ within that region. While the resulting predictive PDF is theoretically correct, errors in the specification of $f_\Theta(\theta)$ can result in large errors in $f_X(x)$.

Covariance among variables

When dealing with more than one variable, uncertainties in one may be associated with uncertainties in another. That is, the uncertainties may not be independent. Consider estimating the "cohesion" and "friction" parameters of a Mohr–Coulomb strength envelope. If the slope of the envelope, ϕ, is mistakenly estimated too high, then for the line to fit the data, the intercept, c, will have to be too low. The reverse is true if the slope is too low. Thus, uncertainties about the slope and intercept are associated with one another. Dependencies among events or among the uncertainties in estimates can be critical to obtaining proper numerical results in reliability analyses. To complicate matters, these dependencies can also be subtle to identify and difficult to estimate.

Independence means that knowing whether or not one event obtains in no way changes the probability of a second event. Described mathematically, independence requires,

$$\begin{aligned} Pr(A \text{ and } B) &= Pr(A)Pr(B|A) \\ &= Pr(A|B)Pr(B) \\ &= Pr(A)Pr(B) \end{aligned} \tag{11-29}$$

in which the probability of *A-and-B* is the probability of the joint occurrence, and the conditional probabilities $Pr(B|A)$ and $Pr(A|B)$ are the probabilities that one event obtains given that it is known that the second one does. Thus, independence means that the conditional probabilities equal the marginal probabilities, $Pr(A|B)=Pr(A)$ and $Pr(B|A)=Pr(B)$.

Causes of covariation and dependence

Probabilities associated with separate events or uncertainties can be made dependent through a number of circumstances.

Causal dependence means that one event physically causes another; for example, liquefaction-induced settlement may directly lead to overtopping of an embankment, thus the liquefaction event and the overtopping event are not be independent of one another. If liquefaction settlement occurs, the probability of overtopping is enhanced.

Probabilistic correlation means that two uncertainties may share a common dependence on a third uncertainty, as in the case of the low-density soil lenses in the example above. The realisation that the low-density soil lenses exist (or do not exist) simultaneously changes the probabilities of both liquefaction cracking and of overtopping.

Spatial or temporal autocorrelation means that two uncertainties depend on the spatial or temporal realisation of some third uncertainty which itself exhibits stochastic dependence in space or time. The performances of two sections of a long levee may depend on soil engineering properties in the naturally occurring valley bottom, which, when modelled as a stochastic (aleatory) process, exhibit a long wavelength of correlation in space; thus adjacent sections will exhibit similar settlements or factors of safety against strength instability.

Statistical correlation means that two uncertainties are simultaneously estimated from a given set of data and, therefore, are influenced by a common sampling variability error. In soil mechanics, a common – and mostly overlooked – statistical correlation is that between soil cohesion and friction angle, which being regression parameters, are negatively correlated.

Covariance and the correlation coefficient

The most common measure of dependence among uncertain quantities is the correlation coefficient. This measures the degree to which one uncertain quantity varies linearly

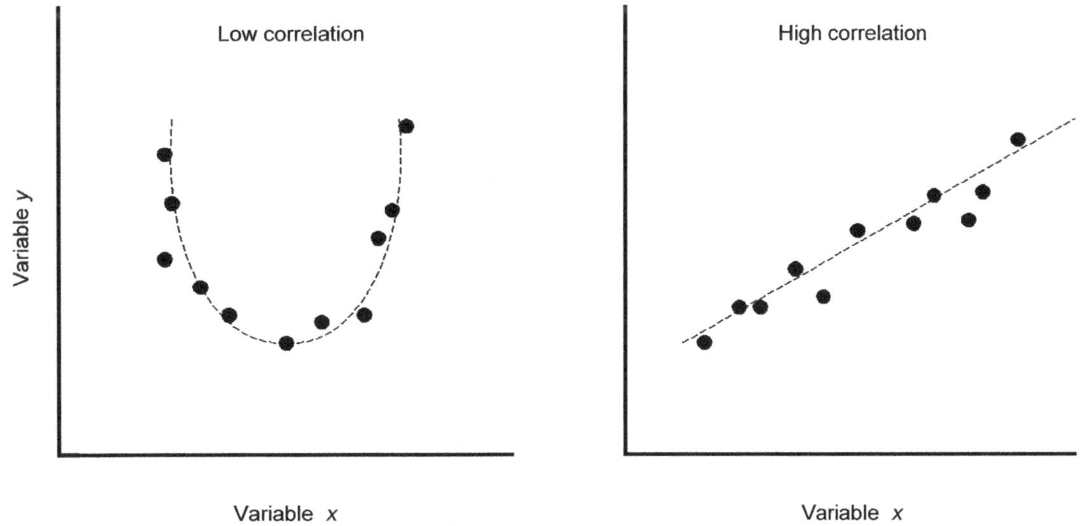

Figure 11-10. Correlation measures the degree of linear dependence between variables

with another uncertain quantity. Note, the notion of linearity here is important: two uncertain quantities may be deterministically related to one another, yet have negligible correlation if the relationship is strongly non-linear (Figure 11-10).

The correlation coefficient for two uncertain quantities x and y is defined as the ratio of the covariance of x and y to the square root of the product of the variances of x and y (i.e. the product of the standard deviations of x and y):

$$\rho_{xy} = \frac{Cov(x, y)}{\sqrt{Var(x)Var(y)}}$$

$$= \frac{E[(x-\mu_x)(y-\mu_y)]}{\sqrt{E[(x-\mu_x)^2]E[(y-\mu_y)^2]}} \tag{11-30}$$

The correlation coefficient varies within [–1, +1], with the higher bound implying a strict linear relation of positive slope and the lower bound a strict linear relation of negative slope. The higher the magnitude, the more closely the data fall on a straight line. Zero correlation coefficient implies no (linear) association between x and y. The correlation coefficient is equivalent to a normalised product moment of inertia in solid mechanics. It expresses the degree to which two parameters vary together. The correlation coefficient is non-dimensional because deviations of x and y are measured in the same units as their respective means.

Choice among distribution models

Another common problem in fitting distribution or other models to data is the comparison among models. For example, are the data better fit by a logNormal vs a gamma PDF, or are the data better fit by alternative aeromechanics models such as different constitutive relations?

As a generalisation, suppose there are k alternative models, $\{M_1, \ldots, M_k\}$. For each model there is some density function of the set of observations, $f_i(\mathbf{z}|\boldsymbol{\theta}_i)$, but with unknown parameters, $\boldsymbol{\theta}_i$. The subscript refers to the model. The *a priori* density function over the parameters is, $f(\boldsymbol{\theta})$, which could be non-informative (i.e. uniform in some sense). Then, for any given model, the marginal PDF of the observations, **z**, over the possible values of the parameters, is:

$$f_i(\mathbf{z}) = \int f(\theta_i) f(\mathbf{z} \mid \theta_i) d\theta_i \qquad (11\text{-}31)$$

and the *a posteriori* PDF of the parameters for each model, given the data is:

$$f_i(\theta_i \mid \mathbf{z}) = \frac{f(\theta_i) f(\mathbf{z} \mid \theta_i)}{\int f(\theta_i) f(\mathbf{z} \mid \theta_i) d\theta_i} \qquad (11\text{-}32)$$

If the *a priori* probabilities of the models, respectively, are $P(M_i)$; then applying Bayes' theorem, the a posteriori probabilities, given the data, **z**, are

$$Pr(M_i \mid \mathbf{z}) = \frac{Pr(M_i) f_i(\mathbf{z})}{\sum_i Pr(M_i) f_i(\mathbf{z})} \qquad (11\text{-}33)$$

Consider the hypothesis test that a set of n data are modelled with a Normal distribution vs a Uniform distribution. Berger and Pericchi (2001) shows that the marginal PDFs of the observations for these two distributions (among others), are:

$$f(\mathbf{z} \mid M_N) = \frac{\Gamma[(n-1)/2]}{2\pi^{(n-1)/2}\sqrt{n}\left[\sum_j (z_j - \bar{z})^2\right]^{(n-1)/2}} \qquad (11\text{-}34)$$

$$f(\mathbf{z} \mid M_U) = \frac{1}{n(n-1)[z_{max} - z_{min}]^{(n-1)}} \qquad (11\text{-}35)$$

in which the subscripts, N and U, refer to the Normal and Uniform distributions.

As an example, $n=31$ data points were generated from a [0, 1] Uniform random number generator by adding sets of four values each. By the central limit theorem, as the number of variables in the sum increases, the PDF should approach Normality. On the other hand, the un-summed variables are by definition Uniform. The resulting statistics were: $n=31$, $\bar{z}=2.02$, $\Sigma(z_j - \bar{z})^2=3.10$, $z_{max}=2.87$, and $z_{min}=1.06$. Thus, the respective marginal PDFs become, $f(\mathbf{z} \mid M_N)=7.16\text{E-}10$ and $f(\mathbf{z} \mid M_U)=2.00\text{E-}11$. The *a posteriori* probability for the Normal hypothesis is then, $P(M_N)=0.97$. Repeating the experiment for sums of two uniform variables led to $n=31$, $\bar{z}=1.08$, $\Sigma(z_j - \bar{z})^2=4.60$, $z_{max}=1.74$, and $z_{min}=0.16$. For this case, the posterior probability for the Normal hypothesis was reduced to $P(M_N)=0.60$. For the case of one Uniform variable (by definition, generated from a Uniform PDF), the corresponding statistics were, $n=31$, $\bar{z}=0.39$, $\Sigma(z_j - \bar{z})^2=2.12$, $z_{max}=0.96$, and $z_{min}=0.03$; and the corresponding *a posteriori* probability for the Normal hypothesis was, $P(M_N)=2 \times 10^{-5} \rightarrow 0$.

11.4 Quantified probability assignments

The fundamental concepts of statistical estimation and inference are well described, and in greater detail than possible here, in many texts. For example, Benjamin and Cornell (1970) and Ang and Tang (1975) provide readable introductions to basic statistical concepts in the context of civil and environmental engineering; while Feller (1967, 1971) provides a general introduction of substantial depth. The present discussion presents an overview of pertinent points.

Exploratory data analysis

Engineering data on soil or rock mass properties are usually scattered. Graphical and simple probabilistic methods are useful in summarising this scatter so that a better understanding of the data – and of the corresponding uncertainties associated with engineering performance – can be developed.

Histograms and frequency diagrams

The most common way to present data is in a histogram, which graphs the number of measurements falling within specific intervals of value as a vertical bar (Figure 11-11). Thus, a histogram is sometimes called a bar chart. The height of the bar above each interval shows the number of measured values within the interval. The sum of the heights of the bars equals the total number of measurements. The histogram divides data into fixed intervals. The choice of intervals is arbitrary, but they should be of uniform width and have convenient end points. If too many intervals are chosen, the general picture of relative frequencies will be jagged and, conversely, if too few intervals are chosen the general picture will be blurred. A frequency distribution is constructed from a histogram by dividing each vertical bar by the total number of measurements. This gives the relative frequency of observed values in each interval as a decimal fraction. The cumulative frequency distribution sums the number of data less than or equal to a particular value.

A histogram gives a summary view of variation. The shape suggests whether the data have a central tendency and, if so, where along the x-axis the bulk of the data is concentrated. The width suggests the dispersion or scale of variation. Some histograms have one point of concentration and are thus called unimodal. Others have more than one and are called multimodal. Usually, hydrological and geotechnical data have unimodal distributions. Multimodal distributions may indicate an inhomogeneous mixture of data from different soil types, hydraulic regimes, construction procedures, or sites. The histogram also shows whether the variation is symmetric or asymmetric, that is, whether high and low variations are evenly balanced and whether variations from the central tendency are more frequent on one side than on the other.

The copper porphyry tailings data on the RHS of Figure 11-11 are not symmetric about a peak frequency, but are skewed with a long upper tail. The highest frequency occurs near the lower end of the scale, at about 2 psi (13.8 kPa), and while frequencies decline on both sides of the peak, they do so more slowly on the upper side. Such distributions are said to be skewed.

To construct a histogram, a simple procedure can be used: (1) divide the data range into about five to ten intervals of constant width; (2) count the number of data having values within each interval; and (3) plot this number as a vertical bar. Five to ten intervals are used because this number typically allows a sufficient number of data in each interval for the observed frequencies to vary smoothly, yet provides adequate definition to the shape of the distribution. If larger numbers of data are available, larger numbers of intervals might be used. For small numbers of data, a convenient rule-of-thumb for choosing the number of intervals is:

$$k = 1 + 3.3 \log_{10} n \qquad (11\text{-}36)$$

in which k = the number of intervals, and n = the number of data values (Sturges, 1926). Usually it is convenient to specify interval boundaries to one fewer decimal place than that to which the data are measured, avoiding the problem of where to place values falling directly on an interval boundary. When this is not possible, a consistent procedure should be adopted for deciding how to count data that fall directly on a boundary. Since the choice of number of intervals can affect visual interpretation of data scatter, it is sometimes useful to construct more than one histogram, using a different number of intervals for each.

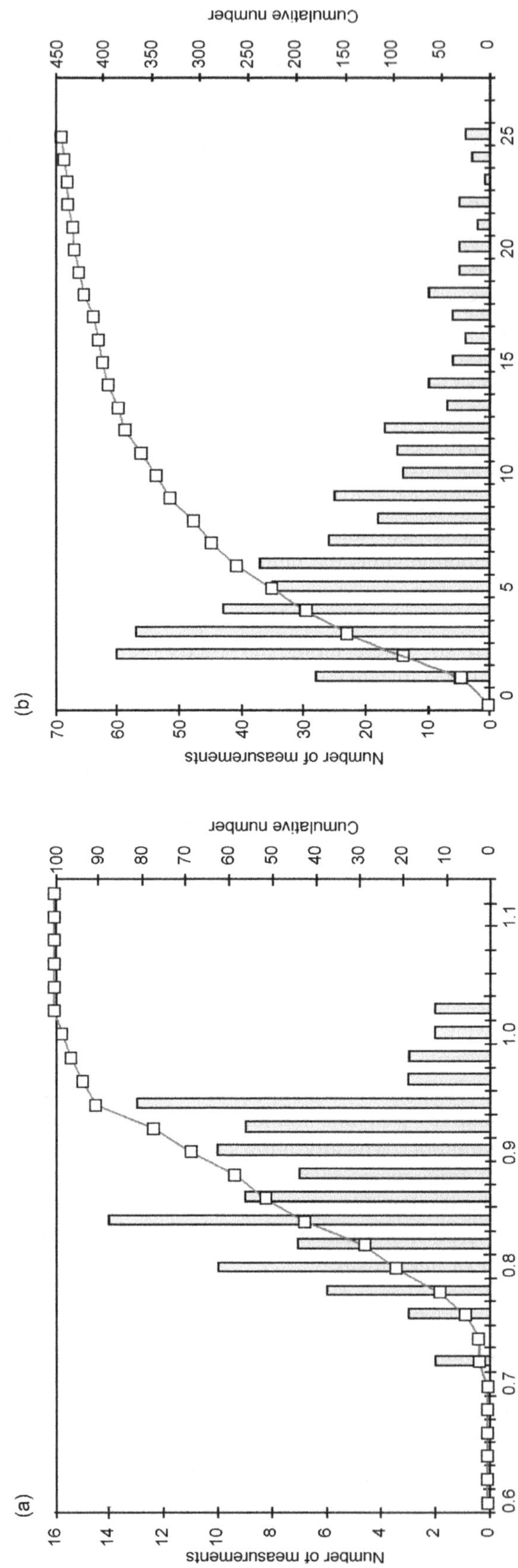

Figure 11-11. Variability of soil property data as shown in histograms: (a) liquidity index; (b) Cone penetration resistance

A cumulative distribution is constructed by summing relative frequencies starting at the lower end of the data and proceeding toward the upper end. The cumulative frequency distribution gives the fraction of measurements less than or equal to a particular value:

$$F_X(x_0) = \textit{fraction of measurements} \leq x_0 \tag{11-37}$$

An empirical cumulative frequency distribution is found by plotting the data against fraction-less-than. The plotting positions in a cumulative frequency distribution are usually taken as:

$$F(x_m) = \frac{m}{N+1} \tag{11-38}$$

in which $F(x_m)$ is the cumulative frequency of the mth point, with $m=1$ being the lowest value, and N being the total number of points (McCuen and Snyder, 1986). Sometimes the plotting points are taken as $F(x_m)=m/N$, for all but the largest value, which is then plotted as in equation 11.3. An advantage of the cumulative distribution is that the data do not need to be grouped into an arbitrary number of intervals. A disadvantage is that the shape of the distribution is not as readily apparent as in the histogram.

Summary statistics

Frequency distributions are convenient representations of data for visual inspection, but statistics of distribution characteristics are often more useful for calculations or for setting standards. The term statistic refers to any mathematical function of a set of measured data, but the most useful usually have to do with central tendency and dispersion of the data.

The common measures of central tendency are the mean, median and mode. The mean of a set of n data $x=\{x_1, \ldots, x_n\}$, sometimes denoted $\bar{x}$, is the arithmetical average:

$$\bar{x} = \frac{1}{n}\sum_{i=1}^{n} x_i \tag{11-39}$$

The mean is the centre of gravity of the frequency distribution along the x-axis. For example, the mean of the liquidity index data above is 0.87, while that of the cone penetration data is 7.13. The median of the set of data, denoted $x_{0.5}$, is that value of x_n for which half the data are less and half more. Correspondingly, the cumulative distribution evaluated at the median is 0.5:

$$F_X(x_{0.5}) = 0.5 \tag{11-40}$$

The median is the midpoint of the data when listed in increasing or decreasing order. Common practice in the case of an even number of data is to define the median as intermediate between the two middle data. The mode of a set of data is the most common data element or interval, i.e. the peak. For theoretical distributions the mode is usually obvious, but in sample data it can be erratic or unstable depending on the vagaries with which data appear. For the liquidity index data the mode is 0.84, although it might nearly have been 0.94.

Common measures of dispersion are the standard deviation, range, and inner quartiles of the frequency distribution. The standard deviation is the root-mean-square (rms) value of the data, the range is the spread between maximum and minimum values, and the inner quartiles bracket the middle 50% of the data. The standard deviation of a set of data $\mathbf{x}=\{x_1, \ldots, x_n\}$, denoted s_x, is,

$$s_x = \sqrt{\frac{\sum_{i=1}^{n} (x_i - \bar{x})^2}{n}} \tag{11-41}$$

in which $\bar{x}$=the mean of the data. Sometimes the denominator $(n-1)$ rather than (n) is used to correct a statistical bias, due to the fact that the mean, too, must be estimated from the same data. The coefficient of variation of a set of data is defined as the standard deviation divided by the mean:

$$\Omega_X = \frac{s_x}{\bar{x}} \tag{11-42}$$

which expresses relative dispersion. The variance of a set of data, sometimes denoted $Var(x)$ or σ_x^2, is the square of the standard deviation:

$$Var(X) = \frac{1}{n} \sum_{i=1}^{n} (x_i - \bar{x})^2 \tag{11-43}$$

In many statistical calculations the variance is a more convenient measure than the standard deviation. The variance is the moment of inertia of the frequency distribution about $\bar{x}$. The range of a set of data, denoted r, is the difference between the largest and smallest values:

$$r_x = |x_{max} - x_{min}| \tag{11-44}$$

The range has poor statistical properties in that it is sensitive to extreme values in a data set; however, it is easily evaluated and therefore often useful. The inner quartiles of a set of data, denoted $x_{0.25}$ and $x_{0.75}$, are the data values for which one-quarter of the data are smaller and one-quarter larger, respectively. The quartiles may be found from the cumulative distribution as:

$$F(x_{0.25}) = 0.25 \text{ and } F(x_{0.75}) = 0.75 \tag{11-45}$$

The inter-quartile range, denoted $r_{0.5}$:

$$r_{0.5} = (x_{0.75} - x_{0.25}) \tag{11-46}$$

is less influenced by extreme values than is the range itself, but more troublesome to compute.

Unplanned data

Statistical approaches to estimating relationships are based on correlations, and thus are limited by the degree to which calibrating data are appropriate to the population of dams for which predictions are sought. That is, the normal limitations of sampling theory pertain to the calibrating data. For example, in establishing base rates of dam failure from historical observations, one should think of the set of dams that have failed in the past as a sample from some population. Conceptually, from this sample we are trying to estimate the characteristics of a sampled population using statistical theory. However, the sample has not been taken according to some *ex ante* plan, but comes about according to which dams happened to have failed. Obviously, that is always the case with historical data, but it means that the set of failures is not a random sample of dams and may disproportionately include small and poorly engineered or poorly maintained structures, often operated by other than competent authority. The objective, on the other hand, is to estimate the characteristics of a target population of dams that are the subject of the risk analysis. In contrast to the sampled population, these are likely to be modern, well-engineered dams. In principle, we can estimate the characteristics of the sampled population from characteristics of the observations, and the reliability of these estimates can be quantified with statistics. But, the characteristics of the target population can only be estimated from the characteristics of the sampled population by judgement, and the reliability of these estimates can only be determined subjectively. This is a conceptual limit to which judgement must be brought to bear in interpreting the results of a statistical analysis.

Probability models

Mathematical models of probability distributions are commonly used in engineering practice either because: (1) reasoning from first principles suggests that certain natural phenomena should exhibit certain mathematical regularity; or (2) computational convenience combined with empirical verification supports the use of simplifying mathematical forms with which to describe relative probabilities.

Probability models for initiating events: flood, earthquakes, and piping

The most common places that statistical estimates are used is in estimating probabilities of initiating events such as extreme floods or earthquakes. These are usually expressed as annual exceedance probabilities (AEP): the probability that an event equal-to or larger-than some magnitude occurs in any given year. An example is the familiar flood-frequency curve.

Statistical models used for different categories of initiating events can be quite different due to the distinctive nature of these events. It is the nature of events which dictates how statistical models are – or can be – constructed. Typically, three events are considered as plausible initiating events which may lead to a failure in dam safety studies. These are:

- floods;
- earthquake ground shaking;
- internal erosion (piping).

Sometimes piping is categorised as design and construction flaws, although piping failure can occur in a dam without flaws, for example, as a result of an earthquake causing displacement and discontinuity in the core or filter. However, the piping failure mode may be present from the moment of the dam design or construction because of an inherent flaw, which may remain dormant or undetectable until other contributing factors reach certain states. Although these other factors are necessary to conclude the failure, they could not by themselves cause the failure to occur. Therefore, it is the flaw that initiates the failure. It is quite clear that this is a chance event and any statistical model needed for derivation of probabilities has to be quite different from that for floods and earthquakes.

In complex, multi-reservoir systems, miscalculation (human error) or flawed operation of the system combined with unavailability or only partial availability of discharge facilities may lead to the same effect as flood, that is to overtopping and consequently to a failure of the dam. Again, this can be probably captured through part of an analysis that starts with floods, but in real case analysis this might be accounted as a fourth initiating event, that is operation or human error. This fourth potential initiating event in itself also requires different statistical models for assigning probabilities.

Floods

Probabilistic analysis of floods dates back many decades and has been primarily concerned with statistical analysis of flow data (annual maximum stream flow) for establishing design floods. Common practice has been to fit a statistical distribution (model) to a sample of flood data and then use the model to estimate required quantiles. These quantiles have been traditionally referred to as floods with given return periods. Although the literature on the subject of statistical procedures and techniques available is abundant, Maidment's (1993) handbook provides an excellent source of information on techniques based on classical approach. Flood frequency curves obtained from such analyses have been used to make inferences about the magnitude of floods with the return periods greater than the periods of record.

It is known, that a data sample of size n, in the absence of *a priori* distributional assumption can furnish information only about exceedance probabilities greater than approximately $1 / n$. Of course, one may fit a distribution to a sample of size 20 and use it to estimate 100-year or 10 000-year floods. The user must recognise, however, that such a model is based on extrapolation and that it begs the question of the form of the tail.

Until the early 1990s the prevailing "rule of thumb" was that "with a reasonable sample length" extrapolation can be extended to a 100-year or "if necessary" to a 200-year return period (Interagency Advisory Committee on Water Data, 1982). In 1983, Yevjevich discussed the problem of limits of extrapolation of frequency curves and indicated that US federal agencies recommended a 500-year return period as the limit. Most recently, with the emergence of risk assessment for dam safety, there has been growing demand for probabilities of floods significantly less frequent than the 200-year flood. The Institution of Engineers, Australia (1972) recommends 2000 years as a credible limit of extrapolation. The Joint Report of Utah State University and Bureau of Reclamation (1999) contains a guide for various limits depending on the kind of data used for analysis (single-site, paleo – flood, or regional) with an interval of 40 000 to 100 000 years as the limit for the combination of all of the above. Although the limits are characterised as "credible" there is no scientific (i.e. mathematical) theory supporting the claim. Credibility could have been established if exact, or at least asymptotic confidence limits accounting for all sources of uncertainty in quantile estimates had been determined but that at the present is not possible.

The major obstacle in improving the quality of estimates is limited length of record. One of the ways to resolve this problem has been multi-site or regional flood frequency analysis. The general concept behind the regional approach is that because more information is available than in a single-site analysis, there is an opportunity to achieve better accuracy in model fitting. Although it is true that regional analysis may result in substantial increase of the sample size, it also creates certain difficulties in observing some of the premises constituting the foundation of the approach. In effect, the gains from increased sample size available for analysis are to a large extent negated by the need to relax the conditions that were required to make the data pooling available. Two essential assumptions required for data pooling which cause the biggest problem are homogeneity (identical distributions at all pooled sites) and statistical independence. However, whether the relaxation is done in the interest of mathematical convenience or simply out of the mathematical necessity, it is irrelevant since it essentially invalidates the entire model. Inability to demonstrate that the assumptions were fulfilled would lead to the same conclusion.

Another approach of sample augmentation in flood frequency is the inclusion of historical and paleo-flood information. Historical information includes additional evidence of floods that exceeded a particular stage prior to the period of systematic recording. Paleo-flood information comes from "the study of movements of water and sediment in channels before the time of continuous hydrological records or direct measurements" (Costa, 1988). Statistical approaches applied for analysis of augmented samples are generally based on censored-sample analytic techniques.

One of the cornerstones of frequency analyses is the requirement that the observations analysed have to be independent and identically distributed (IID). In practical terms this means that not only all local natural conditions within the watershed (land use, morphology, etc.) but also global conditions in the surrounding areas (climate) should remain unchanged within the sampling period. Unless the changes in local conditions are drastic, available statistical methods of detection are rather ineffective. It is usually not clear whether known and documented changes in local conditions are insufficient to modify the distributions of the analysed random variables or whether the modifications are simply undetectable by current testing methods.

Global changes related mostly to climate conditions may pose a significant problem for the intended use of flood frequency analysis as a source of probabilistic data input into risk assessment. The issue of global climate change is relatively new and although the implications of potential climate changes on flood frequency analysis has been already recognised, there is no mechanism in place (both from theoretical and practical perspective) with which to address this issue. Since the probability issue of floods in the context of risk assessment is strictly an exercise in prediction, the question is relatively simple: can the past flow records represent the future flows when the generating mechanism is changing?

Although unrelated to climate change, recent research activities exploring climate, weather and stream flow interfaces are bringing new information which can directly affect the way flood frequency should be performed. Findings from recent studies (Jain and Lall, 2000, 2001; Pan *et al.*, 1999; Sarma, 1999) indicate that there is a considerable difference in the frequencies of extreme flood events during La Nina and El Nino years:

- Complex, low-frequency variations in climate states may cause inter-annual, inter-decadal and longer-term variations in flood frequencies by changing patterns of atmospheric moisture transport in the flood seasons.
- Slowly varying climate states may affect finite sample flood statistics and that may lead to non-stationarities in flood records.
- Large-scale climate oscillations may impact stability of hydrologic regimes.

If the impact of weather and climate components is taken into account, it is not surprising that there have been attempts to infer probabilistic characterisation of floods through analysis of the forces driving the process, that is, weather phenomena. However, the unresolved issues of independency and homogeneity of records still prevent these attempts from being a successful alternative to classical flow frequency analysis.

Physical modelling of the atmosphere, region and watershed system based on first principles, if considered in the future and viable, may help in better estimates of probabilistic properties of floods via analysis of synthetic time series generated with the help of such models.

Earthquake hazard

Annual exceedance probabilities for earthquake hazards are often divided in two parts: a point process model describing the occurrence of events (earthquakes) in time and space, and a magnitude model describing the severity of events that do occur. The point process of initiating events is often modelled as a stationary Poisson process, with a single parameter, λ, describing the number of occurrences per unit time, usually a year. This model is:

$$f_n(n \mid \lambda) = \frac{\lambda^n e^{-\lambda}}{n!} \tag{11-47}$$

in which n is the number of occurrences, and λ is the rate per unit time. Stationary means that λ is constant over time. Other point process models could also be used to represent the occurrence of initiating events, for example, if one wanted to introduce probabilistic dependence among the event occurrences, a Negative binomial or related model might be used (Lewis, 1972).

The most important assumption of the Poisson distribution is that events occur independently. Thus, if a large storm occurs this year, the probability of a similar storm occurring next year is unaffected: in principle, one could have two $Pr=0.001$ events (1000-year storms) back-to-back. The probability, p_n, of one or more events occurring in a time period, n years, is:

$$p_n = 1 - e^{-n\lambda} \tag{11-48}$$

The exception to independence in the Poisson model is that, for most natural hazards, the length of historical record is short. New information, such as the occurrence of an extreme storm or a large earthquake, affects the statistical estimate of the parameter λ. Thus, the occurrence of an extreme event changes the parameter of the model of the process itself, and consequently influences the estimated probabilities of future events. For example, in constructing Oroville Dam, two presumably 250-year ($p=0.004$) storms and one 1000-year ($p=0.001$) storm occurred during the few years of construction (Gilbert, 2002). With the occurrence of each of these storms the rate parameter λ for the site changed, as did subsequent estimates of the annual exceedance curve.

The usual approach to quantifying the probability distribution for an uncertain model parameter, such as λ, based on statistical data is to start from some uniform or non-informative probability density function (PDF) on the parameter, presumed to exist prior to having observed the data, and then to update that probability distribution using Bayes' theorem. Modelling the occurrence of initiating events as a Poisson process typically presumes the uncertainty in those events to be due to natural variability in time or space. Given the average rate of occurrence specified by the parameter λ, the actual occurrence of initiating events is then a stochastic process. The only place that knowledge uncertainty enters the formulation is in the statistical estimate of the parameter λ itself (which may be far from negligible).

Usually, an event tree analysis presumes stationarity for stochastic initiating events. This means that the random process remains the same over space or time. However, for events such as flood discharges or river stages such an assumption may not be good. Changing upstream conditions in land use and other factors may cause flood frequencies or rating curves to gradually change with time. This means that the probability distributions used to represent random variables, such as peak flood discharge, may change.

The magnitude of an event, given an occurrence, is typically modelled by a common univariate PDF, often an Exponential or a logNormal distribution. The Exponential PDF is of the form, $f(x \mid \theta) = \theta \exp(-\theta x)$, with mean $E(x) = 1/\theta$, and variance $Var(x) = 1/\theta^2$. The logNormal PDF is of the Normal distribution form over the logarithm of x. In either case, the parameters of the distributions are fit to historical records. Convolving the point process model with the magnitude PDF allows for annual exceedance probability to be calculated and plotted as a plot of magnitude vs return period.

Probability models for engineering parameters

In contrast to probability models for initiating events, risk analysis also requires probability models for engineering parameters from which engineering response is calculated. Considerations on choosing distribution shapes for various engineering parameters, and specifically the interplay of the presumed random process models that generate engineering parameters and the corresponding distributional shapes, are considered below. For many problems it is convenient to approximate probability distributions by a mathematical function. Surprisingly, a comparatively small set of mathematical functions can be used to fit a broad range of frequency distributions encountered in practice. Johnson and Kotz (1969, 1970, 1972) and Evans *et al.* (1993) give mathematical equations, ranges, and moments for a variety of commonly used probability mass and density functions.

In the late 19th century, Karl Pearson attempted to systematise families of probability distributions. The system consists of seven solutions to a differential equation, which also approximate a wide range of distributions of different shapes. Gruska *et al.* (1973) describe how the different Pearson curves can be fit to an empirical distribution. This system, using a simple graph to plot the relationship of families of distributions to functions of their moments, is widely used in hydrologic and geotechnical engineering: an example is given by Lumb (1974). Detailed explanation of Pearson's families of distributions is given by Ord (1972).

Johnson *et al.* (1992) describe another system of frequency curves representing transformations of the standard Normal curve. By applying such transformations, a wide variety of non-Normal distributions can be approximated, including distributions which are bounded on either one or both sides (e.g. U-shaped distributions). The advantage of this approach is that once a particular Johnson curve has been fit, the normal integral can be used to compute expected percentage points under the curve. Methods for fitting Johnson curves, so as to approximate the first four moments of an empirical distribution, are described in Hahn and Shapiro (1967).

In many cases there are physical considerations that suggest appropriate forms for the probability distribution function of an uncertain quantity. In such cases there may be cogent reasons for favouring one distributional form over another, no matter the behaviour of limited numbers of observed data.

Success and failures: the Binomial PMF. The number of dam failures in any year is the sum of all the failures of individual dams across the world. For purposes of illustration, assume that these failures are independent from one dam to another; that is the failure of one dam is assumed for the purposes of data modelling in no way to affect potential failures of other dams; and, furthermore, that no common cause initiating event simultaneously affects failures of multiple dams.[1] Presuming independence among the failures of large dams, let the probability of any one dam failing in a particular year be p. The probability that x dams fail in one year is then, p^x. The probability that the remaining $n-x$ dams do not fail is $(1-p)^{n-x}$. Thus, the probability of x dams out of a total on n dams failing is the product of the probability of x failing and the complement $n-x$ not failing, or $p^x(1-p)^{n-x}$. Since there are many different combinations of how x dams can be chosen from a set of n total dams, the probability distribution (probability mass function) of the number of failures of x out of n dams in any given year is:

$$f(x \mid n)=\binom{x}{n} p^x(1-p)^{n-x} \qquad (11\text{-}49)$$

in which:

$$\binom{x}{n}=\frac{n!}{x!(n-x)!} \tag{11-50}$$

is the number of combinations of n things taken x at a time. The mean of x is $E[x]=np$, and the variance is $Var[x]=np(1-p)$. Equation 11-49 is known as the Binomial distribution, and is originally due to correspondence between Pascal and Fermat.

Random Occurrences: Poisson PMF and Exponential PDF. In surveying rock joints in a dam abutment, sampling lines are laid out across outcrops, borings are cored into the abutment, and the number of fractures intersected by the sampling line or boring is recorded. Jointing patterns in rock masses are known to display distinct orientation clusters, but it is often assumed that the location of fractures within clusters are random (Einstein and Baecher, 1983). Making the assumption that joints of a particular cluster are random and independent, the number intersecting a unit length of sampling line follows the binomial PMF, in which x is the number occurring within a given interval, p is the probability of any one fracture occurring within the interval, and n is the total number of fractures. In the field, the total number of fractures n is large and, thus, it is convenient to replace n and p with their product $\lambda=np$, the density of fractures per unit length. Taking the limit of the binomial PMF as $n\rightarrow\infty$ and $p\rightarrow 0$, while keeping their product λ constant, leads to:

$$f(x\,|\,\lambda)=\frac{\lambda e^{-\lambda}}{x!} \tag{11-51}$$

This is known as the Poisson distribution. The mean of x is $E[x]=\lambda$ and the variance $Var[x]=\lambda$. For sampling widths w other than unit length, the density λ is replaced by the density over w, which is λw, to obtain:

$$f(x\,|\,\lambda, w)=\frac{(\lambda w)^x e^{-\lambda w}}{x!} \tag{11-52}$$

for which $E[x]=\lambda w$ and $Var(x)=\lambda w$.

The distribution of spacings among adjacent rock fractures is obtained by noting that, if starting at a given intersection, the next closest fracture is at distance s, then there must be no occurrences of fractures within the interval $(0, s)$. Thus, setting the sampling width equal to s and making it the argument of the distribution, and setting $x=0$, one obtains:

$$f(s\,|\,\lambda)=\lambda e^{-\lambda s} \tag{11-53}$$

known as the Exponential distribution. For example, Priest and Hudson (1976) have used this model for rock fractures surveys. The mean of s is $E[s]=1/\lambda$ and the variance $Var[s]=1/\lambda^2$ (Figure 11-12).

Adding variables: the Normal PDF. In calculating the limiting equilibrium strength of a potential failure arc through an earthen embankment, the contributing strengths of a number of segments of the arc passing through different zones are added together to estimate total resistance. The contributing strengths of each of these segments is known only with some uncertainty, say up to a mean and a variance. What distributional form is appropriate for total resistance? Figure 11-13 shows the normalised sums of uniformly distributed random variables in which one, two, five and ten variables,

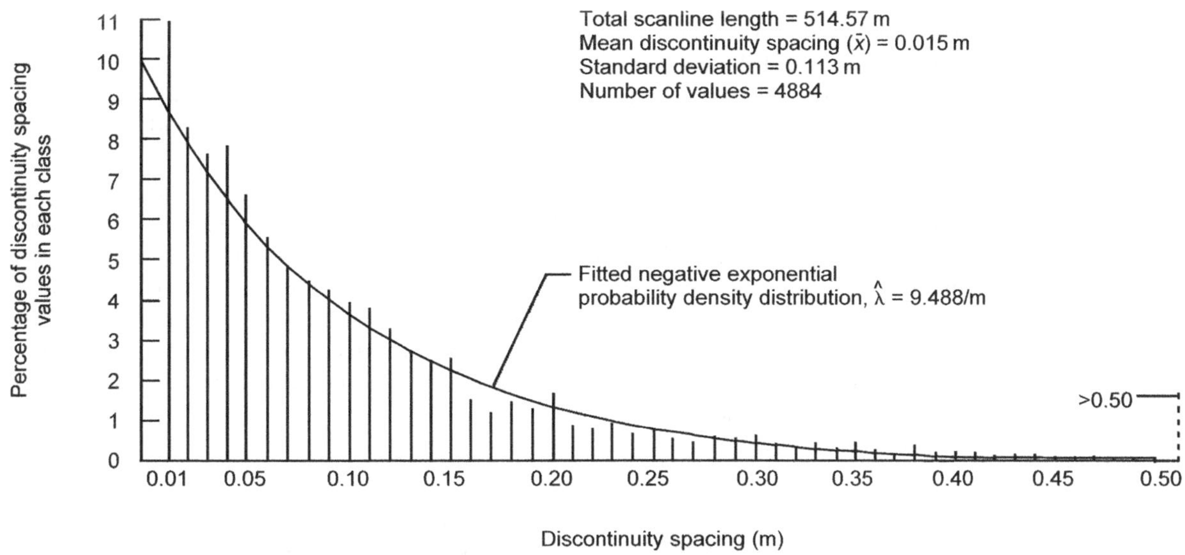

Figure 11-12. Exponential distribution (Priest and Hudson, 1976)

respectively, are summed, and then divided by the number of variables in the sum. Thus, the plot labelled $N=1$ shows 100 realisations of a random variable distributed as:

$$f_x(x) \sim U(0, 1) = \begin{cases} 1 & for\ 0 \leq x \leq 1 \\ 0 & otherwise \end{cases} \tag{11-54}$$

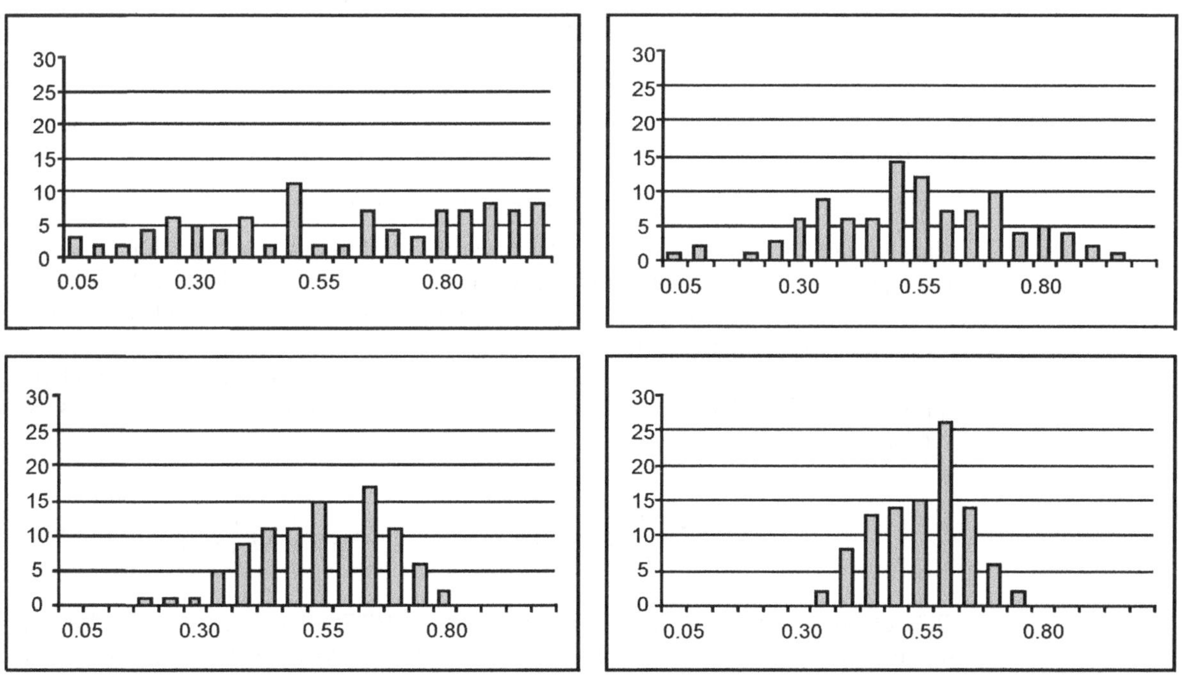

Figure 11-13. Normalised sums of uniformly distributed random variables, involving N = 1, 2, 5, and 10 variables, respectively. Note the trend toward Normality ('bell-shape') as the number of variables in the sum increases

The plot labelled $N=2$ shows 100 realisations of a random variable that is the sum of two uniformly distributed variables, divided by 2. That is, $x_2=(x_i+x_j)/2$, in which the x_i and x_j are independently and identically distributed (IID) as equation 11-56. Similarly for the plots labelled $N=5$ and $N=10$; they are the sums of five and ten uniform variables, normalised by 5 and 10, respectively.

As N increases, the distributions of Figure 11-13 become more concentrated and more bell-shaped. In fact, as N becomes large, the distribution of the sum of independent random variables asymptotically approaches a Normal distribution, almost regardless of the distributions of the underlying variables. This is reflected in the central limit theorem (CLT) (Feller, 1967), which states that the distribution of the sum of N random variables approaches Normality as N becomes large.[2] Thus, for uncertainties, such as the average strength across a large failure arc which averages random variations, the Normal PDF is an appropriate model.

The Normal PDF is:

$$f_X(x\,|\,\mu,\,\sigma)=\frac{1}{\sigma\sqrt{2\pi}}\exp\left\{-\frac{1}{2}\left(\frac{x-\mu}{\sigma}\right)^2\right\} \tag{11-55}$$

in which μ is the mean and σ^2 the variance. As can be seen from the exponent, the PDF is symmetric about a mode at μ, and falls off quickly as x deviates from the mean. The Normal PDF is usually tabulated or calculated numerically using its Standard Normal form, with mean zero and variance one, $N(0, 1)$ (Benjamin and Cornell, 1970).

Multiplying variables: the logNormal distribution. In the same way that calculating limiting equilibrium strengths involves the sum of random variables, other calculations involve the product of random variables. For example, in calculating certain material properties a series of modifying terms is sometimes applied to measurements. These modifying terms may have to do with known testing and measurement biases, model uncertainties, and the like, resulting in a term of the form:

$$z=k_1k_2k_nx \tag{11-56}$$

Taking the logarithm of each side leads to an equation involving the sum of logarithms:

$$\log z=\log k_1+\log k_2+\ldots+\log k_n+\log x \tag{11-57}$$

Thus, since $\log z$ is the sum of the logarithms of the k and x, with only loose assumptions on the distributions of the logarithms of the k and x, the distribution of log z should approach Normality (and hence z approach *logNormality*) as n becomes large. If the mean and standard deviation of the data themselves are μ and σ, respectively, and the means of the logarithms of the data are ζ and λ, respectively, the following relations apply and are useful in working with logNormal distributions (Aitchison and Brown, 1969):

$$\zeta^2=\ln\left(1+\frac{\sigma^2}{\mu^2}\right)\quad \lambda=\ln(\mu)-\frac{1}{2}\zeta^2\quad \mu=\exp\left(\lambda+\frac{1}{2}\zeta^2\right) \tag{11-58}$$

Tails of probability distributions: Extreme Value PDFs. The peak annual flow in a stream is the largest discharge to occur in any of the weeks in the year. Presume, for the purpose of discussion, that peak weekly flows are independent of one another (this is clearly not the case for daily flows, since storms typically last more than one day). If the peak weekly discharge, say, can be modelled by a Normal PDF, what is the appropriate distributional form of the peak annual discharge? In essence, the question reduces to the

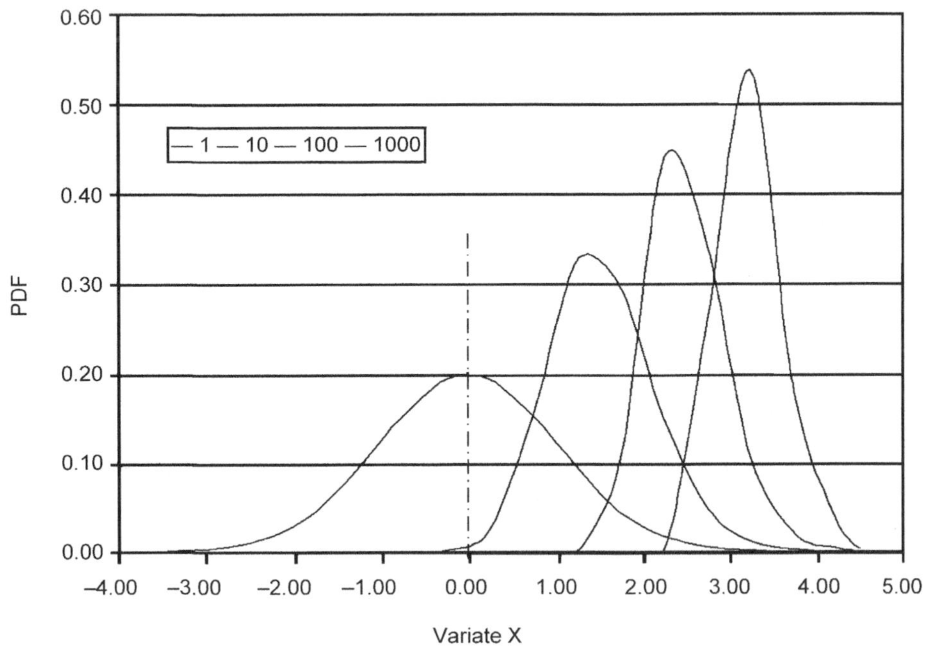

Figure 11-14. The PDFs of the largest value within samples of 1, 10, 100 and 1000 drawn from a Normal parent

distribution of the maximum value within samples of size 52 drawn from a Normal population (Figure 11-14). In concept, this question is the same as that of the weakest link of a chain, which can be thought of as the minimum value within samples of a size reflecting the number of links in the chain. Such problems are said to involve the distribution of extreme values.

Let the population from which peak weekly discharges are drawn be $f_{Qw}(q)$, in which Q_w is peak weekly discharge. For the peak annual discharge to be less than or equal to some value q_o, all the 52 weekly peak discharges must be less than q_o. Thus:

$$F_{Q_{max}}(q \mid n=52)=F_Q^{52}(q \mid n=1) \tag{11-59}$$

In the more general case, in which the largest of n uncertain quantities is sought, the cdf of the maximum becomes:

$$F_{Q_{max}}(q \mid n)=F_Q^n(q) \tag{11-60}$$

and, after taking the derivative of each side, the PDF of the maximum is:

$$F_{Q_{max}}(q \mid n)=\frac{d}{dq}F_{Q_{max}}(q \mid n)=nF_{Q_{max}}^{n-1}(q)f_Q(q) \tag{11-61}$$

Figure 11-14 shows the PDFs of the largest value within samples of 1, 10, 100 and 1000 drawn from a Normal parent distribution. As would be expected, as the sample size becomes larger, the mean value of the largest observation within the sample also becomes larger, but the variance becomes smaller. The classical reference on the statistics of extreme values is Gumbel (1954, 1958).

The shape of the extreme value is sensitive to the shape of the tails of the parent distribution. The PDF of the largest value within a sample is sensitive to the shape of the upper tail of the parent PDF, and likewise the PDF of the smallest value is sensitive to the lower tail. For a Normal distribution these are symmetric, but this is not the case for all parent distributions. Gumbel-classified extreme value distributions into three categories, depending on their asymptotic behaviour as $n \rightarrow \infty$. The type-I limiting distribution arises for the largest variable from a parent distribution with an exponentially decaying upper tail, that is, an upper tail that falls off as:

$$F_x(x)=1-\exp(-g(x)) \tag{11-62}$$

in which $g(x)$ is an increasing function of x. For example, the Normal, gamma, and Exponential distributions are all of this type. Gumbel showed that for large n this distribution approaches:

$$F_X(x)=\exp(-e^{-\alpha(y-u)}) \tag{11-63}$$

$$f_X(x)=\alpha\exp(-\alpha(y-u)-e^{-\alpha(y-u)}) \tag{11-64}$$

The parameters of the distribution are typically estimated from observed data, the presumption being that the extreme variable arises from a large but perhaps unknown number n of, say, large stream discharges during the course of a year.

The type-II limiting distribution arises for the largest variable from a parent distribution with an upper tail that falls off as:

$$F_X(x)=1-\beta\left(\frac{1}{x}\right)^k \tag{11-65}$$

Gumbel showed that for large n this type-II distribution approaches:

$$F_X(x)=e^{(-u/y)^k} \tag{11-66}$$

$$f_X(x)=\frac{k}{u}\left(\frac{u}{y}\right)^{k-1} \tag{11-67}$$

The type-III limiting distribution arises for either the largest or smallest variable from a parent distribution with a limited tail, that is, a tail that falls off as:

$$F_X(x)=1-c(w-x)^k \tag{11-68}$$

in which $x\leq w$ and $k>0$. In practice, this distribution is most often used to model smallest values, having a lower tail of the form $F_X(x)=c(w-\varepsilon)^k$ for $x\geq\varepsilon$, where ε is the lower limit for x (Benjamin and Cornell, 1970). For example, the gamma distribution is of this form. Gumbel showed that for large n this type-III distribution approaches:

$$F_X(x)=1-\exp\left[-\left(\frac{x-\varepsilon}{u-\varepsilon}\right)^k\right] x\geq\varepsilon \tag{11-69}$$

$$f_X(x)=\frac{k}{u-\varepsilon}\exp\left[-\left(\frac{x-\varepsilon}{u-\varepsilon}\right)^{k-1}\right]\exp\left[-\left(\frac{x-\varepsilon}{u-\varepsilon}\right)^k\right] x\geq\varepsilon \tag{11-70}$$

Inferences from data

This section builds on Section 11-3 to continue the discussion of the use of Bayesian methods for inferring probability distributions from observed data. The Bayesian approach leads to statements of probability directly over parameter values, models and other states of nature, as needed for probability assignments in risk analysis.[3] In most cases, this is the preferred approach to making probabilistic inferences from data. More complete results can be found in standard texts on Bayesian statistics (e.g. O'Hagen, 1994).

Inferring probability distribution functions

This section provides two examples of inferences of probability distributions from data, first for a discrete-valued probability mass function (Poisson PMF), and then for a continuous-valued probability density function (Normal PDF).

Poisson process. Section 11-3 illustrated the use of Bayes' theorem by reference to a simple Binomial process involving independent events, each with a common probability of occurrence, π. The Poisson process with parameter λ is related to the binomial, in that it describes the occurrence of independent events at random point in time or space. The Poisson PMF has the form,

$$fN(n \mid \lambda) = \frac{(\lambda t)^n e^{-\lambda t}}{n!} \tag{11-71}$$

in which n is the number of occurrences, and λ is the rate of occurrence; the mean is $E[n] = \lambda t$ and variance is $Var[n] = \lambda t$. Thus, the likelihood of the parameter λ for a set of observations $\mathbf{n} = \{n_1, \ldots, n_k\}$ in periods $\mathbf{t} = \{t_1, \ldots, t_k\}$ is:

$$L(\lambda \mid n) \propto \prod_i (\lambda t_i)^{n_i} e^{-\lambda t_i} \tag{11-72}$$

The non-informative prior distribution of λ is:

$$f_\lambda(\lambda) \propto \lambda^{-1} \tag{11-73}$$

The informative, conjugate prior distribution of π is of the gamma form with parameters (a, b):

$$f_\lambda(\lambda \mid a, b) = \lambda^{a-1} \exp(-b\lambda) \tag{11-74}$$

and the posterior PDF on π is of the same form, but with parameters, $a' = a + \Sigma n$ and $b' = b + \Sigma t$,

$$f_\lambda(\lambda \mid \mathbf{n}, \mathbf{t}) \propto \lambda^{(a + \Sigma n_i) - 1} e^{-(b + \Sigma t_i)\lambda} \tag{11-75}$$

As an example, consider that five events have been observed in ten years. The uniform prior PDF on λ^{-1} and the posterior PDF are shown in Figure 11-15. The modal (most probable) value of λ is found by setting the derivative of Equation (11-89) to zero, giving $\hat{\lambda} = 0.4$. The outcome, that the mode is located at 0.4 rather than the average of the sample (events/year = 0.5), derives from the prior PDF on λ. For the case of uniform prior PDF on λ, the posterior mode is 0.5, the same as the mode of the likelihood function. No matter how large the sample size, there will always remain a $1/\lambda$ difference in the posterior PDF between the cases of the two prior PDFs.

Normal process. The most common sampling encountered in practice is from Normally distributed populations. The familiar Normal process with parameters (μ, σ) has the form,

$$f_X(x \mid \mu, \sigma) = \frac{1}{\sigma\sqrt{2\pi}} \exp\left\{\frac{-(x - \mu)^2}{2\sigma^2}\right\} \tag{11-76}$$

in which x is a continuous variable with domain $(-\infty, +\infty)$; the mean is $E[x] = \mu$ and $Var[x] = \sigma^2$. Thus, the likelihood of the parameters (μ, σ) is:

$$L(\mu, \sigma \mid x) \propto \frac{1}{\sigma^n} \prod_i \exp\left\{\frac{-(x_i - \mu)^2}{2\sigma^2}\right\} \tag{11-77}$$

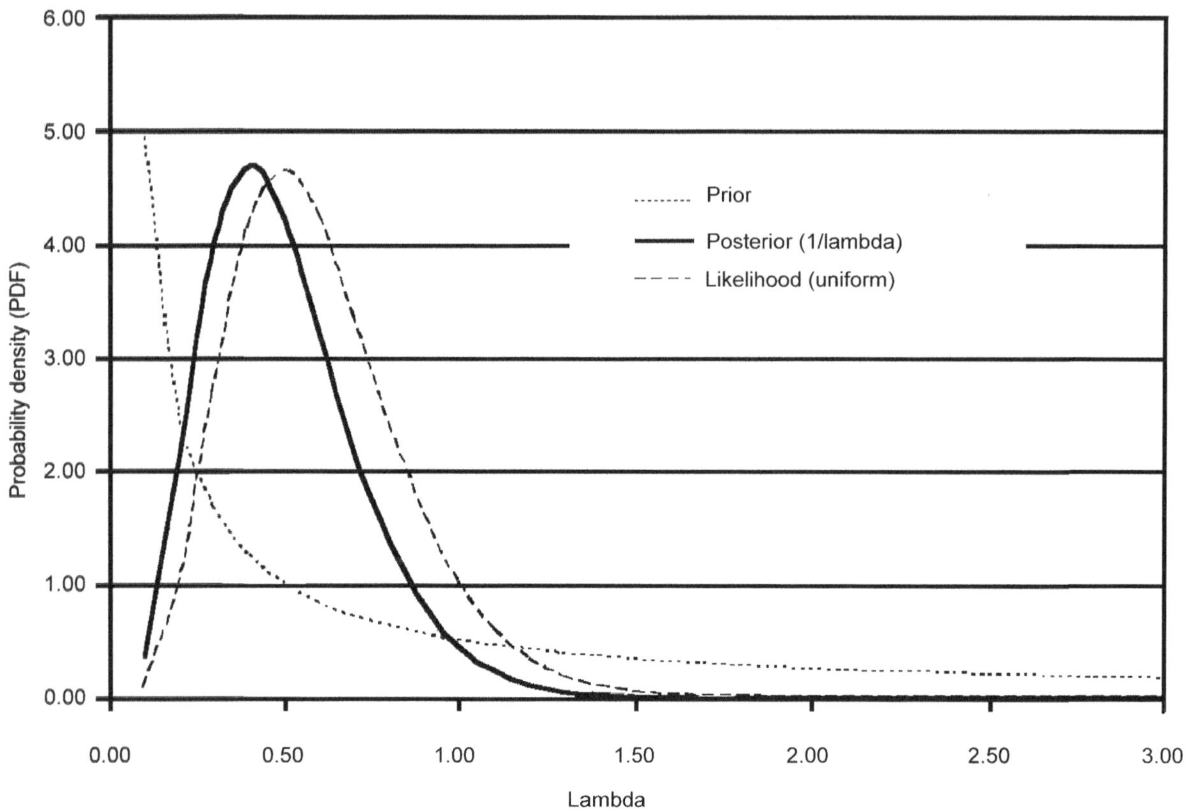

Figure 11-15. Prior and posterior PDFs for Poisson sampling

for a set of observations $\mathbf{x}=\{x_1, \ldots, x_n\}$. The non-informative prior distribution of (μ, σ) is:

$$f_{\mu,\sigma}(\mu, \sigma) \propto \sigma^{-1} \tag{11-78}$$

The informative, conjugate prior distribution of (μ, σ) is of the Normal-Inverted gamma form with parameters (m', d', n', ν'),

$$f(\mu, \sigma | m', d', n', \nu') \propto \left[\exp\left\{-\frac{1}{2\sigma^2} n'(\mu - m')^2\right\}\right]\left[\frac{1}{\sigma}\exp\left\{-\frac{1}{2\sigma^2} d'\nu'\right\}\left(\frac{1}{\sigma^2}\right)^2\right] \tag{11-79}$$

That is, the prior distribution of μ conditional on σ is Normal, and the marginal prior distribution of σ is inverted-gamma. The posterior PDF of μ, σ is of the same form, with parameters (Raiffa and Schlaifer, 2000),

$$\begin{aligned} m'' &= \frac{1}{n''}(n'm' + nm) \\ d'' &= \frac{1}{\nu''}[(\nu'd' + n'm'^2) + (\nu d + nm^2) - n''m''^2] \\ n'' &= n' + n \\ \nu'' &= \nu' + \nu + 1 \end{aligned} \tag{11-80}$$

in which,

$$m = \frac{1}{n} \sum_i x_i = \bar{x}$$

$$d = \frac{1}{\nu} \sum_i (x_i - m)^2 \qquad (11\text{-}81)$$

$$\nu = n - 1$$

For the data of Figure 11-16, this yields the joint posterior PDF shown as a contour plot. The ovals in the figure show the contours of the PDF of the mean and standard deviation estimated from the data.

Among other advantages of the Bayesian approach is that it is computationally stronger than the traditional frequentist approach based on the sampling distributions of estimators. Indeed, sampling distributions of some sample statistics, such as the correlation coefficient, are quite difficult mathematical problems. For example, the sampling distribution for estimators of the bivariate regression problem when both variables contain measurement errors remains unsolved, despite the fact that this is among the most common problems encountered in practice. In contrast, Bayesian computations are based solely on the likelihood function, which is usually obvious by inspection. Jeffreys (1983) was able to exploit this capability of Bayesian methods relatively directly to derive strong results, in contrast to the mathematical hurdles Fisher was forced to overcome deriving the corresponding traditional sampling theory results (Jaynes, 2003). As a case in point, the straightforward Bayesian solution to the problem of bivariate regression with error in both variables is given by Zellner (1971).

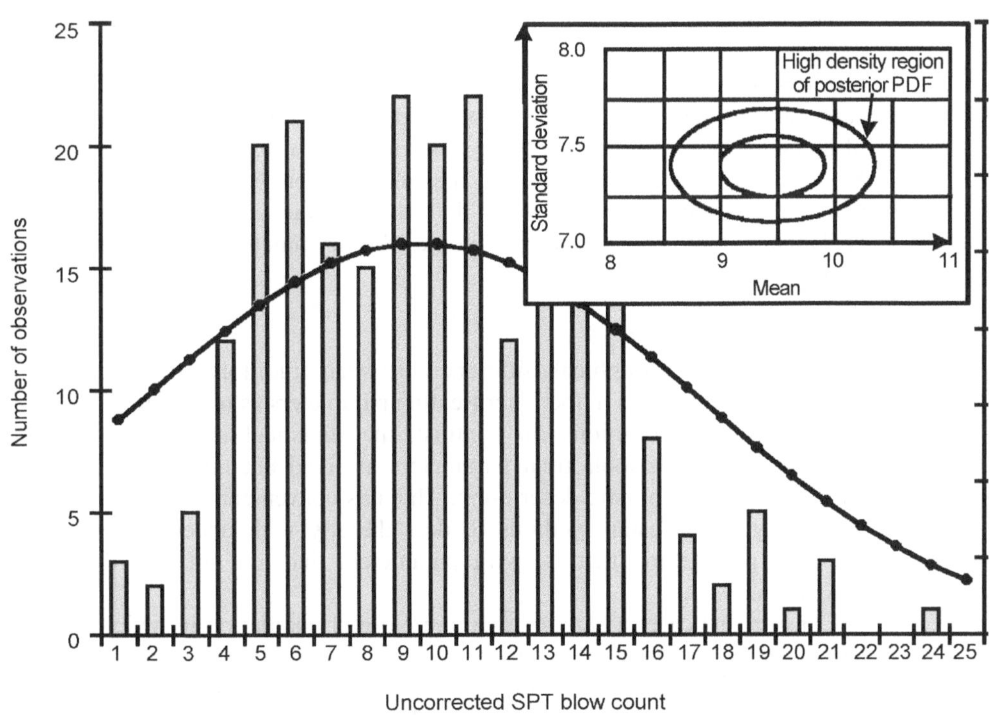

Figure 11-16. Inference for a Normally distributed variable

Sequential use of Bayes' theorem

If data occur in time, or in any event may be divided into parts, such that the observations $\mathbf{z}$ may be decomposed as $\mathbf{z}=\{\mathbf{x}_1, \mathbf{z}_2\}$, in which $\mathbf{z}_1$ and $\mathbf{z}_2$ are themselves vectors of data, then Bayes' theorem can be applied in two or more stages to obtain the same result as if all the data $\mathbf{z}$ were used to update uncertainty about a set of parameters all in one stage.

Having observed $\mathbf{z}_1$, the PDF of a set of parameters θ can be updated using Bayes' theorem to obtain:

$$f_\Theta(\theta|\mathbf{z}_1)=\frac{f_\Theta(\theta)L(\theta|\mathbf{z}_1)}{f(\mathbf{z}_1)} \tag{11-82}$$

in which the normalising constant $N=\int_\theta f(\theta)L(\theta|\mathbf{z}_1)d\theta=f(\mathbf{z}_1)$ is simply the marginal distribution of the observations. Now, having observed $\mathbf{z}_2$, the PDF of a set of parameters θ can be updated once again using Bayes' theorem to obtain:

$$\begin{aligned} f_\Theta(\theta|\mathbf{z}_1, \mathbf{z}_1)&=\frac{f_\Theta(\theta|\mathbf{z}_1)L(\theta, \mathbf{z}_1|\mathbf{z}_2)}{f(\mathbf{z}_2|\mathbf{z}_1)} \\ &=\frac{\dfrac{f_\Theta(\theta)L(\theta|\mathbf{z}_1)}{f(\mathbf{z}_1)}L(\theta, \mathbf{z}_1|\mathbf{z}_2)}{f(\mathbf{z}_2|\mathbf{z}_1)} \\ &=\frac{f_\Theta(\theta)L(\theta|\mathbf{z}_1, \mathbf{z}_2)}{f(\mathbf{z}_1, \mathbf{z}_2)} \end{aligned} \tag{11-83}$$

Thus, the two-stage updating process is equivalent to updating directly from $f_\theta(\theta)$ using the combined observations. In a similar way, the data may be divided further to update sequentially for each observation, as might be the case with data that are observed serially. Note, of course, that the likelihood of $\mathbf{z}_i$ at a particular stage is conditioned on all the preceding data $\mathbf{z}_j$, $j<i$.

The notion of sufficiency, introduced earlier, is easily demonstrated by reference to the staged updating of uncertainties according to equation 11-83. If a set of data $\mathbf{z}_2$ are uncorrelated with the uncertain parameter at issue, that is, $L(\theta|\mathbf{z}_2)=f(\mathbf{z}_2)$, then those data carry no information about θ (they are said to be uninformative of θ). The second stage of updating by $\mathbf{z}_2$ will not affect the posterior PDF of θ, and thus, $f_\Theta(\theta|\mathbf{z}_1, \mathbf{z}_2)=f_\Theta(\theta|\mathbf{z}_1)$. Since the data $\mathbf{z}_2$ are irrelevant to the inference on θ, they need not be made; it is sufficient to observe the data which do have an influence on the inference of θ, namely, $\mathbf{z}_1$. Thinking more generally of $\mathbf{z}_1$ as some arbitrary function of the sample data, $\mathbf{z}_1$ is said to be a sufficient statistic.

Nuisance parameters

The principal philosophical difference between traditional statistical inference and Bayesian inference lies in the latter treating parameters as uncertain quantities admitting of probability distributions. In contrast, the statistical inference of the Fisher and Neyman–Pearson schools treat parameters as fixed but unknown and, therefore, not admitting of probabilistic description. This has convenient consequences, among them the ability of Bayesian methods to deal directly with uncertainties associated with nuisance parameters; that is, with parameters that are not directly important to a prediction.

Consider the case of sampling from a Normal population in which both the mean μ and standard deviation σ are uncertain; but, assume that for purposes of analysis or decision-making, only uncertain in the mean is pertinent. For example, we may wish to predict the average density of an earth fill used in a dyke. Using a Bayesian approach, one may simply integrate out over the PDF of uncertainty in the nuisance parameter, here σ, to obtain the marginal distribution of the parameter needed for decision-making, here μ. This is not permissible within the Fisher or Neyman–Pearson approaches, since in those approaches a probability distribution over the parameters μ, σ is undefined.

The procedure for eliminating the nuisance parameter(s) is, first, to update the probability distribution on the suite of parameters based on the observed data. Thus:

$$f_\mu(\mu, \sigma | \mathbf{z}) \propto f_{\mu,\sigma}(\mu, \sigma) L(\mathbf{z} | \mu, \sigma) \tag{11-84}$$

Then, uncertainty in the nuisance parameter(s) can be integrated out:

$$f_\mu(\mu | \mathbf{z}) = \int_\sigma f_\mu(\mu, \sigma | \mathbf{z}) d\sigma \tag{11-85}$$

In a similar way, the uncertainty in a quantity x, marginal of uncertainty in the parameters upon which its PDF depends, can be found by integrating out the uncertainty in the parameters. Thus, if x has PDF $f_x(x | \mu, \sigma)$, the distribution of x marginal of uncertainty in its mean μ and standard deviation σ is:

$$f_x(x) = \int_\mu \int_\sigma f_x(x | \mu, \sigma) f_{\mu,\sigma}(\mu, \sigma) d\mu d\sigma \tag{11.86}$$

This is said to be the predictive distribution of x (Chapter 7). Predictive distribution results for a variety of cases can be found in Zellner (1971), Aitcheson and Dunsmore (1975), and Box and Tiao (1992).

Regression analysis

It is often the case that soil properties or other variables are related to one another, as illustrated in Figure 11-17, in which a gradual increase in maximum past pressure in a clay formation is observed with depth. Since the undrained strength of the clay increases with pre-consolidation pressure, this information can be used in a risk analysis to model

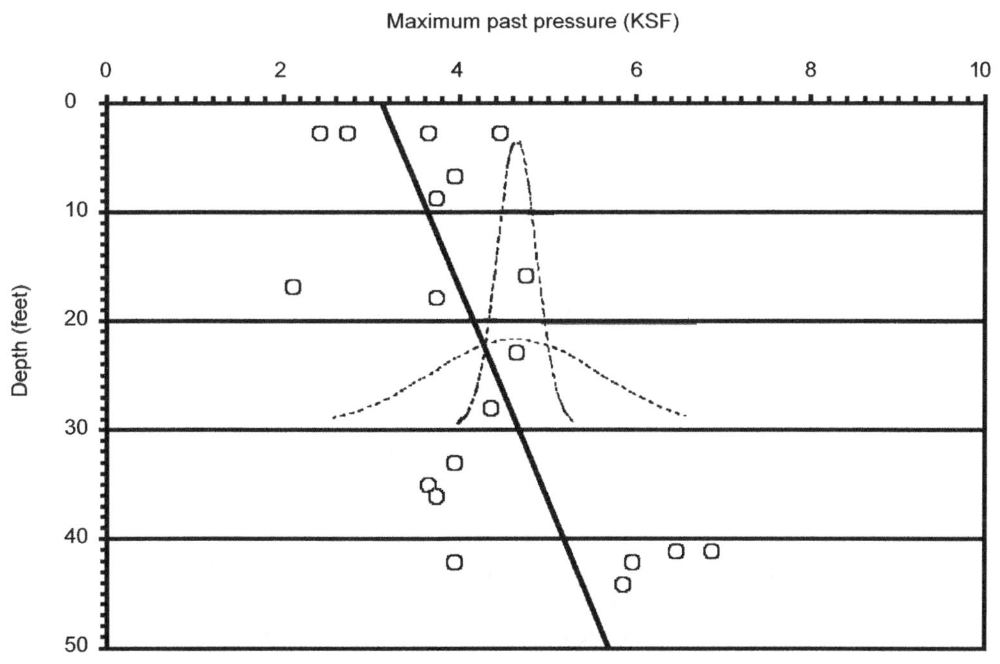

Figure 11-17. Vertical profile showing maximum past pressure and field vane strengths of clay with depth (Baecher and Ladd, 1998)

the increasing strength of the clay as one moves deeper in the deposit. To do so, however, necessitates a quantitative relationship between depth and maximum past pressure.

The correlation coefficient between depth and maximum past pressure is $\rho=0.41$. What we need, however, is a mathematical relationship that captures this correlation, and allows a probabilistic prediction to be made of maximum past pressure for a known depth.

The (univariate) linear regression model has the form:

$$y_i=\beta_1+\beta_2 x_i+u_i \tag{11-87}$$

in which $\mathbf{y}=\{y_1,\ldots,y_n\}_i$ are dependent observed variables, $\mathbf{x}=\{x_1,\ldots,x_n\}$ are independent variables, $\beta=\{\beta_1,\beta_2\}$ are scalar constants usually called regression parameters (intercept and slope), and $\mathbf{u}=\{u_1,\ldots,u_n\}$ are random errors. The assumptions are made that: (1) the $\mathbf{u}$ are Normally and independently distributed with zero-mean and a common variance, σ^2; and (2) the $\mathbf{x}$ are either fixed, deterministic variables, or they are random variables independent of $\mathbf{u}$ (see, e.g., Zellner, 1971).

The likelihood of the observations y conditioned on $\mathbf{x}$ and the parameters of the regression model is:

$$L[\mathbf{y}|\mathbf{x},\beta_1\beta_2\sigma]=\prod_{i=1}^{n} N[\mathbf{y}|\mathbf{x},\beta_1\beta_2\sigma]$$

$$\propto\frac{1}{\sigma^n}\exp\left[-\frac{1}{2\sigma^2}\sum_{i=1}^{n}(y_i-\beta_1+\beta_2 x_i)^2\right] \tag{11-88}$$

From Bayes' theorem, the posterior PDF of the regression parameters, starting from the non-informative prior $f(\beta_1,\beta_2,\sigma)\propto 1/\sigma$, is:

$$f(\beta_1,\beta_2,\sigma|\mathbf{x},\mathbf{y})\propto\frac{1}{\sigma^{n+1}}\exp\left[-\frac{1}{2\sigma^2}\sum_{i=1}^{n}(y_i-\beta_1+\beta_2 x_i)^2\right] \tag{11-89}$$

from which the marginal posterior PDFs are:

$$f(\beta_1,\beta_2|\mathbf{x},\mathbf{y})=\int_0^\infty f(\beta_1,\beta_2,\sigma|\mathbf{x},\mathbf{y})d\sigma$$

$$\propto[\nu s^2+n(\beta_1-\hat{\beta}_1)^2+2(\beta_1-\hat{\beta}_1)(\beta_2-\hat{\beta}_2)\Sigma x_i+(\beta_2-\hat{\beta}_2)^2\Sigma x_j^2]^{-n/2} \tag{11-90}$$

$$f(\sigma|\mathbf{x},\mathbf{y})\propto\frac{1}{\sigma^{\nu+1}}\exp\left(-\frac{\nu s^2}{2\sigma^2}\right)$$

in which, $\nu=n-2$, and

$$\hat{\beta}_1=\bar{y}-\hat{\beta}_2\bar{x}$$
$$\hat{\beta}_2=\frac{\Sigma(x_i-\bar{x})(y_i-\bar{y})}{\Sigma(x-\bar{x})^2} \tag{11-91}$$
$$s^2=\nu^{-1}\Sigma(y_i-\hat{\beta}_1-\hat{\beta}_2 x_i)^2$$

The derivation of these and related results is presented by Zellner (1971) Box and Tiao (1992), O'Hagan (1994), and others.

Note the strong dependence between β_1 and β_2 (Figure 11-18). Since both parameters depend on the same observational data, they are not independent. The same is true for σ in its relation to β_1 and β_2, although this cannot be seen in the marginal distribution. A similar correlation arises in estimates of the Mohr–Coulomb strength parameters,

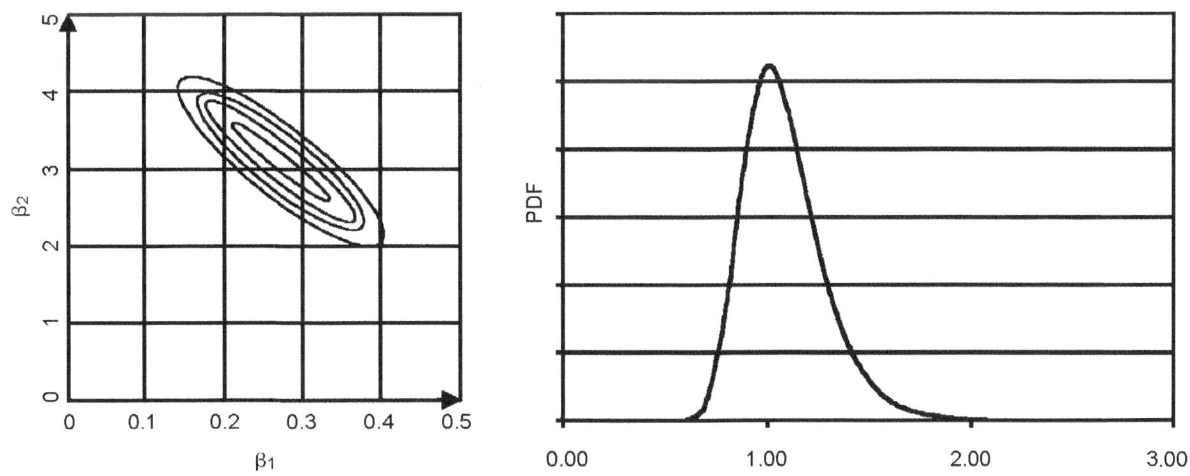

Figure 11-18. Posterior PDF of regression parameters estimated from maximum past pressure data. LHS shows slope and intercept; RHS shows standard deviation of residual

c, ϕ, which are negatively correlated when estimated from measurements. Many hydrologic parameter estimates also show the same property; indeed, any curve fitting estimation has this property.

The predictive distribution of maximum past pressure at a particular depth is found by integrating the conditional PDF with given regression coefficients over the uncertainty in the regression coefficients:

$$f(y_i|x_i) = \iiint N[y_i|x_i, \beta_1, \beta_2, \sigma] f(\beta_1, \beta_2, \sigma) d\beta_1 d\beta_2 d\sigma \qquad (11\text{-}92)$$

which has the form of the Student-t distribution with $\nu = n-2$ degrees of freedom (O'Hagan, 1994; Zellner, 1971), in which:

$$\frac{y_i - (\hat{\beta}_1 + \hat{\beta}_2 x_i)}{s\left(\frac{1}{n} + \frac{x_i}{S_{xx}} + 1\right)^{1/2}} \sim t_\nu \qquad (11\text{-}93)$$

in which $S_{xx} = \Sigma(x_i - \bar{x})^2$. The predictive distribution on the mean of y for a given x is also Student-t, with mean $(\hat{\beta}_1 + \hat{\beta}_2 x_i)$, spread $s\left(\frac{1}{n} + \frac{x_i}{S_{xx}} + 1\right)^{\frac{1}{2}}$, and degrees of freedom ν. Aitchison and Dunsmore (1975) discuss the problem of updating an estimate of regression relationships (in their case, "calibrations") to account for site specific information. Figure 11-17 shows results for the problem of maximum past pressure.

Reliability models

Engineering analysis deals with the relationship between the demands that a system must meet and its capacity to do so. Both the demands and capacity may be uncertain. This section summarises current approaches to engineering reliability modelling, and its application to probability assignment in dam safety risk analysis.

Demands, capacity and reliability

The demand to which an engineering system is exposed is Q. The available capacity is R. In this context "demands" and "capacity" are taken, in a broad sense, to include not only forces and stresses but also seepage, settlement or other design considerations. The values of both R and Q are uncertain, so these variables have mean values, variances and covariances, as well as full PDFs describing their uncertainty.

The margin of safety, M, is the difference between the capacity and the demand:

$$M = R - Q \tag{11-94}$$

It follows that the mean of M is:

$$\mu_M = \mu_R - \mu_Q \tag{11-95}$$

in which μ indicates the expected value; and the variance of M is:

$$\sigma_M^2 = \sigma_R^2 + \sigma_Q^2 - 2\rho_{RQ}\sigma_R\sigma_Q \tag{11-96}$$

in which σ indicates the standard deviation.

The reliability index, β, is defined as:

$$\beta = \frac{\mu_M}{\sigma_M} = \frac{\mu_R - \mu_Q}{\sqrt{\sigma_R^2 + \sigma_Q^2 - 2\rho_{RQ}\sigma_R\sigma_Q}}, \tag{11-97}$$

which describes the distance of the mean margin of safety from its critical value ($M=0$) in units of standard deviation. If the demand and capacity are uncorrelated, the correlation coefficient is zero, and:

$$\beta = \frac{\mu_M}{\sigma_M} = \frac{\mu_R - \mu_Q}{\sqrt{\sigma_R^2 + \sigma_Q^2}} \tag{11-98}$$

Figure 11-19 shows plots of typical probability distributions of R and Q. Figure 11-20 shows the resulting probability distribution of M. The probability of failure must be the probability that M is less than zero, which is the shaded area in the figure. The probability of failure is the intercept of the cumulative distribution function with the vertical axis at $M=0$.

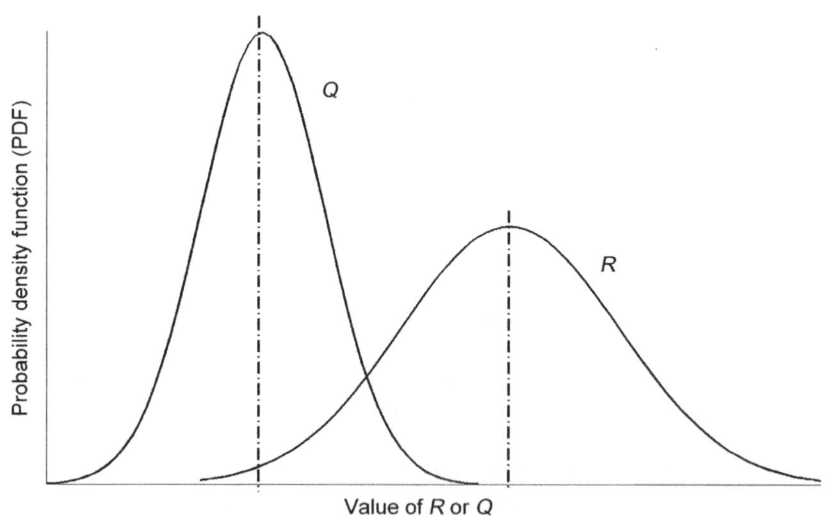

Figure 11-19. Uncertainties in demand and capacity

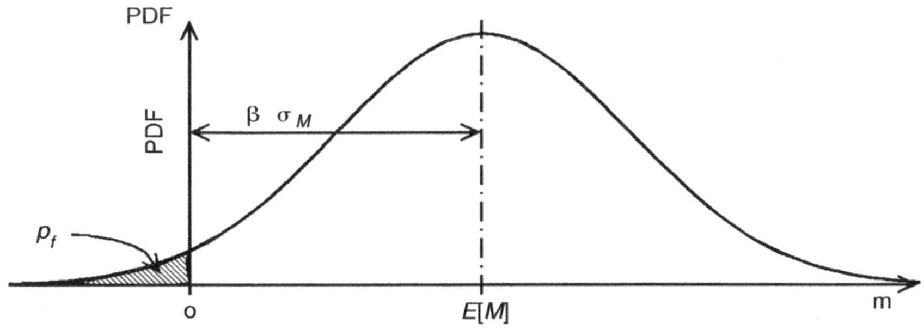

Figure 11-20. Probability distribution of margin of safety

In the special case that R and Q are Normally distributed, M is Normally distributed as well. Thus, the reliability index, β, which normalises M with respect to its standard deviation, is a standard Normal variate, usually designated Z.

Engineers are more accustomed to working with the factor of safety, F, defined by

$$F = \frac{R}{Q} \tag{11-99}$$

Failure occurs when $F = 1$, and a reliability index is defined by

$$\beta = \frac{E[F] - 1}{\sigma_F} \tag{11-100}$$

Calculations of the reliability index are more difficult when it is expressed in terms of the factor of safety because F is the ratio of two uncertain quantities while M is their difference. To avoid this problem, some workers have assumed that R and Q are logNormally distributed (that is, the logarithms of R and Q are Normally distributed) so that the logarithm of their ratio becomes the difference between their logarithms. Then, the formulation is identical to the equations above, but numerical results for a given problem will be different. Also, in estimating the statistical parameters of R and Q from experimental and field data, calculations must be made on the logarithms of the data rather than the arithmetic values.

Results for different distributions of the performance function

Either M or F can describe the performance of a system, so either can be called the performance function. The relation between β and the probability of failure is not unique but depends on the shape of the probability distribution of the performance function (Baecher and Christian, 2003). For most of the range of the reliability index the assumption of a Normal distribution is conservative and that, for values of β less than about 2, there is little difference between the probabilities of failure calculated using a range of PDFs for either M or F. For very small values of β the probability of failure is actually slightly smaller for the Normal distribution than for the others. These results suggest that it is reasonable to assume a Normal distribution in the absence of further information and that the assumption will probably overestimate the probability of failure.

A further argument in favour of using the Normal distribution to relate β to the probability of failure is the central limit theorem, a fundamental result of probability theory. It states that, for a wide variety of conditions, the distribution of the sum of a large number of random variables converges to a Normal distribution. As long as there is a reasonably large number of variables, no one variable dominates, and the variables are not highly dependent on each other, the theorem applies with only weak conditions, regardless of the distributions of the individual variables. An extension of this argument is that geological and hydrological phenomena are determined by the combined contributions of a large number of small effects and, therefore, that the distribution of the overall effect ought to be Normal. A different argument is that these combinations are not the *sum of* individual effects but their *product*, in which case the resulting distribution should reflect the sum of their logarithms and be logNormally distributed. Most workers would probably agree that it is reasonable to assume that M is Normally distributed. Other distributions apply to some geotechnical variables. For example, probability of exceeding a particular seismic acceleration is often represented by an exponential function. Gamma and beta distributions are appropriate for other parameters. There is more controversy over the appropriate distribution for F. However, it seems reasonable and simple, to assume that F is Normally distributed unless demonstrated otherwise. The sometimes voiced argument that Normal distributions are inappropriate because they allow negative values is usually moot, since the probabilities of negative realisations are usually vanishingly small.

Approximations in reliability calculations

Most practical applications of reliability analysis in dam safety studies involve approximations. One should expect that different approximations will give different answers. In particular, the calculation of the probability of failure from the reliability index usually implies an assumption that the margin or factor of safety is Normally distributed, and this is seldom the case. Therefore, it is good practice to compare results among approaches.

The question to be answered in most reliability analyses as used in dam safety studies, is, given some set of probabilistic descriptions of loads and capacities, what is the probability of a specified adverse outcome (i.e. "failure", however defined) occurring? This question is normally approached by propagating the uncertainties in loads and capacities through an engineering model to determine the probability distribution (or low-order moments) of the performance function, and then integrating to determine the probability that the performance is unsatisfactory. For example, the PDF or mean and variance of, say, the margin of safety is calculated from the joint PDF of the loads and capacities, and the area under the resulting $f(M)$ for $M<0$ determines the probability of failure. This is called a derived distribution or error propagation approach.

First-order second-moment method

The treatment of error propagation starts by recognising that the result of calculations can be considered a function g of the input uncertainties, evaluated at some point $(x_1, x_2, \ldots, x_n)$ (Taylor, 1997):

$$g = g(x_1, x_2, \ldots, x_n). \qquad (11\text{-}101)$$

If there is only one independent variable and the value of g is known for some value of X, say $\bar{X}$, then the value of g can be found for any other value x by using a Taylor series approximation:

$$g(x) = g(\bar{X}) + \frac{1}{1!}(x-\bar{X})\frac{dg}{dx} + \frac{1}{2!}(x-\bar{X})^2\frac{d^2g}{dx^2} + \frac{1}{3!}(x-\bar{X})^3\frac{d^3g}{dx^3} + \ldots \qquad (11\text{-}102)$$

This equation is asymptotically exact. In practical applications, x is chosen to be near $\bar{X}$, so higher order terms become small and the series can be truncated after only a few terms. For more than one input variable, this approximation becomes:

$$g(x_1, x_2, \ldots, x_n)=g(\mu_{X_1}, \mu_{X_2}, \ldots, \mu_{X_n})+\frac{1}{1!}\sum_{i=1}^{n}(x_i-\mu_{X_i})\frac{\partial g}{\partial x_i}$$

$$+\frac{1}{2!}\sum_{i=1}^{n}\sum_{j=1}^{n}(x_i-\mu_{X_i})(x_j-\mu_{X_j})\frac{\partial^2 g}{\partial x_i \partial x_j}$$

$$+\frac{1}{3!}\sum_{i=1}^{n}\sum_{j=1}^{n}\sum_{k=1}^{n}(x_i-\mu_{X_i})(x_j-\mu_{X_j})(x_k-\mu_{X_k})\frac{\partial^2 g}{\partial x_i \partial x_j \partial x_k}+\ldots \quad (11\text{-}103)$$

in which the partial derivatives are taken at $\mu_{X}, \mu_{X}, \ldots$.

The first-order second-moment (FOSM) method uses the first (linear) terms of a Taylor series expansion of the performance function to estimate the expected value and variance (first two moments) of the performance function. The first order terms reduce to:

$$g(x_1, x_2, \ldots, x_n)\approx g(\mu_{X_1}, \mu_{X_2}, \ldots, \mu_{X_n})+\sum_{i=1}^{n}(x_i-\mu_{X_i})\frac{\partial g}{\partial x_i} \quad (11\text{-}104)$$

As this is a linear function, the mean of the function equals the function of the means, thus,

$$\mu_g \approx g(\mu_{X_1}, \mu_{X_2}, \ldots, \mu_{X_n}) \quad (11\text{-}105)$$

This is the intuitive result, however, it remains an approximation, and introduces error when $g(\mathbf{x})$ is non-linear.

The variance of g is,

$$Var[g]=\sigma_g^2=E[(g-\mu_g)^2] \quad (11\text{-}106)$$

which as a first-order approximation becomes:

$$\sigma_g^2 \approx \sum_{i=1}^{n}\sum_{j=1}^{n}\rho_{X_iX_j}\sigma_{X_i}\sigma_{X_j}\frac{\partial g}{\partial x_i}\frac{\partial g}{\partial x_j}$$

$$=\sum_{i=1}^{n}\sigma_{X_i}^2\left(\frac{\partial g}{\partial x_i}\right)^2+\sum_{i=1}^{n}\sum_{j\neq i}^{n}Cov(X_i, X_j)\frac{\partial g}{\partial x_i}\frac{\partial g}{\partial x_j} \quad (11\text{-}107)$$

Neither result depends on the forms of the distributions of the variables or the functions, provided the distributions and functions are regular and smooth.

Equation 11-104 involves partial derivatives. Sometimes it is possible to differentiate $g(\mathbf{x})$ exactly, but more often than not the function is too complex to allow a closed-form solution, or may be implicit. The easiest way around this problem is by means of central differences. The function g is evaluated with each of the variables set at its mean value; this yields the estimate of μ_g. Then the partial derivative for each variable in turn is found by incrementing the variable by a small amount, numerically finding the difference between the resulting values of g, and dividing the difference by the increment. The optimal choice is an increment as small as possible without unacceptable rounding error. The larger the increment, the larger the error in the estimate of the derivative. It has been suggested that the increment in each variable should be equal to the standard deviation of that variable. This is not a good idea for it degrades the approximation.

If g is a linear function of the variables $\mathbf{x}$, the first order approximations are exact. If g is a product function of the variables x_i, the first order approximations become,

$$g = a_1 x_1 \cdot a_2 x_2 \times \ldots \times a_n x_n = \prod_{i=1}^{n} a_i x_i \qquad (11\text{-}108)$$

$$\mu_g = \prod_{i=1}^{n} a_i \mu_{X_i} \qquad (11\text{-}109)$$

$$\sigma_g^2 \approx \sum_{i=1}^{n} \prod_{\substack{j=1 \\ j \neq i}}^{n} \mu_{X_i}^2 \;\; \sigma_{X_i}^2 \qquad (11\text{-}110)$$

Equation 11-108 is exact, but equation 11-110 is approximate. In the case that the variables are mutually independent, the exact expression for the relations among the coefficients of variation is:

$$1 + \Omega_g^2 = (1 + \Omega_{X_1}^2)(1 + \Omega_{X_2}^2) \ldots (1 + \Omega_{X_n}^2) = \prod_{i=1}^{n} (1 + \Omega_{X_i}^2) \qquad (11\text{-}111)$$

Finally, when there are two correlated variables, the mean value of their product is:

$$\mu_g = \mu_{X_1}\mu_{X_2} + \rho_{X_1X_2}\sigma_{X_1}\sigma_{X_2} \qquad (11\text{-}112)$$

A variety of other simple results for special cases can be found in Baecher and Christian (2003).

Other approaches to derived distributions

A number of other approaches to the derived distribution problem have been used in dam safety risk analyses. The most common are Rosenblueth's (1975) point estimate method, and Hasofer and Lind's (1974) first-order reliability method. The Monte Carlo method, which is widely used for a number of calculational purposes, is considered separately in the following section.

The point estimate method. Rosenblueth (1975) proposed a simple and elegant method of obtaining the moments of the performance function by evaluating the performance function at a set of specifically chosen discrete points. One of the disadvantages of the original method is that it requires that the performance function be evaluated $2N$ times, and this can become a very large number when the number of uncertain parameters is large. Recent modifications reduce the number of evaluations to the order of $2N$ but introduce their own complications. This method is not further considered here, but is described by Harr (1987). Recent extensions are given by Christian and Baecher (1999, 2002).

The Hasofer–Lind method. Hasofer and Lind (1974) proposed an improvement on the FOSM method based on a geometric interpretation of the reliability index as a measure of the distance in dimensionless space between the peak of the multivariate distribution of the uncertain parameters and a function defining the failure condition. The acronym FORM, for first-order reliability method, is often referred to this method. The method has gained broad acceptance in structural reliability practice, and is,

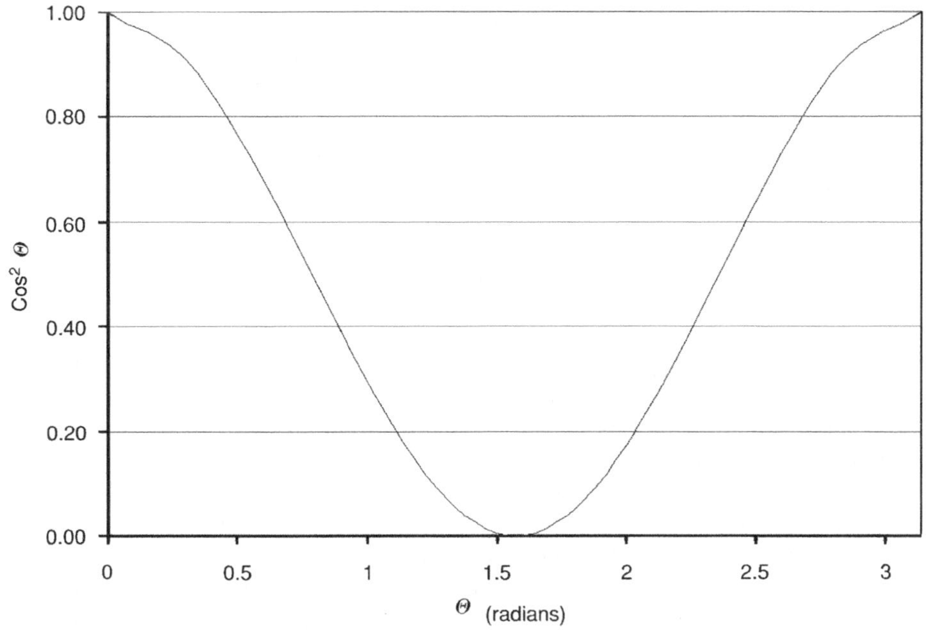

Figure 11-21. Function $\cos^2 2$ *to be integrated between* 0 *and* π *by Monte Carlo simulation. Integral is area under the curve*

perhaps, the most common current way of calibrating partial safety factors in load and resistance factor design codes. This method is not further considered here.

Monte Carlo method

The Monte Carlo simulation method has come to dominate event-tree-based risk analysis for dam safety studies. In this approach the analyst creates a large number of sets of randomly generated values for the uncertain parameters and numerically computes the performance function for each set. The statistics of the resulting set of values of the function can be computed and β or p_F calculated directly. The method has the advantage of conceptual simplicity, but it can require a large set of values of the performance function to obtain adequate accuracy. Furthermore, the method does not give insight into the relative contributions of the uncertain parameters that is obtained from other methods. The computational effort can be reduced significantly by using statistical techniques known as variance reduction schemes, and these should be employed whenever possible.

Basic considerations of random simulation

The applications of Monte Carlo methods can be divided into two types: simulation of processes that are random and, solution of problems that are deterministic but that can be solved by simulation with random variables. The focus in dam safety is the former, but a simple example of a deterministic solution is instructive.

Consider a definite integral of the function $\cos^2\theta$ between 0 and π, for which the solution is known to be:

$$I = \int_0^{\pi} \cos^2\theta \, d\theta = \pi/2 \tag{11-113}$$

This integral can also be approximated to any degree of accuracy by random simulation. Imagine that the function is plotted and enclosed in a box as in Figure 11-21. The area of the box is π. The integral is the area under the curve. We now generate N points with random x, y coordinates. If a point falls in the area under the curve, we assign it a value "1", if not, "0." The integral is found by averaging the values of the points. The exact answer is $\frac{1}{2}$.

Clearly, whether or not the point falls in the area of the curve is a Bernoulli process. Thus, while the expected result of the calculation is $\frac{1}{2}$, the variability of the result has a variance that is simply that of the mean of a Bernoulli process sampled with N observations:

$$Var[\hat{I}]=A^2 Var[\hat{p}]=A^2\frac{1}{N}p(1-p)=\frac{I}{N}(A-I) \tag{11-114}$$

This is a simple but pure example of solving a mathematical problem with simulation. The purpose of using simulation to solve a problem involving random processes is to generate similar solutions for the derived distribution problem so, in concept, there is not much difference between the two.

In repeated sampling, the variability of the estimator of the integral decreases as the square root of the number of trials. This is a slow rate of convergence, and much of the modern work on Monte Carlo methods has been directed at either reducing the error in the sampling process or decreasing the number of trials necessary to achieve a desired accuracy. Many researchers have investigated the question of how many replications must be made with random simulation to achieve a given accuracy; see, for example, Rubinstein (1981), Morgan *et al.* (1990) or Fishman (1996).

In most cases, one is less interested in estimating the error than in determining how many trials must be made to achieve a desired accuracy. Using Chebychev's inequality, Fishman (1996) shows that the probability of the error between the estimate and actual value of p is bounded by:

$$Pr\left[\frac{|\hat{p}-p|}{\sigma_z}\geq\beta\right]\leq\frac{1}{\beta^2} \tag{11-115}$$

in which σ is the standard deviation of the error $(\hat{p}-p)$ (i.e. β is the number of standard deviations of error). Thus, Fishman demonstrates that the number of trials needed to achieve a probability that the absolute error $Pr[|\hat{p}-p|<\varepsilon]=1-\delta$, is approximately,

$$N_c \approx p(1-p)/(\delta\varepsilon^2) \tag{11-116}$$

in which the subscript c indicates the Chebychev approximation.

Simulation of random processes

The principal application of Monte Carlo simulation is to study the behaviour of random processes. These are problems in which the input is stochastic. Monte Carlo simulation is particularly effective when the process is nonlinear or involves many uncertain inputs, which may be distributed differently from one another.

To perform such a study, one generates a random value for each uncertain variable and performs the calculations necessary to yield a solution for that set of values. This gives one sample of the process outcome. The trials are repeated many times, giving many samples. With a large number of runs it is possible to study the outputs statistically to obtain moments (means, variances), exceedance probabilities and other statistical descriptions. Regardless of the number of stochastic inputs, each run gives one observation of the process. Hence, increasing the number of stochastic input variables does not increase the number of runs for a given level of accuracy. The technique is one of repeated sampling, and methods of statistical inference are used to establish the accuracy and precision of results.

An important thing to note, is that Monte Carlo simulation depends on reliable random numbers. Generating random numbers is an art, and one method will not work for all circumstances. There is a substantial literature on the subject, and anyone contemplating an important study involving random numbers is advised to be satisfied that the random number generators to be used are appropriate to the job. This topic is otherwise outside the scope of the present volume.

Variance reduction methods

A Monte Carlo application may require that a large number of random points be used. Variance reduction methods attempt to reduce this computational burden by increasing the precision of the statistical inferences made from the results of random simulation. These methods use known information about the problem to increase efficiency. The methods concentrate calculations on that part of the sampling domain where the most information can be gained about a process. Some improvements are relatively simple, for example, exploiting symmetries, but others are more subtle.

Importance sampling starts with the observation that, if we are going to sample randomly, we should distribute the points to put most of them in the region that contains information and to waste as few as possible. For example, in using random simulation to find probabilities of failure or other performance variables, the central problem is to find the integral of a function $g(x)$:

$$I=\int g(x)dx \tag{11-117}$$

where x may be a vector, the integration can be multidimensional, and the limits of integration are appropriate to the specific problem. Assume that we can find a probability distribution function $f(x)$ concentrated in the region of interest, for example, near the limiting state. Then:

$$I=\int \frac{g(x)}{f(x)}f(x)dx=E\left[\frac{g(x)}{f(x)}\right] \tag{11-118}$$

By generating points from the distribution $f(x)$, the Monte Carlo estimate of the integral is:

$$I\approx\frac{1}{n}\sum_{i=1}^{n}\frac{g(x_i)}{f(x_i)} \tag{11-119}$$

the optimal choice of $f(x)$ is (see Rubinstein, 1981, Press *et al.*, 1992, Fishman, 1996):

$$f(x)=\frac{g(x)}{I} \tag{11-120}$$

Thus, a good choice of $f(x)$ is one that closely approximates the function to be integrated in the significant region (e.g. the tail of a distribution near a limit state). The trick is to determine an $f(x)$ without expending more effort than is saved by the increased efficiency of the simulation, and a variety of methods are available for doing this, but are outside the scope of the present document.

Stratified sampling is widely used in statistical sampling, and it can be applied to Monte Carlo simulation as well. Whereas importance sampling is an attempt to place most of the sampling points in regions that contribute most to the integral, stratified sampling tries to place the points so that more of them will be found in regions where the variance of the function $g(x)$ is largest. Consider that the sampling uses N points in

m regions, and that the probability that the random variable lies in the region i is P_i. If the standard deviation of the process in region i is σ_i, the number of sampling points to be used in region i should be:

$$N_i = N \frac{P_i \sigma_i}{\sum_{j=1}^{m} P_j \sigma_j} \tag{11-121}$$

A common procedure is to select intervals so that the P_i's are all the same. Then the distributions are weighted according to the variances in each interval. A special case occurs when $P_i = 1/m$ and $N_i = N/m$. Rubinstein (1981) calls this systematic sampling. It can also be shown that, when one point is used in each interval, the most efficient place for that point is in the middle of the interval. Although stratified sampling can reduce the number of sampling points, if there are D variables to be sampled and each is to be sampled in m regions, the total number of simulation points becomes m^D, which can grow into a very large number. To avoid this, the Latin hypercube method may be employed. For each variable the m sampling points are placed in independent random order. Then m samplings are made by choosing the first randomised value for each variable, the second randomised value for each variable, and so on until m sets of randomised variables have been chosen. This ensures that each value is used once and that their combination is randomised.

11.5 Judgement and probability

An issue in risk analysis for dam safety is the extent to which a detailed analytical procedure is needed within which to coalesce judgements about the probability of dam failures. One view holds that the collective judgement of engineers, structured within a process of debate, yields as good an assessment of probabilities as can be obtained by detailed analysis. Another view holds that purely intuitive reasoning leading to collective judgements that are absent an analytical framework provides no assurance of probability assignments that are either internally consistent or in line with engineering fundamentals. The resolution to this contrast of views depends, in part, on the use to which a risk analysis is to be put, for example, if the intent is to guide further safety investigations, or if the intent is to make specific remedial action decisions on the basis of it. The procedural details of subjective probability elicitation protocols are discussed at some length in Chapter 7. This section focuses on the nature and use of judgemental probabilities.

The nature of judgement

Judgement is critical to the success of practical engineering activities, yet what people mean by judgement differs from one person to another. For example, Hartford and Stewart (2002) have noted, "To preserve the credibility of judgement in engineering practice it is necessary to define what constitutes the process of 'exercising judgement,' and who is qualified to engage in the process in a precise manner." Ferguson (1992) argues for the centrality of intuitive judgement in engineering design, and several well-known dam engineers have discussed the importance of judgement to the design and construction of dams (Casagrande, 1965; Peck, 1969, 1980; Whitman, 1984).

The *Oxford English Dictionary* says that, "judgement is the formation of an opinion or notion concerning something by exercising the mind upon it; an opinion, or estimate". This is somewhat broad for engineering purposes. Within the dam safety community, there are at least two schools of thought on what constitutes judgement: (1) One holds that judgement reflects a base of knowledge held by a person or persons, and manifests in quantitative estimates of probabilities or other parameters as a reflection of intuition; (2) The other holds that judgement reflects an analytical process of reasoning, and manifests in quantitative estimates as a reflection of logic. These contrasting views of judgement reflect an age-old rivalry between the mathematical and the intuitive mind (Berlin and Hardy, 1980; Berlin *et al.*, 1979). Pascal (1966) thought the two ways of thinking to be incompatible within the same person. Nonetheless, each approach is supportable and internally consistent, and each relates to the other but they differ in their practical implications.

Within the school of thought that holds judgement to reflect an intuitive process, judgement is seen as based on the recognition of patterns in the world from which correlates can be identified. This is sometimes referred to as the "lens theory" and is attributed to Brunswik's (1952) research on perception and cognition (Hammond, 1996). Brunswik theorised that people do not directly perceive the essence of an object or a situation, but rather perceive a set of implicit cues about it which may be ill-defined. Such cues are statistically related in a person's mind, whether consciously or not, with the essential aspects one is attempting to draw a judgement about, and these statistical relationships are learned from experience.

Within the school of thought that holds judgement to reflect an analytical process, judgement is seen as based on reasoning from observations, known facts and physical principles; wherein, reasoning means to determine or conclude something by logical thinking. Reasoning is similar to mathematical argumentation, but with verbal statements and relationships rather than symbolic ones. An important quality of the reasoning approach to judgement, in the eyes of its proponents, is the paper trail of evidence it leaves to justify conclusions that are reached.

Within the normal enterprise of risk analysis, an opinion, in contrast to a judgement, is a belief held with confidence but not substantiated by positive knowledge, proof or explicit reasoning. This contrasts, and should not be confused, with the means of the term in the legal arena, where an opinion is a formal statement by an adjudicative body of the legal reasons and principles for a set of conclusions.

Coherence, correspondence

The philosophical world distinguishes between two types of truth or judgements: coherence and correspondence (Hammond, 1996). The coherence theory of judgement focuses on whether an individual's judgemental process is internally consistent. The correspondence theory focuses on whether an individual's judgements have empirical accuracy.

The cognitive basis of intuitive judgement is poorly understood. Brunswik's model, which has been the basis of later work by Hammond (1996), Cooksey (1996), and others – and which has been cited by Parkin (2000), Vick (2002), and others – is based on perception, specifically the perception of attributes which Brunswik calls cues. Such cues are statistically related to objects and situations based on experience. When faced with a new object or situation, the individual perceives a relatively limited number of cues (Brunswik (1952) speculates that the number seldom exceeds seven) from among the almost limitless possibilities, and from these draws conclusions. The cues tend to be complex, thus they may not be interpreted in the same way each time they are perceived, or they may be perceived differently each time. Different people, presumably, perceive and interpret cues in different ways, presumably place different weights on them and combine them in different ways and, thus, may come to different conclusions about the same object or situation. Brunswik posits that most intuitive judgements can be modelled as weighted averages of cues and that the redundancy among cues ensures that conclusions are not highly sensitive to the exact values of the weights, suggesting a robustness of judgement.

Hammond (1996) and Cooksey (1996) combined Brunswik's model of cues and intuitive judgement with reasoning and calculation to form cognitive continuum theory.

Cognitive continuum theory holds that intuitive judgements should be evaluated by the correspondence between the weighted average of the cues perceived about an object or situation, on the one hand, and the critical attributes of the real object or situation they reflect, on the other. If these two correspond, then the judgement is said to be valid. In contrast, reasoning or calculation should be evaluated by the coherence of the model produced. If the parts of the model form an internally consistent totality, then the reasoning or calculation is said to be valid. The correspondence of this logically sound model to physical reality is of secondary importance.

Cognitive continuum theory further holds that people do not – or cannot – simultaneously think in both correspondence mode (intuitive judgement) and coherence mode (reasoning and calculation), but rather flip back and forth between these two cognitive processes. People form intuitive judgements, then subject those judgements to reasoning and calculation, and then take the results back into an intuitive correspondence mode and so on, and so on. Hammond calls this, "quasi-rational cognition". In solving a difficult problem, one might first look to hunches, intuitive guesses or premonitions, and then subject whatever arises to analytical thought. When the analysis becomes bogged down, one might go the other way and seek hunches about the analysis. This is something akin to Karl Popper's (1968) hypo-deductive view of the scientific method, in which an hypothesis is developed intuitively but then tested deductively.

Parkin (2000) notes that, in practical applications, analytical cognition is both more highly accurate on average than is intuitive judgement, yet sometimes can be wildly inaccurate. This is unsurprising. When underlying assumptions are more or less correct, analytical cognition can be both accurate and precise, but when those same assumptions are incorrect, the conclusions based on analysis can be widely off. Large errors were sometimes made in the analytical mode, but research suggests that they are less frequent in the intuitive mode (Hammond, 1996).

The confidence that people, both professionals and the public, place in a prediction appears to be related to the degree of internal consistency that manifests in the arguments or model underlying the prediction. For example, this has been observed in legal proceedings, among other places, where juries seem to reconcile conflicting evidence by constructing stories based on prior experiences (Pennington and Hastie, 1993), the more complete and consistent the story is, the more it tends to be believed. Thus, because intuitive judgements often lack the clearly stated internal consistency of analytical cognition, they are sometimes underestimated in empirical value.

Engineering judgement vs regular judgement

Is "engineering judgement" distinct from any other kind? People have a great deal of confidence in human intuition; this is not limited to geotechnical engineers, although the geotechnical literature makes a great deal to do about engineering judgement with little discussion of what the notion means. There seems tacit agreement that readers understand the notion, yet a cursory review of the civil engineering literature provides little comfort.

To some writers, engineering judgement is the basic tenet of responsible professional decision-making, without which even the term "professional" is suspect (Peck, 1980; Vick, 2002); to others, engineering judgement is little more than guesswork, a flimsy substitute for logic and reasoning (Hartford, 2000; NRC, 1999a). Although too ambitious a task for the current volume, it would be helpful to better understand the nature and limitations of engineering judgement in practice.

While engineering judgement is raised to transcendent heights within the geotechnical community, it is often questioned by policy makers and the public. Despite the benefits provided to society by modern constructed facilities, adverse environmental consequences and other unfavourable impacts of those same facilities are much on people's minds. To an extent, engineers are blamed for these adverse consequences and the paternalistic sentiment embedded in the notion of engineering judgement – that professionals know best – is out of fashion. In recent years, US federal government agencies, such as the Nuclear Regulatory Commission, have even discouraged the use of the term engineering judgement in their deliberations.

Even though public opinion polls consistently rank engineers among the most trusted occupational groups, that same public increasingly questions whether engineering judgement can be relied on to answer questions of public safety. Perhaps safety can be improved by more public involvement, regulatory oversight and explicit proof of engineers' contentions. Perhaps engineering truth is no more than a social construct, lacking objective validity, the way deconstructionist theory in recent years has viewed science itself (Hacking, 1999). This is to say that, the elevated position to which engineering judgement is raised by the geotechnical community is by no means echoed by the rest of society.

The role of engineering judgement in routine design, for example, in selecting pile driving criteria or in dimensioning a retaining wall, is relatively accepted, but its role in making decisions with potentially catastrophic consequences, for example in dam safety or environmental remediation, is another matter. The public and regulatory authorities that represent public interests require more than assurances of confidence from experts to put fear to rest. Modern history has examples of technological failures about which engineering experts had expressed great confidence beforehand: the Tay Bridge, *Titanic*, Saint Francis Dam, Malpasset Dam, the spaceshuttle *Challenger* among others. To say to the public – or even to owners and clients – "trust us", is to miss the mark by a wide margin.

Problem solving and cognition

So, what is engineering judgement, how does it relate to decision-making and risk management in geotechnical practice, and how does it relate to probabilistic risk and reliability? In a book dedicated to this topic, Parkin (2000) notes that "judgement is informed by experience, expertise, reasoning and analysis". Such judgement may be brought to bear quickly, or only after a long period of deliberation. It may be the result of one person's considerations, or of a collaborative process involving many people, possibly with differing professional expertise.

We know a great deal about the psychological factors that come into play when people attempt to estimate risks subjectively. This understanding has developed over the past 30 years in the work of Kahneman, Slovic and Tversky (1982) and many others (Chapter 7). We similarly know something about the psychology of judgement, although the present volume is not the place to delve into this topic. This understanding has developed in the field of social judgement theory growing out of the work of Brunswik (1952) and others, out of various fields of philosophy, for example in the work of Nozick (1993) on rationality, and to some extent in the growing literature of knowledge management, for example in the work of Prusak (Davenport and Prusak, 1998) and others.

The field of social judgement theory views judgement as the ability to arrive at conclusions based on attributes of an object or situation. Margolis (1987) identifies seven progressive levels of cognitive function, which Parkin suggests may be thought of as different ways that people process experience and reach decisions on actions. These are: (1) simple feedback; (2) pattern recognition; (3) learning; (4) choice; (5) intuitive judgement; (6) reasoning; and (7) calculation.

Simple feedback is involuntary stimulus–response of the nervous system; driving a car and hand-eye coordination involve simple feedback. Pattern recognition is the integration of stimuli from the natural world into recognised configurations; butterflies migrate thousands of miles based on patterns of temperature, wind speed and other things. Learning is the process by which a lasting change in potential behaviour occurs as a result of practice or experience – the pattern of cues is remembered and abstracted from the past; in response to the stimulus of a cabinet door opening and a can of food being opened, a pet dog will run to its bowl. Choice arises when multiple stimulus–response relations have been learned; the dog may flee danger or stand and fight when someone tries to take its bowl away.

Feedback, pattern recognition, learning and choice are cognitive processes we share with many other animals. Judgement, reasoning, and calculation are not. Intuitive judgement has to do with being able to see with the mind's eye what might happen if specific actions are taken in the external world and, thereby, to engage in mental simulation to find an answer to a problem. Pattern recognition is brought within the process of cognition. Data, experience, representations and patterns are compared within the mind until a fit is found with the problem at hand. Margolis (1987) says that this is not the sort of thing that "very plausibly requires logic, calculation, following out of formal rules, or even verbal reasoning of the most informal sort". It is the sort of thing that is at the centre of craftsmanship, and even our cousins in the primate world evidence judgement.

Reasoning and calculation lie beyond intuitive judgement, and are the stuff of human intelligence. Reasoning is the explicit use of language to facilitate judgement. It is among the reasons that language skills are so important to engineering. Without language there is no reasoning and, thus, reasoning is a human development of relatively recent vintage, since language is a development of the late Pleistocene. Language provides the abstract symbolic framework within which people move beyond intuitive judgement. Syntax, the rules whereby words or other elements of language are combined to form grammatical sentences, is the logical structure within which, to use Parkin's phrase, the "language equivalent of physical reality," can be manipulated as one might manipulate an engineering model. Semantics, the meanings and classifications of what words signify, provides the elements to be manipulated. So, reasoning is a rationally structured way to draw conclusions about the world and to choose among alternate courses of action. The emphasis here is on rationally structured – in contrast to intuitive – which is the substance of judgement.

It is a small step, conceptually, from reasoning to calculation. Calculation replaces the informal logic of reasoning with the formal logic of mathematics. Semantic names are replaced by abstract variables, and the disciplines of symbolic logic and mathematics brought to bear on uncovering relationships or patterns and reaching conclusions. This is the "engineering modelling" with which so much of the university curriculum is obsessed. For most people, there is little intuitive about calculation, and calculation and reasoning are tightly bound.

The place of engineering judgement

Among other things, this book embraces two apparently contradictory positions. First, applications of probability to engineering problems yield the most satisfactory results when probability is based on a degree of belief. That is, the most important aspects of probability are fundamentally subjective. Second, engineers, along with other experts, are usually excessively optimistic about their knowledge of important parameters and especially about the uncertainty in that knowledge. In fact, however, the two are not contradictory. Much of our uncertainty is subjective, and we are often not very good at estimating that uncertainty. One of the things that distinguishes a professional in any field from a dilettante is the degree to which he or she is able to assimilate experience and extrapolate from it to new situations. In effect, what is important is not the amount of experience but the amount of evaluated experience.

Properly applied, engineering judgement reflects accumulated and evaluated experience. The term should not be used as an excuse for avoiding analysis or investigation. As the quotations in the earlier sections reveal, Terzaghi's success depended in large part on a willingness to examine a problem deeply before rendering an opinion (Peck, 1980). Engineers who have invoked "engineering judgement" without thought when asked to justify an opinion have done themselves and the profession a disservice and tended to put the phrase into disrepute.

Who is an "expert"?

It is common to distinguish between two types of expertise: (1) Substantive expertise, means that an expert's judgements are likely to be accurate and precise. On average, the judgements are correct, in the sense that reasonably high probabilities are assigned to observed outcomes; (2) normative expertise, means that an expert's judgements are calibrated to observed occurrences. In a series of judgements, the observed frequencies of occurrence match the probabilities assigned, even if those probabilities are relatively diffuse.

In as much as one's state of knowledge is an intrinsic property of subjective probability, it is sometimes assumed that better values will be elicited from those with the most advanced technical understanding of the state-of-the-art, but this is not necessarily so. Substantive knowledge bears no general relation to the normative "goodness" of a probability judgement, and it is necessary but not sufficient for the kind of expertise that subjective probability assessment requires.

Klein (1998) shows that experts are distinguished from novices by the following: (1) ability to recognise patterns; (2) ability to detect anomalies or deviations from these patterns; (3) ability to track all important information, draw inferences from it and project it forward in time; (4) use of mental simulation in interpreting the operation of a process; (5) ability to make fine discriminations and detect subtle differences; and (6) awareness of their own limitations. The relationship of many of these skills to limiting heuristic and bias effects is apparent, and they provide a model for the kinds of abilities that probability elicitation should seek to exploit.

Geotechnical practice is virtually unique among engineering disciplines in its reliance on knowledge and understanding derived from field performance case histories, and this tradition provides powerful support in formulating subjective probability judgements. Detailed review of applicable field performance case histories is so fundamental as to represent a mandatory addition for geotechnical application of elicitation procedures. Case-history information is used in several ways. Occasionally, structures of a specific class are so similar and numerous that performance statistics can be invoked as base-rate frequency data (Johansen *et al.*, 1997), but this is comparatively rare. More commonly there are a handful of well-documented case histories that represent the "type case" for a particular failure mode, such as Lower San Fernando Dam in relation to seismic liquefaction flow slides for example. These illustrate the conditions, performance characteristics, and elements of failure processes that pattern recognition requires. The availability heuristic, however, would caution against exclusive reliance on failures alone, and case histories for non-failure conditions help identify negative cues or missing factors that differentiate failure from non-failure patterns. Near-failure case histories may provide the most useful information in this respect by identifying conditions or circumstances responsible for terminating what might otherwise have become a failure sequence. The careful study and wise use of case histories of all kinds almost inevitably embodies all of the desirable characteristics of expertise previously enumerated.

Expert panels

Although most of the literature on judgemental probability assignment treats an individual expert, in most dam safety studies a panel of engineers or experts is formed to provide oversight and guidance. Each member from such a panel comes from a different set of experiences, may have different impressions of the current problem, and is likely to provide a probability assignment that is somewhat or even considerably different from those provided by other experts. The risk analyst, while presumably valuing all this diversity, in the final analysis needs a single probability or probability distribution to use in the risk analysis. So, somehow these differing probabilities or probability distributions need to be reconciled. That is the core issue in speaking of consensus or aggregate probability assignments: how to forge a single assignment out of a set of differing ones?

Choosing expert panels

The choice of experts is the most important step in determining success or failure of the expert elicitation process. Depending on personality, experience and technical background, individual experts may serve different functions. These individuals can be categorised into distinct types (Cooke, 1991):

- Resources – individuals with special knowledge of a problem, technology, site or project.
- Proponents – individuals with a particular and strongly argued point of view about a technology, issue or project.
- Evaluators – individuals with open mind, willing to objectively judge the credibility of different hypotheses, points of view or opinions.
- Facilitators – individuals versed in the technical issues who are also willing to lead discussions and manage interactions among other experts.

To be successful, an expert panel usually needs to include an evaluator and a facilitator. Resources and proponents are typically used to elucidate specific issues, to give depth in specific areas; they are typically not used to give balance and breadth.

It is important in considering potential experts that individuals be willing to be objective, commit time and interact with others in a professional manner. Panel members who cannot be counted on to commit to these terms harm the process more than help it.

Combining probability assignments

When a panel of experts is used to develop judgemental probabilities, the difficulty must be faced of how to aggregate the possibly differing views of the members of the panel into a single probability, a distribution of probability (PDF or PMF) over some variable to be estimated, or a higher-order probability (i.e. a probability distribution over probability assignments). Clearly, this is not an easy task, especially if the differences within the panel over probability assignments reflect fundamentally different interpretations of a physical situation. In the latter case, the experts not only differ on the numerical value of the probability assignment, but also on *why* the assignment was made the way that it was. This section does not attempt to solve the problem of probability aggregation but to survey approaches that appear to have been useful in other applications.

There are two general approaches to aggregating judgemental probabilities; (1) forming a consensus among the experts by means of discussion; and (2) combining individual assignments through some mathematical procedure. Cooke (1991), Rowe (1992), and Clemen and Reilly (2001) provide summaries of these two sets of methods.

A few general principles for combining probability assignments apply no matter which approach is taken. First, in principle, whatever the method of combining probability assignments it should not be structured so as to create an incentive for an expert to bias his or her responses in order to influence the group outcomes. That is, the method should, in principle, be structured to make "gaming" the system hard to do. It should be structured to create an incentive for the expert to assign the probabilities that best reflect his or her opinion. This is often difficult to achieve in practice.

Second, in combining probability assignments, each expert should have the same *ex ante* ability to influence a group outcome. In other words, each expert should be treated equally. Consensus schemes, that give a more articulate or more aggressive expert subtle advantages to influence group conclusions, should be avoided. Similarly, weighting schemes or other mathematical approaches that give more importance to some experts over others (e.g. more senior experts), should be avoided. Research suggests that more senior experts are not necessarily better at assigning probabilities that are better calibrated to reality than are more junior ones, although one might hope that the more senior experts displayed more definitive probabilities (i.e. on average, nearer zero and one), but this, too, is not supported by research (Edwards and Tversky, 1967).

Cooke (1991) identifies several desirable qualities for a consensus process to obtain group probabilities. Reproducibility, means that the details of the process of soliciting and computing group probability assignments should be documented to the extent needed to make them reproducible by peers. Accountability, means that the experts, the opinions of each, and the sources of evidence or support should be identified for reference by others who may be evaluating the outcomes. Empirical control, means that probability assignments should be, if possible, testable by empirical controls. Neutrality, means that the method of gathering and combining probability assignments should encourage experts not to bias their responses (not to "game"). Fairness, means that experts are treated equally during the process and in processing numerical results.

Consensus probabilities or distributions intend to reflect the combined judgement of a group or panel. From a philosophical view, degrees of belief are unique to the individual; they are not properties of a group of people. Nonetheless, the approach to consensus probabilities is to draw on the combined expertise of the members of a group, and to have the individuals within the group jointly explore the implications of information or experiences in arriving at a probability assignment that all, or most, can agree to. Vick (2002) provides an extended discussion of group process to develop consensus probabilities. These group processes have been described by Snizek and Henry (1990) as revision and weighting, which is self-descriptive, with combinations of probability assignments being made by negotiation or a form of self-weighting within the group.

Many people have had the experience that discussing a technical problem within a group setting, enjoying the differing perspectives of people with differing backgrounds and expertise, can lead to a better informed opinion on a particular problem. Research suggests that this group interaction may motivate people to do a better job (Rowe, 1992), and decision science research suggests that the accuracy of group probability assignments may be more accurate than those of individuals (Snizek and Henry, 1990). This often requires the use of a "facilitator" to smooth the process and help people come to an agreement (Vick 2002).

Mathematical methods of combining probability assignments vary from the simplistic to the complex. At the simplest level, individual probability assignments can just be treated as unbiased observations from a sample. Individual point assignments of probability can be plotted in a histogram as any other data, and a frequency or probability distribution assigned to them. Assignments of probability distributions, say, to an engineering parameter, can be integrated over to obtain an aggregate distribution. Refinements to this sampling approach might include imprecision in the individual probability assignments, based on self-weighting or evaluation by a third party.

Genestand and Zidek (1986) provide an overview of various methods for combining probability assignments, along with example weighting schemes. Among these are weighted combinations of the arithmetic average, geometric average, harmonic average, maximum value, minimum value and a generalised weighted average. Ultimately, the problem with all these weighting schemes is that someone must play the role of "super-decision-maker" and choose relative weights for the experts. This is also a problem with the informational approach described in the following paragraphs.

Beginning in the 1970s, an informational approach to combining probability assignments grew up around the work of Lindley (Lindley *et al.*, 1979), Winkler (1968), Morris (1974, 1977, 1983) and others; and the work in many guises continues to the present (Gelfand *et al.*, 1995). This approach has earlier roots in wartime intelligence (Kent, 1949). Conceptually, this use of judgement is not dissimilar from the use of engineering models, which also generate probabilistic outputs that the engineer must interpret based both on prior suspicions and on an appraisal of the model. The work of Gelfand *et al.* (1995), for example, deals with model selection, but the concepts are closely related. This means that the large literature on model selection is also directly pertinent to the evaluation of judgement (Berger, 1993; Stein 1999).

For the single expert, let $f_e(x|y)$ be the probability assignment summarised in a subjective probability distribution on an uncertain quantity, x, which might be an event, parameter, or some other value; and which in turn is conditioned on a state of information indicated by y. The analyst whose task it is to incorporate the expert's judgement in a risk analysis, also starts with some prior probability distribution on the same uncertain quantity, $f(x|z)$, which precedes and is independent of the assessment obtained from the expert. The conditioning information, z, in this case may or may not be the same as y. Having learned of the expert's judgement, the analyst's updated probability distribution on x is found from a simple application of Bayes' theorem. In this way, the expert's judgement is used as any other information would be. For an expert panel, the likelihood is replaced by a joint likelihood, entailing covariances, and so forth, but remains conceptually straightforward.

Endnotes

1. Of course, this assumption of independence is not strictly true even for large dams and may not be even approximately true for small dams. The failure of a dam upstream may directly cause the failure of a dam down stream by releasing a flood wave that over-tops the lower dam. When the Austin, Pennsylvania Dam failed in 1911, it washed out seven dams downstream. Similarly, extreme weather events may cause multiple dam failures across a region. When Hurricane Floyd struck North Carolina in 1999 it led to the failure of 36 dams in 44 North Carolina counties.
2. The central limit theorem is not without constraints upon the underlying or component distributions. For example, Kaufman (1963) has shown in the context of oil reserve estimation that the sum of logNormal variables does not satisfy these conditions, and thus the distribution of reserve estimates involving logNormally distributed pool volumes is more complex than they would be if the CLT applied to them.
3. In civil engineering practice, one often encounters frequentist sampling inferences (e.g. confidence intervals) used as if they were probability distributions on estimated parameters. This is conceptually incorrect. Confidence intervals describe the behaviour of estimators (i.e. mathematical summaries of sample outcomes), given presumptions about the values of population parameters, not uncertainty bounds on the parameters being estimated.

References

Aitchison, J. and Brown, J. A. C. (1969) *The Lognormal Distribution: with Special Reference to its Uses in Economics*, Cambridge University Press, Cambridge.

Aitchison, J. and Dunsmore, I. R. (1975) *Statistical Prediction Analysis*, Cambridge University Press, Cambridge, Eng and New York.

Al-Futaisi, A. and Stedinger, J. R. (1999) "Hydrologic and economic uncertainties and flood-risk project design", *Journal of Water Resources Planning and Management*, 125(6), 314–324.

Allais, M. (1953) "La psychologie de l'home rationnel devant le risque: critique des postulats et axiomes de l'école Américaine", *Econometrica*, 21(4), 503–546.

Allais, M. and Hagen, J. (1979) *Expected utility hypotheses and the Allais Paradox*, Reidel, Dordrecht.

Almeida, A. B. and Viseu, T. (1996) "Dams and valley safety. A present and future challenge", *Proceedings, Workshop on Dams and Safety Management of Downstream Valleys*, Balkema, Lisbon.

Alpert, M. and Raiffa, H. (1982) "A progress report on the training of probability assessors", in: *Judgement Under Uncertainty, Heuristics and Biases*, by D. Kahneman, P. Slovic, and A. Tversky (eds), Cambridge University Press, Cambridge, pp. 294–306.

Ang, A. H.-S. and Tang, W. H. (1975) *Probability Concepts in Engineering Planning and Design*, Wiley, New York.

Apt, S. R., Whittler, R. J., Taylor, A. and Love, D. J. (1992) "Human stability in a high hazard flood zone", *Water Resources Bulletin*, 25(4), 881–890.

Armstrong, E. L., Mermel, T. W. and Engineering Foundation (US) (1975) *Converting Existing Hydro-electric Dams and Reservoirs into Pumped Storage Facilities*, Engineering Foundation Conference, Franklin Pierce College, Rindge, NH, August 18–23, 1974, American Society of Civil Engineers, New York.

Assaf, H. and Hartford, D. N. D. (2002) "A virtual reality approach to public protection and emergency preparedness planning in dam safety analysis", *Proceedings of the Canadian Dam Association Conference*, Victoria, Canada.

Assaf, H., Hartford, D. N. D. and Cattanach, J. D. (1997) "Estimating dam breach flood survival probabilities", *ANCOLD Bulletin No. 107*, December.

Baecher, G. B. (1972) "Site exploration: A probabilistic approach", PhD, Massachusetts Institute of Technology, Cambridge, MA.

Baecher, G. B. and Christian, J. T. (2003) *Reliability and Statistics in Geotechnical Engineering*, Wiley, London.

Baecher, G. B. and Ladd, C. C. (1998) "Formal observational approach to staged loading", *Transportation Research Board Annual Meeting*.

Baecher, G. B., Pate, E. M. and de Neufville, R. (1979) "Risk of dam failure in benefit/cost analysis", *Water Resources Research*, 16(3), 449–456.

Barnett, V. (1999) *Comparative Statistical Inference*, Wiley, Chichester and New York.

Bayes, T. (1763) "An essay toward solving a problem in the doctrine of chances", *Philosophical Transaction of the Royal Society (London)*, 53, 370–418.

BC Hydro (1997) "Hugh Keenleyside Dam Deficiency investigations: earthfill dam–risk assessment", BC Hydro, Burnaby.

Bechteler. W. and Broich, K. (1993) "Computational analysis of the dam-erosion problem", *Advances in Hydro-Science and Engineering*, 1, 723–728.

Bedford, T. and Cooke, R. M. (2001) *Probabilistic Risk Analysis: Foundations and Methods*, Cambridge University Press, Cambridge and New York.

Benjamin, J. R. and Cornell, C. A. (1970) *Probability, Statistics, and Decision for Civil Engineers*, McGraw-Hill, New York.

Berger, J. O. (1993) *Statistical Decision Theory and Bayesian Analysis*, Springer-Verlag, New York.

Berger, J. O. and Pericchi, L. (2001) "Objective Bayesian model selection: introduction and comparison", in: Lahiri, P. (ed.), *Model Selection*, Lecture Notes Monograph Series, Vol. 38, Institute of Mathematical Statistics, Beachwood, OH. pp. 135–207.

Berlin, I. and Hardy, H. (1980) *Against the Current : Essays in the History of Ideas*, Viking Press, New York.

Berlin, I., Hardy, H. and Kelly, A. (1979) *Russian Thinkers*, Penguin Books, New York.

Bernoulli, J. (1713) *Jacobi Bernoulli ars conjectandi: opus posthumum: accedit tractatus de seriebus infinitis, et epistola gallice scripta de ludo pilae reticularis*, Impensis Thurnisiorum Fratrum, Basileae.

Bernstein, P. L. (1996) *Against the Gods : The Remarkable Story of Risk*, Wiley, New York.

Bier, V. M. (1997) "An overview of probabilistic risk analysis for complex engineered systems", *Fundamentals of Risk Analysis and Risk Management*, V. Molak, (ed.), Lewis Publishers, Boca Raton.

Bodine, B. R. (undated) *Users Manual for FLOW SIM 1, Numerical Method for Simulating Unsteady and Spatially Varied Flow in Rivers and Dam Failures*, US Army Corps of Engineers, Southwestern Division, Dallas, TX.

Bowles, D. S. and McClelland, D. M. (1999) "Loss of life estimation: what can we learn from case histories". *ANCOLD Bulletin No. 113*, 75–91.

Bowles, D. S. and McClelland, D. M. (2000) *Estimating Life Loss for Dam Safety Risk Assessment*, Institute for Dam Safety Risk Assessment. Utah State University, Logan, UT.

Box, G. E. P. and Tiao, G. C. (1992) *Bayesian Inference in Statistical Analysis*, Wiley, New York.

Brinded, M. (2000) "Perception vs analysis. How to handle risk." *The 2000, Lloyd's Register Lecture*, Royal Academy of Engineering, London.

British Standards Institution (BSI) (1991) BS 5760: Part 5: 1991 Reliability of systems equipment and components – Guide to failure modes, effects and criticality analysis (FMEA and FMECA).

Brown, C. A. and Graham, W. J. (1988) "Assessing the threat to life from dam failure", *Water Resources Bulletin*, 2(4), 6.

Brunswik, E. (1952) *The Conceptual Framework of Psychology*, University of Chicago Press, Chicago.

Buckle, H. T. (1858) *History of Civilization in England*, D. Appleton and Company, New York.

Budnitz, R. H., Apostolakis, G., Boore, D. M., Culff, L. S., Coppersmith, K. J., Cornell, C. A. and Morris, P. A. (1998) "Use of technical expert panels: applications to probabilistic seismic hazard analysis", *Risk Analysis*, 18(4), 463–469.

Bühlmann, H. (1970) *Mathematical Methods in Risk Theory*, Springer, Berlin, Heidelberg and New York.

Bury, K. V. and Kreuzer, H. (1985) "The assessment of the failure probability for a gravity dam", *Water Power and Dam Construction*, 37.

Bury, K. V. and Kreuzer, H. (1986) "The assessment of risk for a gravity dam", *Water Power and Dam Construction*, 38(2).

Canadian Standards Association (CSA) (1991) *Risk Analysis Requirements and Guidelines*, Q636-91.

Canadian Standards Association (CSA) (1993) *Guidelines and Requirements for Reliability Analysis Methods*, Q636-93.

Casagrande, A. (1965) "The role of the 'calculated risk' in earthwork and foundation engineering", *Journal of the Soil Mechanics and Foundations Division, ASCE*, 91(SM4), pp. 1–40.

Chambers, W&R, Ltd (1972) Chambers Twentieth Century Dictionary, A. M. Macdonald (ed.), W&R Chambers, Ltd, London.

Christian, J. T. and Baecher, G. B. (1999) "The point estimate method as numerical quadrature", *Journal of Geotechnical and Geoenvironmental Engineering, ASCE*, 125(7).

Christian, J. T. and Baecher, G. B. (2002) "The point-estimate method with large numbers of variables". *International Journal of Numerical and Analytical Methods in Geomechanics*, 26(15), 1515–1529.

Christofano, E. A. (1965) *Method of Computing Erosion Rate for Failure of Earthfill Dams*, US Bureau of Reclamation, Denver, CO.

Clausen, L. and Clark, P. B. (1990) "The development of criteria for predicting dambreak flood damages using modelling of historical dam failures, in: *International Conference on River Flood Hydraulics* by W. R. White (ed.), John Wiley, and Hydraulic Research Ltd: Chichester, pp. 369–380.

Clemen, R. T. and Reilly, T. (2001) *Making Hard Decisions with Decision Tools*, Duxbury/ Thomson Learning, Pacific Grove, CA.

Cooke, R. M. (1991) *Experts in Uncertainty: Expert Opinion and Subjective Probability in Science*, Oxford University Press, New York.

Cooke, R.M. and Goossens, L. H. J. (1999) *Procedures Guide for Structured Expert Judgement.* Report EUR 18820, Brussels, Luxembourg.

Cooksey, R. W. (1996). *Judgment Analysis: Theory, Methods, and Applications*, Academic Press, San Diego, CA.

Costa, J. E. (1988) "Floods from dam failures", in: *Flood Geomorphology*, V. R. Baker, R. C. Kochel and P. C. Patton (eds), John Wiley, New York, pp. 439–463.

Cox, D. R. and Lewis, P. A. W. (1966) *The Statistical Analysis of Series of Events*, Methuen, London.

Dasgupta, P., Sen, A. K. and Marglin, S. A. (1972) *Guidelines for Project Evaluation*, United Nations, New York.

Daston, L. (1988) *Classical Probability in the Enlightenment*, Princeton University Press, Princeton, NJ.

Davenport, T. H. and Prusak, L. (1998) *Working Knowledge, How Organizations Manage What they Know*, Harvard Business School Press, Boston, MA.

David, F. N. (1962) *Games, Gods and Gambling; The Origins and History of Probability and Statistical Ideas from the Earliest Times to the Newtonian Era*, Hafner Pub. Co., New York.

De Finetti, B. (1937) "Foresight: its logical laws, its subjective sources", in: *Studies in Subjective Probability*, by H. Kyburg and H. Smokler (eds), Wiley, New York, pp. 93–158.

De Finetti, B. (1972) *Probability, Induction and Statistics: The Art of Guessing*, Wiley, London and New York.

De Finetti, B. (1990) *Theory of Probability: A Critical Introductory Treatment*, Interscience Publishers, Chichester and New York.

DeKay, M. L. and McClelland, G. H. (1991) *Setting Decision Thresholds for Dam Failure Warnings: A Practical Theory-Based Approach.* Center for Research on Judgement and Policy. University of Colorado, Boulder, CO.

DeKay, M. L. and McClelland, G. H. (1993) "Predicting loss of life in cases of dam failure and flash flood", *Risk Analysis*, 13(2), 193–205.

Duesenberry, J. S. (1958) *Business Cycles and Economic Growth*, New York, pp. 14–15.

Eccleston, C. H. (2001) *Effective Environmental Assessments*, Lewis Publishers, Boca Ratan.

Eckstein, O. (1958) *Water-resources Development: The Economics of Project Evaluation*, Harvard University Press, Cambridge, MA.

Edwards, W. (1975) "Personal probabilities of probabilities", in: *Theory and Decision*, by J. Marschak (ed.), pp. 136–139.

Edwards, W. and Tversky, A. (1967) *Decision Making: Selected Readings*, Penguin, Harmondsworth.

Einstein, H. H. and Baecher, G. B. (1983) "Probabilistic and statistical methods in engineering geology, part one: exploration", *Rock Mechanics and Rock Engineering*, 16(1), 39–72.

Einstein, H. H., Labreche, D. A., Markow, M. J. and Baecher, G. B. (1976) "Decision analysis applied to rock tunnel exploration", *Engineering Geology*, 12(2), 143–161.

Ellsberg, D. (1961) "Risk, ambiguity, and the savage axioms", *Quarterly Journal of Economics*, 75, 643–669.

Environment Canada (1996) *ERA of Priority Substances Under the Canadian Environmental protection Act Draft 2.0*, Environment Canada, Ottawa.

Evans, M., Hastings, N. and Peacock, B. (1993) *Statistical Distributions*, Wiley, New York.

Evans, S. G. (1986) "The maximum discharge of outburst floods caused by the beaching of man-made and natural dams", *Canadian Geotechnical Journal*, 23(1), 385–387.

Feller, W. (1967) *An Introduction to Probability Theory and its Applications*, Wiley, New York.

Feller, W. (1971) *An Introduction to Probability Theory and its Applications*, Wiley, New York.

Ferrell, W. R. and McGoey, P. J. (1980) "A model of calibration for subjective probabilities", *Organizational Behaviour and Human Performance*, 26, 32–53.

Fischhoff, B., Slovic, P. and Lichtenstein, S. (1997) "Knowing with certainty: the appropriateness of extreme confidence", *Journal of Experimental Psychology*, 3(4), 552–564.

Fishburn, P. C. (1964) *Decision and Value Theory*, Wiley, New York.

Fisher, R. A. (1921) "On the probable error of a coefficient of correlation deduced from a small sample", *Metron*, 1, 1–32.

Fisher, R. A. (1925a) "Theory of statistical estimation", *Proceedings of the Cambridge Philosophical Society*, 26, 528.

Fisher, R. A. (1925b) *Statistical Methods for Research Workers*, Oliver & Boyd, London.

Fisher, R. A. and Bennett, J. H. (1971) *Collected Papers of R. A. Fisher*, University of Adelaide, Adelaide.

Fishman, G. S. (1996) *Monte Carlo: Concepts, Algorithms, and Applications*, Springer-Verlag, New York.

Folayan, J., Höeg, K. and Benjamin, J. (1970) "Decision Theory Applied to Settlement Predictions", *Journal Soil Mechanics and Foundation Engineering*, 96(4), 1127–1141.

Foster, M., Fell, R. and Spannagle, M. (2000) "The statistics of embankment dam failures and accidents", *Canadian Geotechnical Journal*, 37(5), 1000–1024.

Fread, D. L. (1977) "The development and testing of a dam-break flood forecasting model", *Dam-Break Flood Modeling Workshop*, Washington, DC, 164–197.

Fread, D. L. (1984) "A breach erosion model for earthen dams", *Specialty Conference on Delineation of Landslides, Flash Flood, and Debris Flow Hazards in Utah*, Logan, 30, UT.

Fread, D. L. (1987) "BREACH: an erosion model for earth dam failures", *HRL-193*, Hydrologic Research Laboratory, National Weather Service, Silver Spring, CO.

Fread, D. L. (1988) "The NWS DAMBRK model: theoretical background/user documentation", *HRL-256*, Hydrologic Research Laboratory, National Weather Service, Silver Spring, CO.

Fread, D. L. (1993) "NWS FLDWAV model: the replacement of DAMBRK for dam-break flood prediction", *10th Annual Conference*, Kansas City, 177–184.

Fread, D. L. (1998) "Dam-breach modeling and flood routing: a perspective on present capabilities and future directions", in: *Dam-Breach Modeling and Flood Routing: A Perspective on Present Capabilities and Future Directions*, US Dept of Agriculture, Stillwater, OK.

Fread, D. L. and Harbaugh, T. E. (1973) "Transient hydraulic simulation of breached earth dams", *Journal of the Hydraulics Division, ASCE*, 99(HY1), 139–154.

Fread, D. L. and Hsu, K. S. (1993) "Applicability of two simplified flood routing methods: level-pool and Muskingum-Cunge", *ASCE National Hydraulic Engineering Conference*, San Francisco, July 25–30, 1564–1568.

Fread, D. L., Lewis, J. M. and Wiele, S. M. (1985) *BREACH: An Erosion Model for Earthen Dam Failures*, Hydraulic Research Laboratory, National Weather Service, Silver Spring, CO.

Fread, D. L., Lewis, J. M. and Wiele, S. M. (1991) "The NWS simplified dam bread flood forecasting model", Hydrologic Research Laboratory, National Weather Service, Silver Spring, CO.

Froehlich, D. C. (1995) "Peak outflow from breached embankment dam", *Journal Water Resource Planning and Management, ASCE*, 121(1), 90–97.

Gál, T., Stewart, T. J. and Hanne, T. (1999) *Multicriteria Decision Making: Advances in MCDM Models, Algorithms, Theory, and Applications*, Kluwer Academic, Boston, MA.

Gelfand, A. E., Mallick, B. K. and Dey, D. K. (1995) "Modeling expert opinion arising as a partial probabilistic specification", *Journal of the American Statistical Association*, 90(430), 598–604.

Gènestand, C. and Zidek, J. (1986) "Combining probability distributions: a critique and an annotated bibliography", *Statistical Science*, 1(1), 114–135.

Gettys, C. M. C., Steiger, J. H., Kelly, C. W. and Peterson, C. R. (1973) "Multiple-stage probabilistic information processing", *Organizational Behavior and Human Performance*, 10.

Gigerenzer, G. (1989) *The Empire of Chance: How Probability Changed Science and Everyday Life*, Cambridge University Press, Cambridge and New York.

Gigerenzer, G. (1991) "How to make cognitive illusions disappear: beyond 'heuristics and biases'", *European Review of Social Psychology*, 2, 83–115.

Gilbert, P. (2002) Personal communication.

Gleick, J. (1988) *Chaos: Making a New Science*, Penguin, New York.

Good, I. J. (1962) "Subjective probability as the measure of a non-measurable set", in: *Logic, Methodology and Philosophy of Science*, by E. Nagel, P. Suppes and A. Tarski (eds), Stanford University Press, Stanford.

Good, I. J. (1983) "Some history of the hierarchical Bayesian methodology", in: *Good Thinking: The Foundations of Probability and its Applications*, by I. J. Good (ed.), University of Wisconsin Press, Minneapolis.

Graham, W. J. (1999) "A procedure for estimating loss of life caused by dam failure", *DSO-99-06*, US Department of the Interior, Bureau of Reclamation, Denver, CO.

Gruetter, F. and Schnitter, N. J. (1982) "Analytical risk assessment for dams", *14th International Congress on Large Dams*, Rio De Janeiro, 611–625.

Gruner, E. C. (1967) "The safety of reservoirs", *World Dams Today (Japan Dam Association)*, 104–109.

Gruska, G. F., Mirkhani, K. and Lamberson, L. R. (1973) *Point Estimation in Non Normal Samples: Using Pearson Frequency Curves, and Burr Cumulative Functions*, Chevrolet Product Assurance Center, Warren, MI.

Gumbel, E. J. (1954) *Statistical Theory of Extreme Values and Some Practical Applications: A Series of Lectures*, US Govt Print Office, Washington, DC.

Gumbel, E. J. (1958) *Statistics of Extremes*, Columbia University Press, New York.

Haack, S. (1996) *Deviant Logic, Fuzzy Logic: Beyond the Formalism*, University of Chicago Press, Chicago.

Hacking, I. (1975) *The Emergence of Probability*, Cambridge University Press, Cambridge.

Hacking, I. (1990) *The Taming of Chance*, Cambridge University Press, Cambridge and New York.

Hacking, I. (1999) *The Social Construction of What?*, Harvard University Press, Cambridge, MA.

Hacking, I. (2001) *An Introduction to Probability and Inductive Logic*. Cambridge University Press, Cambridge.

Hagen, V. K. (1982) "Re-evaluation of design floods and dam safety", *14th ICOLD Congress*, Rio de Janeiro.

Hahn, G. J. and Shapiro, S. S. (1967) *Statistical Models in Engineering*, John Wiley, New York.

Haimes, Y. Y. (1990) *Hierarchical Multiobjective Analysis of Large-scale Systems*, Hemisphere, New York.

Hall, N. (1993) *Exploring Chaos : A Guide to the New Science of Disorder*, W. W. Norton, New York.

Hammond, K. R. (1996) *Human Judgment and Social Policy: Irreducible Uncertainty, Inevitable Error, Unavoidable Injustice*, Oxford University Press, New York.

Harr, M. E. (1987) *Reliability-based Design in Civil Engineering*, McGraw-Hill, New York.

Harrington, H. J. (1991) *Business process improvement: the breakthrough strategy for total quality, productivity, and competitiveness*, McGraw-Hill, New York.

Harris, G. W. and Wagner, D. A. (1967) *Outflow From Breached Earth Dams*, University of Utah, Salt Lake City, UT.

Hartford, D. N. D. (2002) "Commentary on dam failure consequence (life safety) analysis", *Dam Safety Interest Group Meeting*, Denver, CO.

Hartford, D. N. D. (2000) "Judged values and value judgements in dam risk assessment: a personal perspective", *ANCOLD Bulletin*, No. 114, 78–86.

Hartford, D. N. D. (2001) Personal communication.

Hartford, D. N. D. and Stewart, R. A. (2002) "Risk assessment and the safety case in dam safety decisions", *Proceedings of the 12th Biennial Conference of the British Dams Society*, 4–8 September, Thomas Telford. 2002.

Hasofer, A. M. and Lind, N. C. (1974) "An exact and invariant first-order reliability format", *Journal of the Engineering Mechanics Division, SCE*, 100(1), 111–121.

Hendron, A. J. and Patton, F. D. (1985) *The Vaiont Slide: A Geotechnical Analysis Based on New Geologic Observations of the Failure Surface*, US Army Corps of Engineers, Waterways Experiment Station, Vicksburg, MS.

Henrion, M. and Fischhoff, B. (1986) "Assessing uncertainty in physical constants" *American Journal of Physics*, 54(9), 791–798.

Hogarth, R. (1975) "Cognitive processes and the assessment of subjective probability distributions", *Journal of the American Statistical Association*, 70(35), 271–294.

Hogarth, R. (1987) *Judgment and Choice,* Wiley, New York.

Howard, R. A. (1988) "Uncertainty about probability: a decision analysis perspective", *Risk Analysis*, 8, 91–98.

Howard, R. A. and Matheson, J. E. (1981) *Principles and Applications of Decision Analysis*, Strategic Decision Group. Menlo Park, CA.

Howson, P. and Urbach, P. (1991) *Scientific Reasoning – the Bayesian Approach,* Open Court Publishers, La Salle, IL.

Hume, D. and Smith, A. (1963) *An Enquiry Concerning Human Understanding, and Selections from a Treatise of Human Nature: With Hume's Autobiography and a Letter from Adam Smith*, Open Court Publishers, La Salle, IL.

Hynes, M. and Vanmarcke, E. (1976) "Reliability of embankment performance predictions", *Proceedings Engineering Mechanics Division Specialty Conference*, University of Waterloo Press, Ontario.

Institution of Engineers, A. S. D, Australasian Institute of Mining and Metallurgy–Sydney Branch and Australian Geomechanics Society. Sydney Group (1972) *Symposium on Monitoring of Dams, Sydney, 11th October, 1972: symposium*, Australian Geomechanics Society, Sydney.

Interagency Advisory Committee on Water Data (1982) *Guidelines for determining flood flow frequency; Bulletin 17B*, US Department of Interior, US Geological Survey, Office of Water Data Coordination, Reston, VG.

International Electrotechnical Commission (IEC) (1985) IEC 812, Analysis Techniques for System Reliability–Procedure for Failure Mode Effects Analysis.

Jain, S. and Lall, U. (2000) "Magnitude and timing of annual maximum floods: trends and large-scale climatic associations for the Blacksmith Fork River, Utah", *Water Resources Research*, 36(12).

Jain, S. and Lall, U. (2001) "Floods in a changing climate: does the past represent the future?" *Water Resources Research*, 37(12).

Jansen, R. B. (1980) *Dams and Public Safety*, US Dept of the Interior Water and Power Resources Sevices, Denver, CO.

Jaynes, E. T. (2003) *Probability Theory: The Logic of Science*, Cambridge University Press, New York.

Jeffreys, H. (1948) *Theory of Probability*, Clarendon Press, Oxford University Press, Oxford.

Jennings, R. H. (1974) *Historic and Modern Practices in Reliability Engineering,* Aerojet Nuclear Company, Idaho Falls, ID.

Johansen, P., Vick, S. and Rikartsen, K. (1997) "Risk analysis of three Norwegian rockfill dams", *Hydropower '97*, Balkema, Rotterdam, 431–442.

Johnson, N. L. and Kotz, S. (1969) *Discrete Distributions*, Houghton Mifflin, Boston, MA.

Johnson, N. L. and Kotz, S. (1970) *Continuous Univariate Distributions*, Houghton Mifflin, New York.

Johnson, N. L. and Kotz, S. (1972) *Distributions in Statistics: Continuous Multivariate Distributions*, John Wiley, New York.

Johnson, N. L., Kotz, S. and Kemp, A. W. (1992) *Univariate Discrete Distributions*, Wiley, New York.

Johnson, P. A. (1999) "Fault tree analysis of bridge failure due to scour and channel instability", *Journal of Infrastructure Systems*, 5(1), 35–41.

Jones, R. B. (1995) *Risk-based Management, A Reliability-centered Approach*, Gulf Professional Publishing.

Jonkman, S. M., van Gelder, P. H. A. J. M. and Vrijling, J. K. (2002) "Loss of life models for sea and river floods", *Flood Defence 2002, Proceedings of the 2nd International Symposium on Flood Defences, Beijing*, Science Press, New York, 196–206.

Kahneman, D. and Tversky, A. (1982) "Subjective probability: a judgment of representativeness", in: *Judgment under Uncertainty: Heuristics and Biases*, by P. Slovic and A. Tversky (eds), Cambridge University Press, New York, pp. 32–47.

Kaplan, S. and Garrick, B. J. (1981) "On the quantitative definition of risk", *Risk Analysis*, 1(1), 11–27.

Karvonen, T., Hepojoki, A., Huhta, H.-K. and Louhio, A. (2000) "The use of physical models in dam-break analysis", *RESCDAM Final Report*, Helsinki University of Technology, Helsinki, Finland, 11 December.

Kaufman, G. M. (1983) *Statistical Decision and Related Technologies in Oil and Gas Exploration*, Prentice Hall, Englewood Cliffs, NJ.

Keeney, R. L. and Raiffa, H. (1993) *Decisions with Multiple Objectives: Preferences and Value Tradeoffs*, Cambridge University Press, Cambridge and New York.

Kendall, M. and Stuart, A. (1977) *The Advanced Theory of Statistics*, C. Griffin, London.

Kent, J. T. (1949) *Military Intelligence*, Princeton University Press, Princeton.

Keren, G. (1994) "The rationality of gambling: gamblers' conceptions of probability, chance and luck", in: *Subjective Probability*, by G. Wright and P. Ayton (eds), Wiley, New York.

Keynes, J. M. (1921) *A Treatise on Probability*, Macmillan, London.

Keynes, J. M. (1937) "The general theory", *Quarterly Journal of Economics*, LI (February), 209–233.

Klein, G. A. (1998) *Sources of Power: How People Make Decisions*, MIT Press, Cambridge, MA.

Klinke, A. and Renn, O. (2002) "A new approach to risk evaluation and management: risk-based, precaution-based, and discoursed strategies", *Risk Analysis*, 22(6).

Knight, F. H. (1921) *Risk, Uncertainty and Profit*, Houghton Mifflin Company, Boston and New York.

Koriat, A., Lichtenstein, S. and Fischhoff, B. (1980) "Reasons for confidence", *Journal of Experimental Psychology: Human Learning and Memory*, 6, 107–118.

Kreuzer, H. and Bury, K. V. (1984) "A probability-based evaluation of the safety and risk of existing dams", *International Conference on Safety of Dams*, Coimbra, Portugal.

Lafitte, R. (1993) "Probabilistic risk analysis of large dams", *International Water Power and Dam Construction*, 45, 13–16.

Lambert, H. E. (1983) "Systems safety analysis and fault tree analysis", Lawrence Livermore National Laboratory, Report UCID-16238, May.

Laplace, P. S. and Dale, A. I. (1814) *Philosophical Essay on Probabilities*, A. I. Dale, (translator), Dover Publications, Inc., New York.

Lee, R., Hu, P. S., Neal, D. M., Ogles, M. R., Sorensen, J. H. and Trumble, D. A. (1986) "Predicting loss of life from floods", draft report prepared by Oak Ridge National Laboratories US DoE for the Institute for Water Resources, USACE, Oak Ridge, TN.

Lehner, P. E., Laskey, K. G. and Dubois, D. (1996) "An introduction to issues in higher order uncertainty", *IEEE Transactions on Systems, Man, and Cybernetics – Part A: Systems and Humans*, 26(3), 289–293.

Leopold, L. B., Clarke, F. E., Henshaw, G. G. and Balsley, J. R. (1971) "A procedure for evaluating environmental impact", *Survey Circular 645*, USGS, Washington, DC.

Leveson, N. (1995) *SafeWare: System Safety and Computers*, Addison-Wesley, Reading, MA.

Lewis, P. A. W. (1972) *Stochastic Point Processes: Statistical Analysis, Theory, and Applications*, Wiley-Interscience, New York.

Liao, S., Veneziano, D. and Whitman, R. V. (1988) "Regression models for evaluating liquefaction probability", *Journal of Geotechnical Engineering, ASCE*, 114(4), 389–411.

Lichtenstein, S., Fischhoff, B. and Phillips, L. (1982) "Calibration of probabilities: the state of the art to 1980", in: *Judgment Under Uncertainty: Heuristics and Biases*, by P. Slovic and A. Tversky (eds), Cambridge University Press, New York, pp. 306–334.

Lichtenstein, S. and Newman, J. R. (1967) "Empirical scaling of common verbal phrases associated with numerical probabilities", *Psychon Science*, 9, 563–564.

Lind, N. C. and Hartford, D. N. D. (2000) "Probability of human instability in a flooding: a hydrodynamic model", in: *Applications of Statistics and Probability to Geotechnical and Structural Engineering*, by R. E. Melchers and M. G. Stewart (eds), Balkema, Rotterdam, p. 1151.

Lindley, D. (1987) "The probabilistic approach to the treatment of uncertainty in artificial intelligence and expert systems", *Statistical Scientist*, 2(1), 17–24.

Lindley, D. V. (1971) *Bayesian Statistics: A review*, Society for Industrial and Applied Mathematics, Philadelphia, PA.

Lindley, D. V., Tversky, A. and Brown, R. V. (1979) "On the reconciliation of probability assessments", *Journal of the Royal Statistical Society, A*, 142(2), 146–180.

Lou, W. C. (1981) "Mathematical modeling of earth dam breaches", PhD, Colorado State University, Fort Collins, CO.

Luce, R. D. and Suppes, P. (1965) *Foundations of Measurement*, Academic Press, New York.

Lumb, P. (1974) "Application of statistics in soil mechanics", in: *Soil Mechanics: New Horizons*, by I. K. Lee (ed.), Newnes-Butterworth, London, pp. 44–112

Maass, A. (1962) *Design of Water-Resource Systems: New Techniques for Relating Economic Objectives, Engineering Analysis, and Governmental Planning*, Harvard University Press, Cambridge, MA.

Macchione, F. and Sirangelo, B. (1988) "Study of earthdam erosion due to overtopping", in: Starosolszky, O. and Melder, O. M. (eds), *Hydrology of Disaster*, James & James, London.

Maidment, D. R. (1993) *Handbook of Hydrology*, McGraw-Hill, New York.

Margolis, H. (1987) *Patterns, Thinking, and Cognition: A Theory of Judgment*, University of Chicago Press, Chicago, IL.

Marshall, K. T. and Oliver, R. M. (1995) *Decision Making and Forecasting: With Emphasis on Model Building and Policy Analysis*, McGraw-Hill, New York.

Martz, H. F. and Waller, R. A. (1982) *Bayesian Reliability Analysis*, Wiley, New York.

Matheron, G. (1989) *Estimating and Choosing: An Essay on Probability in Practice*, Springer-Verlag, Berlin and New York.

McCann, M. W. (1995) "Unpublished data. National performance of dams program", Center on the Performance of Dams, Stamford, CT.

McCann, M. W. (1997) "Risk assessment for seismic prioritization of dams", *Notes for a Seminar Presented to the USACE, Office of the Chief of Engineers*, Washington, DC.

McCann, M. W., Franzini, J. B., Kavazanjian, E. and Shah, H. C. (1985) "Preliminary safety evaluation of existing dams", Federal Emergency Management Agency, Stamford University, Stamford, CT.

McClelland, A. G. R. and Bolger, F. (1994) "The calibration of subjective probabilities: theories and models 1980–1994", in: *Judgmental Forecasting*, by G. Wright and P. Ayton (eds), Wiley, Chichester and New York, pp. 453–482.

McCormick, N. J. (1981) *Reliability and Risk Analysis: Methods and Nuclear Power Applications*, Academic Press, New York.

McCuen, R. H. and Snyder, W. M. (1986) *Hydrologic Modeling: Statistical Methods and Applications*, Prentice Hall, Englewood Cliffs, NJ.

Macdonald, A. M. (ed.) (1972) *Chambers Twentieth Century Dictionary*, W & R Chambers Ltd, Toronto.

McDonald, L. (2002) "Probabilities–discussion on considerations", *Working Paper*, Dam Safety Interest Group.

McQuaid, J. (2002) "Risk assessment – its development and some current priorities", *Proceedings of the 12th Biennial Conference of the British Dams Society*, Thomas Telford, London.

Merriam-Webster (2000) *WWWebster Dictionary*, Merriam-Webster, Incorporated, New York, http://www.m-w.com/

Meyer, M. A. and Booker, J. M. (1990) "Eliciting and analyzing expert judgment, a practical guide", *NUREG/CR-5424*, US Nuclear Regulatory Commission, Washington, DC.

Mishan, E. J. (1988) *Cost-benefit Analysis: An Informal Introduction*, Unwin Hyman, London and Boston.

Morgan, M. G. and Henrion, M. with a chapter by Small, M. (1990) *Uncertainty: A Guide to Dealing with Uncertainty in Quantitative Risk and Policy Analysis*, Cambridge University Press, Cambridge and New York.

Morris, P. A. (1974) "Decision analysis expert use", *Management Science*, 20(9), 1233–1241.

Morris, P. A. (1977) "Combining expert judgments: a Bayesian approach", *Management Science*, 23(7), 679–693.

Morris, P. A. (1983) "An axiomatic approach to expert resolution", *Management Science*, 29(1), 24–32.

Mosleh, A. and Bier, V. M. (1996) "Uncertainty about probability: a reconciliation with the subjectivist viewpoint", *IEEE Transactions on Systems, Man, and Cybernetics–Part A: Systems and Humans*, 26(3), 1083–1090.

Mosleh, A., Bier, V. M. and Apostolakis, G. (1987) "Methods for the elicitation and use of expert opinion in risk assessment: Phase I, a critical evaluation and directions for future research", *NUREG/CR-4962*, US Nuclear Regulatory Commission, Washington, DC.

Multiagency Task Force (1972) "Guidelines for implementing principles and standards for multi-objective planning of water resources", US Bureau of Reclamation, Washington, DC.

Murphy, A. H. and Winkler, R. L. (1974) "Probability forecasts: a survey of national Weather Service forecasters", *Bulletin of the American Meteorological Society*, 55, 1449–1453.

Murphy, A. H. and Winkler, R. L. (1977a) "Can weather forecasters formulate reliable forecasts of precipitation and temperature?", *National Weather Digest*, 2, 2–9.

Murphy, A. H. and Winkler, R. L. (1977b) "The use of credible intervals in temperature forecasting: Some experimental results", in: *Decision Making and Change in Human Affairs*, by H. Jungerman and G. de Zeeuw (eds), Reidel, Dordrecht.

National Institute for Standards and Technology (NIST) (1998) *Engineering Statistics Handbook*, NIST, Washington, DC.

National Research Council. (1996) *Understanding Risk: Informing Decision in a Democratic Society*, National Academies Press, Washington, DC.

National Research Council (1995) *Flood Risk Management and the American River Basin: An Evaluation*, National Academies Press, Washington, DC.

National Research Council (1999a) *Improving American River Flood Frequency Analyses*, Committee on American River Flood Frequencies, National Academies Press, Washington, DC.

National Research Council (1999b) *The Impacts of Natural Disasters: A Framework for Loss Estimation*, National Academies Press, Washington, DC.

National Research Council (2000) *Risk Analysis and Uncertainty in Flood Damage Reduction Studies*, National Academies Press, Washington, DC.

Newman, J. R. (1966) *The World of Mathematics: A Small Library of the Literature of Mathematics from A'h-mosé the Scribe to Albert Einstein*, Simon and Schuster, New York.

Neyman, J. and Pearson, E. S. (1967) *Joint Statistical Papers*, University of California Press, Berkeley,CA.

NIBS (2003) *HAZUS Program Overview*, National Institute for Building Sciences, Washington, DC.

Nozick, R. (1993) *The Nature of Rationality*, Princeton University Press, Princeton, NJ.

O'Hagan, A. (1994) *Kendall's Advanced Theory of Statistics*, Edward Arnold, London.

Ord, J. K. (1972) *Families of Frequency Distributions*, Hafner Pub. Co., New York.

Orkin, M. (1999) *What are the Odds? Chance in Everyday Life*, W. H. Freeman.

Pan, Z., Segal, M., Arritt, R. W., Chen, T.-C. and Weng, S.-P. (1999) "Effects of quasi-stationary wave anomalies on the regional hydraulic cycle in the continental United States", in: Bottschalk *et al.* (eds), *Proceedings of the IUGG Intl Symposium, Hydraulic Extremes: Understanding, Predicting, Mitigating*, IAHS Publication No. 255, UK Environment Agency, London.

Papoulis, A. (1991) *Probability Random Variables, and Stochastic Processes*, McGraw Hill, New York.

Parkin, J. (2000) *Engineering Judgement and Risk*, Thomas Telford, London.

Parr, N. M. and Cullen, N. (1998) "Risk management and reservoir maintenance", *Journal of the Institution of Water and Environmental Management*, 2, 587–593.

Pascal, B. (1966) *Pensaes*, Penguin Books, Harmondsworth.

Pate-Cornell, M. E. (1984) "Fault trees *vs* event trees in reliability analysis", *Risk Analysis*, 4(3), 177–186.

Pearl, J. (1988) *Probabilistic Reasoning in Intelligent Systems: Networks of Plausible Inference*, Morgan Kaufmann Publishers, San Mateo, CA.

Pearl, J. (2001) *Causality*, Cambridge University Press, New York.

Pearson, K. and Pearson, E. S. (1978) *The History of Statistics in the 17th and 18th Centuries Against the Changing Background of Intellectual, Scientific, and Religious Thought: Lectures by Karl Pearson given at University College, London, During the Academic Sessions, 1921–1933*, C. Griffin, London.

Peck, R. B. (1969) "Advantages and limitations of the observational method in applied soil mechanics", *Geotechnique*, 19(2), 171–187.

Peck, R. B. (1980) "Where has all the judgment gone?", *Canadian Geotechnical Journal*, 17, 584–590.

Peirce, C. S. (1998) *Chance, Love, and Logic: Philosophical Essays*, University of Nebraska Press, Lincoln, NB and London.

Pennington, N. and Hastie, R. (1993) "A theory of explanation-based decision making", in: *Decision Making in Action: Models and Methods*, by G. A. Klein *et al.* (eds), Ablex Publishing, Norwood, NJ.

Petrascheck, A. W. and Sydler, P. A. (1984) "Routing of dam break floods", *International Water Power and Dam Construction*, 36, 29–32.

Phillips, L. D. and Edwards, W. (1966) "Conservatism in a simple probability inference task", *Journal of Experimental Psychology*, 72, 346–357.

Pitbaldo, R. and Turney, R. (1996) *Risk Assessment in the Process Industries*. Institution of Chemical Engineers.

Pólya, G. (1954) *Mathematics and Plausible Reasoning*, Princeton University Press, Princeton, NJ.

Ponce, V. M. and Tsivoglou, A. J. (1981) "Modeling of gradual dam breaches", *Journal of the Hydraulics Division, ASCE*, 107(HY6), 829–838S.

Popper, K. R. (1968) *The Logic of Scientific Discovery*, Harper & Row, New York.

Porter, T. M. (1986) *The Rise of Statistical Thinking, 1820–1900*, Princeton University Press, Princeton, NJ.

Porter, T. M. (1995) *Trust in Numbers: The Pursuit of Objectivity in Science and Public Life*, Princeton University Press, Princeton, NJ.

Pratt, J. W., Raiffa, H. and Schlaifer, R. (1995) *Introduction to Statistical Decision Theory*, MIT Press, Cambridge, MA.

Press, W. H. (1997) *Numerical Recipes in C: The Art of Scientific Computing*, Cambridge University Press, Cambridge and New York.

Press, W. H., Teukolsky, S. A., Vetterling, W. T. and Flannery, B. P. (1992) *Numerical Recipes in FORTRAN: The Art of Scientific Computation*, Cambridge University Press, Cambridge.

Priest, S. D. and Hudson, J. A. (1976) "Discontinuity spacings in rock", *International Journal of Rock Mechanics and Mining Science*, 13, 135–148.

Raiffa, H. (1968) *Decision Analysis: Introductory Lectures on Choices Under Uncertainty*, Random House, New York.

Raiffa, H. and Schlaifer, R. (1971) *Applied Statistical Decision Theory*, Wiley, New York.

Ramsey, F. P. and Braithwaite, R. B. (1931) *The Foundations of Mathematics and Other Logical Essays*, K. Paul Trench Trubner & Co Ltd, London.

Rasmussen, N. C. (1975) *Reactor Safety Study: An Assessment of Accident Risk in US Nuclear Power Plants*, WASH 1400, NUREG 75/104, US Nuclear Regulatory Commission.

Reagan, R., Mosteller, F. and Youtz, C. (1989) "Quantitative meanings of verbal probability expressions", *Journal of Applied Psychology*, 74(3), 433–442.

Reiter, P. (2001) "Loss of life caused by dam failure, the RESCDAM LOL method", P R Water Consulting Ltd, *Summary of the Final Report to the RESCDAM Project.*

Rosenblueth, E. (1975) "Point estimates for probability moments", *Proceedings, National Academy of Science*, 72(10), 3812–3814.

Rossini, F. A. and Porter, A. L. (1983) *Integrated Impact Assessment*, Westview Press, Boulder, CO.
Rowe, G. (1992) "Perspectives on expertise in the aggregation of judgments", in: *Expertise and Decision Support*, by G. Wright and F. Bolger (eds), Plenum Press, New York, pp. 155–180.
Rowe, W. D. (1977) *An Anatomy of Risk*, Wiley, New York.
Rubinstein, R. Y. (1981) *Simulation and the Monte Carlo method*, Wiley, New York.
Salmon, G. M. and Hartford, D. N. D. (1995) "Lessons from the application of risk assessment to dam safety", *Australian National Committee on Large Dams Bulletin*, 101, 54–67.
Salmon, W. C. (1967) *The Foundations of Scientific Inference*, University of Pittsburgh Press, Pittsburgh.
Salmon, W. C. (1998) *Causality and Explanation*, Oxford University Press, New York.
Sarma, A. A. L. N., Kumari, B. P. and Srinivas, S. (1999) Proc. of the IUGG Intl. Symposium (eds), "Studies on Hydrological Extremes" *Hydrological Extremes: Understanding, Predicting, Mitigating, IAHS Publication No. 255.*
Savage, L. (1954) *The Foundations of Statistics*, Wiley, New York.
Schachter, R.D. (1986) Evaluating influence diagrams, *Operations Research*, 34(6), 871–882.
Seaver, D. A., von Winterfeldt, D. and Edwards, W. (1978) "Eliciting subjective probability distributions on continuous variables", *Organizational Behavior and Human Performance*, 21, 379–391.
Serafim, J. L. (1984) *Safety of Dams: Proceedings of the International Conference on Safety of Dams, Coimbra/23–28 April 1984*, A. A. Balkema, Rotterdam and Boston.
Shannon, C. E. (1948) "The mathematical theory of communication", *Bell System Technical Journal*, July–October.
Sherbot, D. M. J. (1997) "Hugh Keenleyside Dam, dam breach ecological risk assessment (Draft)", BC Hydro, Burnaby.
Singh, K. P. and Snorrason, A. (1984) "Sensitivity of outflow peaks and flood stages to the selection of dam breach parameters and simulation models", *Journal of Hydrology*, 68, 295–310.
Singh, S. (1997) *Fermat's Enigma*, Viking, New York.
Singh, V. P. (1996) *Dam Breach Modelling Technology*, Kluwer Academic Publishers, Dordrecht.
Singh, V. P. and Scarlatos, P. D. (1988) "Analysis of gradual earth-dam failure", *Journal of Hydraulic Engineering*, 114(1), 21–42.
Sivia, D. S. (1996) *Data Analysis: A Bayesian Tutorial*, Oxford University Press, Oxford.
Slovic, P., Fischhoff, G. and Lichtenstein, S. (1977) "Behavioral decision theory", *Annual Review of Psychology*, 28, pp. 1–39.
Slovic, P., Fischhoff, B. and Lichtenstein, S. (1982) "Facts versus fears: understanding perceived risk", in: *Judgment Under Uncertainty: Heuristics and Biases*, by P. Slovic and A. Tversky (eds), Cambridge University Press, New York, pp. 463–489.
Smith, J. F. and Kida, T. (1991) "Heuristics and biases: expertise and task realism in auditing", *Psychological Bulletin*, 109, 472–485.
Smith, R. P. (2000) "Risk, uncertainty and ambiguity in engineering design decision making", *The Open Workshop on Decision-Based Design: Origin, Status, Promise, and Future, NSF*, State University of New York at Buffalo.
Sparks, R. E. (1996) "Ecosystem effects: positive and negative outcomes", in: *The Great Flood of 1993*, by S. Changnon (ed.), Westview Press, Boulder, CO, pp. 132–162.
Stedinger, J. R., Heath, D. C. and Thompson, K. (1996) "Risk assessment for dam safety evaluation: hydrologic risk", *96-R-13*, US Army Corps of Engineers, Institute for Water Resources, Alexandria, VA.
Stein, M. L. (1999) *Interpolation of Spatial Data: Some Theory for Kriging*, Springer, New York.
Stevens, S. S. (1951) "Mathematics, measurements, and psychophysics", in: *Handbook of Experimental Psychology*, by S. S. Stevens (ed.), Wiley, New York, pp. 1–49.
Stewart, R. A. (2000) *Dam Risk Management – The ICOLD Lecture Proceedings. GeoEng 2000*, Balkema.
Stigler, S. M. (1986) *The History of Statistics*, Harvard University Press, Cambridge, MA.

Sturges, H. A. (1926) "The choice of a class interval", *Journal of the American Statistical Association*, 21, 65–66.

Swain, A. D. and Guttman, H. E. (1983) *Handbook of Human Reliability Analysis with Emphasis on Nuclear Power Plant Applications*, US Nuclear Regulatory Commission, NUREG/CR-1278, August.

Swain, A. D. (1987) "Accident sequence evaluation program human reliability analysis procedure", *NUREG/CR-4772, SAND86-1996*, Sandia National Laboratories, Albuquerque, NM, February.

Taylor, J. R. (1997) *An Introduction to Error Analysis*, University Science Books, Sausalito, CA.

Terzaghi, K., Peck, R. B. and Mesri, G. (1996) *Soil Mechanics in Engineering Practice*, Wiley, New York.

Tolstoy, I. (1981) *James Clerk Maxwell: A Biography*, Canongate, Edinburgh.

Turney, R. and Pitbaldo, R. (1996) *Risk Assessment in the Process Industries*, Institute of Chemical Engineers, Rugby, Warwickshire.

US Army Corps of Engineers (1992) *Guidelines for Risk and Uncertainty Analysis in Water Resources Planning*, Volume 1, Principles with technical appendices, 92-R-1, Volume 2, Example cases," *92-R-2*, Institute of Water Resources, Fort Belvoir, VA.

US Bureau of Reclamation (1989) *Policy Procedures for Dam Safety Modification Decision Making*, US Department of the Interior, Denver, CO.

USEDA (1998) "Guidelines for ecological risk assessment", *Federal Register*, 63(93): 26846–26934.

US Department of Defense (1980) *Mil-Std-1629A*, Military standard procedures for performing a Failure Modes, Effects and Criticality Analysis, Washington, DC.

US Water Resources Council (1983) *Economic and Environmental Principles and Guidelines for Water and Related Land Resources Implementation Studies*, US Government Printing Office, Washington, DC.

USACE, C-E-Y (1995) "Hydrologic engineering requirements for flood damage reduction studies", *EM 1110-2-1419*, US Army Corps of Engineers, Washington, DC.

USACE, IfWR (1988) "National economic development procedures manual – urban flood damage", *IWR Report 88-R-2*, US Army Corps of Engineers, Fort Belvoir, VA.

USACE (1996) "Risk-based analysis for flood damage reduction studies", *EM 1110-2-1619*, US Army Corps of Engineers, Washington, DC.

USACE (1997) "Metropolitan region of Louisville, Kentucky, Beargrass Creek Basin, final feasibility analysis report", *EM 1110-2-1419*, US Army Corps of Engineers, Louisville District, Louisville, KT.

US Department of the Interior (1987) "Techniques to measure damages to natural resources", *Publication No. PB88-100136*, National Technical Information Service, Springfield, VA.

US Nuclear Regulatory Commission (1975) "Reactor safety study: an assessment of accident risks in US commercial nuclear power plants", *WASH 14 00 (NUREG-75/014)*, US Nuclear Regulatory Commission, Washington, DC.

US Nuclear Regulatory Commission (1981) *Fault Tree Handbook*, NUREG-0492, US Nuclear Regulatory Commission, Washington, DC.

USEPA (1998) "Guidelines for ecological risk assessment", *Federal Register*, 63(93), 26846–26924.

Utah State University and Bureau of Reclamation (1999) "A framework for characterizing extreme floods for dam safety risk assessment", Utah State University and US Department of the Interior, Bureau of Reclamation, Denver, CO.

Veneziano, D. (1994) "The earth, engineers, and education", *A symposium in honor of Robert V. Whitman*, Massachusetts Institute of Technology, Cambridge, pp. 102–124, MA.

Venn, J. (1866) *The Logic of Chance. An Essay on the Foundations and Province of the Theory of Probability, with Especial Reference to its Application to Moral and Social Science*, Macmillan, London and Cambridge.

Vick, S. G. (1997) "Dam safety risk assessment: new directions", *Water Power and Dam Construction*, 49(6).

Vick, S. G. (2002) *Degrees of Belief: Subjective Probability and Engineering Judgement*, ASCE, Reston, VG.

Vick, S. G. and Bromwell, L. F. (1989) "Risk analysis for dam design in karst", *Journal of the Geotechnical Engineering Division*, 115(6), 819–835.

Vick, S. G. and Stewart, R. A. (1996) "Risk analysis in dam safety practice", *Uncertainty in the Geologic Environment*, ASCE Geotechnical Special Publication No. 58. Madison, pp. 586–603, WN.

Von Mises, R., Neyman, J., Sholl, D. and Rabinowitch, E. I. (1939) *Probability, Statistics and Truth*, Macmillan, New York.

Von Thun, L. (1984) "Application of decision analysis techniques in dam safety evaluation and modification", *International Conference on Safety of Dams*, Coimbra, Portugal.

Von Thun, L. (1996) "Risk assessment of Nambe Falls Dam", *Uncertainty in the Geologic Environment*, Madison, WN, 604–635.

Von Thun, L. (1999) "Risk analysis for Navajo Dam, US Bureau of Reclamation", Technical Memorandum NZ-8311-3, Denver, CO.

Wahl, T. L. (1997) "Predicting embankment dam breach parameters – a needs assessment", *XXViith IAHR Congress*, San Francisco.

Wald, A. (1971) *Statistical Decision Functions*, Chelsea Pub. Co., Bronx, NY.

Walder, J. S. and O'Connor, J. E. (1997) "Methods of predicting peak discharge of floods caused by failure of natural and constructed earth dams", *Water Resources Research*, 33(10), 2337–2348.

Warner, M. L. and Preston, E. H. (1974) "A Review of Environmental Impact Assessment Methodologies", Environmental Protection Agency, Washington, DC.

Wetmore, J. N. and Fread, D. L. (1984) *The NWS simplified dam bread flood forecasting model for desk-top and hand-held microcomputers*, Federal Emergency Management Agency, Washington, DC.

Whitman, R. V. (1984) "Evaluating calculated risk in geotechnical engineering", *ASCE Journal of Geotechnical Engineering*, 110(2), 145–188.

Whitman, R. V. (2000) "Organizing and evaluating uncertainty in geotechnical engineering", *Journal of the Geotechnical and Geoenvironmental Engineering Division, ASCE*, 126(7), 583–593.

Winkler, R. L. (1967) "Assessment of prior distributions in a Bayesian analysis", *Journal of the American Statistical Association*, 62, 776–800.

Winkler, R. L. (1968) "Combining probability distributions from dependent information sources", *Management Science*, 27(4), 479–488.

Winkler, R. L. and Murphy, A. H. (1968) "'Good' probability assessors", *Journal of Applied Meteorology*, 7, 751–758.

Winkler, W. F. and Murphy, A. H. (1973) "Experiments in the laboratory and the real world", *Organizational Behaviour and Human Performance*, 10, 252–270.

Wurbs, R. A. (1987) "Dam-breach flood wave models", *Journal of Hydraulic Engineering*, 113(1), 29–46.

Yegian, M. K. and Whitman, R. V. (1978) "Risk analysis for ground failure by liquefaction", *Journal of the Geotechnical Engineering Division, ASCE*, 104(7), 921–928.

Yevjevich, V. (1983) "Extremes in hydrology", in: *Statistical Extremes and Applications*, by J. T. De Oliveira (ed.), Reidel, Dordrecht.

Zadeh, L. A. (1977) *Theory of Fuzzy Sets*, Electronics Research Laboratory College of Engineering, University of California, Berkeley, CA.

Zellner, A. (1971) *An Introduction to Bayesian Inference in Econometrics*, Wiley, New York.

Glossary and interpretations

Recently, the International Commission on Large Dams (ICOLD) agreed on a wide range of definitions and interpretations of risk analysis terms for use in risk assessment for dam safety, to be published in the forthcoming ICOLD Bulletin on Risk Assessment in Dam Safety Management. Earlier drafts of Part I of this book, which included a draft Glossary of Terms, served as a foundation for the ICOLD Bulletin. The authors of the ICOLD Bulletin extended this Glossary of Terms to include a wide range of risk assessment terms. Selected terms relevant to risk analysis as agreed by ICOLD together with other accepted terms are presented below. Owners and others, who intend to use definitions other than those described herein, are encouraged to precisely define their intended meaning in their documents to reduce the chance of misinterpretation and confusion.

TERM	**INTERPRETATION**
Annual Exceedance Probability (AEP)	The estimated probability that an event of specified magnitude will be equalled or exceeded in any year.
Conditional probability	The probability of an outcome, given the occurrence of some event.
Consequences	Impacts in the downstream, as well as other, areas resulting from failure of the dam or its appurtenances.
Cumulative distribution function (CDF)	The integral of the probability density function calculated in the direction of increasing values of the random variable.
Dam-break affected zone	That zone of flooding where the changes in depth and velocity of flooding due to dam-break are such that there is potential for loss of life or damage to property or the environment.
Dam-break analysis	An analysis that provides an estimation of downstream flooding effects resulting from dam failure. The analysis includes a dam-break analysis and the routing of the dam-break hydrograph through the downstream channel and areas that would be inundated.
Deterministic	Describing a process with an outcome that is always the same for a given set of inputs.

TERM	INTERPRETATION
Direct losses	Losses that result from contact with floodwaters.
Effects	In the context of failure modes analysis (FMEA and FMECA), this term refers to the consequences for the functioning of a system, such as a dam, of a failure at some point within the system. If the system boundary is the dam, these consequences would be distinguished from the remote consequences, which may also be the result of the same failure, such as loss of life and property damage due to a dam-break wave downstream of the dam.
Event tree analysis	An event tree analysis is an inductive analysis process that utilises an event tree graphical construct that shows the logical sequence of the occurrence of events in, or states of, a system following an initiating event.
Expected value	The average or central tendency of a random variable. Specifically in relation to risk analysis, the product of the probability of an event and of its consequences, aggregated over all possible values of the variable.
Extreme event	Event, which has a very low annual exceedance probability (AEP).
Failure mechanism	A mechanism described by the physical processes and states that must occur in accordance with natural laws for the failure mode to progress from failure mode initiation (cause) through to the realisation of ultimate failure effect of interest.
Failure of dam	In the general case, the inability of a dam system, or part thereof, to function as intended. In the context of dam safety, failure is generally confined to issues of structural integrity, and in some contexts to the special case of uncontrolled release of the contents of a reservoir through collapse of the dam or some part of it.
Failure mode	Away that failure can occur, described by the means by which element or component failures must occur to cause loss of the sub-system or system function.
Failure mode and effects analysis (FMEA)	An inductive method of analysis where particular faults or initiating conditions are postulated and the analysis reveals the full range of effects of the fault or the initiating condition on the system. FMEA can be extended to perform what is called failure modes, effects and criticality analysis (FMECA). In a FMECA, each failure mode identified is ranked according to the combined influence of its likelihood of occurrence and the severity of its consequences.
Fault tree analysis	A systems engineering method for representing the logical combinations of various system states and possible causes which can contribute to a specified event (called the top event).
Fragility curve	A function that defines the probability of failure as a function of an applied load level. A particular form of the more general system response.

TERM	**INTERPRETATION**
Frequency	A measure of likelihood expressed as the number of occurrences of an event in a given time or in a given number of trials.
Hazard	Threat; condition, which may result from either an external cause (e.g. earthquake, flood, or human agency) or an internal vulnerability, with the potential to initiate a failure mode. A source of potential harm or a situation with a potential to cause loss.
Human factors	Human factors refer to environmental, organisational and job factors, and human and individual characteristics which influence behaviour at work in a way which can affect safety.
Human Reliability Analysis	A structured and systematic way of estimating the probability of human error in specific tasks, and the consequences of those errors.
Human Reliability Assessment (HRA)	A structured and systematic way of estimating the probability of human error in specific tasks, the consequences of those errors, the tolerability of the risks and the needed control measures.
Indirect losses	Indirect losses are the difference between total losses and *direct losses*.
Individual risk	The increment of risk imposed on a particular individual by the existence of a hazardous facility. This increment of risk is an addition to the background risk to life, which the person would live with on a daily basis if the facility did not exist or, in the context of dam safety, if the dam did not fail.
Joint probability	The probability that two or more variables will assume certain values simultaneously or within particular time intervals.
Likelihood	Used as a qualitative description of probability and frequency.
Population at risk	All those persons who would be directly exposed to floodwaters within the dam-break affected zone if they took no action to evacuate. See also *dam-break affected zone*.
Probabilistic	A description of procedures, which are based on the application of the laws of probability. Probabilistic analysis takes explicit account of the random variations in natural and other events and properties.
Probability	A measure of the degree of confidence in a prediction, as dictated by the evidence, concerning the nature of an uncertain quantity or the occurrence of an uncertain future event. It is an estimate of the likelihood of the magnitude of the uncertain quantity, or the likelihood of the occurrence of the uncertain future event. This measure has a value between zero (impossibility) and 1.0 (certainty). *There are two main interpretations:* ***Statistical*** *– frequency or fraction.* *The outcome of a repetitive experiment of some kind like flipping coins. It includes also the idea of <u>population variability</u>. Such a number is called an "objective" probability because it exists in*

TERM	INTERPRETATION
	the real world and is in principle measurable by doing the experiment. ***Subjective probability*** – *Quantified measure of belief, judgement, or confidence in the likelihood of an outcome, obtained by considering all available information honestly, fairly, and with a minimum of bias.* *Subjective probability is affected by the state of understanding of a process, judgement regarding an evaluation, or the quality and quantity of information. It may change over time as the state of knowledge changes.*
Probability density function	A function describing the relative likelihood that a random variable will assume a particular value in contrast to taking on other values.
Qualitative risk analysis	An analysis, which uses word form, descriptive or numeric rating scales to describe the magnitude of potential consequences and the likelihood that those consequences will occur.
Quantitative risk analysis	An analysis based on numerical values of the potential consequences and likelihood, the intention being that such values are a valid representation of the actual magnitude of the consequences and the probability of the various scenarios which are examined.
Random variable	A quantity, the magnitude of which is not exactly fixed, but rather the quantity may assume any of a number of values described by a probability distribution.
Reliability	Likelihood of successful performance of a given project element. It may be measured on an annualised basis or for some specified time period of interest or, for example, in the case of spillway gates, on a per demand basis. Mathematically, Reliability = (1 − Probability of failure).
Reliability theory	A theory that enables assessment of the probability of failure in terms of the uncertainties in capacity (or resistance) and demand (or load).
Risk	Measure of the probability and severity of an adverse effect to life, health, property, or the environment. In the general case, risk is estimated by the combined impact of all triplets of scenario, probability of occurrence and the associated consequence. In the special case, average risk is estimated by the mathematical expectation of the consequences of an adverse event occurring (that is, the product of the probability of occurrence and the consequence, combined over all scenarios).
Risk analysis	The use of available information to estimate the risk to individuals or populations, property or the environment, from hazards. Risk analyses generally contain the following steps: scope definition, hazard identification, and risk estimation.

TERM	**INTERPRETATION**
	Consistent with the common dictionary definition of analysis, viz. "*A detailed examination of anything complex made in order to understand its nature or to determine its essential features*", risk analysis involves the disaggregation or decomposition of the dam system and sources of risk into their fundamental parts.
Risk assessment	The process of making a decision recommendation on whether existing risks are tolerable and present risk control measures are adequate, and if not, whether alternative risk control measures are justified or will be implemented. Risk assessment incorporates the risk analysis and risk evaluation phases.
	Consistent with the common dictionary definition of assessment, viz. "*To analyse critically and judge definitively the nature, significance, status or merit of . . . [risk]*", risk assessment is a decision-making process, often sub-optimal between competing interests, that results in a statement that the risks are, or are not, being adequately controlled. Risk assessment involves the analysis, evaluation and decision about the management of risk and all parties must recognise that the adverse consequences might materialise and owners will be required to deal effectively with consequences of the failure event.
Risk evaluation	The process of examining and judging the significance of risk. The risk evaluation stage is the point at which values (societal, regulatory, legal and owners) and value judgements enter the decision process, explicitly or implicitly, by including consideration of the importance of the estimated risks and the associated social, environmental, economic, and other consequences, in order to identify and evaluate a range of alternatives for managing the risks.
Risk identification	The process of determining what can go wrong, why and how.
Root cause	The fundamental reason for the failure. The fundamental beginning of the failure can be determined through scientific analysis. This does not exclude the possibility of failure modes referring to root causes, this will be the case if the failure modes analysis is carried out at the most fundamental level of the system.
Scenario	A unique combination of states, such as: • initiating event; • concurrent wind state; • prior reservoir storage state; • gate operating state; • failure mode; • downstream and tributary concurrent flows; and • factors that would determine the presence of people below a dam at the time of failure. A scenario defines a suite of circumstances of interest in a risk assessment. Thus there may be loading scenarios, failure scenarios or downstream flooding scenarios.

TERM	**INTERPRETATION**
Sensitivity analysis	An analysis to determine the rate at which an output parameter varies, given unit change in one or more input parameters.
State	A specified value or range of a variable, that is used to describe a situation of interest in a risk assessment study. As examples, there may be a wind velocity state, a peak inflow flood discharge state or a spillway gate operating state.
System	A system is a defined entity that consists of identifiable, interacting discrete elements. It is an orderly arrangement of these elements (e.g., area within spatial boundaries, structures, mechanical and electrical equipment items, and operators) designed to show the interactions between the various elements in the performance of the system function. For simplicity, the general term system will be used without distinction between various levels of systems.
System response	How a dam responds, expressed as a conditional probability of failure, to a given scenario of applied loads and concurrent conditions. See also *fragility curve.*
Uncertainty	Now used to describe any situations without sureness, whether or not described by a probability distribution. In the context of dam safety, uncertainty can be attributed to (i) inherent variability in natural properties and events, and (ii) incomplete knowledge of parameters and the relationships between input and output values.
Warning time (WT)	In relation to methods for estimating loss of life from dam failure, the time from first issue of a warning to the population at risk (PAR) to the time of arrival of dangerous flooding at the PAR.

Index

Note: Figures and Tables are indicated by *italic page numbers*, glossary by **emboldened numbers**, notes by suffix 'n'

Printed and bound by CPI Group (UK) Ltd, Croydon, CR0 4YY
16/09/2026
14963404-0001